Computer Graphics

Computer Graphics

Theory and Practice

Jonas Gomes
Luiz Velho
Mario Costa Sousa

CRC Press
Taylor & Francis Group
Boca Raton London New York

CRC Press is an imprint of the
Taylor & Francis Group, an **informa** business

AN A K PETERS BOOK

CRC Press
Taylor & Francis Group
6000 Broken Sound Parkway NW, Suite 300
Boca Raton, FL 33487-2742

© 2012 by Taylor & Francis Group, LLC
CRC Press is an imprint of Taylor & Francis Group, an Informa business

No claim to original U.S. Government works

Printed in the United States of America on acid-free paper
Version Date: 20120320

International Standard Book Number: 978-1-56881-580-0 (Hardback)

Library of Congress Cataloging-in-Publication Data

Gomes, Jonas.
    Computer graphics : theory and practice / Jonas Gomes, Luis Velho, Mario Costa Sousa.
      p. cm.
    Summary: "Computer Graphics: Theory and Practice provides complete and integrated coverage of the subject, including geometric modeling, graphics interface, and visualization. It focuses on conceptual aspects of computer graphics, covering fundamental mathematical models as well as the inherent problems encountered in the implementation of the models. The approach assumes only a fundamental knowledge of calculus and linear algebra and provides the basis for an introductory course. The text offers a global conceptual view of the field and an understanding of its main problems. For each problem, solution strategies are compared and presented in algorithmic form and complete working implementations are provided in the C language. The complete C source code for the implementation of all the algorithms can be accessed from the books website. Together, the book and electronic portion enable readers to understand and practice with the basic techniques involved in the implementation of a 3D graphics system"-- Provided by publisher.
    Summary: "This book focuses on conceptual aspects of computer graphics, covering the fundamental mathematical models as well as the inherent problems encountered in the implementation of those models. The approach assumes only a basic knowledge of calculus and linear algebra and provides the basis for an introductory course. It is accompanied by online materials that offer opportunities to practice concepts presented in the book"-- Provided by publisher.
    Includes bibliographical references and index.
    ISBN 978-1-56881-580-0 (hardback)
     1. Computer graphics. 2. Microcomputers--Programming. 3. Image processing--Mathematics. 4. C (Computer program language) I. Velho, Luiz. II. Costa Sousa, Mario, 1967- III. Title.

    T385.G649 2011
    006.6--dc23   2011030312

Visit the Taylor & Francis Web site at
http://www.taylorandfrancis.com

and the CRC Press Web site at
http://www.crcpress.com

To Solange and Daniel
—J.G.

To Noni and Alice
—L.V.

To Patricia and Marianna
—M.C.S.

Contents

Foreword

Depiction is about creating a signal that tickles our visual systems in a way that is just good enough for us to extract meaning from it. Such a signal itself need not be visual, for there is now strong evidence that nonvisual signals such as suggestion, storytelling, recall, imagination, and visualization in the original sense of the word all can deeply engage our visual systems. Artists have understood this for millennia. They have developed ingenious techniques for visual depiction using various media, combining insights into both phenomenology and the power of suggestion.

Computer graphics is about giving visual depiction a systematic computational underpinning. Here, we immediately run into some difficult questions. What do we mean by *systematic*? What are the dependencies? What is fundamental and unchanging and what is ephemeral? How do we separate one from the other?

While such questions plague most fields, they are particularly acute in computer graphics. As a field it is barely a half-century old and combines the development of many original techniques with extensive adoption of ideas from other fields. It has undergone an extremely high rate of adaptation and growth, so that some techniques are older than the field itself, while others have only been around for a year or two. This can give rise to inconsistencies in representations, processes, and workflow. The field has also been steered in some ways by its most successful applications, including animated films, special effects, visual simulation, electronic games, and design and manufacture. As a result, it is not always clear how relevant certain aspects of computer graphics are to new applications. That said, what a remarkable success computer graphics has been as a mathematical science, an engineering discipline, and digital medium!

How then should one introduce the concepts of computer graphics? As a medium, we have seen a wide array of books on the use of certain graphics systems, and on the practice and aesthetics of graphical depiction. As a technology, there are numerous books providing instruction on the use of specific application programming interfaces (APIs) and programming development environments to build graphical applications. What has remained problematic is how to define and explain the fundamental concepts of computer graphics, knowing full well that ultimately these concepts must be relevant both to further research in the field as well as to practical applications. To this end, some authors have advocated programming-based approaches that rely on specific languages or APIs. Others have removed the reliance on specific implementation environments and instead advocate algorithmic approaches. But then, what concepts are key to the development of algorithms, and how are those concepts to be stated outside of the algorithms themselves?

This book by Jonas Gomes, Luiz Velho, and Mario Costa Sousa has clear priorities: first, explain the concepts of computer graphics precisely, but not pedantically, using basic mathematics; second, explore the mathematical implications of these concepts by constructing models of graphical processes that are seen as fundamental; third, after those models are understood, exploring their algorithmic formulation. It thus means that in reading this book you will try to understand before you build, and the exercises in each chapter reinforce that discipline. After going through this book myself, I particularly enjoyed the early treatment of topics such as projective mappings and color spaces, as they informed many topics later on.

Ultimately, this book helps to expose what we do best in computer graphics. It isn't merely in our ability to create beautiful images, or in our ability to make things work very quickly. We of course help to do both of those things. It is instead in our ability to create an ever-growing set of *visual models*, to simulate or prove properties about those models so as to explore their capabilities, and to map those models onto usable technology so that we may all better express ourselves visually.

—Eugene Fiume, University of Toronto

About the Cover

"The Liquid Dark Side of the Moon"

Simplicity itself, a jet black $12'' \times 12''$ square with a line drawing of a luminous white prism at its center. A thin beam of white light penetrates the left side of the prism at an angle and exits on the right, split into a fanned spectrum of glowing color.

My name is Dan Abbott and I work as part of a compact but busy design collective StormStudios, based in London, England. You may or may not be familiar with our work, but chances are you've stumbled across the image I describe above as the 1973 cover graphic to Pink Floyd's gazillion-selling "The Dark Side of the Moon" album. Of course the same graphic elements were already well rooted in the collective conscious well before 1973, thanks to the work of our old friend Isaac Newton, and reproduced in a thousand and one school science textbooks.

That the prism landed on the cover of Pink Floyd's seventh album was due to the efforts of my esteemed colleague and tormentor Storm Thorgerson, who at that time co-helmed influential sleeve design company Hipgnosis with Aubrey 'Po' Powell. Hitherto, Storm and Po's designs for Pink Floyd had been exclusively photographic in nature, but the band requested something graphic by way of a change. Hipgnosis rustled up seven exciting new designs and much to their surprise the band voted unanimously for the one with the prism. Storm claims he tried to talk them out of it, but their minds were all made up. Thus ends the fable of "How the Prism Got Its Album" and magically leapt from textbook to record racks worldwide.

Two decades later in 1993 history started to repeat itself—traditional practice in the rock 'n' roll universe. The Dark Side of the Moon was re-released in shiny, all new digitally remastered, twentieth anniversary CD form. So Storm decided to "remaster" the cover too, replacing the 1973 drawing with a photo of real light being refracted through a real-life glass prism. What could be more honest than that? Funnily enough, few fans seemed to notice the switcheroo, which I think might tell you something about the power of the basic setup of the image.

Ten years later still and it was suggested that the design be tweaked once again for the thirtieth anniversary re-release on SACD (which we were reliably informed was the absolutely definitive audio format of the future). Thirtieth anniversaries are very significant for all triangular life forms, so how could we refuse? So we built a four-foot square stained glass window to the exact proportions of the original design, and photographed it. "Hmm, maybe this idea's got legs after all" we thought. In the following years we created several further homages to the original design: a prism made of words for a book cover, a prism painted a-la Claude Monet, a Lichtenstein-esque pop art number, and rather curiously, a prism created entirely with fruit for a calendar (this probably came about after someone joked about calendars being made from "dates").

To execute the above-mentioned "Fruity Side of the Moon," we built a large wooden tray with each line of the design being a walled-off section, keeping all the dates, raisins, cranberries, apricots, oranges, and baby lemons in their right and proper positions. It was then photographed from above. I can't remember if we ate the contents afterwards, but shoots are hungry work so it's very likely. Later, one of us (might've been Pete, might've been Storm) inspected the empty tray and had the bright idea that colored paint or ink poured into the various sections might make yet another cool photo. The tray was quickly modified with any leaky corners made watertight, and the relevant hue of paint was poured into each section. The effect was smooth, glossy, and rather pleasing to the eye.

Then, the unplanned started to occur. The separate areas of paint began slowly but surely to bleed into each other. But rather than becoming a hideous mess the experiment began to take on a whole new dimension, and we experienced something of a eureka moment. We started helping the migrating paint go its own sweet way. A swish here, a couple of drips there, and soon the previously rather rigid composition began to unravel into a wild psychedelic jungle. Areas of leaking paint expanded into impressive swirling whorls and delicate curlicues of color, stark and vibrant against their black backdrop. Fine and feathery veins of pigment unfurled like close-ups of a peacock's plumage or like NASA photos of the gigantic swirls in Jupiter's atmosphere. Blobs and bubbles emerged organically bringing to mind Pink Floyd's early liquid light shows. Detail was crisp and went on and on, a feast for the eyes and seriously entertaining for us. All the time, our intrepid photographer Rupert was poised a few feet above, dangling with his camera from a gantry, snapping frame after frame. Our magic tray had done most of our work for us, and we christened the process "controlled random." All that remained was for us to select a couple of shots for use—a nigh-on-impossible task given the multitude of beautiful frames we'd captured.

And so we come to the most recent stop on our prismatic journey. A few months ago we received an email from Mario Costa Sousa. He had spied "Liquid DSoM" (as we came to call it) on our website and politely enquired as to whether he and his fellow authors might use it as the cover for their new computer graphics textbook. Our first response was a friendly "yes" followed by fairly patronizing words to the effect of, "But Mario dear, do you realize that we created this for real, that it's not computer generated in any way?" Mario, clearly a man with his head screwed on the right way round, calmly explained that it was just what was needed.

First off, the basic image of the prism diffracting a beam of light is central to light and color theory and a truly crucial element in computer graphics. Second, the controlled randomness of the paint as it flows in specific, distinct directions reflects algorithmic modeling techniques often used in computer graphics, particularly in procedural image synthesis. Third, they enjoyed the idea of featuring a hand-created real life image on the front of a computer graphics textbook, implying that a technical reader might gain valuable insights into the theory and practice of computer graphics by observing real-world phenomena. And fourth, I suspect the authors may also be Pink Floyd fans, but we'll leave that for another day.

How appropriate then, that our design, an image that some might say was cribbed from a school textbook, should wind up through a variety of fairly exotic twists and turns, back on the cover of a textbook. Nothing random about that, eh?

— Dan Abbott, StormStudios
London, December 2011

Preface

This book has been used for various years in an introductory graduate-level course at the Institute of Pure and Applied Mathematics (IMPA), Rio de Janeiro, as part of the joint graduate program with the Catholic University of Rio de Janeiro, PUC-Rio, in computer graphics. This material has also been used in recent years at a senior undergraduate/first-year graduate level course in computer graphics in the Department of Computer Science at the University of Calgary. Many students of mathematics, engineering, and computer science have attended these courses at both IMPA, PUC-Rio, and the University of Calgary. The results have strengthened our conviction of the importance of emphasizing mathematical models in teaching computer graphics. This is especially true for students interested in pursuing more advanced studies: the important problems at the knowledge frontier in computer graphics involve nonelementary aspects of mathematical modeling.

This textbook has its fundamental roots in a publication by Jonas Gomes and Luiz Velho, *Computação Gráfica*, Volume 1, IMPA, 1998 (in Portuguese). Various chapters have been rewritten, other chapters have been carefully reviewed, new chapters have been added, and exercises have been included in order to cover the core material usually offered in an introductory course in computer graphics at the upper undergraduate or first-year graduate level. The book uses a problem-based learning approach in the sense that its fundamental goal is to provide a broad conceptual view of the main problems in computer graphics and to provide a framework for their solution. The content and exposition were elaborated in order to avoid the need for complementary texts at the fundamental computer graphics level. Prerequisites for this book include calculus, linear algebra, and basic topology and data structures.

As this is an introductory textbook, no previous knowledge of computer graphics is required, although the conceptual approach of the book requires that the reader be familiar with some concepts in continuous and discrete mathematics. This conceptual approach also allows this book to be adopted in more advanced courses with the appropriate complements. To facilitate its use, we included a list of additional topics at the end of each chapter.

It is important to highlight that the mathematical models of computer graphics only blossomed as a result of the various graphics and images produced on a computer screen, making the implementation of those models an inherent problem of the area. For this, we provide as supplemental material a complete e-book dedicated to this subject: *Design and Implementation of 3D Graphics Systems*. In our coursework programs we used this book and related notes to emphasize the implementation aspects.

Acknowledgments

Various colleagues collaborated on the initial volume from 1998 that gave origin to this book. Paulo Roma Cavalcanti gave us a great incentive for materializing this project. Paulo not only taught the course and created a set of initial notes, but also provided the very early preliminary reviews. Luiz Henrique de Figueiredo did a detailed and thorough review of some of the chapters and produced some of the illustrations that appear in the text, all properly credited. Many thanks to Margareth Prevot (IMPA, VisGraf Lab) who collaborated in the production of various images used in the text. We also thank everyone who allowed us to use figures from their works, all properly acknowledged in this book.

Various other colleagues read the preliminary versions of various chapters, saved us from some pitfalls, and gave us valuable suggestions. Among them, we can highlight Antonio Elias Fabris, Romildo José da Silva, Cícero Cavalcanti, Moacyr A. Silva, Fernando W. da Silva, Marcos V. Rayol Sobreiro, Silvio Levy, and Emilio Vital Brazil. We thank all sincerely. We also thank Jamie McInnis, Sarah Chow, and Patricia Rebolo Medici for their reviews and suggestions and for carefully editing and proofreading the book.

We sincerely thank Alice Peters for her dedication to this book project. We are honored to have the foreword in this book by Eugene Fiume and thank him for his inspiring words.

We are very grateful to Storm Thorgerson and Dan Abbott from StormStudios for giving us permission to use their original art photography "The Liquid Dark Side of the Moon" as our book cover. Many thanks to everyone else from StormStudios who helped to produce this art piece: Peter Curzon, Rupert Truman, Lee Baker, Laura Truman, Jerry Sweet, Charlotte Barnes, and Nick Baker. We would like to thank Dan Abbott very much for also describing "How the Prism Got Its Cover" as part of this book. Many thanks to Kara Ebrahim for working in the final cover layout design and production and to Dan Abbott for his valuable design suggestions.

The project of writing this book has been facilitated by the fruitful teaching and research environments of the computer graphics laboratory at IMPA (Visgraf Lab), and both the Department of Computer Science and the Interactive Reservoir Modeling and Visualization Group (*Illustra*Res/iRMV)/Computer Graphics Research Lab at the University of Calgary. Our sincere thanks go to all their members for their constant support. Finally, we sincerely appreciate the support from NSERC/Alberta Innovates Technology Futures (AITF)/Foundation CMG Industrial Research Chair program in Scalable Reservoir Visualization.

— Jonas Gomes, Luiz Velho, and Mario Costa Sousa
Rio de Janeiro and Calgary, December 2011

1 Introduction

Defining a research area is a difficult task and often an impossible one. The best definition is often pragmatic: what are the main problems to be solved, and how can they best be approached? In the case of computer graphics, the fundamental problem is ultimately that of transforming data into images:

As a result, computer graphics is commonly defined as a group of methods and techniques for transforming data into images displayed through a graphics device.

In this book, we will take the fundamental problem of computer graphics as our framework for studying the subject. We will divide it into subproblems and develop the theory and the mathematical models needed to solve each of them. Understanding and solving these subproblems will allow us to obtain a solution to the main problem.

The line between open and solved problems is often blurry in applied mathematics. In pure mathematics, new solutions to a solved problem do not necessarily amount to innovations contributing to scientific progress. By contrast, in applied mathematics, different solutions to the same problem generally follow from the use of new models, and may be greatly preferable from the viewpoint of practical applications.

This being an introductory book, we will generally use simple mathematical models, accessible to an undergraduate in the exact sciences who has taken basic linear algebra and some multivariable calculus. Some knowledge of data structures and the theory of algorithms is also useful. We stress, however, that our treatment of the subject is broad enough to be widely applicable, independently of the mathematical models being used.

1.1 Data, Images, and Computer Graphics

From its early days, the goal of computer graphics has been to allow the visualization of information. There is virtually no limitation on the source and nature of such information, and today computer graphics finds applications in virtually all fields of human activity: design and research of every sort, medicine, finance, entertainment … the list is endless. Despite the variety of applications, there is a conceptual core of shared techniques and methods, which can usefully be grouped into subdisciplines, based on the nature of the inputs and outputs (see Figure 1.1).

❑ *Geometric modeling* treats the problem of describing and structuring geometric data on the computer.

❑ *Image synthesis*, also known as *rendering*, involves manipulating data generated by a geometric modeling system to obtain an image that can be displayed on a graphics output device such as a monitor or printer.

❑ In *image processing*, the input is itself an image, to be modified in some way; the output is the processed image. Typical examples include colorizing, enhancing details, or combining images, as in the processing performed on the image stream from a satellite.

❑ *Image analysis*, more broadly known as *computer vision*, has the goal of extracting geometrical, topological, and physical information about the objects depicted in an image. Such techniques are very important, for example, in robotics—allowing robots to "see"—and in applications where real and synthetic scenes must be combined. Thus, while rendering focuses on the generation of images, computer vision treats the problem of interpreting them.

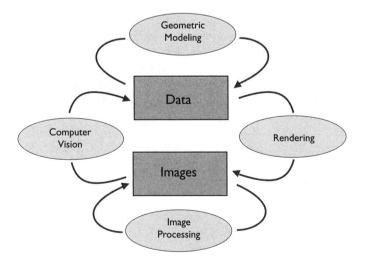

Figure 1.1. The four broadest subdisciplines of computer graphics.

1.1.1 Motion

Once we add the time dimension to computer graphics, things get even more interesting. Figure 1.2 indicates the analogous conceptual subdisciplines that take into account the time evolution of data and images.

- *Motion modeling* or *specification* deals with the modeling and description of moving objects in a scene. This includes both motion itself and elements such as path specification, merging and splitting of objects, and appearance changes.

- *Motion visualization* or *animation* translates the object and scene description into a sequence of *frames* (images), collectively known as *video*. Video can be stored in a variety of formats for subsequent visualization.

- *Video processing* is the manipulation of an animation sequence.

- *Motion analysis* is the part of image analysis that deals with obtaining information about a dynamic scene from the sequence of images that depict it.

1.1.2 Graphics Objects

We can extend this four-part scheme to other areas of computer graphics, drawing diagrams similar to Figures 1.1 and 1.2. But this repetitive process is an indication that we can introduce broader concepts that would allow us to merge all these diagrams into a more unified view of computer graphics. The key concept, to be introduced later, is that of a *graphics object*. Once that is done, we will be able to relate that concept to the four broad

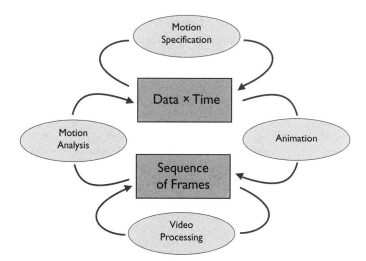

Figure 1.2. Subdisciplines of computer graphics, as applied to systems in motion.

realms exemplified in the last two diagrams: modeling, rendering, processing, and analysis. The notion of a graphics object, which will be introduced later on, must be broad enough to include geometric models, images, animation, video, etc.

1.1.3   What This Book Covers

This book covers the basics of geometric modeling and image processing, plus a more detailed treatment of rendering, particularly 3D surface visualization. The reader will also find an introduction to hierarchy animation and the parameterization of rigid motions in Euclidean space. Computer vision is mentioned in various parts of the book, but a deeper study of it lies outside our scope.

It should be stressed that, although we have drawn a distinction between modeling, rendering, and so on, most computer graphics applications require several or all of these subdisciplines to act in a unified way. For example, one or more images of a certain terrain captured by a satellite can be used to obtain a 3D reconstruction of the terrain. After colorization (shading), the image can be mapped onto the 3D model of the terrain, and the model can be rendered from many different angles. Similarly, medical applications using computer graphics require the seamless integration of a multitude of techniques from image processing, rendering, and computer vision. The combined use of techniques from these subdisciplines is what gives computer graphics its immense power, and has opened up whole new research areas.

1.2   Applications of Computer Graphics

As mentioned, computer graphics finds applications today in practically every field of knowledge. It makes a vital contribution especially when one needs to visualize objects that are still under design, or that cannot be seen directly, or that lie beyond our 3D reality.

In the first case, an object is modeled and virtually constructed on the computer. Such electronic prototypes can be manipulated and used in simulations to obtain information about a real object that is still in an early phase of design.

The visualization of objects that cannot be seen may include the rich subject of molecular modeling, where atomic-scale structures can be simulated, visualized, and analyzed, leading to the synthesis of new compounds and the prediction of the properties of existing ones. It may also include computerized tomography, of great value in medicine, materials science, and other areas. The rendering of physical and astronomical data collected in portions of the electromagnetic spectrum not visible to the eye (e.g., infrared, ultraviolet, x-rays) may also be included in this list.

Finally, computer graphics frees us from the confines of a 3D universe ruled by the laws of mechanics and Euclidean geometry. On the computer, one can visualize multidimensional objects and study the evolution of systems subject to laws different from those of the physical universe. This ability is extensively explored in mathematics, statistics, and physics.

Computer graphics applications can be grouped into three main areas: computer-aided design and manufacturing, data and motion visualization, and human-computer interaction.

In *computer-aided design*, or CAD, computer graphics allows the creation, representation, and analysis of models during the design phase, making it possible to visualize and try solutions not yet physically realized. Computer graphics can also be useful in the actual manufacturing process—down to the creation of the final product itself, in the case of desktop publishing or the machining of parts using numerically controlled tools. Computer-aided manufacturing is closely integrated with computer-aided design, hence the common abbreviation CAD/CAM.

The idea that "an image is worth a thousand words" motivates a plethora of computer graphics applications related to *data* and *motion visualization*. The computer becomes a tool that makes a fast qualitative analysis of complex data possible. In particular, the large-scale use of computer graphics for scientific visualization has become prevalent in the last two decades and is now indispensable in many branches of pure and applied scientific research.

Early *human-computer interaction* was entirely character-based, but today anyone who uses a computer or digital device typically does so through a graphical interface, and advances in computer graphics have made human communication with computers vastly easier and more pleasant. The most common paradigm is that of WIMP-based interfaces (window, icon, menu, pointing device) such as MS-Windows and Mac-OS, but other interfaces continue to evolve both for the general public and for specialized applications: iconographic languages, accessible systems for visually impaired users, navigation systems, and many other categories. We will close this section with a brief discussion of multimedia, a major source of applications of computer science today.

1.2.1 Multimedia

Information reaches us through different channels, such as sound, images, and text. The effectiveness of communication can be enhanced by integrating these channels to exploit the advantages and specific perceptual qualities of each, while ensuring that they act in concert and reinforce each other. This is what lies behind the notion of multimedia, or the simultaneous use of several media to convey information in a unified way (Figure 1.3).

The basic challenges of multimedia, above and beyond those associated with each medium, are *representation* (how the different information channels can be coherently encoded), *control* (including synchronization of the components of the information flow), and *storage* (how to keep and retrieve the information in the different forms in which it will be used).

One can reflect on the way these challenges were solved in a successful early example of multimedia: the motion picture with soundtrack. The audio waveform was encoded as an image that ran parallel to the frames (photographic images) along the film, and was converted back into sound by an appropriate transducer.

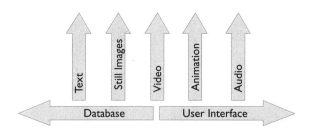

Figure 1.3. Conceptual model of a multimedia system.

Today, numerous types of digital format have been developed to address the needs of representation and storage. As for control, multimedia systems can be *local* or *distributed*. In local systems, control resides on a single computer only. In distributed systems, several computers control the system through network communication.

Computer graphics makes an important contribution to digital multimedia systems. This can take several forms, of which we highlight user interfaces, image synthesis, animation, and electronic publishing of text, images, and video. The presence of computer graphics in multimedia systems is so pervasive that some of it goes unnoticed to anyone but a computer graphics specialist.

1.3   The Four-Universe Paradigm

In applied mathematics, we need to model a variety of objects. For best results, we must create a hierarchy of abstractions and apply the most appropriate mathematical models to each abstraction level. In the case of computer graphics, a good abstraction paradigm consists of establishing four universes (sets): the physical, the mathematical, the representation, and the implementation universes:

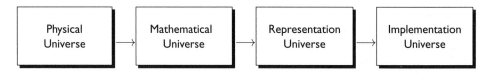

This is called the *four-universe paradigm*. The *physical universe P* contains the real-world objects we intend to study and model; the *mathematical universe M* contains an abstract description of those physical objects; the *representation universe R* is made up of symbolic, finite descriptions associated to objects of the mathematical universe; and in the *implementation universe I*, we associate the descriptions in the representation universe to data structures needed for computer manipulation.

Thus, to study a real-world phenomenon or object by computer, we first associate to it a mathematical model, then a finite representation of this model, which in turn is susceptible to computer implementation. The last step isolates the representation or discretization stage from the particulars of the programming language to be used in the implementation.

Example 1.1 (Numerical representation). Consider the problem of measuring objects of the physical world. For each object, we wish to associate a number representing its length, area, or volume, relative to a chosen unit.

In the mathematical universe, we associate a real number to each measurement. Rational and irrational numbers, respectively, correspond to objects that are commensurable and incommensurable with the unit of measurement adopted.

To represent the measurements, we must choose a discretization of the real numbers, such as the commonly used floating-point representation. Since all real numbers are then represented by a finite set of rational numbers, the notion of incommensurability does not exist in the representation universe. An implementation of the real numbers using floating-point representation can be made using the IEEE standard. For an introductory discussion to these topics, see [Higham 96]. ❑

This simple example illustrates one of the most vexing problems we face in computational mathematics, and therefore in computer graphics: moving from the mathematical to the representation universe generally involves loss of information. Here the possibility of incommensurability is lost altogether. It is necessary to be mindful of this loss of information at all times; indeed, much ingenuity is spent on minimizing its effects.

1.3.1 Interaction between Problems and Paradigm

Based on the four-universe paradigm, we can pose several general problems in our area of study:

❑ the definition of the elements of the mathematical universe M,

❑ the relation between the universes P, M, R, and I,

❑ the definition of the representation methods of M in R,

❑ the study of the properties of the various representations of M in R, and

❑ the conversion between different representations.

Once we define the elements of universe M, more specific problems can be posed, possibly with the creation of abstraction sublevels, in a process similar to the *top-down method* in structured programming.

The hierarchization of abstraction levels allows us to encapsulate the problems of each level and so reach a better description of these problems and their subsequent solutions, much as in object-oriented programming. The four-universe paradigm will be used on several occasions throughout the book. In each case, the paradigm will be particularized to the given area in order to elucidate that area's problems.

1.4   Example Models: Terrains and 2D Images

We now present two important example applications of the four-universe paradigm. These examples illustrate how different physical objects can be described by the same mathematical models.

1.4.1   Terrain Modeling

Consider the problem of storing on the computer the topography of a plot of land (for instance, a mountain). This can be done by using a height map: we establish a certain reference level and, for each point on the surface of the land, we consider its elevation.

In the mathematical universe, the height map defines a function $F \colon U \subset \mathbb{R}^2 \to \mathbb{R}$, $z = f(x, y)$, where (x, y) are the coordinates on the plane and z is the corresponding height, or elevation. Geometrically, the terrain is described by the graph of the height function f:

$$G(f) = \{(x, y, f(x, y))\}.$$

Figure 1.4(a) shows the graph of the height function of a portion of the Aboboral Mountains in the state of São Paulo, Brazil [Yamamoto 98].

How can we represent the mathematical model of the terrain? In other words, how can we discretize this function? If U is a rectangle $[x_{\min}, x_{\max}] \times [y_{\min}, y_{\max}]$, a simple method consists in taking partitions of the x- and y-axes,

$$P_x = \{x_0 < x_1 < \cdots < x_n\} \quad \text{and} \quad P_y = \{y_0 < y_1 < \cdots < y_m\}$$

(where $x_0 = x_{\min}$, $x_n = x_{\max}$, $y_0 = y_{\min}$, $y_m = y_{\max}$), and forming the Cartesian product of the finite sets so defined. This gives a grid of points (x_i, y_j), $i = 0, 1, \ldots, n$,

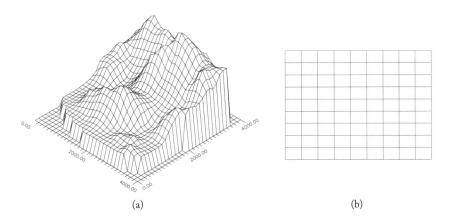

(a)   (b)

Figure 1.4. (a) Graph of the terrain function [Yamamoto 98]. (b) Associated grid.

$j = 0, 1, \ldots, m$, in the domain of the function. At each grid vertex (x_i, y_j), we take the value of the function $z_{ij} = f(x_i, y_j)$. We then have represented the terrain by a matrix of elevation values (z_{ij}). This is called the *sampling representation*.

A particularly simple case is that of *uniform sampling*, where the grid coordinates satisfy $x_{i+1} = x_i + \Delta x$ and $y_{i+1} = y_i + \Delta y$ for certain fixed Δx and Δy (and all i, j). Uniform sampling generates a *uniform grid* on the plane; the cells $[x_{i+1} - x_i] \times [y_{i+1} - y_i]$ are congruent rectangles, as illustrated in Figure 1.4(b). In this case, an implementation can easily be done using a matrix data structure, each element of the matrix indicating the elevation at the corresponding point.

1.4.2 2D Image Modeling

We now turn to the problem of representing a 2D, black-and-white image on the computer, say a photograph, regarded as a physical object in its own right. We thus have a support set—a rectangular piece of paper—and a certain darkness (gray tone) associated to each point of that support. We can associate to each gray tone a number in the interval $[0, 1]$, where 0 represents black and 1 represents white.

The rectangular support of the image is represented by a rectangular subset $U \subset \mathbb{R}^2$ of the plane. Therefore, the mathematical model of a black-and-white image is a function $f \colon U \subset \mathbb{R}^2 \to \mathbb{R}$, $z = f(x, y)$, associating to each point (x, y) the value z of the corresponding gray tone. This function is called an *image function*. Thus, we can describe the image using the graph of the image function. In Figure 1.5, we show an image and the graph of the corresponding function.

Given that the mathematical model of an image is the same as that of a terrain, we can use the uniform sampling representation described in the previous section. Thus, two completely different objects from the physical universe, a terrain and a black-and-white image, can be modeled by the same type of object in the mathematical universe: a real function of two variables, $f \colon U \subset \mathbb{R}^2 \to \mathbb{R}$.

Chapter 6 will be devoted to the study of images (including color images) and their representation. We will also study the problem of describing and representing models more complex than a terrain.

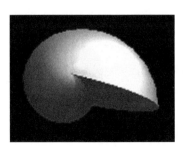 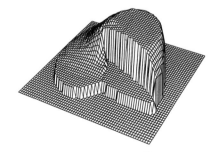

Figure 1.5. Image and graph of the image function.

1.5   Reconstruction

In a variety of situations, it is useful to be able to invert the representation process, deriving an object in the mathematical universe from a discrete representation. This is called *recon-struction* (see Figure 1.6). Reconstruction is important, for instance, in converting from one representation of an object to another, when we wish to work in the continuous domain to minimize computational error, in the ultimate visualization of an object (on a computer screen, for example), or when a graphics object is originally specified in the representation universe.

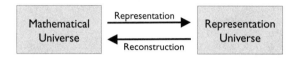

Figure 1.6. Representation and reconstruction.

It is clear from Example 1.1 that, given a representation in R of an object in M, the result of reconstruction is generally some other object in M that merely approximates the object of origin. In special cases, normally occurring only when the mathematical universe is populated with discrete objects to begin with, a representation may be *exact* or *lossless*, meaning that there is a reconstruction method that always yields the same object of origin; in other words, the reconstruction arrow is an exact inverse to the representation arrow in Figure 1.6. Exact representations are rare; most representations are *approximate* or *lossy*.

In Section 1.4 we introduced the sampling representation, in connection with terrain and 2D image models. The reconstruction operation in this case is called *interpolation*, and it amounts to constructing a continuous function $U \to \mathbb{R}$ given its values z_{ij} at finite num-ber of points of U. There are many variants: linear interpolation, Lagrange interpolation, and so on. They each lead to different approximations of the original function. Figure 1.7 shows two different reconstructions of the Aboboral Mountain data of Figure 1.4.

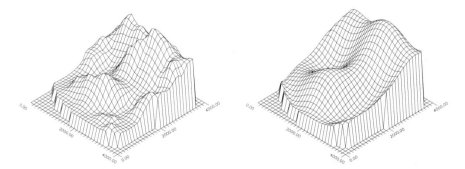

Figure 1.7. Two reconstructions of the same object [Yamamoto 98], using different sampling recon-struction methods and parameters. Compare with Figure 1.4.

1.6   A Practical Problem

We next illustrate the four-universe paradigm as it applies to a practical example: creating a computer system to guide the cutting of plane metal sheets into polygonal shapes.

In the physical universe, the objects to be modeled are flat pieces of metal as in Figure 1.8(a). In the mathematical universe, these shapes correspond to polyhedra. However, we can simplify the model by ignoring the thickness of the sheet, using instead planar polygonal regions, as in Figure 1.8(b).

Recall that a *closed polygonal curve* is a sequence of straight line segments $P_i P_{i+1}$, $i = 1, \ldots, n$, such that $P_{n+1} = P_1$, $P_i \neq P_j$ for $1 \leq i, j \leq n + 1$, and two segments only intersect each other at a common vertex. The points P_i are the *vertices* and the segments $P_i P_{i+1}$ are the *edges* of the polygonal curve.

Our problem now is to represent closed polygonal curves; that is, we must devise a finite description for a polygonal curve. Different representations are possible, of which we will mention and contrast two of them.

The simplest representation consists in listing the finitely many vertices P_1, P_2, $\ldots$, P_n of the polygonal curve in terms of its coordinates in $\mathbb{R}^2$; see Figure 1.9(a). We call this the *vertex list representation*.

A second representation is obtained by observing that the shape of a polygonal curve is completely determined if we know the lengths ℓ_i of the sides and the internal angles θ_i at each vertex; see Figure 1.9(b). We can then list the values $\ell_1, \theta_1, \ell_2, \theta_2, \ldots, \ell_n, \theta_n$. This is called the *internal angle representation*.

Observe that this second representation only determines the *shape* of the polygonal curve, not its placement or orientation in the plane. To obtain vertex coordinates we would need to fix the position of, say, vertex P_0, and the direction of, say, edge $P_0 P_1$. In other words, the internal angle representation determines the curve only up to a translation and a plane rotation. Because it does not depend on a particular coordinate system (unlike the vertex representation), we say the internal angle representation is *intrinsic*.

From the implementation point of view, both representations are equally easy; they can be implemented by list structures. What other advantages and disadvantages might there be to the two representations?

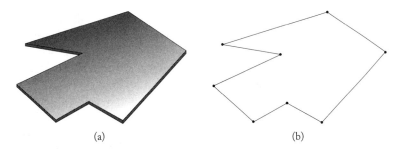

(a)   (b)

Figure 1.8. Different representations of same object. (a) Polyhedron. (b) Planar polygonal region.

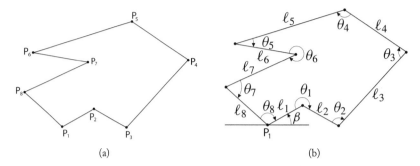

Figure 1.9. Representations of a planar polygonal curve. (a) Vertex list. (b) Internal angle.

The vertex representation is robust with respect to transformations of the plane. To apply a transformation T to the polygonal curve $C = (P_1, P_2, \ldots, P_{n+1})$, we apply T to each vertex, thus obtaining a list $T(C) = (T(P_1), T(P_2), \ldots, T(P_{n+1}))$, which serves as the representation of another polygonal curve. (Of course, unless T preserves straight lines, the edges of C don't necessarily map to edges of $T(C)$ under T.)

The representation by internal angles may not be so robust. If the transformation involves only rigid motions and scaling, there is no problem; the θ_i remain the same, and the ℓ_i are all multiplied by the same factor—and remain unchanged in the case of a rigid motion. But for an arbitrary transformation T of the plane, it is not obvious how the lengths and angles are affected. In fact, the simplest way to compute the representation of the transformed curve may be to calculate the vertices, apply T and convert back to the internal angle representation. Exercise 19 of this chapter (page 17) deals with the problem of converting between these two representations.

1.7 Image Making: The Physical and Mathematical Universes

Consider the process of photographing a constructed scene, say a still life or an image for a product catalog. Many factors contribute toward the product image. The photographer must choose where the photoshoot will take place: outdoors? in an indoor setting? in a professional studio? The objects in the scene must have been created somehow, either by natural processes of by their manufacturer. They must be artfully arranged to form the scene. They should be lit properly, generally with multiple light sources for the best effect. A camera is then used to record the image, after appropriate settings are chosen. Whether it is a film camera or a digital one, various steps are likely still needed after the shoot: development, printing, and perhaps toning and retouching in the former case; color correction, scaling, retouching, and finally rendering in the latter.

In the creation of an image in computer graphics, most of these factors from the real world have analogs in the mathematical universe. For example:

- The physical environment corresponds to the choice of a mathematical model of space, such as $\mathbb{R}^3$.

- The creation of the scene's objects is paralleled by the use of mathematical models to build the objects in the virtual scene.

- The placement of objects in the scene has as its counterpart the use of transformations in space.

- Physical lighting is replaced by mathematical illumination models with physical significance.

- The recording of the image corresponds to a transformation projecting the (generally 3D) scene to a plane (virtual camera).

- The photographic image is directly modeled in the mathematical universe, as briefly discussed in Section 1.4.2.

- As in the case of film and digital photographs, the synthesized image may undergo further processing steps—in this case mathematical transformations—that will lead to the final result.

1.7.1 What This Book Covers, Revisited

We use these parallels, together with the four-universe paradigm, as a guide to introduce the contents of the rest of this book, of which we now give a brief overview.

First, note the role of the ambient space and the transformations of that space in the scene. The geometry of space is fundamental to the study of computer graphics, and for this reason the next chapter is devoted to the notion of geometry, or rather of *geometries*. There we investigate and answer an important question: what is the geometry most appropriate to the practice of computer graphics?

Two further chapters cover specific geometric material needed later: Chapter 3 recapitulates changes between coordinate systems, and Chapter 4 studies the parameterization of the space of 3D rotations, a subject that is likely to be new to most students, and is not covered in detail in the literature.

Photography and indeed vision are only possible due to the presence of light, and a key characteristic of light is its frequency spectrum, which manifests itself perceptually as color. The color of an object in a scene is the result of absorption and emission of visible light, which often bounces off several objects until it goes through the lens and hits our retina, or the camera's photographic film or photosensitive back. In Chapter 5 we take the first step toward understanding color in computer graphics: namely, the mathematical modeling of the space of colors. The interaction of light and color with objects is left to later chapters.

Chapter 6 revisits the mathematical modeling of images and introduces basic notions of image processing.

We next turn to 2D and 3D objects (Chapters 7 and 8, respectively). Real-world objects have their mathematical counterpart in the notion of *graphical objects*, which are introduced, together with techniques used to manipulate them, in these two chapters.

The creation of mathematical models of physical objects is facilitated by the use *hierarchies*, discussed in Chapter 9. There and in Chapter 10, which is devoted to geometric modeling, we discuss the creation and structuring of data on the computer and the factors involved in choosing the best representation of a given object from the physical world.

After having discussed the objects of our virtual world and defined the concept of image, we proceed in Chapter 11 to study the problem of photographing our *virtual scene*. We introduce the notion of the virtual camera, the mathematical counterpart of a real photographic camera.

In subsequent chapters, we cover several operations and notions relevant to image synthesis and processing: clipping (Chapter 12), visibility (Chapter 13), illumination (Chapters 14 and 19), rasterization (Chapter 15), mapping techniques, including textures (Chapter 16), and image composition (Chapter 17). Chapter 18 provides an introduction to radiometry and photometry, an important subject for the understanding of illumination, providing complementary and background material for Chapters 5, 12, and 19.

1.8   Comments and References

The goal of this book is to give students, or anyone new to computer graphics, a global conceptual view of the field and an understanding of its main problems. Implementation issues are the subject of a companion volume, *Design and Implementation of 3D Graphics Systems*. Both books are used in computer graphics courses regularly offered at the Instituto de Matemática Pura e Aplicada in Rio de Janeiro and at the University of Calgary. Additional course-related material, particularly on implementation, can be found at http://www.crcpress.com/product/isbn/9781568815800.

The prerequisites for reading this book were briefly discussed on page 1. Elementary notions of topology are also useful. We assume some familiarity with graphics devices; a detailed discussion of such devices can be found in [Foley et al. 96] and [Gomes and Velho 02, Chapter 10].

More details on the four-universe paradigm discussed in this chapter and used throughout this book can be found in [Gomes and Velho 95].

Exercises

1.   Generalize the diagram of Figure 1.1 for the case of any graphics object.

2.   Find out something about *virtual reality* and how computer graphics is used in it. Try to formulate a description or classification, along the lines of this chapter, for the applications of computer graphics to virtual reality. Do the same for *mixed reality*.

3.   Compare and contrast visualization, image processing and computer vision. What are some areas where each of these disciplines finds applications?

4.   The *weighted average* of n points $p_1, \ldots, p_n$ in $\mathbb{R}^n$, with weights $w_1, \ldots, w_n \in \mathbb{R}$, is the point

$$p = \sum_{i=1}^{n} w_i p_1, \quad \text{where } w_1 + \cdots + w_n = 1.$$

Show that the weighted average is the minimum point of the function $f \colon \mathbb{R}^n \to \mathbb{R}$ defined by

$$f(q) = \frac{1}{2} \sum_{i=1}^{n} w_i |q - p_i|^2.$$

How could we use this notion of weighted average to describe an algorithm for terrain reconstruction?

5.   Give examples of applications involving

   (a) visualization and image processing,

   (b) visualization and computer vision,

   (c) visualization, computer vision, and image processing.

6.   Describe ten computer graphics applications in different areas of knowledge.

7.   Discuss the use of the four-universe paradigm in other areas of science. Establish, in each case, the abstraction levels and the nature of the elements at each level.

8.   Discuss the following: in a computer, every object is discrete. How is the notion of reconstruction useful, if a computer model, at any level, cannot be continuous? Or, perhaps, *what does a continuous object mean in the computer?*

9.   Give examples of an exact and an approximate representation.

10. Give an example of a terrain that cannot be represented by the functional model introduced in this chapter.

11. Our terrain model only considers the surface of the terrain. Describe a terrain model that takes into account its subsurface. Indicate at least one application where such a model can be useful.

12. Consider a function $f \colon \triangle ABC \to \mathbb{R}$, defined on a triangle $\triangle ABC \subset \mathbb{R}^2$. Knowing the values $f(A)$, $f(B)$, and $f(C)$ of f at the vertices of the triangle, describe a method to find an approximate linear reconstruction of f agreeing with f at the vertices. (Hint: remember barycentric coordinates.)

13. Consider a function $f \colon \square ABCD \to \mathbb{R}$ defined on a quadrilateral $\square ABCD \subset \mathbb{R}^2$. Knowing the values $f(A)$, $f(B)$, $f(C)$, and $f(D)$ of f at the vertices of the quadrilateral, is it possible to find an approximate linear reconstruction of f agreeing with f at the vertices? Why or why not?

14. Continuing the previous exercise, describe a method to obtain an approximate quadratic reconstruction of f. (Hint: think of bilinear interpolation.)

15. Extend the previous exercise to functions defined on a cube in space. What is the degree of the approximating polynomial?

16. Topographical data for a terrain are typically obtained through elevation measurements at a number of irregularly distributed points. Those measurements provide a representation of the height function as a list of points (x_i, y_i, z_i), where x_i and y_i are the horizontal coordinates of the ith point and z_i is the elevation there. This is called a *scattered data representation*.

   (a) Describe a method to obtain an approximate reconstruction of the terrain from that representation.

   (b) How can we obtain a uniform representation of the terrain starting from a scattered representation?

17. In a topographic survey of a terrain, it is very common to draw *elevation curves*.

   (a) Define precisely the notation of an elevation curve in our functional model of a terrain.

   (b) Discuss how to calculate the elevation curves of a terrain starting from a scattered representation.

   (c) Repeat the previous item for uniform sampling.

18. In a topographical survey, elevations relative to sea level (elevation 0) were determined by uniform sampling to have the values shown in Figure 1.10. Construct a polygonal approximation to the area that lies above sea level.

19. Define precisely representations of polygonal curves based on

   (a) edge lengths and edge orientations (angles formed by the edges with the x-axis),

   (b) edge vectors.

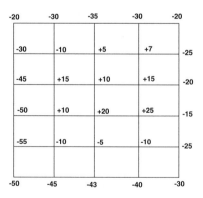

Figure 1.10. Topographic survey for Exercise 18.

Investigate and describe conversion procedures among the four representations discussed so far for polygonal curves: those in Section 1.6 and the two just introduced.

20. (a) Which of the four representations of polygonal curves in Exercise 19 is best suited to the study of deformations of polygonal curves, as physically modeled for instance by elastic or flexible rods?

(b) Which of them is best suited to the polygonal curves defining the outlines of characters, for use in optical character recognition (OCR) software?

21. Consider the problem of implementing on a computer a system for modeling and visualizing metal disks. Analyze this problem using the four-universe paradigm.

22. Consider a disk of radius r rotating uniformly around a point O on the plane. Give a detailed description of the disk motion using the four-universe paradigm, considering the representations of the disk and the motion.

2

Geometry

The key problem of computer graphics is to transform data into images. Roughly speaking, the data describe geometric objects that represent, or *model*, physical objects. Geometry, therefore, plays an important role in the methods and techniques used in computer graphics.

With this in mind, this chapter is not meant as a review of geometry. Its goal is to address a wider question: what is the right geometry for computer graphics?

Here we are using the word "geometry" in the sense of a coherent set of idealized objects, transformations, and rules. Selecting the appropriate type of geometry helps pose and solve problems correctly, both conceptually and from the implementation point of view.

Geometric transformations are of particular significance. They are essential in the creation and manipulation of geometric objects, in their visualization, and in the post-processing of the resulting images. To help answer our question, then, we list some of the characteristics the right geometry should have:

1. The objects in the geometry, such as points, vectors, lines, and planes, should have simple and well-defined semantics to allow consistent operations.

2. The transformations in the geometry should

   ❏ preserve the geometry's objects,

   ❏ possess a simple representation, to allow an easy, efficient, and consistent computer implementation, and

   ❏ suffice for the manipulation of the objects we wish to create and visualize.

2.1 What Is Geometry?

This is a difficult question to answer in a few words. Several types of geometry exist, and several ways of defining them. We will describe briefly three common methodologies used for defining a geometry: the axiomatic method, the coordinate method, and the transformation groups method.

2.1.1   The Axiomatic Method

In the axiomatic method, we introduce a space, or set of points in the geometry; the objects of the geometry, such as lines and planes; and a set of basic properties, called *axioms*, that the objects must satisfy. After that, we deduce other geometric properties in the form of theorems. This method was introduced by Euclid, a Greek mathematician from the third century BC, to define what we now call Euclidean geometry.

The set of axioms must be *consistent* (must not lead to a logical contradiction) and *complete* (must be enough to prove all the desired properties). Also useful is *independence* (no axiom should be derivable as a consequence of the others). The controversy over the independence of Euclid's fifth axiom is well known: it lasted 2,000 years and led to the discovery of non-Euclidean geometries.

The axiomatic method has great power to synthesize; it allows the common properties of many distinct spaces and objects to be subsumed into a single set of axioms.

From the computational point of view, the axiomatic method lends itself to the automatic demonstration of theorems, for example, through the so-called *logical framework approach* (LFA). However, the axiomatic method has the disadvantage of not determining a representation of the geometry in the computer.

2.1.2   The Coordinate Method

The coordinate method, also known as analytic geometry, was introduced by the French mathematician and philosopher René Descartes (1596–1650). It consists of defining a coordinate system in the geometry space such that both the objects of the geometry and the geometric properties (axioms and theorems) are translated into mathematical equations. The method allows an analytical approach for a great diversity of geometries, from Euclidean geometry on the plane to Riemannian geometry in differentiable manifolds.

A coordinate system introduces redundancy and arbitrariness: the coordinates (x, y, z) of a point in 3D space, say, indicate the distance from the point to three coordinate planes, and so depend on the choice of a coordinate system. We say they are not *intrinsic* to the geometric object (in this case a point). Thus, whenever we define a notion or property using coordinates, we need to show that it does not depend on the coordinate system used. Still, the coordinate method is very convenient computationally, if one chooses a correct representation of the coordinate system:

The representation of objects is dependent on the coordinate system, so it is difficult to develop automatic methods for semantic checks on geometric properties.

2.1.3   The Transformation Group Method

The transformation group method was formally introduced by the German mathematician Felix Klein (1849–1925), who was inspired by the principle of overlapping, used implicitly by Euclid in the demonstration of some of his geometric theorems. In this method, a geometry consists of a space S (the points in the geometry) and a group G of transformations of S, that is, one-to-one maps $g \colon S \to S$ that are continuous in both directions. Being a group, G satisfies the following properties:

❏ **associativity**: given $g, h, l \in G$, we have $(gh)l = g(hl)$;

❏ **identity element**: there exists $e \in G$ such that $ge = eg = g$ for every $g \in G$;

❏ **inverse elements**: for every $g \in G$, there exists $g^{-1} \in G$ such that $gg^{-1} = g^{-1}g = e$.

In this formulation, a *geometric object* is a subset of S. A *geometric property* is a property of geometric objects that is invariant under the action of G. Thus, if a geometric object O has property P, all images $g(O)$, for $g \in G$, must have property P. Two geometric objects O_1 and O_2 are said to be *congruent* if there exists an element $g \in G$ such that $g(O_1) = O_2$. When we want to emphasize the group G, we use the prefix G, and speak of G-congruence, G-properties, and so on.

Klein's approach can be summarized as follows:

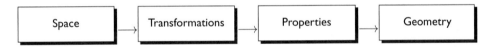

From Klein's point of view, the study of the geometry defined by a group G consists of identifying the properties invariant under G, describing G-congruence classes, and determining the relation between congruence classes and properties.

Example 2.1 (Plane Euclidean geometry). The best-known example, and one that inspired Klein, is *plane Euclidean geometry*, where the space S is a plane and the transformation group G is generated by rotations, translations, and reflections of the plane (G is called the set of *rigid motions* of the plane). Distance is preserved by all these transformations, and constitutes a property in this geometry.   ❑

The transformation group approach is mathematically very elegant and general, and has been enormously influential in the development of geometry. It has the advantage of allowing one to relate different geometries on the same space, by considering the relationship between the respective groups. In the transformation group approach there are no axioms, only theorems.

From the computational point of view, we should look for a representation of the space S and group G, so as to be able to implement models of the geometry:

In computer graphics, transformations are also associated with the motion of objects in space. For instance, the motion of a rigid body in its surrounding space can be characterized by changes in position and orientation. Our model will be that much more robust if those transformations are part of the transformation group of the geometry.

We will adopt the transformation group approach in our efforts to find the most appropriate geometry for computer graphics.

2.2   Transformations and Computer Graphics

Geometric transformations are related to two key concepts in computer graphics: coordinate changes and the deformation of objects in space.

Changing coordinates. Coordinate systems are used to formulate problems analytically. Through a coordinate system, we can calculate positions, velocities, and other measurements associated with physical objects, as discussed in the next chapter. A change of coordinates between two systems is performed by a transformation in space.

Deforming objects in space. In addition to rigid motions—on the plane, say, or in 3D space, as already discussed—one can consider *nonrigid deformations* of objects. Whereas rigid motions keep unaltered the distances between points (and for this reason are also called *isometries*), nonrigid deformations can change the distances between points internal to the objects being moved.

2.3   Euclidean Geometry

Euclidean geometry describes the space of our everyday experience extraordinarily well. People design buildings, build cars and put satellites into orbit using Euclidean geometry.

One generally makes these calculations using coordinates, but the end result does not depend on the coordinates used. We will follow a similar strategy: we introduce Euclidean geometry by means of its transformation group, but define the transformations on a very familiar space, $\mathbb{R}^n$—using coordinates, so to speak.

The space of points of Euclidean geometry will therefore be

$$\mathbb{R}^n = \{(x_1, \ldots, x_n) \; ; \; x_i \in \mathbb{R}\}. \tag{2.1}$$

We denote points in this space by boldface letters: $\mathbf{x}, \mathbf{u}, \mathbf{v}$, etc. In $\mathbb{R}^n$ there is a notion of addition, given by

$$(x_1, x_2, \ldots, x_n) + (y_1, y_2, \ldots, y_n) = (x_1 + y_1, x_2 + y_2, \ldots, x_n + y_n),$$

and multiplication by a scalar (i.e., a real number), given by

$$\lambda\,(x_1, x_2, \ldots, x_n) = (\lambda\,x_1, \lambda\,x_2, \ldots, \lambda\,x_n).$$

These make $\mathbb{R}^n$ into a *vector space*, and the points of $\mathbb{R}^n$ are called also vectors when we want to stress the vector space character of $\mathbb{R}^n$. The point $\mathbf{0} = (0, 0, \ldots, 0)$ plays a special role in the vector space. But in our everyday space there are no special points, nor is the notion of addition of points meaningful. In other words, the vector space structure of $\mathbb{R}^n$ is not actually part of Euclidean geometry; we just use it as a stepping stone. This can be compared with the remarks in Section 2.1.2.

2.3.1 Linear Transformations

Transformations preserving the vector space structure of $\mathbb{R}^n$ are called *linear*. That is, a linear transformation $L\colon \mathbb{R}^n \to \mathbb{R}^n$ is characterized by the properties

$$L(\mathbf{u} + \mathbf{v}) = L(\mathbf{u}) + L(\mathbf{v}) \quad \text{and} \quad L(\lambda\,\mathbf{u}) = \lambda\,L(\mathbf{u}), \tag{2.2}$$

for every $\mathbf{u}, \mathbf{v} \in \mathbb{R}^n$ and $\lambda \in \mathbb{R}$. In particular, L must fix the origin.

For computations, we use the well-known matrix representation of linear transformations. If $L\colon \mathbb{R}^n \to \mathbb{R}^n$ is linear and $\mathbf{e}_1 = (1, 0, 0, \ldots, 0)$, $\mathbf{e}_2 = (0, 1, 0, \ldots, 0)$, ..., $\mathbf{e}_n = (0, 0, 0, \ldots, 1)$ are the elements of the standard basis of $\mathbb{R}^n$, we define n vectors

$$\mathbf{a}_1 = L(\mathbf{e}_1) = (a_{11}, a_{21}, a_{31}, \ldots, a_{n1}),$$
$$\mathbf{a}_2 = L(\mathbf{e}_2) = (a_{12}, a_{22}, a_{32}, \ldots, a_{n2}),$$
$$\vdots$$
$$\mathbf{a}_n = L(\mathbf{e}_n) = (a_{1n}, a_{2n}, a_{3n}, \ldots, a_{nn}).$$

We now associate to L the matrix L_e whose columns are, in this order, the vectors $\mathbf{a}_1, \mathbf{a}_2, \ldots, \mathbf{a}_n$:

$$L_e = \begin{pmatrix} a_{11} & a_{12} & \cdots & a_{1n} \\ a_{21} & a_{22} & \cdots & a_{2n} \\ \vdots & \vdots & \ddots & \vdots \\ a_{n1} & a_{n2} & \cdots & a_{nn} \end{pmatrix}. \tag{2.3}$$

As L is linear, it follows that $L(\mathbf{x}) = L_e\mathbf{x}$ for any vector $\mathbf{x} \in \mathbb{R}^n$, where, on the right-hand side, the juxtaposition denotes the product of matrices and $\mathbf{x}$ is written as a column vector, that is, an $n \times 1$ matrix. (We will use this convention throughout: a vector written to the right of a matrix is a column vector. Some authors use the notation $\mathbf{x}^{\mathrm{T}}$ to make this explicit, but that is unnecessary since no other interpretation is possible.)

We have shown that the value of a linear transformation at a point can be obtained by left-multiplying the point (vector) by the matrix associated to the transformation. Conversely, if A is a matrix of order n, we define a transformation $L \colon \mathbb{R}^n \to \mathbb{R}^n$ by

$$L(\mathbf{x}) = A\mathbf{x} = \begin{pmatrix} a_{11} & a_{12} & \cdots & a_{1n} \\ a_{21} & a_{22} & \cdots & a_{2n} \\ \vdots & \vdots & \ddots & \vdots \\ a_{n1} & a_{n2} & \cdots & a_{nn} \end{pmatrix} \begin{pmatrix} x_1 \\ x_2 \\ \vdots \\ x_n \end{pmatrix}. \tag{2.4}$$

It is easy to verify that L is linear and that $L_e = A$. Therefore, we have a one-to-one correspondence between linear transformations of $\mathbb{R}^n$ and the set of $n \times n$ matrices. The importance of this correspondence lies in that it preserves the operations on the two spaces. That is, the composition of linear transformations corresponds to the product of matrices, and the sum of linear transformations corresponds to the sum of matrices. In symbols,

$$(T \circ L)(\mathbf{x}) = T(L(\mathbf{x})) = (T_e L_e)\mathbf{x},$$
$$(T + L)(\mathbf{x}) = T(\mathbf{x}) + L(\mathbf{x}) = (T_e + L_e)\mathbf{x}.$$

Therefore, we have a good representation for the space of linear transformations in $\mathbb{R}^n$. Computationally, the manipulation of linear transformations boils down to performing matrix operations.

2.3.2 Orthogonal Transformations, Isometries, and the Euclidean Group

The next ingredient we need is a metric, or notion of distance, in $\mathbb{R}^n$. We start with the *inner product* (also known as dot product)

$$\langle \mathbf{u}, \mathbf{v} \rangle = \sum_{i=1}^{n} u_i v_i, \tag{2.5}$$

where $\mathbf{u} = (u_1, \ldots, u_n)$ and $\mathbf{v} = (v_1, \ldots, v_n)$. We can now define two important notions:

❑ The *length* (or *norm*) of a vector u is $\|\mathbf{u}\| = \sqrt{\langle \mathbf{u}, \mathbf{u} \rangle}$.

❑ The *angle* θ between two nonzero vectors $\mathbf{u}$ and $\mathbf{v}$ is

$$\cos \theta = \frac{\langle \mathbf{u}, \mathbf{v} \rangle}{\|\mathbf{u}\| \, \|\mathbf{v}\|}.$$

Geometrically, the inner product is the measure of the projection of vector $\mathbf{x}$ over $\mathbf{y}$, weighted (multiplied) by the length of $\mathbf{y}$. See Figure 2.1(a). Equivalently, it is the projection of vector $\mathbf{y}$ over $\mathbf{x}$, weighted by the length of $\mathbf{x}$.

The distance $d(\mathbf{x}, \mathbf{y})$ between two points $\mathbf{x}$ and $\mathbf{y}$ in $\mathbb{R}^n$ is defined as the length of the difference vector: $d(\mathbf{x}, \mathbf{y}) = \|\mathbf{y} - \mathbf{x}\|$. See Figure 2.1(b). Clearly, $d(\mathbf{x}, \mathbf{y}) = d(\mathbf{y}, \mathbf{x})$. We have now made $\mathbb{R}^n$ into a metric space, that is, a space with a notion of distance.

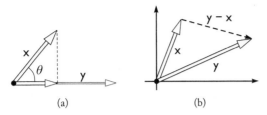

Figure 2.1. (a) The inner product of **x** and **y** is equal to the length of **y** times the length of the projection of **x** in the direction of **y**. (b) The distance from **x** to **y** is the length of **y** − **x**.

A linear transformation $T \colon \mathbb{R}^n \to \mathbb{R}^n$ is called *orthogonal* if it preserves the inner product; that is, if

$$\langle T(\mathbf{u}), T(\mathbf{v}) \rangle = \langle \mathbf{u}, \mathbf{v} \rangle \quad \text{for all } \mathbf{u}, \mathbf{v} \in \mathbb{R}^n.$$

Orthogonal transformations preserve lengths and distances; thus they are *isometries*. Isometries move points and objects around, but maintain all metric relations between them.

There are other isometries besides orthogonal transformations, of course: for instance, the *translation* $T \colon \mathbb{R}^2 \to \mathbb{R}^2$ defined by $T(\mathbf{u}) = \mathbf{u} + \mathbf{t}$, for a fixed $\mathbf{t} \in \mathbb{R}^n$. (Translations—other than the trivial one, when $\mathbf{t} = \mathbf{0}$—cannot be linear transformations since they do not preserve the origin of $\mathbb{R}^n$.)

Two objects $\mathcal{O}_1$ and $\mathcal{O}_2$ in space are *congruent* if an isometry $T \colon \mathbb{R}^n \to \mathbb{R}^n$ exists such that $T\mathcal{O}_1 = \mathcal{O}_2$.

Euclidean geometry (in n dimensions) *is the geometry of isometries of* $\mathbb{R}^n$. Euclidean space is the set $\mathbb{R}^n$, considered not as a vector space, but as something acted on by isometries. The vector space structure was just the scaffolding. The isometries are what count. Or, put yet another way: the group of transformations of Euclidean geometry, in the sense of Klein, is the group of isometries of the metric space $\mathbb{R}^n$.

So we naturally want to know what this group is. It can be shown that a (not necessarily linear) transformation $T \colon \mathbb{R}^n \to \mathbb{R}^n$ is an isometry if and only if it is of the form

$$T(\mathbf{u}) = L(\mathbf{u}) + \mathbf{t} \quad \text{for all } \mathbf{u}, \tag{2.6}$$

where L is a fixed orthogonal linear transformation and $\mathbf{t}$ is a fixed vector. In geometric terms, an isometry is the composition of an orthogonal linear transformation with a translation. (Orthogonal transformations of $\mathbb{R}^n$ are precisely those isometries for $\mathbf{t} = 0$, that is, those that preserve the origin.)

Lines, planes, and other "flat pieces" of Euclidean space are the images of vector subspaces of $\mathbb{R}^n$ under isometries. (Recall that vector subspaces are those lines, planes, and so on *that contain the origin*. By applying all isometries of $\mathbb{R}^n$—or just all translations, in fact—to these objects, we obtain all lines and planes in $\mathbb{R}^n$.)

Equation (2.6) gives a working representation of Euclidean isometries. (We can also write it as $T(\mathbf{u}) = L_e \mathbf{u} + \mathbf{t}$, where L_e is the matrix of L.) This representation is not as easy to work with as one would like. For example, suppose we have two Euclidean isometries in the form (2.6), say $T(\mathbf{u}) = L(\mathbf{u}) + \mathbf{t}$ and $T'(\mathbf{u}) = L'(\mathbf{u}) + \mathbf{t}'$. If we are interested in composing the two, we can write

$$(T \circ T')(\mathbf{u}) = T(T'(\mathbf{u})) = L(T'(\mathbf{u})) + \mathbf{t} = L(L'(\mathbf{u}) + \mathbf{t}') + \mathbf{t}$$
$$= L(L'(\mathbf{u})) + L(\mathbf{t}') + \mathbf{t} = (L \circ L')\mathbf{u} + (L(\mathbf{t}') + \mathbf{t}).$$

That is, the expression of $T \circ T'$ in the form (2.6) has a linear part $L \circ L'$ that is just the product of the linear parts of the component isometries, but the translation part is messy: $L(\mathbf{t}') + \mathbf{t}$. It would be much nicer if the algebra of Euclidean isometries were as simple as that of linear transformations of $\mathbb{R}^n$. This can in fact be achieved, as we shall see in the next section.

The trick is to place our n-dimensional Euclidean space inside a larger space, and look at those linear transformations of the larger space that map the smaller space onto itself. We'll describe this in more detail next.

2.4 Affine Geometry

We have seen that Euclidean space is, in some sense, $\mathbb{R}^n$ minus the vector space structure. What exactly does this mean? One way to make this clearer is to choose a different model of Euclidean geometry, one that is not a vector space to begin with. A *model* of a geometry is simply a set of points $\mathcal{P}$, concretely defined somehow, where all the features of the geometry apply: in this case distances, isometries, lines, planes...[1]

We will take as our alternate model the subset $\mathcal{P}$ of $\mathbb{R}^{n+1}$ given by

$$\mathcal{P} = \{(x_1, \ldots, x_n, 1) \, ; \, x_i \in \mathbb{R}\}.$$

(See Figure 2.2.) This set is not a vector space; if we add two points in it as vectors in $\mathbb{R}^{n+1}$, the result is not in $\mathcal{P}$ (its last coordinate is 2). However, we can immediately transfer to $\mathcal{P}$ all the paraphernalia already defined for $\mathbb{R}^n$, via the correspondence $(x_1, \ldots, x_n) \leftrightarrow (x_1, \ldots, x_n, 1)$. Some observations are worth making:

❏ The metric defined on $\mathcal{P}$ by transfer from $\mathbb{R}^n$ is the same as the metric it has as a subset of the Euclidean space $\mathbb{R}^{n+1}$.

❏ Points in $\mathcal{P}$ correspond to vector lines in $\mathbb{R}^{n+1}$ that are not parallel to $\mathcal{P}$. (Given a vector line ℓ in $\mathbb{R}^{n-1}$, take its intersection with $\mathcal{P}$; conversely, given a point in $\mathcal{P}$, take the line joining it to the origin of $\mathbb{R}^{n+1}$.)

❏ Lines in $\mathcal{P}$ (defined by transfer from $\mathbb{R}^n$) correspond to vector planes in $\mathbb{R}^{n+1}$ that are not parallel to $\mathcal{P}$.

---

[1]More generally, $\mathcal{P}$ can be an abstract set acted on by a transformation group isomorphic to the group defining the geometry. For instance, affine spaces, which we define in a concrete way below, can be defined abstractly by starting from a pair $(\mathcal{P}, \mathcal{V})$, where $\mathcal{P}$ is a set on which the additive group of a vector space $\mathcal{V}$ acts faithfully and transitively. The transformations of $\mathcal{P}$ that are compatible with the action of $\mathcal{V}$, in an appropriate sense, are defined to be affine transformations. We will not need this much generality.

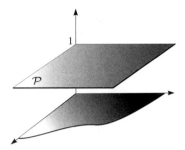

Figure 2.2. Model of n-dimensional Euclidean space as an affine hyperplane of $\mathbb{R}^{n+1}$.

❑ An isometry of $\mathcal{P}$ (defined by transfer from $\mathbb{R}^n$) is always the restriction to $\mathcal{P}$ of some linear transformation of $\mathbb{R}^{n+1}$.

This last point deserves elaboration, because it allows the computational simplification referred to at the end of the previous section. Consider first an isometry Λ of $\mathcal{P}$ that comes from an orthogonal transformation L of $\mathbb{R}^n$. That is, if $L(x_1, \ldots, x_n) = (y_1, \ldots, y_n)$, then $\Lambda(x_1, \ldots, x_n, 1) = (y_1, \ldots, y_n, 1)$. If the $n \times n$ matrix of L is L_e, we clearly have

$$\Lambda(x_1, \ldots, x_n, 1) = \left(\begin{array}{c|c} L_e & \begin{matrix} 0 \\ \vdots \\ 0 \end{matrix} \\ \hline 0 \quad \cdots \quad 0 & 1 \end{array} \right) (x_1, \ldots, x_n, 1), \qquad (2.7)$$

so Λ is the restriction to $\mathcal{P}$ of the linear transformation of $\mathbb{R}^{n+1}$ corresponding to the matrix in (2.7). Next, consider the isometry τ of $\mathcal{P}$ that comes from a translation $\mathbb{R}^n$ by the vector $\mathbf{t} = (t_1, \ldots, t_n)$. Then τ acts on $\mathcal{P}$ as the translation by $(t_1, \ldots, t_n, 0) \in \mathbb{R}^{n+1}$. Now, a translation *of all of* $\mathbb{R}^{n+1}$ would not be a linear transformation of $\mathbb{R}^{n+1}$, but we just need a linear transformation that acts as a translation *on part of* $\mathbb{R}^{n+1}$, namely, on our model $\mathcal{P}$. This is easy to find—for example, by considering what the transformation must do to the standard basis of $\mathbb{R}^{n+1}$, given what it does to points in $\mathcal{P}$. The answer is

$$\tau(x_1, \ldots, x_n, 1) = \left(\begin{array}{ccc|c} 1 & & & t_1 \\ & \ddots & & \vdots \\ & & 1 & t_n \\ \hline 0 & \cdots & 0 & 1 \end{array} \right) (x_1, \ldots, x_n, 1), \qquad (2.8)$$

Now, because of the decomposition (2.6), any Euclidean transformation of $\mathcal{P}$ can be obtained by applying first a linear transformation of $\mathbb{R}^{n+1}$ of the form shown in (2.7) and then another linear transformation of the form shown in (2.8). Since the composition of two linear transformations is a linear transformation, we have reached our goal of representing Euclidean isometries as linear transformations (of a vector space one dimension higher).

We remark that although the transformation of $\mathbb{R}^{n+1}$ shown in (2.7) is orthogonal, and so is an isometry on all of $\mathbb{R}^{n+1}$, the same is not true of the linear transformation of $\mathbb{R}^{n+1}$ shown in (2.8) (unless $t_1 = \cdots = t_n = 0$, of course). The latter type of transformation is called a *shear*, and it will play a role in Chapter 11 (see also Exercise 8 on page 48).

2.4.1   Affine Transformations and Affine Space

Something else comes out of the previous discussion. Any linear transformation of $\mathbb{R}^{n+1}$ obtained by multiplying matrices of the forms shown in (2.7) and (2.8) will induce an isometry of $\mathcal{P}$, because a composition of isometries is an isometry. So the set of all such linear transformations—more precisely, the group generated by the transformations whose matrices are given in (2.7) and (2.8)—is exactly the set of linear transformations of $\mathbb{R}^{n+1}$ that map $\mathcal{P}$ to itself *isometrically*.

But why stop there? We can just as well ask what are *all* the linear transformations of $\mathbb{R}^{n+1}$ that map $\mathcal{P}$ to itself, isometrically or not. It is not hard to find the answer. Since such a transformation keeps unaltered the $(n+1)$-st coordinate for points in $\mathcal{P}$, it must do so for points everywhere, by linearity. Therefore the last row of the corresponding matrix must be $(0, \ldots, 0, 1)$; that is, the matrix has the form

$$
\begin{pmatrix}
a_{11} & a_{12} & \cdots & a_{1n} & a_{1\,n+1} \\
a_{21} & a_{22} & \cdots & a_{2n} & a_{2\,n+1} \\
\vdots & \vdots & \ddots & \vdots & \vdots \\
a_{n1} & a_{n2} & \cdots & a_{nn} & a_{n\,n+1} \\
0 & 0 & \cdots & 0 & 1
\end{pmatrix}.
\tag{2.9}
$$

Conversely, all invertible linear transformations of $\mathbb{R}^{n+1}$ whose matrix has this form preserve $\mathcal{P}$. The corresponding transformations of $\mathcal{P}$ are called *affine transformations*.

Note that we can write (2.9) as the product

$$
\begin{pmatrix}
1 & & & a_{1\,n+1} \\
& \ddots & & \vdots \\
& & 1 & a_{n\,n+1} \\
0 & \cdots & 0 & 1
\end{pmatrix}
\begin{pmatrix}
a_{11} & \cdots & a_{1n} & 0 \\
\vdots & \ddots & \vdots & \vdots \\
a_{n1} & \cdots & a_{nn} & 0 \\
0 & \cdots & 0 & 1
\end{pmatrix},
\tag{2.10}
$$

with both matrices invertible since (2.9) is invertible. The matrix on the right is like that in (2.7) (where now $L_e = (a_{ij})_{i,j=1,\ldots,n}$ is not required to be orthogonal), and its invertibility is equivalent to that of L_e. Moreover, if we consider for a moment the model $\mathbb{R}^n$ instead of $\mathcal{P}$ (that is, if we drop the $(n+1)$-st coordinate), the action of this matrix reduces to a linear transformation of $\mathbb{R}^n$, given by multiplication by L_e.

The matrix on the left in (2.10) is exactly of the same form as the one in (2.8), with $t_i = a_{i\,n+1}$; so its action on $\mathcal{P}$ is a translation, from the discussion preceding (2.8), and it is always invertible.

To summarize our work so far: *Affine geometry* (in n dimensions) is the geometry of affine transformations, which are defined as the restrictions to $\mathcal{P} = \{(x_1, \ldots, x_n, 1) \mid x_i \in \mathbb{R}\}$ of those linear transformations of $\mathbb{R}^{n+1}$ that map $\mathcal{P}$ onto itself. If we identify $\mathcal{P}$ with $\mathbb{R}^n$ by dropping the $(n+1)$-st coordinate, a transformation $T \colon \mathbb{R}^n \to \mathbb{R}^n$ is affine if and only if it has the form

$$T(\mathbf{u}) = L(\mathbf{u}) + \mathbf{t} \quad \text{for all } \mathbf{u}, \tag{2.11}$$

where L is a fixed invertible linear transformation of $\mathbb{R}^n$ and $\mathbf{t}$ is a fixed vector.

2.4.2 Points, Vectors, and Subspaces

In both Euclidean and affine spaces, points cannot be added together, nor multiplied by scalars. This is clear in the $\mathcal{P}$ model, which has no origin. (Earlier we identified $\mathcal{P}$ with $\mathbb{R}^n$, with $(0, \ldots, 0, 1)$ corresponding to the origin of $\mathbb{R}^n$; if we had chosen any other point in $\mathcal{P}$ for this matter, defining the rest of the correspondence by subtraction, we would have the same result. For instance, the last few lines of the previous section would be equally true. (You are encouraged to stop and persuade yourself of this with careful arguments.)

By contrast, points in affine space[2] can be *subtracted* from one another, though the result is not a point in the space—it is a vector! More precisely, the *translations* of an affine space form a vector space, and whatever the model, we can regard the difference between two points of the space as the unique translation that takes one to the other. We can also turn this around and write the action of a translation additively:

$$\mathbf{q} = \mathbf{p} + \mathbf{u} \quad \Longleftrightarrow \quad \mathbf{q} - \mathbf{p} = \mathbf{u},$$

where $\mathbf{p}, \mathbf{q} \in \mathcal{P}$ and $\mathbf{u}$ is a translation vector.

Example 2.2 (Parametric equation of a straight line). Given two distinct points $\mathbf{q}_1$ and $\mathbf{q}_2$ in affine space, the vector $\mathbf{q}_2 - \mathbf{q}_1$ can be used to describe the unique line containing $\mathbf{q}_1$ and $\mathbf{q}_2$. We simply add arbitrary multiples of this vector to either point:

$$\mathbf{r}(t) = \mathbf{q}_1 + t(\mathbf{q}_2 - \mathbf{q}_1), \quad t \in \mathbb{R}. \tag{2.12}$$

We can rewrite this expression as $(1-t)\mathbf{q} - 1 + t\mathbf{q}_2$, but we need to be careful about what this means. Neither summand is meaningful as a point in affine space! It is only the whole expression that has meaning. What makes it work is that the coefficients of $\mathbf{q}_1$ and $\mathbf{q}_2$ add up to 1, and so the expression can be rearranged in the form (2.12), which makes sense because $\mathbf{q}_2 - \mathbf{q}_1$ is a vector, which can be multiplied by scalars. ❏

---

[2]From now on, when we mention affine space, we leave it to the reader to check whether the same is true for Euclidean space. Remember that Euclidean space is affine space with more structure—the metric—and of course with fewer structure-preserving transformations. So nonmetric statements, say about lines, translations, or affine combinations, apply equally well to Euclidean space.

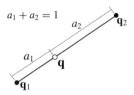

Figure 2.3. Affine combination of points.

In spite of this caveat, the expression $(1-t)\mathbf{q}_1 + t\mathbf{q}_2$ is very useful. It is called an *affine combination* of $\mathbf{q}_1$ and $\mathbf{q}_2$ with coefficients $a_1 = 1-t$ and $a_2 = t$. See Figure 2.3. More generally, we can form the *affine combination* of n points $\mathbf{q}_i \in \mathcal{P}$ with weights a_i whose sum is 1:

$$\sum_{i=1}^{n} a_i \mathbf{q}_i \in \mathcal{P} \quad \Longleftrightarrow \quad \sum_{i=1}^{n} a_i = 1.$$

If each a_i is in $[0, 1]$, the affine combination is also called *convex combination* or *interpolation*.

We can summarize the palette of operations between points in affine geometry by stating that, unlike points in a vector space (i.e., vectors), which can be combined with arbitrary coefficients, points in affine space can be only combined in two situations:

❏ affine combination: when the sum of the coefficients is 1, the result is a point in affine space;

❏ subtraction: when the sum of the coefficients is 0, the result is a vector.

Affine subspaces of $\mathcal{P}$ can be characterized in several equivalent ways, the first two having already been informally mentioned (see bullet list on page 26):

❏ as the intersections with $\mathcal{P}$ of vector subspaces of $\mathbb{R}^{n+1}$,

❏ by transfer of affine subspaces of $\mathbb{R}^n$ (which are the usual lines, planes etc., not necessarily going through the origin; see page 25), or

❏ as sets of all affine combinations of some number of points (just as vector subspaces of a vector space are the sets of linear combinations of some number of points).

Affine transformations, not surprisingly, preserve affine combinations:

$$\sum_{i=1}^{n} a_i = 1 \quad \Longrightarrow \quad T\left(\sum_{i=1}^{n} a_i \mathbf{p}_i\right) = \sum_{i=1}^{n} a_i T(\mathbf{p}_i).$$

In particular, since a line is the set of affine combinations of any two of its points, affine transformations take lines to lines; likewise for higher-dimensional subspaces.

Example 2.3 (Parallelism). If Euclidean geometry can be characterized as the geometry that preserves the usual metric of $\mathbb{R}^n$, so can affine geometry be characterized as the geometry that preserves parallelism.

Two lines in affine space are *parallel* if they can be mapped to one another by a translation. Likewise for two planes, or two affine subspaces of the same dimension. Affine transformations preserve parallelism: if r and s are parallel lines (planes, subspaces) and T is an affine transformation, then $T(r)$ and $T(s)$ are parallel. We leave the justification as an important exercise. ❏

2.4.3 Affine Coordinates

Let $\mathbf{o}$ be a fixed point in affine space and let $\{\mathbf{v}_1, \mathbf{v}_2, \ldots, \mathbf{v}_n\}$ be a basis of the vector space of translations. Any point $\mathbf{p}$ in affine space can be written as $\mathbf{v} + \mathbf{o}$, where $\mathbf{v}$ is a vector, and therefore as

$$\mathbf{p} = c_1\mathbf{v}_1 + c_2\mathbf{v}_2 + \cdots + c_n\mathbf{v}_n + \mathbf{o}, \tag{2.13}$$

with the coefficients c_i uniquely defined. The list $F = (\mathbf{v}_1, \mathbf{v}_2, \ldots, \mathbf{v}_n, \mathbf{o})$, called a *reference frame*, defines a coordinate system in affine space: the point $\mathbf{p}$ is assigned the coordinates $(c_1, c_2, \ldots, c_n, 1)$.

The 1 at the end makes computations easier, as we saw in Section 2.4 for the *standard frame* of $\mathcal{P}$, the one where $\mathbf{o} = \{0, \ldots, 0, 1\} \in \mathcal{P}$ and $\mathbf{v}_i$ is the translation vector in the i-th coordinate direction. The matrix representation defined there for affine maps works for any frame of any affine space (just as a linear transformation of a vector space has a matrix representation in any basis, not just the standard basis). Specifically, if $F = (\mathbf{v}_1, \mathbf{v}_2, \ldots, \mathbf{v}_n, \mathbf{o})$ is a frame and T is an affine transformation, we can express the images of certain points in terms of F:

$$T(\mathbf{o}) = \sum_{i=1}^{n} a_{i\,n+1}\mathbf{v}_i \quad \text{and} \quad T(\mathbf{v}_j + \mathbf{o}) = \sum_{i=1}^{n} a_{ij}\mathbf{v}_i + T(\mathbf{o}).$$

Then, for any point $\mathbf{p}$ of the form (2.13), we have

$$T(\mathbf{p}) = \sum_{j=1}^{n} c_j T(\mathbf{v}_j) + T(\mathbf{o}) = \sum_{i=1}^{n}\sum_{j=1}^{n} a_{ij}c_j\mathbf{v}_i + \sum_{i=1}^{n} a_{i\,n+1}\mathbf{v}_i + T(\mathbf{o})$$

$$= \sum_{i=1}^{n}\left(\sum_{j=1}^{n} a_{ij}c_j + a_{i\,n+1}\right)\mathbf{v}_i + T(\mathbf{o}).$$

Thus the coordinates of $T(\mathbf{p})$ in the reference frame F are given by the matrix product

$$\begin{pmatrix} a_{11} & a_{12} & \cdots & a_{1n} & a_{1\,n+1} \\ a_{21} & a_{22} & \cdots & a_{2n} & a_{2\,n+1} \\ \vdots & \vdots & \ddots & \vdots & \vdots \\ a_{n1} & a_{n2} & \cdots & a_{nn} & a_{n\,n+1} \\ 0 & 0 & \cdots & 0 & 1 \end{pmatrix} \begin{pmatrix} c_1 \\ c_2 \\ \vdots \\ c_n \\ 1 \end{pmatrix}. \tag{2.14}$$

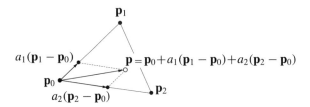

Figure 2.4. Affine combination in a triangle.

Example 2.4 (Affine bases and barycentric coordinates). Consider $n + 1$ points $\mathbf{p}_0$, $\mathbf{p}_1, \ldots, \mathbf{p}_n$ and take the differences $\mathbf{v}_i = \mathbf{p}_i - \mathbf{p}_0$. If $(\mathbf{v}_1, \ldots, \mathbf{v}_n, \mathbf{p}_0)$ is a reference frame, we say that $(\mathbf{p}_0, \mathbf{p}_1, \ldots, \mathbf{p}_n)$ is an *affine basis*.

Any point $\mathbf{p}$ of affine space can be written in the frame $(\mathbf{v}_1, \ldots, \mathbf{v}_n, \mathbf{p}_0)$, say with coordinates $(a_1, \ldots, a_n, 1)$. Then

$$\mathbf{p} = a_1 \mathbf{v}_1 + a_2 \mathbf{v}_2 + \cdots + a_n \mathbf{v}_n + \mathbf{p}_0$$
$$= a_1(\mathbf{p}_1 - \mathbf{p}_0) + \cdots + a_n(\mathbf{p}_n - \mathbf{p}_0) + \mathbf{p}_0 = a_0 \mathbf{p}_0 + a_1 \mathbf{p}_1 + \cdots a_n \mathbf{p}_n,$$

where we have defined $a_0 = 1 - a_1 - \cdots - a_n$. Thus we have expressed p as an affine combination of $\mathbf{p}_0, \mathbf{p}_1, \ldots, \mathbf{p}_n$. We call $(a_0, a_1, \ldots, a_n)$ the *barycentric coordinates* of $\mathbf{p}$ in the affine basis $(\mathbf{p}_0, \mathbf{p}_1, \ldots, \mathbf{p}_n)$. Figure 2.4 illustrates this when $n = 2$. ❏

Note that an affine basis of an n-dimensional space has $n + 1$ points. Fewer points $(\mathbf{p}_0, \mathbf{p}_1, \ldots, \mathbf{p}_m)$ will form an affine basis of the subspace they span—that is, the set of their affine combinations—as long as the differences $\mathbf{p}_i - \mathbf{p}_0$ are linearly independent. An *affine transformation* between two m-dimensional affine subspaces is any transformation that is affine when expressed as a transformation from $\mathbb{R}^m$ to $\mathbb{R}^m$ using affine bases for the two subspaces; the choice of bases does not matter.

2.5   The Geometry of Computer Graphics

Based on our discussion of Euclidean and affine geometries up to this point, we may be tempted to conclude that affine geometry is a good answer to our question, what is the right geometry for computer graphics? Indeed, affine transformations include Euclidean geometry's rigid motions, and in addition can be represented by matrices, which allows a simple computational structure.

We now look at the issue of the viewing transformations. Figure 2.5(a) is an aerial view, from far above, of a straight road on idealized flat ground. Figure 2.5(b) shows the same road as seen from a point nearby. Notice that this process of viewing (photographing) an object is one of the stages within the problem of visualizing data in computer graphics.

Geometrically, the image in Figure 2.5(b) corresponds to a transformation of the objects in (a). It shows straight lines as still straight, but it has a salient feature absent in

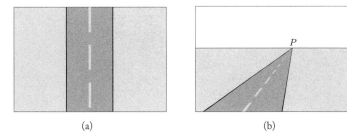

(a)   (b)

Figure 2.5. Picture of a road from different viewpoints. (a) Aerial view straight from above. (b) Perspective view.

affine transformations: the sides of the road, which are parallel lines in (a), are not parallel in (b). Thus the transformation used in the viewing process does not preserve parallelism, and so cannot be an affine transformation (see Example 2.3). So affine geometry does not yet fulfill all our needs, and we must search further.

How can we extend the group of affine transformations so as to include viewing, or perspective, transformations? The answer is projective geometry, which we look at next.

2.6   Projective Space

We use the idea of perspective to motivate the definition of projective space. The geometric transformation involved in viewing a scene is called *central projection*. Consider a point O of Euclidean space $\mathbb{R}^{n+1}$ and a hyperplane $\Pi \subset \mathbb{R}^{n+1}$ such that $O \notin \Pi$ (see Figure 2.6). The conical projection of a point $P \in \mathbb{R}^{n+1}$, $P \neq O$ on the plane Π is the point P' where the line r passing through O and P intersects Π.

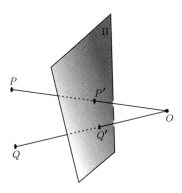

Figure 2.6. Conical projection.

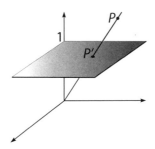

Figure 2.7. The line PP' is a plane in projective space. Its intersection $\mathbb{P}'$ with the affine plane $\mathcal{P}$ is the corresponding point of $\mathcal{P}$.

All points on the line r defined by O and P, with the exception of point O, project to the same point P'. This makes it natural to consider the line r, excluding the point O, as a *projective point*.

Taking O as the origin of $\mathbb{R}^{n+1}$, we define n-*dimensional projective space*, written $\mathbb{RP}^n$, as the set of lines passing through O. To be explicit: each point of projective space $\mathbb{RP}^n$, is a line through the origin in the vector space $\mathbb{R}^{n+1}$. A projective line, naturally enough, is the set of projective points whose lines in $\mathbb{R}^{n+1}$ belong to some plane in $\mathbb{R}^{n+1}$ passing through the origin. Similarly, projective subspaces of dimension m in $\mathbb{RP}^n$, $m < n$, come from vector subspaces of dimension $m + 1$ in $\mathbb{R}^{n+1}$.

You may have heard the statement that *parallel lines meet at infinity*. To really give a sense to this we need a notion of "at infinity," but one thing is clear: in Figure 2.5, the edges of the road, while parallel to Euclidean eyes, appear to converge at the horizon in the perspective view. We now show how projective geometry formalizes this fact.

Points in the model $\mathcal{P}$ of affine space are in correspondence with a certain subset of projective space: namely, the set of projective points whose lines in $\mathbb{R}^{n+1}$ are not parallel to $\mathcal{P}$. Each such line determines a point in $\mathcal{P}$ by intersection (see Figure 2.7). These are called the *affine points* of the projective plane.

Left over are the points of $\mathbb{RP}^n$ that are lines parallel to $\mathcal{P}$. These are called *ideal points* or *points at infinity*. To explain the reason for this name, we start by discussing an important difference between projective and affine geometry: *parallel lines do not exist in projective space*. More precisely, in a projective plane ($n = 2$), two lines always intersect. To see this, recall that lines in projective plane correspond to planes through the origin in $\mathbb{R}^3$. Two such planes always intersect in a line in $\mathbb{R}^3$, and the corresponding projective point is therefore an intersection point of the two original projective lines.

Figure 2.8 shows the situation for two projective lines whose traces in $\mathcal{P}$ are parallel affine lines. The planes $\mathcal{P}_1$ and $\mathcal{P}_2$ in $\mathbb{R}^3$ that define the projective lines intersect in a line r in $\mathbb{R}^3$, which is of course a projective point, but one that lies *outside of affine space*. Thus, each family of parallel lines of the affine plane corresponds to a point at infinity in the projective space, where they all intersect. We can think of them as points on the horizon, where the sides of an infinitely long, straight road converge (Figure 2.5).

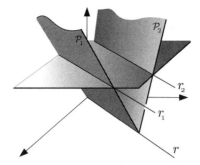

Figure 2.8. Two projective lines whose traces in the affine plane are parallel intersect "at infinity."

The situation in higher dimensions is analogous. In summary, a point at infinity in projective space is a direction of lines in affine space.

2.6.1   Homogeneous Coordinates

To make calculations in projective space, we need to introduce coordinates. Let a point $\mathbf{p}$ be given in $\mathbb{RP}^n$. Any point (i.e., vector) $\mathbf{p}' \in \mathbb{R}^{n+1}$ on the corresponding line through the origin is equally entitled to represent $\mathbf{p}$ (with the exception is the origin itself, of course). All such vectors are multiples of each other. We associate to $\mathbf{p}$ the coordinates in $\mathbb{R}^{n+1}$ of any of these vectors. They are only defined up to multiplication by a nonzero scalar, and so are called *homogeneous coordinates*. In the literature, it is common to write homogeneous these coordinates in brackets:

$$[x_1, \ldots, x_n, x_{n+1}] = \lambda[x_1, \ldots, x_n, x_{n+1}], \quad \lambda \neq 0. \qquad (2.15)$$

Note that a point has 0 for its last coordinate if and only if it is at infinity.

Projective hyperplanes, that is, $(n-1)$-dimensional subspaces, of $\mathbb{RP}^n$ correspond to vector hyperplanes in $\mathbb{R}^{n+1}$, and so are defined by homogeneous linear equations

$$a_1 x_1 + a_2 x_2 + \cdots + a_{n+1} x_{n+1} = 0.$$

For example, lines in the projective plane (the case $n = 2$) are defined by equations $a_1 x_1 + a_2 x_2 + a_3 x_3 = 0$.

2.7   Projective Transformations

What are the transformations of projective space? This question is easier to answer than the corresponding questions for Euclidean and affine spaces. Quite simply, any invertible linear transformation of $\mathbb{R}^{n+1}$ defines a transformation of $\mathbb{RP}^n$, because it takes vector lines in $\mathbb{R}^{n+1}$ to vector lines.

Now recall that the transformations of affine (and Euclidean) space were also expressed as linear transformations of $\mathbb{R}^{n+1}$, using the appropriate models. We see that projective geometry is a natural extension of affine geometry—which is fortunate for us, because after all, affine geometry has many useful properties for computer graphics!

We will not make any distinction between a linear transformation $T\colon \mathbb{R}^{n+1} \to \mathbb{R}^{n+1}$ and the resulting transformation in projective space $\mathbb{RP}^n$.

Note that if $T\colon \mathbb{RP}^n \to \mathbb{RP}^n$ is a projective transformation and $\lambda \in \mathbb{R}$, $\lambda \neq 0$, then by using the linearity of T, we have $(\lambda T)P = T(\lambda P) = T(P)$. In other words, projective transformations are defined up to multiplication by a nonzero scalar.

2.7.1   Anatomy of a Plane Projective Transformation

We now concentrate on the projective plane. A projective transformation $\mathbb{RP}^2 \to \mathbb{RP}^2$ is represented by an invertible matrix M of order 3. Our goal now is to understand the anatomy of this transformation: how does it act on projective points? We will divide M into four blocks:

$$
M = \left(\begin{array}{cc|c} a & c & t_1 \\ b & d & t_2 \\ \hline p_1 & p_2 & s \end{array} \right) = \begin{pmatrix} A & T \\ P & S \end{pmatrix},
$$

where

$$
A = \begin{pmatrix} a & b \\ c & d \end{pmatrix}, \quad P = \begin{pmatrix} p_1 & p_2 \end{pmatrix}, \quad T = \begin{pmatrix} t_1 \\ t_2 \end{pmatrix}, \quad \text{and} \quad S = \begin{pmatrix} s \end{pmatrix}. \tag{2.16}
$$

When P is the null matrix and $s = 1$ this reduces to the form (2.9) of an affine transformation, so we already know what the behavior is in this case, at least for affine points in $\mathbb{RP}^2$; but we will recapitulate it for completeness.

Effect of M. First suppose that P and T have all entries zero and that $s = 1$. In this case, by applying the transformation to a point $(x, y, 0)$ at infinity, we have

$$
\begin{pmatrix} a & c & 0 \\ b & d & 0 \\ 0 & 0 & 1 \end{pmatrix} \begin{pmatrix} x \\ y \\ 0 \end{pmatrix} = \begin{pmatrix} ax + cy \\ bx + dy \\ 0 \end{pmatrix}.
$$

Therefore, the resulting point is also at infinity. The transformation leaves the line at infinity invariant.

On the other hand, if $(x, y, 1)$ is an affine point on the projective plane, its image under the transformation is given by

$$
\begin{pmatrix} a & c & 0 \\ b & d & 0 \\ 0 & 0 & 1 \end{pmatrix} \begin{pmatrix} x \\ y \\ 1 \end{pmatrix} = \begin{pmatrix} ax + cy \\ bx + dy \\ 1 \end{pmatrix}.
$$

This shows that the result is also an affine point. In other words, the affine plane inside the projective plane is also left invariant by the transformation. Moreover, the affine coordinates of the transformed point are given by

$$\begin{pmatrix} a & c \\ b & d \end{pmatrix} \begin{pmatrix} x \\ y \end{pmatrix} = \begin{pmatrix} ax + cy \\ bx + dy \end{pmatrix}.$$

Thus a projective transformation of this restricted form acts as a linear transformation of the vector space $\mathbb{R}^2$. Therefore, the group of projective transformations on the plane contains, in a natural way, the group of linear transformations of the plane (and in particular, the group of Euclidean plane isometries preserving the origin).

Effect of T. Next, let A be the identity matrix of order two, and let $P = (0\,0)$ and $s = 1$. Then the image of point with coordinates $(x, y, 1)$ is

$$\begin{pmatrix} 1 & 0 & t_1 \\ 0 & 1 & t_2 \\ 0 & 0 & 1 \end{pmatrix} \cdot \begin{pmatrix} x \\ y \\ 1 \end{pmatrix} = \begin{pmatrix} x + t_1 \\ y + t_2 \\ 1 \end{pmatrix};$$

that is, the transformation operation on the affine plane is of translation by the vector (t_1, t_2). The reader can check that a point $(x, y, 0)$ at infinite is left invariant.

Effect of S. The effect of the element s, forming the block S of the matrix, is a homothety (scaling) of the affine plane of factor $1/s$, $s \neq 0$. In fact,

$$\begin{pmatrix} 1 & 0 & 0 \\ 0 & 1 & 0 \\ 0 & 0 & s \end{pmatrix} \begin{pmatrix} x \\ y \\ 1 \end{pmatrix} = \begin{pmatrix} x \\ y \\ s \end{pmatrix} = \begin{pmatrix} x/s \\ y/s \\ 1 \end{pmatrix}.$$

In all cases so far, the projective transformation preserves both affine points and points at infinity. Therefore, in essence, the projective transformations do not introduce any novelty. What we showed above can be summarized by saying that the group of projective transformations contains the group of affine transformations (and therefore, the rigid motions of Euclidean geometry).

Effect of P. We will now analyze the block P of matrix M in (2.16). We take A as being the identity matrix, T as null and $s = 1$. By applying the transformation to an affine point with coordinates $(x, y, 1)$, we get

$$\begin{pmatrix} 1 & 0 & 0 \\ 0 & 1 & 0 \\ p_1 & p_2 & 1 \end{pmatrix} \begin{pmatrix} x \\ y \\ 1 \end{pmatrix} = \begin{pmatrix} x \\ y \\ p_1 x + p_2 y + 1 \end{pmatrix}.$$

If $p_1 \neq 0$ or $p_2 \neq 0$, the equation $p_1 x + p_2 y + 1 = 0$ has infinitely many solutions. This shows that some affine points $(x, y, 1)$ are transformed into points $(x, y, 0)$ at infinity.

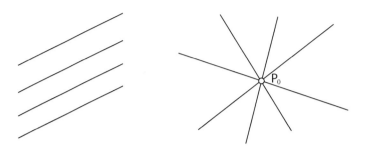

Figure 2.9. A projective transformation may take a point at infinity (intersection point of a family of parallel lines) to an affine point P_0, the vanishing point.

Similarly, by applying the transformation to a point at infinity $(x, y, 0)$, we obtain the result $(x, y, p_1 x + p_2 y)$. Since there are values of (x, y) such that $p_1 x + p_2 y \neq 0$, we conclude that points at infinity may be transformed into affine points.

If a point at infinity is transformed into an affine point P_0, the geometric interpretation is that the family of parallel lines in the direction of that point at infinity is transformed into the family of lines going through point P_0 (Figure 2.9). We call P_0 the *vanishing point* of the transformation.

A *main vanishing point* is a vanishing point corresponding to the direction of one of the coordinate axes in $\mathbb{R}^n$. Thus, projective transformations in two dimensions have up to two main vanishing points. In the notation of (2.16), if $p_1 \neq 0$ and $p_2 = 0$, we only have one main vanishing point, corresponding to the point at infinity $(1, 0, 0)$ in the direction of the x-axis. If $p_1 = 0$ and $p_2 \neq 0$, we likewise only have the main vanishing point corresponding to the y-axis. If both p_1 and p_2 are nonzero, we have two main vanishing points. Figure 2.10 shows a projective transformation of a rectangle with two vanishing points. Notice that the image of the rectangle is a quadrilateral, since projective transformations preserve straight lines.

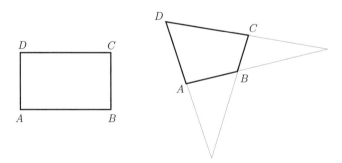

Figure 2.10. Transformation with two vanishing points.

2.7.2   Projective Transformations in Cartesian Coordinates

Consider a projective transformation $T \colon \mathbb{RP}^2 \to \mathbb{RP}^2$ defined in homogeneous coordinates by the matrix

$$\begin{pmatrix} a & b & c \\ d & e & f \\ g & h & i \end{pmatrix}.$$

(Without loss of generality, we can take $i = 1$ in the above equation—why?)

If $z = (x, y, 1)$ is an affine point, we have

$$T(x, y, 1) = \begin{pmatrix} a & b & c \\ d & e & f \\ g & h & i \end{pmatrix} \begin{pmatrix} x \\ y \\ 1 \end{pmatrix} = (ax + by + c, dx + ey + f, gx + hy + i).$$

If $gx + hy + i \neq 0$, that is, the image of $(x, y, 1)$ is not a point at infinity, then we can write

$$T(x, y) = T(x, y, 1) = \left(\frac{ax + by + c}{gx + hy + i}, \frac{dx + ey + f}{gx + hy + i} \right). \tag{2.17}$$

This equation expresses a projective transformation in cartesian coordinates in $\mathbb{R}^2$.

2.7.3   The General n-Dimensional Case

Generalizing to n dimensions the work just done for $n = 2$ is not difficult. The points at infinity of $\mathbb{RP}^n$ form a projective space $\mathbb{RP}^{n-1}$. Parallel hyperplanes in the affine space $\mathbb{R}^n$ ($x_{n+1} = 1$) intersect one another in hyperplanes of the projective space in $\mathbb{RP}^{n-1}$ at infinity. A projective transformation is defined by an invertible linear transformation $T \colon \mathbb{R}^{n+1} \to \mathbb{R}^{n+1}$, represented by an invertible matrix of order $n + 1$. As before, we divide the matrix of this transformation into four blocks:

$$\left(\begin{array}{c|c} n \times n & n \times 1 \\ \hline 1 \times n & 1 \times 1 \end{array} \right).$$

The square block of order n corresponds to linear transformations in the affine space $\mathbb{R}^n$ of $\mathbb{RP}^n$ (regarded as a vector space). The $n \times 1$ block defines translations in $\mathbb{R}^n$, and the $1 \times n$ block defines the n main vanishing points of the projective transformation: a vanishing point for each coordinate direction.

In projective space $\mathbb{RP}^3$, we have the matrix

$$\begin{pmatrix} a_{11} & a_{12} & a_{13} & t_1 \\ a_{21} & a_{22} & a_{23} & t_2 \\ a_{31} & a_{32} & a_{33} & t_3 \\ p_1 & p_2 & p_3 & s \end{pmatrix}.$$

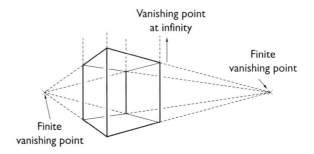

Figure 2.11. Vanishing points in the perspective of a cube.

We recommend that the reader undertake a detailed study of this case, similar to what we did for the 2D case. Figure 2.11 shows a cube in 3D space, projectively transformed with two vanishing points. (In this case, it is common to say that the vanishing point, along the direction where parallelism is preserved, is at infinity.)

2.8   The Fundamental Theorem of Projective Geometry

A *projective basis* in projective space $\mathbb{RP}^n$ is a set of $n + 2$ points such that any subset with $n + 1$ elements is linearly independent (as a set of directions or vectors in $\mathbb{RP}^{n+1}$). The canonical basis of $\mathbb{RP}^n$ is given by

$$
\begin{aligned}
\mathbf{e}_1 &= (1, 0, \ldots, 0, 0), \\
\mathbf{e}_2 &= (0, 1, \ldots, 0, 0), \\
&\vdots \\
\mathbf{e}_n &= (0, 0, \ldots, 1, 0), \\
\mathbf{e}_{n+1} &= (0, 0, \ldots, 0, 1), \\
\mathbf{e}_{n+2} &= (1, 1, \ldots, 1, 1) = \mathbf{e}_1 + \cdots + \mathbf{e}_{n+1}.
\end{aligned}
$$

Theorem 2.5. *Given any projective basis* $\mathbf{a}_1, \ldots, \mathbf{a}_{n+1}$, *there exists a projective transformation* $T \colon \mathbb{RP}^n \to \mathbb{RP}^n$ *such that*

$$
T(\mathbf{e}_i) = \lambda_i \mathbf{a}_i, \quad i = 1, \ldots, n+2, \tag{2.18}
$$

where the λ_i, $i = 1, \ldots, n+2$, *are nonzero scalars.*   ❑

Proof: As the first $n+1$ vectors of the two bases are linearly independent, the first $n+1$ equations in (2.18) define a linear transformation $T \colon \mathbb{R}^{n+1} \to \mathbb{R}^{n+1}$ whose matrix in the bases $\mathbf{e}_1, \ldots, \mathbf{e}_{n+1}$ and $\mathbf{a}_1, \ldots, \mathbf{a}_{n+1}$ is given by

$$\begin{pmatrix} \lambda_1 \mathbf{a}_1 & \lambda_2 \mathbf{a}_2 & \cdots & \lambda_{n+1} \mathbf{a}_{n+1} \end{pmatrix}.$$

(Here each $\mathbf{a}_i$ is considered as an $n \times 1$ column matrix.) Further, the equality $T(\mathbf{e}_{n+2}) = \lambda_{n+2} \mathbf{a}_{n+2}$ in matrix form is

$$\begin{pmatrix} \lambda_1 \mathbf{a}_1 & \lambda_2 \mathbf{a}_2 & \cdots & \lambda_{n+1} \mathbf{a}_{n+1} \end{pmatrix} \begin{pmatrix} 1 \\ 1 \\ \vdots \\ 1 \end{pmatrix} = \lambda_{n+2} \mathbf{a}_{n+2}.$$

or,

$$\begin{pmatrix} \mathbf{a}_1 & \mathbf{a}_2 & \cdots & \mathbf{a}_{n+1} \end{pmatrix} \begin{pmatrix} \lambda_1 \\ \lambda_2 \\ \vdots \\ \lambda_{n+1} \end{pmatrix} = \lambda_{n+2} \mathbf{a}_{n+2}.$$

This equation uniquely defines $\lambda_1, \ldots, \lambda_{n+1}$. Thus the transformation T is defined up to the choice of λ_{n+2}. ❑

Here is an immediate consequence of this theorem:

Corollary 2.6. *Given two projective bases* $\{\mathbf{a}_1, \ldots, \mathbf{a}_{n+2}\}$ *and* $\{\mathbf{b}_1, \ldots, \mathbf{b}_{n+2}\}$, *there exists a unique projective transformation T such that $T(\mathbf{a}_i) = \mathbf{b}_i$.* ❑

Here of course "unique" is to be understood as unique up to a multiplication by a scalar. Geometrically, we have an affine transformation defined by the first $n + 1$ elements of the basis and we have the freedom of choosing the extension of this transformation to the last element. As a projective transformation, T is uniquely determined.

A *projective frame* is a projective basis $\{\mathbf{a}_1, \ldots, \mathbf{a}_{n+2}\}$, such that $\mathbf{a}_{n+2} = \mathbf{a}_1 + \mathbf{a}_2 + \cdots + \mathbf{a}_{n+1}$. The canonical basis of a projective space is a projective frame. The theorem below is easy to prove using Corollary 2.6.

Theorem 2.7 (Fundamental theorem of projective geometry). *Given two projective frames* $\{\mathbf{a}_1, \ldots, \mathbf{a}_{n+2}\}$ *and* $\{\mathbf{b}_1, \ldots, \mathbf{b}_{n+2}\}$, *there exists a unique projective transformation T such that $T(\mathbf{a}_i) = \mathbf{b}_i$.* ❑

Four points in the projective plane define a projective basis if they form a nondegenerate quadrilateral. Since projective transformations are invertible, they map this quadrilateral to another nondegenerate quadrilateral. In short, the fundamental theorem states that a unique projective transformation exists between two quadrilaterals on the plane (see Figure 2.12).

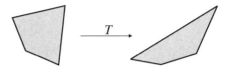

Figure 2.12. A projective transformation takes a nondegenerate quadrilateral to another.

2.8.1   Transformations between Quadrilaterals

We turn to the interesting problem of determining the projective transformation matrix that takes a given quadrilateral to another. The road map for demonstrating the fundamental theorem allows us to perform this calculation. Observe first that it is enough to determine the projective transformation T that takes a unit square $I^2 = [0,1] \times [0,1]$ into an arbitrary square Q. For then the inverse T^{-1} transforms Q in I^2, so given two arbitrary quadrilaterals Q_1 and Q_2, we get the transformations $T_1 \colon I^2 \to Q_1$ and $T_2 \colon I^2 \to Q_2$ and take the composition $T = T_2 \circ T_1^{-1}$ (see Figure 2.13).

The vertices of I^2 are represented by $\mathbf{e}_1 = (1,0,1)$, $\mathbf{e}_2 = (0,1,1)$, $\mathbf{e}_3 = (0,0,1)$, $\mathbf{e}_4 = (1,1,1)$, so finding the transformation T taking I^2 to an arbitrary quadrilateral is the same as determining the projective transformation that maps the canonical projective frame to another arbitrary projective frame. Thus this calculation in effect repeats the demonstration of Theorem 2.5 in the case $n = 2$.

Let $\mathbf{q}_1 = (x_1, y_1, 1)$, $\mathbf{q}_2 = (x_2, y_2, 1)$, $\mathbf{q}_3 = (x_3, y_3, 1)$, $\mathbf{q}_4 = (x_4, y_4, 1)$ be the projective coordinates of the vertices of the square Q (see Figure 2.14). We then have

$$T(\mathbf{e}_1) = \lambda_1 \mathbf{q}_1 = (\lambda_1 x_1, \lambda_1 y_1, \lambda_1),$$
$$T(\mathbf{e}_2) = \lambda_2 \mathbf{q}_2 = (\lambda_2 x_2, \lambda_2 y_2, \lambda_2),$$
$$T(\mathbf{e}_3) = \lambda_3 \mathbf{q}_3 = (\lambda_3 x_3, \lambda_3 y_3, \lambda_3).$$

Therefore, the matrix of T in the canonical basis of $\mathbb{R}^3$ is given by

$$\begin{pmatrix} \lambda_1 x_1 & \lambda_2 x_2 & \lambda_3 x_3 \\ \lambda_1 y_1 & \lambda_2 y_2 & \lambda_3 y_3 \\ \lambda_1 & \lambda_2 & \lambda_3 \end{pmatrix}. \tag{2.19}$$

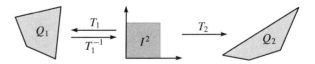

Figure 2.13. Finding the projective transformation that maps a quadrilateral to another.

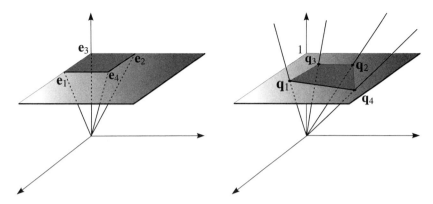

Figure 2.14. Point at infinity transformed into finite point.

Setting $T(\mathbf{e}_4) = \mathbf{q}_4 = (x_4, y_4, 1)$, that is, taking $\lambda_4 = 1$ in the theorem, we have

$$\begin{pmatrix} \lambda_1 x_1 & \lambda_2 x_2 & \lambda_3 x_3 \\ \lambda_1 y_1 & \lambda_2 y_2 & \lambda_3 y_3 \\ \lambda_1 & \lambda_2 & \lambda_3 \end{pmatrix} \begin{pmatrix} 1 \\ 1 \\ 1 \end{pmatrix} = \begin{pmatrix} x_4 \\ y_4 \\ 1 \end{pmatrix},$$

or,

$$\begin{pmatrix} x_1 & x_2 & x_3 \\ y_1 & y_2 & y_3 \\ 1 & 1 & 1 \end{pmatrix} \begin{pmatrix} \lambda_1 \\ \lambda_2 \\ \lambda_3 \end{pmatrix} = \begin{pmatrix} x_4 \\ y_4 \\ 1 \end{pmatrix}.$$

This system admits a single solution, and by finding it we obtain the values of λ_1, λ_2 and λ_3. By replacing these values in (2.19), we obtain the matrix of the transformation T.

Another method. Consider two quadrilaterals on the plane with vertices $P_k = (u_k, v_k)$ and $Q_k = (x_k, y_k)$, $k = 1, \ldots, 3$. To determine the projective transformation T satisfying $T(P_k) = Q_k$, we should solve the matrix equation

$$T(u_k, v_k) = (x_k, y_k), \quad k = 0, 1, 2, 3.$$

Using the expression of T given in the Equation (2.17), with $i = 1$, we can write

$$x_k = \frac{au_k + bv_k + c}{gu_k + hv_k + 1} \implies u_k a + v_k b + c - u_k x_k g - v_k x_k h = x_k,$$

$$y_k = \frac{du_k + ev_k + f}{gu_k + hv_k + 1} \implies u_k d + v_k e + f - u_k y_k g - v_k y_k h = y_k.$$

By varying $k = 0, 1, 2, 3$ in the above equation, we obtain a linear system with eight equations:

$$
\begin{pmatrix}
u_0 & v_0 & 1 & 0 & 0 & 0 & -u_0x_0 & -v_0x_0 \\
u_1 & v_1 & 1 & 0 & 0 & 0 & -u_1x_1 & -v_1x_1 \\
u_2 & v_2 & 1 & 0 & 0 & 0 & -u_2x_2 & -v_2x_2 \\
u_3 & v_3 & 1 & 0 & 0 & 0 & -u_3x_3 & -v_3x_3 \\
0 & 0 & 0 & u_0 & v_0 & 1 & -u_0y_0 & -v_0y_0 \\
0 & 0 & 0 & u_1 & v_1 & 1 & -u_1y_1 & -v_1y_1 \\
0 & 0 & 0 & u_2 & v_2 & 1 & -u_2y_2 & -v_2y_2 \\
0 & 0 & 0 & u_3 & v_3 & 1 & -u_3y_3 & -v_3y_3
\end{pmatrix}
\begin{pmatrix}
a \\ b \\ c \\ d \\ e \\ f \\ g \\ h
\end{pmatrix}
=
\begin{pmatrix}
x_0 \\ x_1 \\ x_2 \\ x_3 \\ y_0 \\ y_1 \\ y_2 \\ y_3
\end{pmatrix} . \tag{2.20}
$$

From the fundamental theorem, we know this system has a unique solution providing the coefficients a, b, c, d, e, f, g, h of the sought projective transformation.

We observe that, in particular cases, the system above is considerably simplified and its solution can be obtained by manual substitution. A case in which this happens is when the quadrilateral of origin is the unit square. In this case we have $u_0 = 0$, $v_0 = 0$, $u_1 = 1$, $v_1 = 0$, $u_2 = 1$, $v_2 = 1$, $u_3 = 0$, $v_3 = 1$. A simple calculation shows that system (2.20) reduces to

$$c = x_0, \quad a + c - gx_1 = x_1, \quad a + b + c - gx_2 + hx_2 = x_2, \quad b + c - hx_3 = x_3,$$
$$f = y_0, \quad d + f - gy_1 = y_1, \quad d + e + f - gy_2 + hy_2 = y_2, \quad e + f - hy_3 = y_3.$$

2.9 Projections and Projective Geometry

We started our discussion of projective geometry in Section 2.5 using the example of perspective. We now return to that example to show that perspective and central projection are, in fact, projective transformations.

2.9.1 Parallel Projection

Given two planes Π and Π' in affine space and a line r not parallel to either, we define the parallel projection $T \colon \Pi \to \Pi'$ in the following way: given $P \in \Pi$, let s be the line passing through the point P and parallel to r. Then $T(P) = s \cap \Pi'$ (Figure 2.15).

When the line r is orthogonal to the plane Π', the projection is called *orthogonal*. It is not difficult to show that parallel projection is an affine transformation from the plane Π onto the plane Π'. In fact, if the planes are parallel, parallel projection defines an isometry between them.

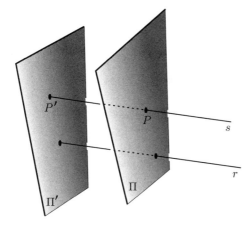

Figure 2.15. Parallel projection from Π to Π'. (Planes are not necessarily parallel.)

2.9.2   Central or Perspective Projection

We now consider the case of central, or perspective, projections—the kind used in our photograph example in Section 2.5. Consider a point O and two projective planes Π and Π' in projective space $\mathbb{RP}^3$ (see Figure 2.16). For every point $P \in \Pi$, the projective line OP intersects the plane Π' at a point P'. We define $T \colon \Pi \to \Pi'$ by setting $T(P) = P'$; see Figure 2.16. The point O is called the *projection center*, and the line OP is a *projection line*.

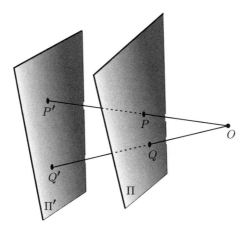

Figure 2.16. Central or perspective projection.

We want to show that T is a projective transformation. We first take a projective transformation L in projective space that takes the projection center O to a point at infinity. All the projection lines are transformed by L into parallel lines of affine space. Therefore, the composition $L \circ T$ is a parallel projection T' between the transformed planes $L(\Pi)$ and $L(\Pi')$. The central projection is then given by $T = L^{-1} \circ T'$. In other words, it is the composition between a parallel projection, which is affine, and the projective transformation L^{-1}, and is therefore a projective transformation.

Projections are important in computer graphics because of the virtual camera transformation. The image of the projection is the plane Π', so the projection can be thought of as a transformation of $\mathbb{R}^3$ in $\mathbb{R}^2$, or, more precisely, of $\mathbb{RP}^3$ in $\mathbb{RP}^2$. From this point of view, a possible more general projection is a projective transformation $T \colon \mathbb{RP}^3 \to \mathbb{RP}^2$ which, in homogeneous coordinates, is given by

$$\begin{pmatrix} y_1 \\ y_2 \\ y_3 \end{pmatrix} = \begin{pmatrix} a_{11} & a_{12} & a_{13} & a_{14} \\ a_{21} & a_{22} & a_{23} & a_{24} \\ a_{31} & a_{32} & a_{33} & a_{34} \end{pmatrix} \begin{pmatrix} x_1 \\ x_2 \\ x_3 \\ x_4 \end{pmatrix}.$$

Using this transformation, we have 11 degrees of freedom to define a virtual camera, and several types of cameras are possible, such as the perspective camera (which uses central projection) and the pinhole, affine, weak-perspective, and orthographic cameras. We will study the perspective camera in detail in Chapter 11.

2.10   Comments and References

An introduction to the methods of projective geometry in computer graphics can be found in [Roberts 66]. There are various sources on projective geometry. However, to our knowledge, there is no book providing an analytical and concise approach to projective geometry stressing its several applications to computer graphics. An attempt in that direction was made in [Penna and Patterson 86], which complements this chapter by providing a large number of examples.

Although for the purposes of this book we came to the conclusion that projective geometry is the right geometry for computer graphics, the truth is that projective geometry, too, has its limitations, especially from the algebraic point of view, as it lacks operations between geometry objects beyond projective transformations. This topic is addressed by *geometric algebra*, a relatively new approach that has not yet become established in courses on algebra and geometry. See [Dorst et al. 07] for a good treatment.

We have not broached the subject of more general spaces called *differentiable varieties*, which contains important topics for computer graphics: particularly the geometry of differentiable curves and surfaces. The computational aspects of geometry were not covered either. Affine geometry has been very well studied from the computational point of view, and is a relevant topic for computer graphics. Computational algebraic geometry, in particular of curves and surfaces, is likewise a rich topic with fascinating ramifications.

Exercises

1. True or false? Justify.

    (a) Linear transformations preserve parallelism.

    (b) Affine transformations preserve parallelism.

    (c) Isometries of Euclidean space preserve angles.

    (d) Affine transformations preserve rectangles.

    (e) Affine transformations preserve perpendicularity.

2. Prove the so-called fundamental theorem of affine geometry: *An affine transformation is completely determined by its values in an affine basis* (see Example 2.4 for this concept). Another, equivalent statement: *if* $(\mathbf{p}_0, \mathbf{p}_1, \ldots, \mathbf{p}_n)$ *and* $(\mathbf{q}_0, \mathbf{q}_1, \mathbf{q}_2 \ldots, \mathbf{q}_n)$ *are affine bases, a unique affine transformation L exists such that $L(\mathbf{p}_i) = (\mathbf{q}_i)$ for $i = 0, 1, \ldots, n$.*

3. Recall that a *convex combination* of points $\mathbf{p}_0, \mathbf{p}_1, \ldots, \mathbf{p}_m$ in affine space is an affine combination of these points with coefficients in $[0, 1]$. If the points are affinely independent (that is, if $\mathbf{p}_1 - \mathbf{p}_0, \ldots, \mathbf{p}_m - \mathbf{p}_0$ are linearly independent vectors), the set σ of convex combinations $\mathbf{p}_0, \ldots, \mathbf{p}_m$ is called an *m-dimensional simplex* and $\mathbf{p}_0, \ldots, \mathbf{p}_m$ are its *vertices*.

    (a) Interpret geometrically simplexes of dimension 1, 2, and 3.

    (b) Show that every simplex σ is a convex set.

    (c) Show that a simplex σ is the convex hull of its vertices.

    (d) For $\mathbf{p} = \sum t_i \mathbf{p}_i \in \sigma$, the coefficients $t_1, \ldots, t_n$ are called the *barycentric coordinates* of p, a notion already introduced in Example 2.4. Interpret barycentric coordinates geometrically for 1D and 2D simplices.

4. If $\mathbf{q}, \mathbf{q}_1$ and $\mathbf{q}_2$ belong to an affine line r, the *affine ratio* is defined by

    $$\left| \frac{\mathbf{q} - \mathbf{q}_1}{\mathbf{q} - \mathbf{q}_2} \right| ,$$

    where the above fraction is to be understood as a ratio of parallel vectors. Thus, if a point q divides a segment $q_1 q_2$ in the ratio $b_2 : b_1$, then

    $$q = \frac{b_1 q_1 + b_2 q_2}{b_1 + b_2}, \quad b_1 + b_2 \neq 0.$$

    Show that affine transformations preserve affine ratios. Conversely, if a transformation of affine space takes lines to lines and preserves the affine ratio, it is an affine transformation.

5. Show that an orthogonal linear transformation preserves the distance between two points in Euclidean space.

6. Consider the linear transformation on the plane defined by

    $$T(x, y) = (3x + y, -2x + 6y).$$

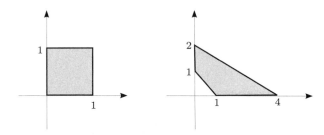

Figure 2.17. Figure for Exercise 7.

(a) Determine its matrix in the basis $\{e_1, e_2\}$, where $e_1 = (1, 0)$ and $e_2 = (0, 1)$.

(b) Determine the matrix of T in the basis $\{(1, 1),\ (-1, 1)\}$.

Consider the different results obtained in items (a) and (b). Do they mean that the correspondence between linear and main transformations is not a one-to-one correspondence? Explain.

7. Consider the unit square and the quadrilateral shown in Figure 2.17. Find the projective transformation matrix that transforms

(a) the square into a quadrilateral,

(b) the quadrilateral into a square.

8. A *vertical shear* of $\mathbb{R}^2$, or *shear along the y-axis*, is the linear transformation defined by the matrix
$$\begin{pmatrix} 1 & 0 \\ a & 1 \end{pmatrix},$$
with $a \neq 0$.

(a) Interpret this transformation geometrically.

(b) Define a shear along the x-axis.

(c) Define shears along the xy-, xz-, and yz-planes in space.

9. Define a projection of an arbitrary vector r on a plane. Find the projection matrix in the basis $\{(1, 0),\ (0, 1)\}$ of the plane.

10. Show, using examples of the $\mathbb{R}^2$ plane, that the composition of geometric transformations is not commutative in general.

11. Determine the matrix of the projective transformation that transforms the quadrilateral on the left in Figure 2.18 to the one on the right.

12. Determine the affine transformation on the plane taking the triangle with vertices $(1, 1)$, $(1, 2)$, and $(3, 3)$ to the equilateral triangle with vertices $(1, 0)$, $(-1, 0)$, and $(0, \sqrt{3})$.

13. Show that a projective transformation is affine if and only if it preserves the relation of parallelism.

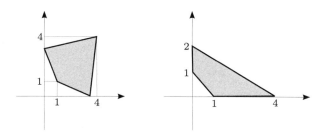

Figure 2.18. Figure for Exercise 11.

14. A transformation $T\colon \mathbb{R}^n \to \mathbb{R}^n$ is a *similarity* if

$$d(P,Q) = r\,\mathrm{d}(T(P),T(Q)),$$

where $r > 0$ is a constant (called a *similarity ratio*). Show that every similarity is an affine transformation.

15. A transformation $T\colon \mathbb{R}^n \to \mathbb{R}^n$ is a *homothety* of center O and ratio $r > 0$, if it satisfies, for every point $P \in \mathbb{R}^n$,

$$\overrightarrow{OT(P)} = r\,\overrightarrow{OP},$$

where r is a constant.

(a) Show that every homothety is a similarity.

(b) Show that a similarity transformation is the composition of a homothety with an isometry.

(c) Write the matrix of a homothety in homogeneous coordinates.

(d) What is the matrix of a similarity in homogeneous coordinates?

16. Given a projective line r on projective space and a point $P \in r$, show that there exists a projective transformation in space taking P to the point at infinity on the line r. Interpret this result geometrically. Is this projective transformation unique?

17. Define what it means for a transformation in $\mathbb{R}^n$ to preserve angles.

(a) Show that an isometry preserves angles.

(b) Give an example of a transformation that preserves angles but is not an isometry.

18. Construct the affine model of projective space of dimension 1, $\mathbb{RP}^1$. Show that, topologically, $\mathbb{RP}^1$ is a circle.

19. If a projective transformation T on the plane is defined by the matrix

$$\begin{pmatrix} a & b & c \\ d & e & f \\ g & h & i \end{pmatrix},$$

show that the inverse of T is given by the matrix

$$\begin{pmatrix} ei-fh & fg-di & dh-ge \\ ch-bi & ai-cg & bg-ah \\ bf-ce & cd-af & ae-bd \end{pmatrix}.$$

(Hint: use the adjunct matrix method to calculate the inverse.)

20. Find the perspective projection matrix with center of projection at the origin in Euclidean space with projection plane $z = d$, $d > 0$.

21. Consider the perspective projection on a plane $z = z_p$ with center of projection at the point $O = (0, 0, z_p+\lambda\mathbf{v})$, where $\mathbf{v} = (v_x, v_y, v_z)$ is a unit vector (see the figure below). Show that the matrix of this projection is given by

$$\begin{pmatrix} 1 & 0 & -\dfrac{v_x}{v_z} & z_p\dfrac{v_x}{v_z} \\ 0 & 1 & -\dfrac{v_y}{v_z} & z_p\dfrac{v_y}{v_z} \\ 0 & 0 & -\dfrac{z_p}{\lambda v_z} & \dfrac{z_p^2}{\lambda v_z} + z_p \\ 0 & 0 & -\dfrac{1}{\lambda v_z} & \dfrac{z_p}{\lambda v_z} + 1 \end{pmatrix}.$$

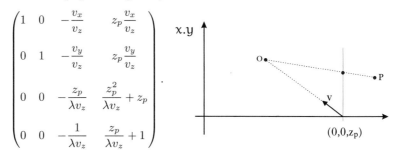

(Hint: use the parametric equation of the projection line to find the projection of a point P.)

22. Obtain the perspective projection matrix whose plane of projection is the plane $z = 0$, and whose center of projection is the point $(0, 0, -d)$, $d > 0$.

23. Assuming the geometry of our universe is projective, show that a traveler walking on a straight line will return to the starting point.

24. This exercise deals with the uniqueness of affine and projective transformations on the plane.

    (a) Let A, B, and C be different points on a line r on the plane, and T an affine transformation of r on another straight line s. Show that the ratio AC/BC is preserved by T.

    (b) Let A, B, and C be different points on a straight line r on the plane, and A', B', and C' be different points on a straight line s. Show that there exists a projective transformation T of r in s, such that $T(A) = A'$, $T(B) = B'$, and $T(C) = C'$.

    (c) Discuss the two previous items in connection with the question of unicity of affine and projective transformations between two lines.

25. Let A, B, C, and D be different points on a straight line r and let $T: r \to s$ be a perspective projection of r on another straight line s (see Figure 2.19).

    Show that

$$\frac{CA/CB}{DA/DB} = \frac{\sin \widehat{COA} \, \sin \widehat{DOB}}{\sin \widehat{COB} \, \sin \widehat{DOA}}.$$

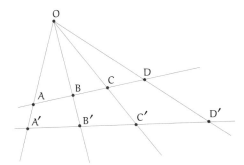

Figure 2.19. Figure for Exercise 25.

Conclude that the ratio on the left-hand side, called the *cross ratio*, is invariant under projective transformations.

26. Select which of these notions are invariant under projective transformations: angle, length, area, incidence, parallelism, orthogonality. Explain each choice briefly.

27. Justify, geometrically, the fact that the set of the conics is invariant under projective transformations. In other words, a projective transformation transforms any conic into another conic: for example, a circle into a hyperbola. How can this be proved analytically (i.e., using coordinate equations)?

28. Let G be a group of transformations of a set S and let H be a subgroup of G. If two objects are H-congruent, then they are G-congruent.

29. Assume that G is a group of transformations of a space S and H is a subgroup of G. If D is a G-property, then D is an H-property.

30. Suppose that T is an isometry of the plane $\mathbb{R}^2$ such that $T(0,0) = (0,0)$ and $T(1,0) = (1,0)$. Then $T(x,y) = (x,y)$, for every $(x,y) \in \mathbb{R}^2$, or $T(x,y) = (x,-y)$ (reflection about the x-axis).

31. Let a, b, and z be complex numbers. Let T be a *similarity transformation* of the complex plane.

    (a) Show that if T preserves orientation, then $T(z) = az + b$, with $a,b,c \in \mathbb{C}$ and a nonzero.

    (b) Show that if T does not preserve orientation, then $T(z) = a\bar{z} + b$, with $a,b,c \in \mathbb{C}$ and a nonzero.

    (c) Express these results using matrices.

32. Given two lists (A,B,C) and (A',B',C'), each with three collinear points, show that, in general, no affine transformation exists transforming one list into another.

33. Given two lists, (A,B,C,D) and (A',B',C',D'), each with four collinear points, show that, in general, no projective transformation exists transforming one list into another.

34. Show that the equation of a quadric in homogeneous coordinates is given by $XCX^{\mathrm{T}} = 0$, where X represents the projective coordinates of a point in space, C is a fixed matrix and superscript ^{T} indicates transposition. Conclude that conics are invariant under projective transformations.

35. Show that projective space $\mathbb{RP}^n$ can be naturally associated to the unit sphere $S^n \subset \mathbb{R}^{n+1}$, with antipodal points identified.

3 Coordinates

In this chapter, we study coordinate systems and change of coordinate transformations, with an emphasis on rectilinear systems. This topic will be important later, in the study of the space of rotations, of hierarchies of articulated objects such as the human body, of color spaces, and of the virtual camera model.

The use of coordinates gives us analytical representations of the objects and of the transformations of a geometry:

However, the representation is dependent on the coordinate system used, so we need a way to change between the two representations, or coordinate systems. A proper choice of a coordinate system often allows a problem to be simplified. Consider the motion of the front wheel of a bicycle: in a coordinate system where the origin is placed on the bicycle's fork, the trajectory of each point of the wheel is a circle. Using a coordinate system attached to the road, those trajectories would be more complex curves, called cycloids and hypocycloids.

For simplicity, we will begin by studying coordinate changes on the affine plane $\mathbb{R}^2$. In this chapter, points of $\mathbb{R}^n$ will be represented by nonbold letters, lowercase or uppercase.

3.1 Affine Transformations and Coordinate Changes

When reading a book, we can select the text that is in the line of sight in two ways: by moving the book or by moving our head. This shows that two points of view exist to interpret how a transformation works: it either moves objects in space (the book), or it changes the coordinate system (our head) leaving the objects fixed.

3.1.1   Transforming Objects

Consider an affine transformation of the plane consisting of a rotation R by an angle θ about the origin $(0,0)$, followed by a translation T by a vector (t_1, t_2). The overall effect of this transformation is given by the product TR, which is another way of denoting the composition $T \circ R$. For an arbitrary point $x = (x_1, x_2) \in \mathbb{R}^2$, we have $TR(x) = T(R(x))$: we rotate x, then translate it.

In matrix form we have

$$R = \begin{pmatrix} \cos\theta & -\sin\theta & 0 \\ \sin\theta & \cos\theta & 0 \\ 0 & 0 & 1 \end{pmatrix}, \qquad T = \begin{pmatrix} 1 & 0 & t_1 \\ 0 & 1 & t_2 \\ 0 & 0 & 1 \end{pmatrix}. \tag{3.1}$$

Therefore, the matrix of the composite transformation TR is

$$TR = \begin{pmatrix} \cos\theta & -\sin\theta & t_1 \\ \sin\theta & \cos\theta & t_2 \\ 0 & 0 & 1 \end{pmatrix}, \tag{3.2}$$

Figure 3.1 shows the effect of such an transformation on a unit square. As there is only one coordinate system involved—here, the standard cartesian coordinates—this is called an *object transformation*. Here is another useful example:

Example 3.1 (Rotation about an arbitrary point). Suppose we want to rotate something by an angle θ about an arbitrary point P on the plane. What is the matrix of this transformation in cartesian coordinates?

A solution method that works in many such situations is to express the desired transformation as a product of simpler transformations. To obtain the desired transformation, we can first apply a translation T_2 that takes the point P to the origin. Then we apply the rotation R about the origin. We finally translate again so the origin is mapped back to P; this is the inverse translation T_2^{-1}.

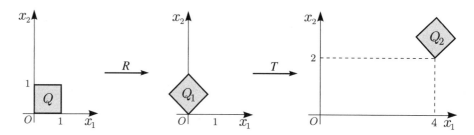

Figure 3.1. Action of the transformation (3.2) on the unit square. The rotation angle θ is $\pi/4 = 45°$ and the translation vector (t_1, t_2) is $(4, 2)$.

If P has coordinates (p_1, p_2), the translation T_2 and its inverse T_2^{-1} are given by

$$T_2 = \begin{pmatrix} 1 & 0 & -p_1 \\ 0 & 1 & -p_2 \\ 0 & 0 & 1 \end{pmatrix} \qquad T_2^{-1} = \begin{pmatrix} 1 & 0 & p_1 \\ 0 & 1 & p_2 \\ 0 & 0 & 1 \end{pmatrix}. \tag{3.3}$$

We already know the matrix for a rotation about the origin; see (3.1). Therefore, the composite final transformation is given by

$$T_2^{-1} R T_2 = \begin{pmatrix} \cos\theta & -\sin\theta & p_1(1-\cos\theta) + p_2\sin\theta \\ \sin\theta & \cos\theta & p_2(1-\cos\theta) - p_2\sin\theta \\ 0 & 0 & 1 \end{pmatrix}.$$

3.1.2 Transforming Reference Frames

Recall from Chapter 2 that a reference frame (loosely speaking, a coordinate system) on the affine plane is defined by a point O and a basis $\{e_1, e_2\}$ of the vector space $\mathbb{R}^2$ (see Figure 3.2(a)). The point O is called the origin. An arbitrary point $P \in \mathbb{R}^2$ is written as $O + \overrightarrow{OP}$, and its coordinates (x_1, x_2) in the given frame are defined by $\overrightarrow{OP} = x_1 e_1 + x_2 e_2$. When the origin is predefined, it is common to think of the frame as just the basis $\{e_1, e_2\}$.

Consider two reference frames $\mathcal{E} = (e_1, e_2, O)$ and $\mathcal{F} = (f_1, f_2, O')$, as in Figure 3.2(b). The fundamental theorem of affine geometry (Exercise 2 on page 47) guarantees there is exactly one affine transformation A taking $\mathcal{E}$ to $\mathcal{F}$; our job now is to find it explicitly. As in Example 3.1, we work in stages. First we find the linear transformation L taking the basis $\{e_1, e_2\}$ to the basis $\{f_1, f_2\}$. Then we find the translation T taking the origin O of the reference frame $\mathcal{E}$ into the origin O' of the reference frame $\mathcal{F}$. The transformation A is then the product $A = TL$.

Assuming the coordinates of the vector $\overrightarrow{OO'}$ in the basis $\{e_1, e_2\}$ to be

$$\overrightarrow{OO'} = t_1 e_1 + t_2 e_2, \tag{3.4}$$

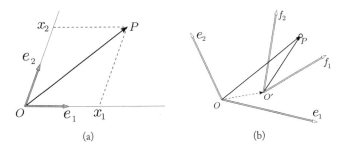

(a)   (b)

Figure 3.2. (a) Coordinate system on the plane; (b) a point in different coordinate systems.

the matrix of T will be given by

$$T = \begin{pmatrix} 1 & 0 & t_1 \\ 0 & 1 & t_2 \\ 0 & 0 & 1 \end{pmatrix}. \tag{3.5}$$

Next, if

$$f_1 = L(e_1) = a_{11}e_1 + a_{21}e_2,$$
$$f_2 = L(e_2) = a_{12}e_1 + a_{22}e_2, \tag{3.6}$$

the transformation L is given by the matrix

$$L = \begin{pmatrix} a_{11} & a_{12} & 0 \\ a_{21} & a_{22} & 0 \\ 0 & 0 & 1 \end{pmatrix}. \tag{3.7}$$

Multiplying we find

$$A = TL = \begin{pmatrix} 1 & 0 & t_1 \\ 0 & 1 & t_2 \\ 0 & 0 & 1 \end{pmatrix} \begin{pmatrix} a_{11} & a_{12} & 0 \\ a_{21} & a_{22} & 0 \\ 0 & 0 & 1 \end{pmatrix} = \begin{pmatrix} a_{11} & a_{12} & t_1 \\ a_{21} & a_{22} & t_2 \\ 0 & 0 & 1 \end{pmatrix}. \tag{3.8}$$

This is called the *transfer matrix* from the reference frame $\mathcal{E}$ to the reference frame $\mathcal{F}$. We will use the notation $A_{\mathcal{E}}^{\mathcal{F}}$ when we need to specify the reference frames. Notice that the first column of this matrix gives the coordinates of the vector $f_1 = L(e_1)$ in the basis $\{e_1, e_2\}$, the second column gives the coordinates of the vector $f_2 = L(e_2)$, and the third column gives the coordinates of the origin $O' = L(O)$ of the new system in the original reference frame system (e_1, e_2, O).

3.1.3 Transforming Coordinates

We now have a transformation or transfer matrix $A_{\mathcal{E}}^{\mathcal{F}}$ that maps the reference frame $\mathcal{E}$ to the reference frame $\mathcal{F}$. How are the coordinates of a point in the reference frame $\mathcal{E}$ related to the coordinates of the same point in $\mathcal{F}$?

We refer again to Figure 3.2(b). In $\mathcal{E}$, the point P is defined by the vector $\overrightarrow{OP}$, and in $\mathcal{F}$, P is defined by the vector $\overrightarrow{O'P}$. Clearly,

$$\overrightarrow{OP} = \overrightarrow{OO'} + \overrightarrow{O'P}. \tag{3.9}$$

To relate the coordinates of P in the two systems, we will write the preceding vector equation using coordinates.

Let (y_1, y_2) be the coordinates of P in the frame (f_1, f_2, O'), and (x_1, x_2) the coordinates of P in the frame (e_1, e_2, O). That is,

$$\overrightarrow{OP} = x_1 e_1 + x_2 e_2, \qquad \overrightarrow{O'P} = y_1 f_1 + y_2 f_2. \tag{3.10}$$

Replacing the expressions of $\overrightarrow{OP}, \overrightarrow{O'P}$ in (3.10), and of $\overrightarrow{OO'}$ in (3.4), in (3.9), we get

$$x_1 e_1 + x_2 e_2 = t_1 e_1 + t_2 e_2 + y_1 f_1 + y_2 f_2.$$

Taking into account the values of f_1 and f_2 in (3.6), we have

$$x_1 e_1 + x_2 e_2 = t_1 e_1 + t_2 e_2 + y_1(a_{11}e_1 + a_{21}e_2) + y_2(a_{12}e_1 + a_{22}e_2)$$
$$= (t_1 + y_1 a_{11} + y_2 a_{12})e_1 + (t_2 + y_1 a_{21} + y_2 a_{22})e_2.$$

We conclude that the change of coordinates (y_1, y_2) of the reference frame $\mathcal{F}$ to the coordinates (x_1, x_2) of the reference frame $\mathcal{E}$, is given by

$$x_1 = t_1 + y_1 a_{11} + y_2 a_{12},$$
$$x_2 = t_2 + y_1 a_{21} + y_2 a_{22},$$

or, using matrices,

$$\begin{pmatrix} x_1 \\ x_2 \\ 1 \end{pmatrix} = \begin{pmatrix} a_{11} & a_{12} & t_1 \\ a_{21} & a_{22} & t_2 \\ 0 & 0 & 1 \end{pmatrix} \begin{pmatrix} y_1 \\ y_2 \\ 1 \end{pmatrix}. \tag{3.11}$$

This is the same matrix as in Equation (3.8). We have just shown an important fact: the transfer matrix $A_{\mathcal{E}}^{\mathcal{F}}$, which expresses the transformation mapping the reference frame $\mathcal{E}$ to the reference frame $\mathcal{F}$, changes the coordinates of a point relative to $\mathcal{F}$ to its coordinates relative to $\mathcal{E}$. Carefully note the reversal to avoid later confusion.

Example 3.2. Let $\mathcal{E}$ be the standard (cartesian) frame of $\mathbb{R}^2$, where $e_1 = e_x$ and $e_2 = e_y$ are the unit vectors in the x and y directions, and $O = (0,0)$. Let $\mathcal{F}$ be the frame with origin $O' = (4, 2)$ and whose basis vectors f_1 and f_2 are obtained from e_1 and e_2 by a counterclockwise $45°$ rotation; see Figure 3.3(a). Since $\sin 45° = \cos 45° = \sqrt{2}/2$, we have

$$f_1 = \frac{\sqrt{2}}{2} e_1 + \frac{\sqrt{2}}{2} e_2, \quad f_2 = -\frac{\sqrt{2}}{2} e_1 + \frac{\sqrt{2}}{2} e_2, \quad O' = 4e_1 + 2e_2.$$

Therefore, the matrix of the transformation T changing the reference frame (e_1, e_2, O) to the reference frame (f_1, f_2, O'), is given by the product

$$A_{\mathcal{E}}^{\mathcal{F}} = \begin{pmatrix} 1 & 0 & 4 \\ 0 & 1 & 2 \\ 0 & 0 & 1 \end{pmatrix} \begin{pmatrix} \sqrt{2}/2 & -\sqrt{2}/2 & 0 \\ \sqrt{2}/2 & \sqrt{2}/2 & 0 \\ 0 & 0 & 1 \end{pmatrix} = \begin{pmatrix} \frac{\sqrt{2}}{2} & -\frac{\sqrt{2}}{2} & 4 \\ \frac{\sqrt{2}}{2} & \frac{\sqrt{2}}{2} & 2 \\ 0 & 0 & 1 \end{pmatrix}. \tag{3.12}$$

Let P be the point having coordinates $(2, 4)$ relative to $\mathcal{F}$. By applying the matrix (3.12), we obtain $(4-\sqrt{2}, 2+3\sqrt{2})$; these are the coordinates of P relative to $\mathcal{E}$. Figure 3.3(b) shows the vectors $\overrightarrow{OP}$ and $\overrightarrow{O'P}$ representing the point P in either system. ∎

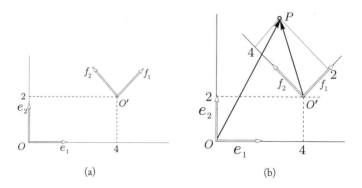

Figure 3.3. (a) Reference frames; (b) change of coordinates.

Inverting a transformation. Recall that the transfer matrix $A_{\mathcal{E}}^{\mathcal{F}}$ represents the transformation taking the reference frame $\mathcal{E}$ to the reference frame $\mathcal{F}$, and it changes the coordinates of a point relative to $\mathcal{F}$ to those relative to $\mathcal{E}$. We next consider the transfer matrix $A_{\mathcal{F}}^{\mathcal{E}}$, which does the same with the roles of $\mathcal{E}$ and $\mathcal{F}$ reversed. Now, applying two matrices in succession is the same as applying their product (more about this soon), so the product $A_{\mathcal{F}}^{\mathcal{E}} A_{\mathcal{E}}^{\mathcal{F}}$ describes the transformation that takes $\mathcal{F}$-coordinates back to $\mathcal{F}$-coordinates. It must be, therefore, the identity matrix. Similarly, $A_{\mathcal{E}}^{\mathcal{F}} A_{\mathcal{F}}^{\mathcal{E}}$ is the identity. This means that the two transfer matrices are inverse to one another:

$$A_{\mathcal{F}}^{\mathcal{E}} = (A_{\mathcal{E}}^{\mathcal{F}})^{-1}.$$

3.1.4 Transforming Equations

Recall that there are two methods of defining a geometric object by means of coordinates: parametric equations and implicit equations. Both types of equations come up often in computer graphics, and there is an important practical difference in the way they behave under coordinate changes, so we will examine the topic carefully in two dimensions.

A familiar example will help to clarify these ideas. A circle of radius 1 about the origin can be described parametrically as the set of points whose coordinates are of the form $x = \cos\theta$, $y = \sin\theta$, for different values of θ. In symbols,

$$C = \{(\cos\theta, \sin\theta) : 0 \leq \theta < 2\pi\}. \tag{3.13}$$

Another way to say this is that we have a map from the interval $[0, 2\pi) \subset \mathbb{R}$ onto the set of interest, the circle, which is inside $\mathbb{R}^2$. This point of view will come up in Section 3.4. The same circle can be defined implicitly as the set of points whose coordinates (x, y) satisfy a well-known quadratic equation:

$$C = \{(x, y) : x^2 + y^2 = 1\}. \tag{3.14}$$

All of this is happening in the standard (cartesian) frame $\mathcal{E} = \{e_1, e_2, O\}$ of $\mathbb{R}^2$. Suppose we want to express the same circle in the frame $\mathcal{F} = \{2e_1, 2e_2, O\}$. Given the $\mathcal{E}$-coordinates of a point of the circle, we know that to get the $\mathcal{F}$-coordinates of the same point we need to apply the transfer matrix $A_{\mathcal{F}}^{\mathcal{E}} = (A_{\mathcal{E}}^{\mathcal{F}})^{-1}$; see the last paragraph of the previous section. In this case

$$A_{\mathcal{F}}^{\mathcal{E}} = (A_{\mathcal{E}}^{\mathcal{F}})^{-1} = \begin{pmatrix} 2 & 0 & 0 \\ 0 & 2 & 0 \\ 0 & 0 & 1 \end{pmatrix}^{-1} = \begin{pmatrix} \frac{1}{2} & 0 & 0 \\ 0 & \frac{1}{2} & 0 \\ 0 & 0 & 1 \end{pmatrix}.$$

Since $A_{\mathcal{F}}^{\mathcal{E}}(\cos\theta, \sin\theta, 1) = \left(\frac{1}{2}\cos\theta, \frac{1}{2}\sin\theta, 1\right)$, we conclude that

$$C = \left\{ \left(\frac{1}{2}\cos\theta, \frac{1}{2}\sin\theta\right) : 0 \le \theta < 2\pi \right\} \quad \text{in } \mathcal{F}\text{-coordinates.}$$

This makes sense; the frame vectors in $\mathcal{F}$ are twice as big, so each point in the circle will have an expression in $\mathcal{F}$ that is twice as small.

Similarly, we can rewrite (3.14) as

$$C = \left\{ \left(\frac{1}{2}x, \frac{1}{2}y\right) : x^2 + y^2 = 1 \right\} \quad \text{in } \mathcal{F}\text{-coordinates.}$$

To get a proper implicit equation we need to get rid of the factors $\frac{1}{2}$ and leave only the variables to the left of the colon; we can do this by making the substitutions $u = \frac{1}{2}x$, $v = \frac{1}{2}y$, so that $x = 2u$, $y = 2v$. Thus, in $\mathcal{F}$-coordinates,

$$C = \{(u, v) : (2u)^2 + (2v)^2 = 1\} = \{(u, v) : 4u^2 + 4v^2 = 1\} = \{(x, y) : 4x^2 + 4y^2 = 1\}.$$

Thus the implicit equation of C in $\mathcal{F}$-coordinates is $4x^2 + 4y^2 = 1$. Note that although it was convenient to introduce new variables u, v to apply the transformation, we ultimately renamed them x, y again, as we are allowed to do with the understanding that these letters now refer to the new coordinates. Especially when there are multiple coordinate changes, as is generally the case in computer graphics, it is pointless to give them all different names; instead we must keep track at any given time of what frame the coordinates refer to.

There is a shortcut to the procedure of introducing new variables to find the new implicit equation. We can write the variables as a vector $(x, y, 1)$, apply the $A_{\mathcal{E}}^{\mathcal{F}}$ to this vector, and then replace each variable in the equation by the corresponding entry in the multiplied vector. In this case,

$$A_{\mathcal{E}}^{\mathcal{F}} \begin{pmatrix} x \\ y \\ 1 \end{pmatrix} = \begin{pmatrix} 2 & 0 & 0 \\ 0 & 2 & 0 \\ 0 & 0 & 1 \end{pmatrix} \begin{pmatrix} x \\ y \\ 1 \end{pmatrix} = \begin{pmatrix} 2x \\ 2y \\ 1 \end{pmatrix},$$

so the recipe calls for replacing x with $2x$ and y with $2y$ in the implicit equation, obtaining $4x^2 + 4y^2 = 1$ in the new coordinates. Before moving on to a more complicated example, we stop to repeat the key difference in the handling of the two types of equations:

❏ To convert a parametric equation such as $(\cos\theta, \sin\theta)$ in $\mathcal{E}$-coordinates to $\mathcal{F}$-coordinates, apply the transfer matrix $A_{\mathcal{F}}^{\mathcal{E}}$ to the *vector of coordinates* (extended by 1 as usual).

❏ To convert an implicit equation such as $x^2 + y^2 = 1$ in $\mathcal{E}$-coordinates to $\mathcal{F}$-coordinates, apply the transfer matrix $A_{\mathcal{E}}^{\mathcal{F}}$ to the *vector of coordinate variables* (extended by 1 as usual), and then replace each variable in the equation by the corresponding entry in the multiplied vector.

Note which transfer matrix is used in each case; the matrix in the implicit equation procedure is inverse to the one in the parametric equation procedure. The reason for this should be carefully considered (review the preceding material if necessary), but it leads to the following: in the parametric equation, the transfer matrix $A_{\mathcal{F}}^{\mathcal{E}}$ is being applied to left of the colon in (3.13). The same transfer matrix can be applied to the left of the colon in (3.14), but this doesn't quite yield an implicit equation yet; the fix is to apply the inverse transfer matrix $A_{\mathcal{E}}^{\mathcal{F}}$ to the *variables* on both sides of the colon, with the net result that we have applied $A_{\mathcal{E}}^{\mathcal{F}}$ to the equation on the right-hand side only.

Example 3.3. We return to the situation of Example 3.2. A line has the implicit equation $x + 2y = 4$, and the parametric equation $(2t, 2-t)$, in cartesian coordinates. What are its equations with respect to $\mathcal{F}$? Try to work it out before reading on.

Implicit equation. Applying the matrix $A_{\mathcal{E}}^{\mathcal{F}}$ of (3.12) to $(x, y, 1)$ gives $\big((x-y)\sqrt{2}/2 + 4,$ $(x+y)\sqrt{2}/2 + 2, 1\big)$. So we must replace x by $(x - y)\sqrt{2}/2 + 4$ and y by $(x + y)\sqrt{2}/2 + 2$ in the implicit equation, obtaining

$$(x - y)\frac{\sqrt{2}}{2} + 4 + 2\left((x + y)\frac{\sqrt{2}}{2} + 2\right) = 4; \quad \text{that is,} \quad (3x + y)\frac{\sqrt{2}}{2} = -4.$$

Parametric equation. For this we need the inverse of $A_{\mathcal{E}}^{\mathcal{F}}$, which is

$$A_{\mathcal{F}}^{\mathcal{E}} = \begin{pmatrix} \frac{\sqrt{2}}{2} & \frac{\sqrt{2}}{2} & -3\sqrt{2} \\ -\frac{\sqrt{2}}{2} & \frac{\sqrt{2}}{2} & \sqrt{2} \\ 0 & 0 & 1 \end{pmatrix}$$

(see the end of Section 3.3.2 for a straightforward way to compute this). Applying this to the vector $(2t, 2 - t, 1)$ and dropping the final 1 gives

$$\left((t - 4)\frac{\sqrt{2}}{2}, (4 - 3t)\frac{\sqrt{2}}{2}\right)$$

for the parametric equation of the line in the frame $\mathcal{F}$.   ❏

3.2   Local and Global Transformations

In computer graphics applications, we generally have a coordinate system that applies to the whole scene, called the *world coordinate system*, and defined by a *global reference frame* such as the standard frame of $\mathbb{R}^3$. Often we also use specific coordinate systems for some objects, because in their own system, the equations involving these objects are simpler. These are called *local coordinate systems*.

A typical example is describing the motion of a rigid object in space. Every possible placement of a rigid body in space is described by an affine frame: roughly speaking, the position in space of a particular point fixed on the object, and a *spatial orientation*, or basis of vectors, also thought of as fixed relative to the object.

Having associated a local reference frame to the rigid body, we can describe the body's instantaneous position by its (moving) reference frame, or more precisely by the matrix that maps the global reference frame to the moving frame, which is also the matrix that converts the moving frame's coordinate system into the global frame's coordinate system. We shall see that it is easy to keep track of this matrix as the object moves, if the motion steps are easily described in the object's frame (like rotation around the body's own axes).

This is because a product of matrices corresponds to a composition of transformations. Given three reference frames $\mathcal{E}$, $\mathcal{E}'$ and $\mathcal{E}''$, the transfer matrices are related by

$$A_{\mathcal{E}}^{\mathcal{E}''} = A_{\mathcal{E}}^{\mathcal{E}'} A_{\mathcal{E}'}^{\mathcal{E}''}. \tag{3.15}$$

Why? Well, $A_{\mathcal{E}}^{\mathcal{E}''}$ converts coordinates relative to $\mathcal{E}''$ to those relative to $\mathcal{E}$. We can do this in two stages: first applying the matrix $A_{\mathcal{E}'}^{\mathcal{E}''}$ that converts coordinates in $\mathcal{E}''$ to coordinates in $\mathcal{E}'$, then the matrix $A_{\mathcal{E}}^{\mathcal{E}'}$ that converts coordinates in $\mathcal{E}'$ to coordinates in $\mathcal{E}$. Recalling that the matrix applied first goes on the right, we obtain (3.15).

But wait! Since $A_{\mathcal{E}}^{\mathcal{E}'}$ is also the matrix expression of the transformation that maps $\mathcal{E}$ to $\mathcal{E}'$, shouldn't $A_{\mathcal{E}}^{\mathcal{E}'}$ be applied first (to map $\mathcal{E}$ to $\mathcal{E}'$) and $A_{\mathcal{E}'}^{\mathcal{E}''}$ next (to map $\mathcal{E}'$ to $\mathcal{E}''$), giving a different rule, $A_{\mathcal{E}}^{\mathcal{E}''} = A_{\mathcal{E}'}^{\mathcal{E}''} A_{\mathcal{E}}^{\mathcal{E}'}$? The answer is no, and for a subtle reason: $A_{\mathcal{E}}^{\mathcal{E}'}$ and $A_{\mathcal{E}'}^{\mathcal{E}''}$ express transformations in different frames. While $A_{\mathcal{E}}^{\mathcal{E}'}$ is the expression of a transformation in the frame $\mathcal{E}$, the other matrix, $A_{\mathcal{E}'}^{\mathcal{E}''}$, expresses a transformation in the frame $\mathcal{E}'$. So the product $A_{\mathcal{E}'}^{\mathcal{E}''} A_{\mathcal{E}}^{\mathcal{E}'}$ is invalid as an attempt to express the composition of the two transformations. The only correct formula is (3.15), after all—which is fortunate, since we know that matrix multiplication is not commutative! (See also Exercise 18 on page 73.)

To make this whole discussion more concrete, suppose we have an initial, global reference frame given by

$$\mathcal{E}^0 = (b_1^0, b_2^0, O^0),$$

and n successive (local) reference frames

$$\mathcal{E}^1 = (b_1^1, b_2^1, O^1), \quad \mathcal{E}^2 = (b_1^2, b_2^2, O^2), \quad \ldots, \quad \mathcal{E}^n = (b_1^n, b_2^n, O^n).$$

Denote by T_{i-1}^i the matrix of the transformation taking the frame $\mathcal{E}^{i-1}$ to the frame $\mathcal{E}^i$, expressed in the reference frame $\mathcal{E}^{i-1}$. That is, $T_{i-1}^i = A_{\mathcal{E}^{i-1}}^{\mathcal{E}^i}$. Using (3.15) repeatedly, we see that the product

$$T = T_0^1 T_1^2 T_2^3 \cdots T_{n-1}^n, \tag{3.16}$$

in that order, is the matrix (in $\mathcal{E}^0$) of the transformation taking $\mathcal{E}^0$ to $\mathcal{E}^n$ or, similarly, the matrix that changes the coordinates of the local reference frame $\mathcal{E}^n$ to the global reference frame $\mathcal{E}^0$.

The order of the product in (3.16) is very convenient for computer graphics. Usually, each transformation T_{i-1}^i is easy to describe in the frame $\mathcal{E}^{i-1}$; it might correspond, for example, to the instruction "turn 90° to the right." And the implementation is easily achieved with a stack data structure, because the transformations are specified in the order $T_0^1, T_1^2, \ldots, T_{n-1}^n$ and are applied in the reverse order: $T_{n-1}^n, T_{n-2}^{n-1}, \ldots, T_1^2, T_2^0$.

Another application of the same principle is when we have several reference frames in a hierarchy (see Chapter 9). Again, the local transformations from one frame to the next are generally easy to obtain; we then multiply them in the right order. We discuss a simple particular case:

Example 3.4 (Robot on the plane). Consider the 2D robot arm shown in Figure 3.4(a). A stem is rigidly attached to the wall and has length d_1. The forearm has length d_2 and rotates about the stem. The hand has length d_3 and rotates about the forearm. Problem: find the coordinates of the point P at the tip of the hand, as the whole assembly moves.

One solution method is a simple trigonometric calculation, which we leave to the reader. We will instead solve the problem using successive coordinate changes, involving the three reference frames shown in Figure 3.4(b): the global reference frame $\mathcal{E}'$, the forearm frame $\mathcal{E}^\infty$, and the hand frame $\mathcal{E}^\in$.

The global, fixed, reference frame $\mathcal{E}'$ is simply the cartesian system. Its origin is at the intersection of the stem with the wall, and its basis is the standard basis of $\mathbb{R}^2$.

The reference frame $\mathcal{E}^\infty$ of the forearm has its origin at the forearm joint and its first basis vector is the unit vector in the direction of the forearm. The second basis vector is the unit vector in the perpendicular direction, forming a right-handed frame.

The reference frame $\mathcal{E}^\in$ of the hand has its origin at the hand-forearm joint and its first basis vector is the unit vector in the direction of the hand. The second basis vector is chosen as in the previous case.

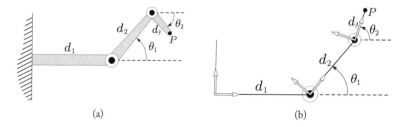

Figure 3.4. (a) A robot arm and (b) the local transformations allowing for its solution.

The transformation from the global reference frame to the reference frame of the hand is given by $T = T_0^1 T_1^2$, where T_0^1 is expressed in the frame $\mathcal{E}'$ and T_1^2 is expressed in the frame $\mathcal{E}_\infty$:

$$\text{Global} \xrightarrow{\;T_1\;} \text{Forearm} \xrightarrow{\;T_2\;} \text{Hand.}$$

The transformation T_0^1 is a translation by the vector $d_1 e_1$, followed by a rotation by an angle θ_1 *about the forearm joint*. Alternatively, we can see it as a rotation by θ_1 *about the global origin* followed by a translation by $d_1 e_1$. The second point of view allows us to immediately write the matrix of T_0^1 (in the frame $\mathcal{E}$) in the form (3.2).

Now here is the beauty of this approach. The second transformation, T_1^2, *has exactly the same form* as T_0^1 because the relationship of the hand to the forearm is the same as that of the arm to the stem (changing d_1 to d_2 and θ_1 to θ_2). Indeed, as shown in Figure 3.4(b), θ_2 measures the angle between the forearm and the hand. In matrix terms,

$$T_0^1 = \begin{pmatrix} \cos\theta_1 & \sin\theta_1 & d_1 \\ -\sin\theta_1 & \cos\theta_1 & 0 \\ 0 & 0 & 1 \end{pmatrix}, \qquad T_1^2 = \begin{pmatrix} \cos\theta_2 & \sin\theta_2 & d_2 \\ -\sin\theta_2 & \cos\theta_2 & 0 \\ 0 & 0 & 1 \end{pmatrix}. \tag{3.17}$$

The final transformation is then given by

$$T = T_0^1 T_2^2 = \begin{pmatrix} \cos(\theta_1+\theta_2) & \sin(\theta_1+\theta_2) & d_1 + d_2\cos\theta_1 \\ -\sin(\theta_1+\theta_2) & \cos(\theta_1+\theta_2) & -d_2\sin\theta_1 \\ 0 & 0 & 1 \end{pmatrix}. \tag{3.18}$$

We now observe that, in the system of the hand, the point P has coordinates $(d_3, 0)$; therefore, to determine the coordinates of the position of point P, it is enough to apply the matrix in (3.18) to the vector $(d_3, 0, 1)$. $\qquad\square$

3.3 Coordinates in Space

We now extend the discussion in Sections 3.1 and 3.2 to higher dimensions, and specifically to three dimensions.

3.3.1 Transforming Reference Frames

As we saw in Section 2.4.3, a reference frame in n dimensions is a list $\mathcal{E} = \{e_1, \ldots, e_n, O\}$, where O is a point called the *origin* and $e_1, \ldots, e_n$ are vectors forming a basis of the vector space $\mathbb{R}^n$. The coordinates of a point P in that frame are the numbers $(x_1, \ldots, x_n)$ such that

$$P = \sum_{j=1}^{n} x_j e_j + O, \quad \text{that is,} \quad \overrightarrow{OP} = \sum_{j=1}^{n} x_j e_j,$$

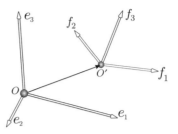

Figure 3.5. Relating two reference frames $\mathcal{E} = \{e_1, e_2, e_3, O\}$ and $\mathcal{F} = \{f_1, f_2, f_3, O'\}$ in space.

Given another reference frame $\mathcal{F} = \{f_1, \ldots, f_n, O'\}$, we want to find the affine transformation mapping the first frame into the second. Its matrix (in $\mathcal{E}$), called the transfer matrix from $\mathcal{E}$ to $\mathcal{F}$, will be denoted by $A_{\mathcal{E}}^{\mathcal{F}}$, as in the 2D case.

We first consider the linear transformation R of the vector space $\mathbb{R}^n$ to itself taking the basis $\{e_1, \ldots, e_n\}$ to the basis $\{f_1, \ldots, f_n\}$, that is, $R(e_i) = f_i$. This transformation is invertible, and its inverse takes $\{f_1, \ldots, f_n\}$ to $\{e_1, \ldots, e_n\}$.

Next we take the translation T of $\mathbb{R}^n$ by the vector $\overrightarrow{OO'}$ taking the origin of the first reference frame to the origin of the second. Figure 3.5 illustrates this in the case $n = 3$.

The composition $TR \colon \mathbb{R}^n \to \mathbb{R}^n$ is the desired affine transformation taking the frame $\{e_1, \ldots, e_n, O\}$ to the frame $\{f_1, \ldots, f_n, O'\}$. In matrix form, we have

$$A_{\mathcal{E}}^{\mathcal{F}} = \begin{pmatrix} 1 & \cdots & 0 & t_1 \\ \vdots & \ddots & \vdots & \vdots \\ 0 & \cdots & 1 & t_n \\ 0 & \cdots & 0 & 1 \end{pmatrix} \begin{pmatrix} a_{11} & \cdots & a_{1n} & 0 \\ \vdots & \ddots & \vdots & \vdots \\ a_{n1} & \cdots & a_{nn} & 0 \\ 0 & \cdots & 0 & 1 \end{pmatrix} = \begin{pmatrix} a_{11} & \cdots & a_{1n} & t_1 \\ \vdots & \ddots & \vdots & \vdots \\ a_{n1} & \cdots & a_{nn} & t_n \\ 0 & \cdots & 0 & 1 \end{pmatrix},$$

which, as before is called the *transfer matrix* from the frame $\mathcal{E}$ to the frame $\mathcal{F}$. Here $(t_1, \ldots, t_n) = \overrightarrow{OO'}$ is the translation vector of T, and the $n \times n$ matrix (a_{ij}) represents the linear map $R \colon \mathbb{R}^n \to \mathbb{R}^n$—that is, the jth column expresses the coordinates of the vector f_j in the basis $\{e_1, \ldots, e_n\}$.

3.3.2 Transforming Coordinates

Given two frames $\mathcal{E}$ and $\mathcal{F}$ related as above and a point P with coordinates $(y_1, \ldots, y_n)$ in $\mathcal{F}$, the coordinates $(x_1, \ldots, x_n)$ of P in $\mathcal{E}$ are given by

$$\begin{pmatrix} x_1 \\ \vdots \\ x_n \\ 1 \end{pmatrix} = \begin{pmatrix} a_{11} & \cdots & a_{1n} & t_1 \\ \vdots & \ddots & \vdots & \vdots \\ a_{n1} & \cdots & a_{nn} & t_n \\ 0 & \cdots & 0 & 1 \end{pmatrix} \begin{pmatrix} y_1 \\ \vdots \\ y_n \\ 1 \end{pmatrix}.$$

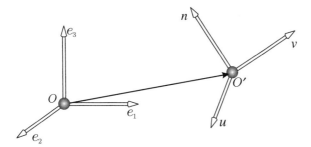

Figure 3.6. Orthonormal reference frames are related by rigid motions.

Example 3.5 (Isometries in $\mathbb{R}^3$). An important special case is that of isometries, or rigid motions, of n-space. Such a transformation preserves lengths and orthogonality, and so maps an orthonormal frame to another orthonormal frame. In this case the a_{ij}'s, which encode the linear part of the transfer matrix $A_{\mathcal{E}}^{\mathcal{F}}$, form an *orthogonal* matrix.

Figure 3.6 shows a typical situation in three dimensions: the global reference frame, still denoted by $\mathcal{E}$, is the standard frame with origin at $(0,0,0)$ and the standard basis $\{e_x, e_y, e_z\}$ of unit vectors in the coordinate directions. It is being mapped to an orthonormal frame $\mathcal{F} = (u, v, n, O')$, with

$$O' = (o_x, o_y, o_z), \quad u = (u_x, u_y, u_z), \quad v = (v_x, v_y, v_z), \quad n = (n_x, n_y, n_z).$$

Note that in this particular example, we use a special notation for correctness: clearly $u_x = a_{11}$, and so on; also, the letter n is chosen because the third vector is often defined as the unit normal to a plane of interest, with coordinates u, v. The numbers o_x, o_y, o_z are the inner products of the vector $\overrightarrow{OO'}$ with the unit vectors e_x, e_y, e_z:

$$o_x = \langle \overrightarrow{OO'}, e_x \rangle, \quad o_y = \langle \overrightarrow{OO'}, e_y \rangle, \quad o_z = \langle \overrightarrow{OO'}, e_z \rangle. \tag{3.19}$$

This is simply because we are expressing the vector in an orthonormal basis.

The transformation has the transfer matrix

$$A_{\mathcal{E}}^{\mathcal{F}} = \begin{pmatrix} u_x & v_x & n_x & o_x \\ u_y & v_y & n_y & o_y \\ u_z & v_z & n_z & o_z \\ 0 & 0 & 0 & 1 \end{pmatrix}.$$

What is the transfer matrix of the inverse transformation? It takes $\mathcal{F}$ to $\mathcal{E}$, so its linear part (the top left 3×3 minor) is obtained just by inversion. This is true in general, but in the orthogonal case it's particularly simple, since as we know the inverse of an orthogonal matrix is the same as its transpose. That is,

$$A_{\mathcal{F}}^{\mathcal{E}} = (A_{\mathcal{E}}^{\mathcal{F}})^{-1} = \begin{pmatrix} u_x & u_y & u_z & u_t \\ v_x & v_y & v_z & v_t \\ n_x & n_y & n_z & n_t \\ 0 & 0 & 0 & 1 \end{pmatrix},$$

with the rightmost column—the translation part—still to be determined. For that we take another look at (3.19), which, as remarked, simply expresses the orthonormality of the basis $\{e_x, e_y, e_z\}$. The basis $\{u, v, n\}$ is also orthonormal, so by reversing roles we see that all we need to do is take the inner products of the vector $\overrightarrow{O'O} = -\overrightarrow{OO'}$ with the basis vectors u, v, n:

$$u_t = -\langle \overrightarrow{OO'}, u \rangle, \quad v_t = -\langle \overrightarrow{OO'}, v \rangle, \quad n_t = -\langle \overrightarrow{OO'}, n \rangle.$$

❑

3.3.3 Application: A Robot Arm in Space

As an application of these ideas, and also those in Section 3.2, we consider a very simple robot arm model in three dimensions, shown in Figure 3.7(a). A rigid arm of length d_1 can rotate around a vertical axis. A rigid forearm is articulated at the end of the arm and rotates around the joint in a vertical plane. Figure 3.7(b) shows the arm in an arbitrary position, parameterized by the two rotation angles θ_1 and θ_2. We wish to set up local frames for the arm and forearm, and find the position of the endpoint P of the forearm as a function of θ_1 and θ_2.

The global frame $\mathcal{E}^0$ has origin O on the vertical axis, and unit vectors in the coordinate directions x, y, z. The frame $\mathcal{E}^1$ of the main arm is obtained from this by a rotation through θ_1 around the z-axis, which is given by the matrix

$$T_0^1 = \begin{pmatrix} \cos\theta_1 & -\sin\theta_1 & 0 & 0 \\ \sin\theta_1 & \cos\theta_1 & 0 & 0 \\ 0 & 0 & 1 & 0 \\ 0 & 0 & 0 & 1 \end{pmatrix}.$$

Next, to obtain the frame of the forearm, we proceed in two steps. First we set up an auxiliary frame $\mathcal{E}^2$ at the arm-forearm joint A, still in the same direction as $\mathcal{E}^1$; in other

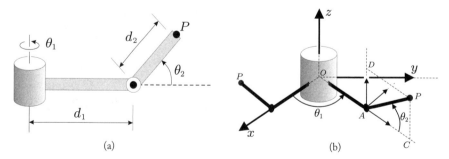

(a)   (b)

Figure 3.7. (a) Arm hierarchy and (b) angles of motion. The figure on the right also shows the global frame $\mathcal{E}^0$ and one of the local frames, $\mathcal{E}^2$, whose origin is at the joint A.

words, $\mathcal{E}^2$ is obtained from $\mathcal{E}^1$ by a translation of length d_1 in the first basis direction, *in the local frame $\mathcal{E}^1$*. The expression of this translation is

$$T_1^2 = \begin{pmatrix} 1 & 0 & 0 & d_1 \\ 0 & 1 & 0 & 0 \\ 0 & 0 & 1 & 0 \\ 0 & 0 & 0 & 1 \end{pmatrix}.$$

The frame $\mathcal{E}_2$ is shown in Figure 3.7(b) at the point A.

Finally, the frame $\mathcal{E}_3$ of forearm (not shown in the figure) is obtained by rotating $\mathcal{E}_2$ around its own second basis vector, through an angle θ_2. The corresponding matrix is

$$T_2^3 = \begin{pmatrix} \cos\theta_2 & 0 & -\sin\theta_2 & 0 \\ 0 & 1 & 0 & 0 \\ \sin\theta_2 & 0 & \cos\theta_2 & 0 \\ 0 & 0 & 0 & 1 \end{pmatrix}.$$

As we know from (3.16), to find the overall transfer matrix taking the global frame $\mathcal{E}_0$ to $\mathcal{E}_3$ we simply need to take the product:

$$T = T_0^1 T_1^2 T_2^3 = \begin{pmatrix} \cos\theta_1\cos\theta_2 & -\sin\theta_1 & -\cos\theta_1\sin\theta_2 & d_1\cos\theta_1 \\ \sin\theta_1\cos\theta_2 & \cos\theta_1 & -\sin\theta_1\sin\theta_2 & d_1\sin\theta_1 \\ \sin\theta_2 & 0 & \cos\theta_2 & 0 \\ 0 & 0 & 0 & 1 \end{pmatrix}.$$

So far, so good. How do we find the coordinates of P in the global frame? In the local frame $\mathcal{E}_3$, the coordinates are simply $(d_2, 0, 0)$, since the first basis vector $\mathcal{E}_3$ points along the forearm. Therefore the coordinates are given by the product

$$T \begin{pmatrix} d_2 \\ 0 \\ 0 \\ 1 \end{pmatrix} = \begin{pmatrix} (d_2\cos\theta_1 + d_1)\cos\theta_1 \\ (d_2\cos\theta_2 + d_1)\sin\theta_1 \\ d_2\sin\theta_2 \\ 1 \end{pmatrix}.$$

Arguably, if we are only interested in the coordinates of P, basic trigonometry might have led to answer more directly, without matrix manipulations. But the advantages of the local frame method become obvious if we consider that the use of matrices *of the same form* allow us to extend the problem further.

For instance, adding a hand (as in Example 3.4) at point P, free to rotate around a "locally vertical" axis—the line at P perpendicular to the forearm AP and contained on the plane ACP of Figure 3.7(b)—gives the linkage full mobility; the tip of the hand will then be able to reach every point in a solid region of space, unlike P (whose position, being parameterized by two angles, can only move along a surface, in this case a torus). To obtain the local frame of the hand from that of the forearm ($\mathcal{E}^3$) we just need to multiply on the right by a translation matrix just like T_1^2, corresponding to the vector AP (that is, a translation by d_2 along the first basis vector of $\mathcal{E}^3$), and then by a translation matrix just like T_0^1, corresponding to a rotation by an angle θ_3 around the new axis at P.

In this way, by composing local transformations of a few basic types, we can model linkages of any complexity with ease. We will return to this topic in Chapter 9.

3.4   Curvilinear Coordinates

So far we have dealt with *affine* coordinate changes only, and especially distance-preserving ones. Often it is useful to work instead with a more general coordinate system. We cannot develop the theory of *curvilinear coordinates* in general, but the basic setup is as follows: a map f is defined from $\mathbb{R}^n$ or some appropriate part $U \subset \mathbb{R}^n$ onto the space of interest S, which might be $\mathbb{R}^n$ as well, or a subset of it, or a space of dimension greater than n, or a subset of it, or something even more abstract. Two very familiar examples are given next.

Example 3.6 (Polar coordinates on the plane). Suppose U is the strip $[0, \infty) \times [0, 2\pi)$ of $\mathbb{R}^2$, also defined as the set of pairs (r, θ) such that $0 \leq r$ and $0 \leq \theta < 2\pi$. The space of interest S is all of $\mathbb{R}^2$. The map f takes $(r, \theta) \in U$ to $(r\cos\theta, r\sin\theta) \in S$. In an ideal situation we would prefer the map $f : U \to S$ to be bijective (one-to-one and onto), but often, in order to cover the whole space of interest, we have to put up with some nonuniqueness; here, all points $(0, \theta) \in U$ map to the origin of S. Alternatively, one could exclude $r = 0$ from the definition of U, and then S is $\mathbb{R}^2$ minus the origin, which is not covered. Or one can allow even more nonuniqueness by defining U with $0 \leq r$ and $0 \leq \theta \leq 2\pi$, so points along the positive x-axis of S can be realized with either $\theta = 0$ or $\theta = \pi$; and many other possibilities, each with its pros and cons.   ❏

Example 3.7 (Latitude and longitude). This time U is the strip $[-\pi, \pi] \times [-\pi/2, \pi/2]$, or, more familiarly, $[-180°, 180°] \times [-90°, 90°]$, and the space of interest is a sphere of radius R in $\mathbb{R}^3$. The first coordinate in U, the *longitude*, will be denoted by θ, and the second, the *latitude*, by ϕ. We choose to place the Greenwich meridian, $\theta = 0°$, on the xz-plane, so our map f takes (θ, ϕ) to the point (x, y, z) defined by

$$x = R\cos\phi\cos\theta, \quad y = R\cos\phi\sin\theta, \quad z = R\sin\phi.$$

See Figure 3.8. Once again there is a trade-off: by allowing ϕ to take the values $90°$ and $-90°$, we can include the poles, but their longitude is arbitrary; likewise by allowing both $\theta = -180°$ and $\theta = 180°$, one meridian is covered twice.

In the context of the visual sphere—say in observing the skies—latitude and longitude are called *elevation* (angle above the horizon) and *azimuth* (angle along the horizon from the direction of the north; east is $90°$).   ❏

We say that the space S is *parameterized* by U, or by f, or by the coordinates on U. The curves on S consisting of points where all parameters but one remain fixed are called

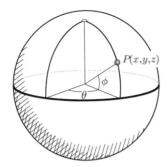

Figure 3.8. Longitude θ and latitude ϕ.

coordinate curves; at least in two dimensions, they help a lot in visualizing the parameterization (Figure 3.8).

For future reference we collect a few more definitions about parameterizations. When the space S is $\mathbb{R}^n$ or a subset of it—which will be most of the time—one can ask whether the map f is continuous, differentiable, etc. Generally, the more of these qualities f has, the better. We will define an *excellent* parameterization, or coordinate system, as one where f is one-to-one and onto, both f and its inverse are continuous, f is differentiable and its partial derivatives are continuous, and the Jacobian matrix of f (matrix of partial derivatives) has full rank, that is, its columns are linearly independent, at all points in the domain. For instance, latitude and longitude give an excellent parameterization of the sphere minus the poles and one meridian; see Example 3.7.

When we insist on excellence, most spaces—even simple ones like a sphere—cannot be parameterized all at once. But a lot can be done with a collection of *local* parameterizations, each covering part, or even most, of the space of interest, and all compatible in an appropriate sense. Such a collection is called an *atlas*: an apt name, because an atlas of the world, if it does not neglect the oceans, is precisely a collection of excellent local parameterizations (by the coordinates of each page) that together cover the whole globe.

Parameterizations can be useful even when they are not excellent. Polar coordinates on the plane are used all the time even though $r = 0$ leads to nonuniqueness. If f is continuous and differentiable, parameter values where the Jacobian matrix of f has full rank are called *regular*, and all others are called *singular*; these are often places where the parameterization fails to be one-to-one, like the line $r = 0$ in the parameter space of polar coordinates, all of whose points map to the origin of $\mathbb{R}^2$, or the line $\phi = 90°$ of latitude-longitude space, which maps to the north pole.

3.5   Comments and References

The subject of curvilinear coordinates and local coordinates is of great importance in many areas of computer graphics, such as modeling, texture mapping, warping, and morphing.

However, a deeper study of it would lead us too far afield, into the realms of differential topology and differential geometry. We will introduce some of the relevant notions in later chapters as needed, starting in the next chapter, where we parameterize the space of rotations of $\mathbb{R}^3$. A good introduction to differential geometry can be found in [do Carmo 75].

The topic of local coordinates and hierarchies, represented by some examples in this chapter, will be approached in more depth in Chapter 9.

Exercises

1. Let the frame $\mathcal{F}$ on the plane be defined by $O = (3, -2)$, $f_1 = 2e_1 + 3e_2$, $f_2 = e_1 + 2e_2$.

   (a) Write down the transfer matrix from the standard (cartesian) reference frame to $\mathcal{F}$.

   (b) If a point has coordinates (x, y) in the frame $\mathcal{F}$, what are its cartesian coordinates?

   (c) If a point has cartesian coordinates (x, y), what are its coordinates in the frame $\mathcal{F}$? Note that the transformation is not orthogonal.

2. Let the frame $\mathcal{F}$ on the plane be obtained from the cartesian reference frame by a counterclockwise rotation about the origin through $135°$. Find the transfer matrices. An ellipse has equation $5x^2 + 6xy + 5y^2 = 1$ in cartesian coordinates; what is its equation in the frame $\mathcal{F}$?

3. The methods for converting equations we saw on page 60 apply without change to problems in higher dimensions. Here are some practice exercises in space. Several objects are given by their equations in cartesian coordinates; convert them to the frame $\mathcal{F}$ with origin $(0, 0, 0)$ and basis vectors $f_1 = -2e_1 + 2e_2 + e_3$, $f_2 = e_2 - 2e_3$, and $f_3 = e_1 - 2e_2 + 2e_3$.

   (a) The plane $x + y + z = 1$.

   (b) The line passing through $(2, -1, 3)$ in the direction $(-2, 3, 1)$.

   (c) The line defined by $x + y = 3$, $y - 2z = 5$.

   (d) The ellipsoid defined by $x^2 + 2y^2 + 3z^2 = 6$.

4. A circle has cartesian equation $x^2 + y^2 + 2x - 3y - 9 = 0$. Find its equation in the coordinates defined by

   (a) the frame whose origin is $(2, 3)$ and whose basis is the standard basis of $\mathbb{R}^2$;

   (b) the frame obtained by reflection of the cartesian frame in the line $x + y = 1$;

   (c) the frame defined by $O = (1, 2)$, $f_1 = e_1 + 2e_2$, $f_2 = e_2 - 2e_1$.

5. An ellipse has cartesian equation $5x^2 + 9y^2 + 6xy = 8$. Write its equation relative to the frame obtained from the standard one by a counterclockwise $45°$ rotation (leaving the origin unchanged). Explain how, by reinterpreting the new equation as one in cartesian coordinates, we have rotated the ellipse $45°$ clockwise.

6. Figure 3.9 shows a line on the plane defined by a point $P = (x, y)$ and a direction vector $v = (\cos \alpha, \sin \alpha)$. Recalling from Exercise 8 in Chapter 2 the definition of a shear along a coordinate axis, give a reasonable definition for a shear along this arbitrary line, with a proportionality constant a. Find the matrix for this linear transformation.

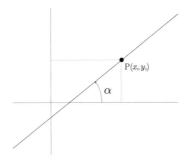

Figure 3.9. Figure for Exercise 6.

7. Find the matrix for the linear transformation L of $\mathbb{R}^3$ that expands distances in the direction of the unit vector $u = (u_1, u_2, u_3)$ by a factor of a, and leaves unchanged all distances in directions orthogonal to it.

    (Hint: first find an auxiliary orthogonal transformation T that takes the unit x-vector to the vector u. You probably do not need to compute all its entries explicitly. Then follow the general method of Example 3.1, or see Exercise 18 below.)

8. Consider a rotation by an angle θ about an axis defined by the unit vector $u = (u_x, u_y, u_z)$ in $\mathbb{R}^3$. Show that the transformation matrix is given by

$$\begin{pmatrix} u_x^2(1-\cos\theta) + \cos\theta & u_x u_y(1-\cos\theta) - u_z\sin\theta & u_x u_z(1-\cos\theta) + u_y\sin\theta \\ u_x u_y(1-\cos\theta) + u_z\sin\theta & u_y^2(1-\cos\theta) + \cos\theta & u_y u_z(1-\cos\theta) - u_x\sin\theta \\ u_x u_z(1-\cos\theta) - u_y\sin\theta & u_y u_z(1-\cos\theta) + u_x\sin\theta & u_z^2(1-\cos\theta) + \cos\theta \end{pmatrix}.$$

    (Hint: same as for the previous exercise.)

9. Find the transformation matrix for a rotation by a $120°$ angle about the axis defined by the unit vector $r = \frac{1}{\sqrt{3}}(1, 1, 1)$. (This of course can be done using the result of the previous exercise, but you might be able to guess the matrix directly by considering what the transformation does to the unit cube $[0, 1]^3$.)

10. We saw in Example 3.5 a special, easy method to invert the matrix associated to a rigid motion of space. Extend the method to homotheties of space (see Exercise 15 in Chapter 2).

11. Formulate the problem of scaling an object in $\mathbb{R}^n$ in terms of (a) finding a transformation matrix in the cartesian system; or (b) changing the coordinate system. Give the corresponding matrices. (Compare with Exercise 5.)

12. Suppose the cylindrical base of the mechanical arm described in Section 3.3.3 moves a certain distance along the diagonal line $z = 0$, $x = y$ in space. What change should be made to the matrix T that expresses the transfer from the global frame to the frame of the forearm? (The global frame remains fixed; its origin no longer coincides with the base of the arm.)

13. As mentioned in Section 3.3.3, now that our matrix machinery is in place, attaching a hand (third segment—see Figure 3.10) to our robot arm is a cinch using the local transformations

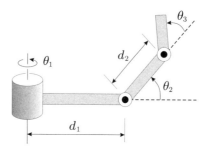

Figure 3.10. Robot arm for Exercise 13.

method. Write out the product of matrices expressing the transfer from the global frame to the frame of the hand; then write the product expressing the position of the tip of the hand. You do not need to compute the products.

14. Show that a counterclockwise rotation through an angle θ about the origin is given by multiplication by $e^{i\theta} = \cos\theta + i\sin\theta$ on the complex plane. Here of course we are thinking of complex numbers $p = x + iy$ as pairs $(x, y) \in \mathbb{R}^2$.

   In general, what transformation of $\mathbb{R}^2$ is given by multiplication by an arbitrary fixed complex number, not necessarily of absolute value 1?

15. Let $z_1 = e^{i\theta_1}$ and $z_2 = e^{i\theta_2}$ be fixed complex numbers of absolute value 1, with θ_1 and θ_2 not far apart. Consider the following three parametric expressions, where t ranges from 0 to 1:

$$R_1(t) = e^{(1-t)i\theta_1 + ti\theta_2}, \qquad R_2(t) = (1-t)z_1 + tz_2, \qquad R_3(t) = \frac{(1-t)z_1 + tz_2}{|(1-t)z_1 + tz_2|}.$$

   Describe the corresponding curves on the complex plane. Interpreting each point in the curve as a linear transformation on the plane (see previous exercise), explain how each of these parameterizations gives a way to interpolate between the two rotations $e^{i\theta_1}$ and $z_2 = e^{i\theta_2}$. Discuss the advantages and disadvantages of each approach.

16. Let the unit sphere in $\mathbb{R}^3$ be parameterized by latitude and longitude (Example 3.7). Fix a point (θ_0, ϕ_0) in the domain U of the parameterization. Find the Jacobian matrix of f at that point.

   When the columns of this matrix are linearly independent, that is, when the matrix has full rank, we also say that f has full rank, or that f is regular, or nonsingular—all this at a given point (θ_0, ϕ_0).) Show that f has full rank unless $\phi_0 = \pm\pi/2$.

   Denote the unit vectors in the θ and ϕ directions on the (θ, ϕ)-plane space by e_θ and e_ϕ. The *image* (or *push-forward*) of e_θ under f is the vector in $\mathbb{R}^3$ given by the first column of the Jacobian; that is, its components are the partial derivatives of the components of f with respect to θ. The image of e_ϕ is the second column.

   These image vectors, denoted by $f_*(e_\theta)$ and $f_*(e_\phi)$, are to be thought of as attached to the point $P = f(\theta_0, \phi_0)$ of the sphere—indeed, they are part of the *tangent plane* to the sphere at that point. (The vectors also depend on (θ_0, ϕ_0), of course, though the notation does not make that explicit.)

If f has full rank at (θ_0, ϕ_0), the vectors $f_*(e_\theta)$ and $f_*(e_\phi)$ are linearly independent, by definition. Then, from the frame $\mathcal{E} = \{e_\theta, e_\phi, (\theta_0, \phi_0)\}$ in U we have obtained a frame $f(\mathcal{E}) = \{f_*(e_\theta), f_*(e_\phi), P\}$ of the tangent space to the sphere at P. What happens to this frame as we get close to the singular points of the parameterization ($\phi = \pm\pi/2$)?

If v is an arbitrary vector on the (θ, phi)-plane, how would the image $f_*(v)$ be defined?

17. Show that the cylinder $S : x^2 + y^2 = 1$ of $\mathbb{R}^3$ can be parameterized by the function $f : \mathbb{R}^2 \to S$ defined by $f(\theta, t) = (\cos\theta, \sin\theta, t)$.

   (a) Show that this parameterization is regular everywhere.

   (b) Define a maximum domain of f, in which it defines a coordinate system in the cylinder (*cylindrical coordinates*).

   (c) Find the image of the frame $\{e_\theta, e_t, (\theta_0, t_0)\}$ under f.

18. Consider the transformation T that maps an affine reference frame $\mathcal{E}'$ to another frame $\mathcal{E}''$. We have denoted the matrix expressing T in the frame $\mathcal{E}'$ by $A_{\mathcal{E}'}^{\mathcal{E}''}$. Show that the matrix expressing T *in another frame* $\mathcal{E}$ is given by $A_{\mathcal{E}}^{\mathcal{E}'} A_{\mathcal{E}'}^{\mathcal{E}''} (A_{\mathcal{E}}^{\mathcal{E}'})^{-1}$. (Hint: call the desired matrix K. Using (3.15) and the subsequent discussion, show that $A_{\mathcal{E}}^{\mathcal{E}'} A_{\mathcal{E}'}^{\mathcal{E}''} = K A_{\mathcal{E}}^{\mathcal{E}'}$.)

Use this to write the formula to convert the matrix of an arbitrary transformation from one frame to another.

4 | The Space of Rotations

The need for 3D rotations is ubiquitous in computer graphics: in animation, in the treatment of linked hierarchies, and in the specification of the virtual camera motion, to mention but a few applications. For this reason it is of great practical importance to have good representations and parameterizations of the space of rotations.

In general, a good representation leads to a good parameterization. The latter, in turn, leads to good interface solutions for the user to specify and manipulate rotations in the computer:

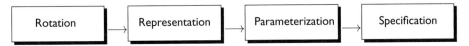

There are at least six major ways in which the description and specification of rotations play a role in computer graphics: interpolation (of keyframes in animation, for instance), direct and inverse kinematics, direct and inverse dynamics, and space-time optimization. A good representation of the space of rotations should satisfy the requirements of these different methodologies, which include the ability to

- ❏ find the orientation of a body in terms of the parameters,

- ❏ calculate the parameters corresponding to a rotation,

- ❏ calculate the kinematical and dynamical magnitudes associated with a rigid body,

- ❏ express and solve equations of motion,

- ❏ interpolate smoothly between rotations, and

- ❏ perform operations between rotations.

As a prelude to the study of 3D rotations, we briefly review the situation in two dimensions, which is much simpler. In a departure from the two previous chapters, in this chapter we will generally be dealing with $\mathbb{R}^2$ and $\mathbb{R}^3$ as vector spaces, and our transformations of $\mathbb{R}^2$ and $\mathbb{R}^3$ will be linear (origin-preserving).

4.1 Plane Rotations

A *rotation* in $\mathbb{R}^2$ is a linear transformation that is orthogonal and positive. *Positive*, or orientation-preserving, means that the determinant of the transformation is positive. (Plane reflections over a line are examples of nonpositive, or orientation-reversing, transformations.) The study of plane reflections is easy, because a rotation is uniquely determined by what it does to a single basis vector, say the vector e_1 of the standard basis. If the rotation takes e_1 to a unit vector b_1, as in Figure 4.1(a), the second basis vector e_2 lands in an entirely predetermined place, b_2, because angles and handedness must be preserved.

As a result, the space of rotations is in a sense just the space of possible images b_1. This space is a circle of radius 1, since b_1 has length 1.

More formally, we have established a correspondence between the space of plane rotations, denoted by $\mathrm{SO}(2)$, and the unit circle. In fact there is more structure than that; in a circle, all points look the same, but the space of rotations is a group (the composition of two rotations is also a rotation) and so it has a distinguished element, the identity.

This is conveniently taken into account if we think of $\mathbb{R}^2$ as the complex plane. Then $e_1 = (1,0) = 1 + 0i = 1$ and $e_2 = (0,1) = 0 + 1i = i$. A rotation that takes e_1 to b_1 acts as multiplication by b_1. The unit circle in the complex plane is a group under multiplication, and is an excellent model, or representation, of $\mathrm{SO}(2)$.

Note that $b_1 = e^{i\theta}$ for some real number $\theta = \arg b_1$, because b_1 has length 1. See Figure 4.1(b). This number is the angle of rotation. (We suggest the reader do Exercise 14 in Chapter 3.) If we add any multiple of 2π to θ, the vector b_1, and therefore the corresponding rotation, is unchanged (recall that $e^{2\pi i} = 1$). Therefore although θ parameterizes the space of rotations, this parameterization is not unique; it is only unique up to multiples of 2π. Uniqueness can be achieved if we restrict θ to an appropriate interval, such as $[0, 2\pi)$, but then other properties are lost: for instance adding two angles can give a result outside the interval.

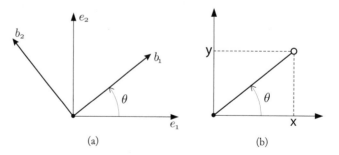

(a)   (b)

Figure 4.1. (a) A rotation in $\mathbb{R}^2$ is characterized by the image of a single vector; (b) the argument $\arg(x + iy)$ is $\theta = \arctan(y/x)$.

A second common representation of $SO(2)$ is by the transformation matrix. As we know, the matrix of a rotation through an angle θ is

$$R(\theta) = \begin{pmatrix} \cos\theta & -\sin\theta \\ \sin\theta & \cos\theta \end{pmatrix}. \tag{4.1}$$

Conversely, an orthogonal 2×2 matrix of positive determinant describes a rotation, and has the form (4.1) for some θ (unique up to multiples of 2π).

Note that the parameter θ in the matrix representation has the same meaning as in the representation by complex numbers.

Using this parameter we can easily interpolate (average) between rotations, simply by interpolating the parameter. If θ_1 and θ_2 are, respectively, the initial and final angles, the weighted average

$$\theta(t) = (1-t)\theta_1 + t\theta_2$$

gives a time-dependent angle varying smoothly between θ_1 and θ_2. The actual rotation can then be achieved by either representation: by multiplication with $e^{i\theta(t)}$, or by the use of the matrix $R(\theta(t))$.

An interesting question is, what happens if θ_1 is replaced by $\theta_1 + 2\pi$, representing the same rotation, but θ_2 is left unchanged? This shows the importance of how one selects parameter values when there is ambiguity, a recurring and important issue in computer graphics. In any given case the choice may be obvious, but formulating and implementing a general rule—for instance, "always choose θ_1 and θ_2 less than $180°$ apart"—is generally difficult, and sometimes impossible.

Note that we do not interpolate the matrices or complex numbers themselves, but rather the parameter. That is, it would not do to take the weighted average

$$(1-t)\begin{pmatrix} \cos\theta_1 & -\sin\theta_1 \\ \sin\theta_1 & \cos\theta_1 \end{pmatrix} + t \begin{pmatrix} \cos\theta_2 & -\sin\theta_2 \\ \sin\theta_2 & \cos\theta_2 \end{pmatrix},$$

which is not even an orthogonal matrix. Nor does the average between $e^{i\theta_1}$ and $e^{i\theta_2}$ represent a rotation. (See Exercise 15 in Chapter 3.)

Determining which representation is better depends on the circumstances. Storing a 2×2 matrix takes more space than storing a complex number, but this seldom matters nowadays. Similarly, multiplying two complex numbers can be faster than multiplying a matrix by a vector, especially in a language that has built-in support for complex numbers. Purely 2D rotations are a minor part of computer graphics processing time, and the need for converting between complex numbers and the matrix representation used in other parts of the code may cancel the speed advantage of the complex number representation.

4.2 Introduction to Rotations in Space

The study of rotations in three dimensions is more complex than in two, and will occupy us for the rest of the chapter. The material necessarily involves a lot of mathematics; readers tempted to skip the calculations should nonetheless strive to follow the text, to grasp the key ideas. This will allow them to understand the details later on, when needed in practice.

Here again, we define a *rotation* to be a linear transformation of $\mathbb{R}^3$ that is orthogonal (hence distance-preserving) and positive. Rotations form a subgroup of the group of all linear transformations of $\mathbb{R}^3$—why?

We denote the group of rotations of $\mathbb{R}^3$ by $\mathrm{SO}(3)$. A rotation R is of course determined by what it does to the standard basis $\{e_1, e_2, e_3\}$ of $\mathbb{R}^3$. Suppose R takes the standard basis to the basis $\{b_1, b_2, b_3\}$, which is orthonormal by definition. The matrix of R has as its columns the expressions of b_1, b_2, b_3 in the standard basis.

Unlike the case of $\mathbb{R}^2$, it is not enough to know b_1, or the first column of the matrix: there are many rotations that take e_1 to b_1. Indeed, given one such rotation, we can then apply any rotation about the axis determined by b_1: the composition is still a rotation taking e_1 to b_1.

However, if both b_1 and b_2 are known, b_3 is determined. This is the cross product familiar from physics: $b_3 = b_1 \times b_2$. (Similarly, $b_1 = b_2 \times b_3$ and $b_2 = b_3 \times b_1$.)

So one way to understand the space $\mathrm{SO}(3)$ is to study the space of possibilities for b_1, and the space of possibilities for b_2 once a choice of b_1 is made. The space of choices for b_1 is just the 2D sphere S^2, the set of vectors of length 1:

$$S^2 = \{(x, y, z) \in \mathbb{R}^3 : x^2 + y^2 + z^2 = 1\}.$$

(Compare with the 2D case, where the space of choices of b_1 was a circle.) Once b_1 is chosen, the choices for b_2 narrow down to a circle, since b_2 must be perpendicular to b_1. And as mentioned, b_3 is then fully determined.

This already tells us that $\mathrm{SO}(3)$ is intrinsically a space of dimension three,[1] in the sense that it can be parameterized by three numbers: two to describe the position of b_1 on the unit sphere—say the latitude and longitude discussed in Exercise 3.7 of Chapter 3—and a third parameter to account for where b_2 lies around the circle where it is allowed to be, with 0 defined in some way. However, there are several complications that make things harder than in the 2D case:

❏ Rotations in three dimensions do not commute. Make your hand flat and turn it horizontally by $90°$, then vertically (say around a north-south line) by $90°$. Now, starting again from the same position, make the same turns in reverse order. Your hand will end up in a different position each time.

❏ Composition has a complicated effect on parameters. For plane rotations, composing (multiplying) rotations is as simple as adding rotation angles. In three dimensions, with the choice of parameters just described or any other choice of three parameters, the relation between the parameters of two rotations and those of their composition is cluttered.

---

[1] This can be proved rigorously as follows. $\mathrm{SO}(3)$ can certainly be represented as a subset of $\mathbb{R}^3 \times \mathbb{R}^3$ using the vectors b_1 and b_2. These vectors must satisfy three constraints: $\langle b_1, b_1 \rangle = 1$, $\langle b_2, b_2 \rangle = 1$, and $\langle b_1, b_2 \rangle = 0$. Solving three algebraic equations in a 6D space gives a 3D subspace provided the matrix of partial derivatives of the equations has full rank everywhere, and this can be shown to be the case.

❏ For plane rotations we were able to find a coordinate that avoids singularities (though not ambiguity). But already for the sphere, singularities cannot be avoided in a global parameterization—see Exercise 16 in Chapter 3 for an example—and this is equally true for SO(3). Any set of three coordinates covering the whole space of rotations will sometimes behave badly—and it is not just that a given rotation may have more than one set of coordinates, but rather that one or more parameters can completely collapse, similar to the situation in the poles of the sphere. (What is the longitude of the pole?)

In practice, then, it is necessary to always work with more than one parameterization. In fact there are many in common use, each with advantages and disadvantages, but before plunging into their study, it will be useful to take a look at some *intrinsic properties* of rotations, those that do not depend on a coordinate system.

4.3   Axis and Angle of Rotation

Leonhard Euler (1707–1783), one of the greatest mathematicians of all time, first published in 1776 a proof of an important fact in geometry and mechanics: every motion of the sphere around its center fixes some diameter. By "a motion of the sphere around its center" is meant a transformation of S^2 preserving distances and orientation, such as those determined by the linear transformations of $\mathbb{R}^3$ that we have called rotations.

Euler's result shows that our rotations are indeed rotations in the sense of turning about an axis, or fixed line. Moreover, this axis is well defined, except in the case of a "trivial rotation," the identity. More precisely:

Theorem 4.1. *Suppose R is an element of* SO(3). *There is a positive orthonormal basis* $\{u, v, w\}$ *of $\mathbb{R}^3$ in which the matrix of R has the form*

$$\begin{pmatrix} 1 & 0 & 0 \\ 0 & \cos\theta & -\sin\theta \\ 0 & \sin\theta & \cos\theta \end{pmatrix}, \tag{4.2}$$

where $\theta \in [0, \pi]$. The number θ is uniquely defined by this property and is called the angle of rotation of R. The element u of the basis is also uniquely defined, unless $\theta = 0$ (in which case u can be any unit vector) or $\theta = \pi$ (in which case u is defined only up to a sign). When $\theta \neq 0$, the line containing u is called the axis of rotation of R. (Unless $\theta = \pi$ we can be more precise and use u to define what is called an oriented axis.*)* ❏

This gives us a good intuitive picture of individual rotations. It also suggests a parameterization of SO(3): two parameters, or degrees of freedom, are used to select the unit vector u on S^2, and the third is the rotation angle θ.

This looks attractive until we realize that this parameterization is singular precisely at the identity, where $\theta = 0$ but u can be any vector at all. In Section 4.7 we will see how this defect can be remedied by working in a 4D space, the field of *quaternions*. The quaternion representation is intimately connected with the axis/angle representation, but is in many ways more elegant.

Example 4.2 (Converting from matrices to axis/angle and vice versa). The representation by axis and angle is intrinsic, but it is not clear how one can calculate with it. Given two rotations, with axes u_1 and u_2 and angles θ_1 and θ_2, what are the resulting axes and angles?

The matrix for a rotation with a given axis and angle was given in Chapter 3, Exercise 8. The axis vector u was specified in the standard basis of $\mathbb{R}^3$, but any orthonormal basis would do. So one can choose a basis, convert the rotations to matrix form and multiply.

The remaining question is, given a rotation matrix, how can we find its axis and angle? Linear algebra comes to the rescue: find the matrix's eigenvectors and eigenvalues. The eigenvalues of an orthogonal matrix have absolute value 1, and are either real or occur in complex conjugate pairs. Since there are three eigenvalues, they must all be real, or one real and one conjugate pair. The product of the eigenvalues is the determinant of the matrix, which is 1. Together these conditions imply that at least one eigenvalue is 1, and the other two are conjugate complex numbers of absolute value 1, that is, $e^{i\theta}$ and $e^{-i\theta}$ for some $\theta \in [0, \pi]$. This θ is the angle of rotation. (If $\theta = 0$ all three eigenvalues equal 1 and the matrix is the identity; if $\theta = \pi$ two eigenvalues are -1). As for the axis—an eigenvector with eigenvalue 1 is fixed by the rotation, and therefore is an axis vector.

In principle, then, we can do algebra with the axis/angle representation. But the matrix algebra is cumbersome. Again, quaternions will come to the rescue in Section 4.7.   ❏

4.4   Parameterizations by Three Rotation Angles

The most common parameterizations of the space of rotations are, in one way or another, based on the composition of three rotations about coordinate axes. We start by considering how we might parameterize rotations *near the identity*, those for which the angle of rotation is small.

4.4.1   Yaw, Pitch, and Roll

Sailors gave names to the three types of rotation a ship can undergo in response to the waves: to *roll* is to tilt sideways, about the vessel's long axis; to *pitch* is to tilt forwards and backwards, so the bow and stern alternately rise and fall; and to *yaw* is to turn around a vertical axis.

The same names were borrowed for aviation, and in that context are illustrated in Figure 4.2. Note that the three types of rotation are rotations about coordinate axes *relative to the aircraft*: the x-axis being the fore-and-aft direction, the y-axis sideways and the z-axis up and down. For an arbitrary object we may not have natural principal axes as in navigation, but we can still, and will, attach three local cartesian axes x, y, z to the object of interest. The world cartesian axes will be called X, Y, Z.

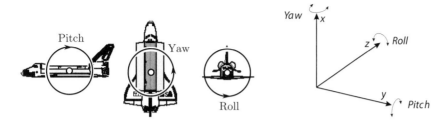

Figure 4.2. The three degrees of rotational freedom of an aircraft.

The rotation matrices for pure rolling, pitching and yawing motions are given by

$$R_x(\alpha) = \begin{pmatrix} 1 & 0 & 0 \\ 0 & \cos\alpha & -\sin\alpha \\ 0 & \sin\alpha & \cos\alpha \end{pmatrix}, \quad R_y(\beta) = \begin{pmatrix} \cos\beta & 0 & \sin\beta \\ 0 & 1 & 0 \\ -\sin\beta & 0 & \cos\beta \end{pmatrix}, \quad R_z(\gamma) = \begin{pmatrix} \cos\gamma & -\sin\gamma & 0 \\ \sin\gamma & \cos\gamma & 0 \\ 0 & 0 & 1 \end{pmatrix},$$

where α, β and γ are the roll, pitch and yaw angles, respectively. (The yaw angle is also known as the *deflection angle*, and the pitch angle as the *steepness angle*.)

Now suppose that we apply a rotation about the z-axis by γ, then one about the y-axis by β, and finally one about the x-axis by α. The overall result is described by the matrix

$$R_z(\gamma)R_y(\beta)R_x(\alpha), \tag{4.3}$$

the order of the product being a consequence of our use of the object-based coordinates x, y, z (see Section 3.2). It is plausible—and can readily be proved[2]—that any rotation near the identity can be written in a unique way in the form (4.3) for small α, β, γ. Thus (4.3) provides a good parameterization of the space of rotations near the identity, one that is fairly easy to visualize.

Note that the matrices $R_x(\alpha)$, $R_y(\beta)$, $R_z(\gamma)$ do not commute with one another in general. But a simple calculation shows that they almost commute when the rotation angles are small. For example, the departure from commutativity for $R_x(\alpha)$ and $R_z(\gamma)$ is given by the difference

$$R_x(\alpha)R_z(\gamma) - R_z(\gamma)R_x(\alpha) = \begin{pmatrix} 0 & (\cos\alpha - 1)\sin\gamma & -\sin\alpha\sin\gamma \\ (\cos\alpha - 1)\sin\gamma & 0 & (\cos\gamma - 1)\sin\alpha \\ \sin\alpha\sin\gamma & (\cos\gamma - 1)\sin\alpha & 0 \end{pmatrix}.$$

The upper left and lower right entries are of *second order* in the angles (and the other nonzero entries are of third order).

---

[2]For example, consider the derivatives

$$\left.\frac{dR_x(\alpha)}{d\alpha}\right|_{\alpha=0} = \begin{pmatrix} 0 & 0 & 0 \\ 0 & 0 & -1 \\ 0 & 1 & 0 \end{pmatrix}, \quad \left.\frac{dR_y(\beta)}{d\beta}\right|_{\beta=0} = \begin{pmatrix} 0 & 0 & 1 \\ 0 & 0 & 0 \\ -1 & 0 & 0 \end{pmatrix}, \quad \left.\frac{dR_z(\gamma)}{d\gamma}\right|_{\gamma=0} = \begin{pmatrix} 0 & -1 & 0 \\ 1 & 0 & 0 \\ 0 & 0 & 0 \end{pmatrix}.$$

Since these matrices are linearly independent, they span the (3D) tangent space of SO(3) at the identity.

Thus, although the product of a yaw, a pitch and a roll depends on the order in which each is applied, the dependence is slight when the angles are small. This means we get an equally good parameterization of the space of rotations near the identity if we take the product in a different order, and any two such parameterizations almost coincide for small angles. (That is, we have $R_z(\gamma)R_y(\beta)R_x(\alpha) = R_y(\beta')R_x(\alpha')R_z(\gamma')$ with $\alpha' \approx \alpha$, $\beta' \approx \beta$, and $\gamma' \approx \gamma$, and similarly for any other ordering.) Which ordering is used depends on the problem at hand.

4.4.2 Euler Angles

We will continue with our search for a good representation by describing a local parameterization of the rotation space, widely used in mechanics, and discovered by Leonhard Euler.

Consider the canonical basis $\mathcal{E} = \{e_1, e_2, e_3\}$ of $\mathbb{R}^3$, as shown in Figure 4.3(a). This reference frame defines the Cartesian coordinates (x, y, z) in space. We will apply three consecutive rotations to this reference frame about the coordinate axes z, x, and z of the canonical basis, as described below:

❑ $R_z(\psi)$: Rotation of an angle ψ about the z-axis (see Figure 4.3(b));

❑ $R_x(\theta)$: Rotation of an angle θ about the x-axis (see Figure 4.3(c));

❑ $R_z(\phi)$: Another rotation about the z-axis by an angle ϕ (see Figure 4.3(d)).

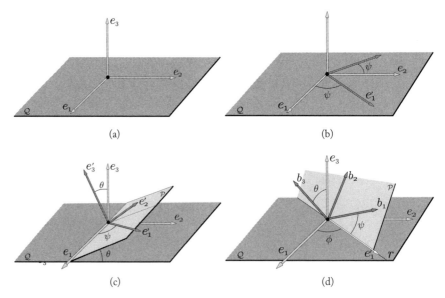

Figure 4.3. Euler's angles: (a) canonical basis $\mathcal{E} = \{e_1, e_2, e_3\}$ of $\mathbb{R}^3$; (b) rotation of angle ψ about the z-axis; (c) rotation of angle θ about the x-axis; (d) rotation of angle ϕ about the z-axis.

The product

$$R_z(\psi)R_x(\theta)R_z(\phi), \tag{4.4}$$

of the above rotations transforms the canonical basis $\mathcal{E}$ in the orthonormal basis $\mathcal{B} = \{\mathbf{b}_1, \mathbf{b}_2, \mathbf{b}_3\}$ shown in Figure 4.3(d).

Now let us assume the orthonormal reference frame $\mathcal{B} = \{\mathbf{b}_1, \mathbf{b}_2, \mathbf{b}_3\}$ is given a priori. Statement: "The reference frame $\mathcal{B}$ can be obtained from the canonical basis $\mathcal{E}$ by three consecutive rotations." In fact, as we showed in Figure 4.3(d), let $\mathcal{P}$ be the plane generated by the vectors $\mathbf{b}_1$ and $\mathbf{b}_2$ of the basis $\mathcal{B}$, and r the intersection straight line between $\mathcal{P}$ and the plane $z = 0$ (r is called *nodal straight line* in mechanics). Notice that the unit vector $\mathbf{e}_3 \wedge \mathbf{b}_3$ points towards the direction of the nodal straight line. (In fact, $\mathbf{e}_3$ and $\mathbf{b}_3$ are normal to the planes $z = 0$ and $\mathcal{P}$, respectively. Therefore, the cross product $\mathbf{e}_3 \wedge \mathbf{b}_3$ should point towards the direction of the intersection straight line between these two planes, which is the nodal straight line.) Let us also indicate by ϕ, the angle between the vector $\mathbf{e}_1$ and the nodal straight line.

Initially we perform a rotation $R_{\mathbf{e}_3}(\phi)$, of an angle ϕ about the e_3 axis, taking the vector $\mathbf{e}_1$ in the vector $\mathbf{e}'_1 = \mathbf{e}_3 \wedge \mathbf{b}_3$. Next, we apply the rotation $R_{\mathbf{e}_3 \wedge \mathbf{b}_3}(\theta)$, by an angle θ about the nodal straight line, taking the vector e_3 in the vector $\mathbf{b}_3$. Finally, we apply the rotation $R_{\mathbf{b}_3}(\psi)$, by an angle ψ about the vector $\mathbf{b}_3$, taking the vector $\mathbf{e}'_1$ in the vector $\mathbf{b}_1$. Of course, this sequence of rotations

$$R_{\mathbf{e}_3}(\phi)R_{\mathbf{e}_3 \wedge \mathbf{b}_3}(\theta)R_{\mathbf{b}_3}(\psi), \tag{4.5}$$

transforms the canonical basis in the basis $\mathcal{B}$.

Notice that the transformation in (4.4) is equal to the one in (4.5), because both coincide in the canonical basis $\mathcal{E}$. However, they apply the rotations in reverse order and with different axes. In the first case, the rotations are always performed about the axes of the Cartesian (global) system; In the second case, the rotations are applied consecutively in local coordinate systems (except for $R_{\mathbf{e}_3}(\phi)$, of course).

We know that a rotation R in space is determined by the positive orthonormal basis $\mathcal{B} = \{\mathbf{b}_1, \mathbf{b}_2, \mathbf{b}_3\}$. (The rotation is given by the linear transformation taking the canonical basis $\mathcal{E} = \{\mathbf{e}_1, \mathbf{e}_2, \mathbf{e}_3\}$ in the basis $\mathcal{B}$). Therefore, we showed above the following result: a rotation $R \in SO(3)$ can be obtained by three consecutive rotations about the coordinate axes.

This result is owed to Leonhard Euler, and the angles (ϕ, θ, ψ) are called *Euler's angles*. They constitute a parameterization of the space $SO(3)$. This parameterization is broadly used in the study of rigid bodies dynamics in $\mathbb{R}^3$.

4.4.3 Euler's Angles and Matrices

We are now going to describe a rotation matrix parameterized by Euler's angles. Indicating the rotation angles about the $\mathbf{e}_1$, $\mathbf{e}_2$ and $\mathbf{e}_3$ axes by α, β and γ, respectively, gives us the rotation matrices at each axis: performing the product of these three matrices, in the order $R_z(\gamma)R_y(\beta)R_x(\alpha)$ (which is the order used to obtain Equation (4.5)), we obtain the parameterization $R(\alpha, \beta, \gamma)$ of the rotation matrix,

$$\begin{pmatrix} \cos\beta\cos\gamma & \cos\gamma\sin\alpha\sin\beta - \cos\alpha\sin\gamma & \cos\alpha\cos\gamma\sin\beta + \sin\alpha\sin\gamma \\ \cos\beta\sin\gamma & \cos\alpha\cos\gamma + \sin\alpha\sin\beta\sin\gamma & -\cos\gamma\sin\alpha + \cos\alpha\sin\beta\sin\gamma \\ -\sin\beta & \cos\beta\sin\alpha & \cos\alpha\cos\beta \end{pmatrix}. \qquad (4.6)$$

which is called *Euler's matrix*. Notice that R defines an application $R\colon \mathbb{R}^3 \to \mathbb{R}^9$.

4.4.4 Singularities and Euler's Angles

Euler's angles do not constitute a global parameterization of the rotation space. The reason being, the space $SO(3)$ is compact, and therefore it cannot be parameterized by an open set of $\mathbb{R}^3$ (in the same way a sphere $S^2 \subset \mathbb{R}^3$ does not admit a global parameterization by an open subset on the plane). In this way, any attempt of extending Euler's angles to cover the whole rotation space, leads to the creation of singularities in the parameterization. These singularities are regions in the parameterization domain, in which we do not have the three degrees of freedom in the rotation matrix (see the discussion of singularity in the parameterization of the sphere in Chapter 3).

   We can explicitly describe singularities of the parameterization by Euler's angles by taking $\beta = \pi/2$ in the parameterization $R(\alpha, \beta, \gamma)$ of Equation (4.6). That is, by performing a rotation of $90°$ about the y-axis (pitch), we obtain the parameterization

$$\begin{aligned} R\left(\alpha, \frac{\pi}{2}, \gamma\right) &= \begin{pmatrix} 0 & \cos\gamma\sin\alpha - \cos\alpha\sin\gamma & \cos\alpha\cos\gamma + \sin\alpha\sin\gamma & 0 \\ 0 & \cos\alpha\cos\gamma + \sin\alpha\sin\gamma & \cos\alpha\sin\gamma - \cos\gamma\sin\alpha & 0 \\ -1 & 0 & 0 & 0 \end{pmatrix} \\ &= \begin{pmatrix} 0 & \sin(\alpha - \gamma) & \cos(\alpha - \gamma) & 0 \\ 0 & \cos(\alpha - \gamma) & \sin(\alpha - \gamma) & 0 \\ -1 & 0 & 0 & 0 \end{pmatrix}. \end{aligned}$$

We see that, despite having two degrees of freedom in the parameter space, we only have one degree of freedom in the parameterization. This is because the parameterized matrix only depends on the difference between the angles. (This phenomenon is similar to the singularity in the parameterization of the sphere that we saw in Chapter 3.)

   It is easy to intuitively understand the singularities of the parameterization of $SO(3)$ using Euler's angles. In fact, the angles perform three consecutive and independent rotations in each one of the coordinate axes: first, we rotate about z, then about y, and later about x. The singularities happen as two rotation axes point towards the same direction. This is because the rotations about those two axes are dependent and therefore we loose a degree of freedom.

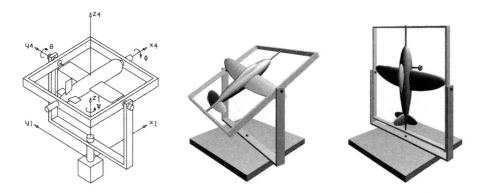

Figure 4.4. Gimbal lock phenomenon. Schematic illustration for rotating an aircraft: by the yaw angle Ψ about the z-axis, by the pitch angle θ about the y-axis and the roll angle φ about the x-axis (left); rotating the aircraft (middle); the yaw gimbal rotates 90 degrees and the aircraft is free to move about the z-axis (right). (*Figures courtesy of Gernot Hoffmann [Hoffmann 03].*)

The gimbal lock phenomenon. The problem of singularities of Euler's angles manifests in practice. The phenomenon is known in aeronautical engineering and mechanical engineering as *gimbal lock*.[3]

Consider the example of gimbal lock in the model shown in Figure 4.4. As the yaw-axis is fixed, after a pitch-axis rotation (leaving the airplane heading in the vertical direction), the roll and yaw axes coincide. In this position, we loose a degree of freedom in the rotation about these two axes.

Of course, by restricting the variation of the Euler's angles, we eliminate the singularities problem. For this reason, the pilot of an airplane does not have the same problems of an animator—fortunately!

4.5 Interpolation of Rotations

We already had the opportunity to discuss, in several occasions, the importance of interpolation methods in computer graphics. They are responsible for the reconstruction of objects starting from a set of samples.

The general problem of interpolating rotations consists on interpolating points in the space $\mathrm{SO}(3)$, and it can be stated as follows: given n rotations $R_0, R_1, \ldots, R_n$, obtain a curve $R\colon [0,1] \to \mathrm{SO}(3)$, of class C^k, $k \geq 0$, and a partition $0 = t_0 < t_1 < \cdots < t_{n-1} < t_n = 1$ in the interval $[0,1]$, such that $R(0) = R_0$, $R(1) = R_1$ and $R(t_i)$ are an approximation of the rotation R_i, for $i = 1, \ldots, n-1$.

---

[3]Gimbal is the name given to the mechanical assembly of equipments requiring a spatial orientation and, for this end, using Euler's angles: for example, gyroscope, optic equipments

In reality, in pure interpolation methods, we require that $R(t_i) = R_i$ for $i = 0, 1, \ldots, n$. The interpolation of rotations play a fundamental role in the *keyframe* animation technique, used in the majority of commercial animation systems. In these systems, the animator specifies the position of the rigid objects (translation+rotation) at some points, the keyframes, and the system computes the interpolation to reconstruct the motion.

An important particular case of the interpolation problem consists on obtaining the interpolation between two rotations R_0 and R_1. As the space holds an affine structure (e.g., the Euclidean space $\mathbb{R}^n$), the natural and simple method for obtaining an interpolation between two objects R_0 and R_1 is a *linear interpolation*, defined as $R(t) = (1-t)R_0 + tR_1$, $t \in [0, 1]$.

In the case R_0 and R_1 are matrices, then linear interpolation makes sense, because the matrix space is a vector space. However, if R_0 and R_1 are rotation matrices, the matrix $R(t) = (1 - t)R_0 + tR_1$, $t \in [0, 1]$, in general, does not represent a rotation. This happens because the rotation space SO(3) is not a linear subspace of the matrix space. (The problem here is similar to a linear interpolation between two points on a sphere: the resulting segment is not contained on the sphere.)

4.5.1   Interpolating Euler's Angles

When we have a parameterization $\varphi \colon U \subset \mathbb{R}^m \to S$ of a space S, the interpolation problem is reduced to interpolate in the parameters space U. In fact, to interpolate n points $\mathbf{p}_1, \ldots, \mathbf{p}_n \in S$, we take the corresponding points $\varphi^{-1}(\mathbf{p}_1), \ldots, \varphi^{-1}(\mathbf{p}_n)$, in the parameter space $U \subset \mathbb{R}^n$, and we obtain an interpolation curve $c \colon [0, 1] \colon \to \mathbb{R}^m$ between these points. The desired interpolation curve is given by the composite $\varphi \circ c \colon [0, 1] \to S$. Therefore, the interpolation problem is reduced to one in $\mathbb{R}^m$, which is broadly covered in the literature.

In particular, the above principle is applied to interpolate rotations using the parameterization by Euler's angles, as shown in Figure 4.5. In this case, if we have two rotations $R_0 = R(\alpha_0, \beta_0, \gamma_0)$ and $R_1 = R(\alpha_1, \beta_1, \gamma_1)$, we obtain a curve c in the parameters space, $c(t) = (\alpha(t), \beta(t), \gamma(t))$, such that $c(0) = (\alpha(0), \beta(0), \gamma(0)) = R_0$ and $c(1) = (\alpha(1), \beta(1), \gamma(1)) = R_1$. The interpolating path in SO(3) is given by the com-

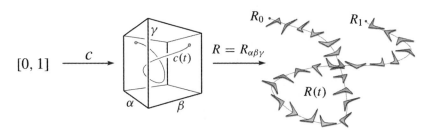

Figure 4.5. Moving a boomerang by interpolating Euler's angles.

posite $R(t) = R(\alpha(t)), R(\beta(t)), R(\gamma(t))$, where $R = R_{\alpha\beta\gamma}$ is the Euler matrix given in the Equation (4.6). (of course, we can use this method to interpolate n rotations.)

The advantage of this method is that the interpolation itself is performed in the $\mathbb{R}^3$ space. In this way, we can interpolate an arbitrary number of rotations using various point interpolation methods described in the literature (e.g., splines, Bezier, Hermite). This method is, of course, applicable to any other parameterization of the rotation space.

The interpolation of Euler's angles is the method used by the majority of keyframe animation systems to specify the orientation of the objects and of the virtual camera. The disadvantages of this method are all those inherent to the parameterization method with Euler's angles, which we will discuss next.

4.6   Commercial Break

The parameterization of the rotation space using Euler's angles is broadly used in the area of animation. A sequence of rotations with Euler's angles is represented by a 3D curve, where each component represents a rotation angle in each axis. Therefore, the animators work by manipulating curves in the $\mathbb{R}^3$ space to modify path, speed, acceleration, etc. In general, the editing is performed over the projection of these curves in the coordinate planes, an interface in which animators are already quite familiar with.

Another advantage of matrices, and in particular of Euler matrices, is that other projective transformations are also represented by matrices, especially translation. However, besides the problem of singularities (gimbal lock), discussed above, the parameterization by Euler's angles has several other inconveniences:

- ❏ Given a rotation matrix, we can have ambiguities in the solution of the inverse problem. What are the axis and rotation angle?

- ❏ The representation by Euler's angles has no unicity because it depends on the order of the axes (anisotropy).

- ❏ The parameterization defines a local coordinate system in the rotation space; In other words, we cannot describe all rotations by only using an Euler coordinate system. Of course, we can use several parameterizations to cover the SO(3) space; However, it is not simple to perform a change of coordinates between two systems of Euler's angles.

- ❏ Finally, specifying rotations using Euler's angles is difficult and not intuitive. Let us see a practical example: hold a football ball, mark a point on it with a pen and rotate it to place the marked point in the vertical position (north pole of the ball). Easy, right? Now try to solve the same problem with three successive rotations about the axes according to Euler's parameterization. In the first solution, in an intuitive and natural way, we used Euler's Theorem, by doing a single rotation about an axis.

The above fact shows that we should look for a representation of $SO(3)$, where each rotation is represented by its axis, together with the rotation angle. For this intuitive representation to be mathematically efficient, we need to introduce a mathematical structure in space constituted by the elements "axis + rotation angle." This will be our goal in the remainder of this chapter.

4.7   Quaternions

Quaternions introduce an algebraic structure in the Euclidean space $\mathbb{R}^4$, similar to the structure of the complex numbers on the $\mathbb{R}^2$ plane. This structure allows to obtain a representation of the rotation space based on the "axis+angle" model to define a rotation.

We denote as $\{\mathbf{1}, \mathbf{i}, \mathbf{j}, \mathbf{k}\}$ the canonical basis of the space $\mathbb{R}^4$: $\mathbf{1} = (1, 0, 0, 0)$, $\mathbf{i} = (0, 1, 0, 0)$, $\mathbf{j} = (0, 0, 1, 0)$ and $\mathbf{k} = (0, 0, 0, 1)$. A point $\mathbf{p} = (w, x, y, z) \in \mathbb{R}^4$ is then written in the form $\mathbf{p} = w + x\mathbf{i} + y\mathbf{j} + z\mathbf{k}$. Amongst the four elements $\mathbf{1}, \mathbf{i}, \mathbf{j}$ and $\mathbf{k}$ of the basis, we define a *product* according to Table 4.1.

Notice that the product is not commutative. We extended the above product to obtain the product $\mathbf{pq}$ between two arbitrary points of $\mathbb{R}^4$ requiring bilinearity. In other words, if $\mathbf{p}_1, \mathbf{p}_2, \mathbf{q}_1, \mathbf{q}_2 \in \mathbb{R}^4$, we have

$$(a\mathbf{p}_1 + \mathbf{p}_2)(b\mathbf{q}_1 + \mathbf{q}_2) = ab\mathbf{p}_1\mathbf{q}_1 + a\mathbf{p}_1\mathbf{q}_2 + b\mathbf{p}_2\mathbf{q}_1 + \mathbf{p}_2\mathbf{q}_2.$$

The $\mathbb{R}^4$ space, with the above multiplicative structure, is called *quaternion space* or simply *quaternions*. Quaternions were discovered by William Hamilton[4] with the goal of obtaining, in space, the same relation between rotations and complex numbers that exists on the plane. The operation of quaternion product is distributive, associative (tedious check), and, highlighting once again, not commutative.

The subspace $\mathbb{R}\,\mathbf{1} = \{(x, 0, 0, 0); x \in \mathbb{R}\}$ is identified with the set of the real numbers, and it is called *scalar space*; the subspace $\mathbb{R}\,\mathbf{i} + \mathbb{R}\,\mathbf{j} + \mathbb{R}\,\mathbf{k} = \{(0, x, y, z); x, y, z \in \mathbb{R}\}$ is naturally identified with the Euclidean space $\mathbb{R}^3$ and it is called *vector space* or *pure quaternion space*. Given a quaternion $\mathbf{u} \in \mathbb{R}^4$, $\mathbf{u} = w + x\mathbf{i} + y\mathbf{j} + z\mathbf{k}$, the number w is called *real part* of $\mathbf{u}$ and it is indicated by $\Re(\mathbf{u})$; the vector $x\mathbf{i} + y\mathbf{j} + z\mathbf{k}$ is called *vector part* of $\mathbf{u}$. If $\mathbf{v} = (x, y, z) \in \mathbb{R}^3$ is a vector, we indicate by $\hat{v}$, the associated pure quaternion: $\hat{\mathbf{v}} = x\mathbf{i} + y\mathbf{j} + z\mathbf{k}$. In this way, every quaternion can be written in the form $\mathbf{p} = w + \hat{\mathbf{v}}$, $w \in \mathbb{R}$, $\mathbf{v} \in \mathbb{R}^3$.

The *conjugated* of a quaternion $\mathbf{u} = w + \hat{\mathbf{v}}$ is defined by $\mathbf{u}^\star = w - \hat{\mathbf{v}}$. In other words, if $\mathbf{u} = w + x\mathbf{i} + y\mathbf{j} + z\mathbf{k}$, then $\mathbf{u}^\star = w - x\mathbf{i} - y\mathbf{j} - z\mathbf{k}$. The conjugation operation satisfies the properties $(\mathbf{u}^\star)^\star = \mathbf{u}$ and $(\mathbf{uv})^\star = \mathbf{u}^\star\mathbf{v}^\star$. Of course, the real part of a quaternion $\mathbf{u}$ is equal to $\Re(\mathbf{u}) = (\mathbf{u} + \mathbf{u}^\star)/2$.

The *norm* of a quaternion $\mathbf{p} = w + x\mathbf{i} + y\mathbf{j} + z\mathbf{k}$ is equal to the norm of the vector (w, x, y, z) in the Euclidean inner product of $\mathbb{R}^4$:

$$|\mathbf{p}| = \sqrt{\langle \mathbf{p}, \mathbf{p} \rangle} = \sqrt{w^2 + x^2 + y^2 + z^2}.$$

---

[4]Sir William Rowan Hamilton (1805–1865), Irish physicist, astronomer, and mathematician.

·	1	i	j	k
1	1	i	j	k
i	i	-1	k	−j
j	j	-k	-1	i
k	k	j	-i	-1

Table 4.1. Multiplication of the quaternions in the basis.

Of course, if $\mathbf{p} = w + \hat{\mathbf{v}}$, then $|\mathbf{p}|^2 = w^2 + |\hat{\mathbf{v}}|^2$. Also, $|\mathbf{p}| = |\mathbf{p}^\star|$. The norm behaves well in relation to the quaternion product:

$$|\mathbf{p}_1\,\mathbf{p}_2| = |\mathbf{p}_1||\mathbf{p}_2|. \tag{4.7}$$

If $|\mathbf{p}| = 1$, we say $\mathbf{p}$ is a *unit quaternion*. Of course, the set of unit quaternions forms the unit sphere $S^3 \subset \mathbb{R}^4$,

$$S^3 = \{(w, x, y, z) \in \mathbb{R}^4; w^2 + x^2 + y^2 + z^2 = 1\}.$$

Besides, for (4.7), the product of two unit quaternions is a unit quaternion. Therefore, S^3, seen as the set of unit quaternions, is a group. (Notice here the analogy with the circle $S^1 \subset \mathbb{R}^2$, which is a group formed by the complex numbers of norm 1.)

The *multiplicative inverse* of a quaternion $\mathbf{u}$ is a quaternion $\mathbf{u}^{-1}$ such that $\mathbf{u}\mathbf{u}^{-1} = \mathbf{u}^{-1}\mathbf{u} = \mathbf{1}$. An immediate calculation shows that every nonnull quaternion $\mathbf{u}$ has a multiplicative inverse, given by

$$\mathbf{u}^{-1} = \frac{\mathbf{u}^\star}{|\mathbf{u}|^2},$$

Notice if $|\mathbf{u}| = 1$, then $\mathbf{u}^{-1} = \mathbf{u}^\star$.

If $\mathbf{q}_1 = a_1 + \hat{\mathbf{v}}_1$ and $\mathbf{q}_2 = a_2 + \hat{\mathbf{v}}_2$, a direct calculation shows that the product of $\mathbf{q}_1$ and $\mathbf{q}_2$ is given by

$$\mathbf{q}_1\mathbf{q}_2 = a_1a_2 - \langle \hat{\mathbf{v}}_1, \hat{\mathbf{v}}_2 \rangle + a_1\hat{\mathbf{v}}_2 + a_2\hat{\mathbf{v}}_1 + \hat{\mathbf{v}}_1 \wedge \hat{\mathbf{v}}_2.$$

We are using the notation $\hat{\mathbf{v}}_1 \wedge \hat{\mathbf{v}}_2$ to indicate $\widehat{\mathbf{v}_1 \wedge \mathbf{v}_2}$. Therefore, if $\mathbf{q}_1$ and $\mathbf{q}_2$ are pure quaternions, that is, $a_1 = a_2 = 0$, then

$$\mathbf{q}_1\mathbf{q}_2 = -\langle \hat{\mathbf{v}}_1, \hat{\mathbf{v}}_2 \rangle + \hat{\mathbf{v}}_1 \wedge \hat{\mathbf{v}}_2.$$

Therefore, the quaternion product synthesizes two types of vector products in $\mathbb{R}^3$: the scalar and cross products (in a way, this let us foresee the great potential of this operation). Notice that if $\mathbf{v}$ is a pure quaternion, $\mathbf{v} \wedge \mathbf{v} = 0$, and by taking $\mathbf{q}_1 = \mathbf{q}_2 = \mathbf{v}$ in the previous equation, then we obtain

$$\mathbf{v}^2 = -\langle \mathbf{v}, \mathbf{v} \rangle = -||\mathbf{v}||^2.$$

In particular, if $\mathbf{v}$ is a unit pure quaternion, we have $\mathbf{v}^2 = -1$.

Every quaternion $q = a + \hat{\mathbf{v}}$ can be written in the form $a + b\hat{\mathbf{u}}$, where $|\mathbf{u}| = 1$ is a unit quaternion. In fact, we can only take

$$\mathbf{u} = \frac{\mathbf{v}}{|\mathbf{v}|}, \quad \text{and} \quad b = |\mathbf{v}|.$$

Observe that the representation of a quaternion in the form $\mathbf{q} = a + b\hat{\mathbf{u}}$, as $|\hat{\mathbf{u}}| = 1$, that is, $\mathbf{u}$ is a pure unit quaternion, we have $\mathbf{u}^2 = -1$. Yet, in this representation, if $\mathbf{q}$ is unit, we have

$$1 = |q|^2 = a^2 + b^2|\hat{\mathbf{u}}|^2 = a^2 + b^2.$$

Therefore, there exists $\theta \in \mathbb{R}$, such that $a = \cos\theta$ and $b = \sin\theta$, and we can write $\mathbf{q} = \cos\theta + \sin\theta\,\hat{\mathbf{u}}$.

Notice the similarity of what we have obtained above with the complex numbers. A complex number has the form $x + iy$ with $i^2 = -1$; Besides, every unit complex number, i.e. of norm 1, can be written in the form $\cos\theta + i\sin\theta$. The difference is that, unlike the complex number i, the unit quaternion $\hat{\mathbf{u}}$, in the representation $p = a + b\hat{\mathbf{u}}$, depends on the quaternion $\mathbf{p}$.

The results here show that quaternions have algebraic properties similar to the ones in complex numbers, with exception for commutativity. Our final test is to try to extend the result involving complex numbers and rotations on the plane to quaternions. This is the subject of the next section.

4.7.1   Representation of Rotations by Quaternions

Given a unit quaternion $\mathbf{q}$, we should define the transformation $R_{\mathbf{q}}\colon \mathbb{R}^4 \to \mathbb{R}^4$ placing $R_{\mathbf{q}}(\mathbf{p}) = \mathbf{pq}$. We have $|R_{\mathbf{q}}(\mathbf{p})| = |\mathbf{p}||\mathbf{q}| = |\mathbf{p}|$, therefore, the transformation $R_{\mathbf{q}}$ is orthogonal. That is, multiplications by unit quaternions, geometrically correspond to rotations in the $\mathbb{R}^4$ space. Now, a natural question comes into play: how do we obtain a representation of SO(3) by quaternions?

A first idea would be to associate, to each quaternion $\mathbf{q} \in \mathbb{R}^3$, the transformation $R_{\hat{\mathbf{q}}}$. Unfortunately, in general, $R_{\hat{\mathbf{q}}} \notin \mathrm{SO}(3)$, that is, the space of pure quaternions, naturally identified with $\mathbb{R}^3$, is not invariant by $R_{\hat{\mathbf{q}}}$. The representation of SO(3) by quaternions is a little subtler as we will see.

Observe that the representation of a unit quaternion $\mathbf{q}$ in the form $\mathbf{q} = \cos\theta + \sin\theta\,\hat{\mathbf{u}}$, with $|\hat{\mathbf{u}}| = 1$, shows this quaternion defining an angle θ and an axis $\hat{\mathbf{u}}$ of the space. It is natural to expect this quaternion to represent a rotation about the $\mathbf{u}$-axis. We will show that, in fact, this happens, and the angle of that rotation is 2θ.

Given a unit quaternion $\mathbf{q}$, we define the transformation $\varphi_{\mathbf{q}}\colon \mathbb{R}^3 \to \mathbb{R}^3$, by placing

$$\varphi_{\mathbf{q}}(\mathbf{v}) = \mathbf{q}\hat{\mathbf{v}}\mathbf{q}^{-1} = \mathbf{q}\hat{\mathbf{v}}\mathbf{q}^{\star}, \qquad \mathbf{v} \in \mathbb{R}^3. \tag{4.8}$$

We will soon demonstrate several properties of the above transformation.

The transformation $\varphi_\mathbf{q}$ is **well defined**. This means $\varphi_\mathbf{q}(\mathbf{v}) \in \mathbb{R}^3$. In fact, it is enough to show $\mathfrak{R}(\varphi_\mathbf{q}(\mathbf{v})) = 0$:

$$
\begin{aligned}
\mathfrak{R}(q\hat{\mathbf{v}}q^{-1}) &= \mathfrak{R}(q\hat{\mathbf{v}}q^\star) \\
&= [q\hat{\mathbf{v}}q^\star + (q\hat{\mathbf{v}}q^\star)^\star]/2 \\
&= [q\hat{\mathbf{v}}q^\star + q\hat{\mathbf{v}}^\star q^\star]/2 \\
&= q[(\hat{\mathbf{v}} + \hat{\mathbf{v}}^\star)/2]q^\star \\
&= q\mathfrak{R}(\hat{\mathbf{v}})q^\star = \mathfrak{R}(\hat{\mathbf{v}})qq^\star = \mathfrak{R}(\hat{\mathbf{v}}) = 0.
\end{aligned}
$$

The transformation $\varphi_\mathbf{q}$ is **linear**. In fact,

$$
\begin{aligned}
\varphi_\mathbf{q}(a\hat{\mathbf{u}} + \hat{\mathbf{v}}) &= q(a\hat{\mathbf{u}} + \hat{\mathbf{v}})q^\star \\
&= qa\hat{\mathbf{u}}q^\star + q\hat{\mathbf{v}}q^\star \\
&= a(q\hat{\mathbf{u}}q^\star) + (q\hat{\mathbf{v}}q^\star) \\
&= a\varphi_\mathbf{q}(\mathbf{u}) + \varphi_q(\mathbf{v}).
\end{aligned}
$$

The transformation $\varphi_\mathbf{q}$ is **orthogonal**. In fact,

$$
|\varphi_\mathbf{q}(\mathbf{v})| = |q\hat{\mathbf{v}}q^\star| = |q||\hat{\mathbf{v}}||q^\star| = |\mathbf{v}|.
$$

The transformation $\varphi_\mathbf{q}$ is **a rotation of $\mathbb{R}^3$**. In fact, as we saw above, it is enough to show that $\varphi_\mathbf{q}$ is positive. In this case, notice that for $\mathbf{q} = 1$ $\varphi_\mathbf{q} = I$, where $I\colon \mathbb{R}^3 \to \mathbb{R}^3$ is the identity transformation. As the space of unit quaternions is connected (because it is the unit sphere in $\mathbb{R}^4$), we conclude, through a continuity argument, that $\varphi_\mathbf{q}$ is positive for every quaternion $\mathbf{q}$.

The **axis of rotation** of $\varphi_\mathbf{q}$ is the quaternion $\hat{\mathbf{u}}$. It is enough to show that $\varphi_\mathbf{q}(\hat{\mathbf{u}}) = \hat{\mathbf{u}}$:

$$
\begin{aligned}
\varphi_\mathbf{q}(\mathbf{u}) &= q\hat{\mathbf{u}}q^\star \\
&= (\cos\theta + \hat{\mathbf{u}}\sin\theta)\,\hat{\mathbf{u}}\,(\cos\theta - \hat{\mathbf{u}}\sin\theta) \\
&= (\cos\theta)^2\hat{\mathbf{u}} - (\sin\theta)^2\hat{\mathbf{u}}^3 \\
&= (\cos\theta)^2\hat{\mathbf{u}} - (\sin\theta)^2(-\hat{\mathbf{u}}) \\
&= \mathbf{u}.
\end{aligned}
$$

The **angle of rotation** of $\varphi_\mathbf{q}$ is 2θ. We will provide a geometric demonstration of this fact, by solving the following problem: given a rotation R of an angle θ about an axis r, how can we obtain the quaternion $\mathbf{q}$ of the representation $\varphi_\mathbf{q}$ in R? To answer this question, consider Figure 4.6, showing the rotation R of point P by an angle β about the OH-axis. The image of point P is the point $P' = R(P)$.

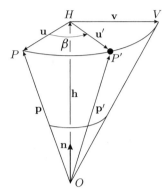

Figure 4.6. Rotation of an angle β about the axis defined by the vector $\mathbf{n}$.

Let us take the positive reference frame $\{\mathbf{u}, \mathbf{v}, \mathbf{n}\}$, where $\mathbf{u} = \overrightarrow{HP}$, $\mathbf{v} = \overrightarrow{HV}$, and $\mathbf{n}$ is the unit vector in the direction of the rotation axis. We also define the vectors

$$\mathbf{h} = \overrightarrow{OH}, \quad \mathbf{p} = \overrightarrow{OP}, \quad \mathbf{p}' = \overrightarrow{OP'}, \quad \text{and} \quad \mathbf{u}' = \overrightarrow{HP'}.$$

with $\mathbf{u} \perp \mathbf{h}$, $\mathbf{h} \perp \mathbf{v}$ and $|\mathbf{v}| = |\mathbf{u}|$. Having $\mathbf{h}$ as the orthogonal projection of $\mathbf{p}$ in the direction $\mathbf{n}$, we have $\mathbf{h} = \langle \mathbf{p}, \mathbf{n} \rangle \mathbf{n}$. It is also easy to verify that

$$\mathbf{p}' = \mathbf{h} + \mathbf{u}'$$
$$\mathbf{u} = \mathbf{p} - \mathbf{h} = \mathbf{p} - \langle \mathbf{p}, \mathbf{n} \rangle \mathbf{n}$$
$$\mathbf{u}' = \cos \beta \, \mathbf{u} + \sin \beta \, \mathbf{v},$$

and besides,

$$\mathbf{v} = \mathbf{n} \wedge \mathbf{u} = \mathbf{n} \wedge (\mathbf{p} - \mathbf{h}) = \mathbf{n} \wedge (\mathbf{p} - \langle \mathbf{p}, \mathbf{n} \rangle) = \mathbf{n} \wedge \mathbf{p}.$$

We then have

$$
\begin{aligned}
R(P) = \overrightarrow{OP'} &= \mathbf{h} + \mathbf{u}' \\
&= \langle \mathbf{n}, \mathbf{p} \rangle \mathbf{n} + \cos \beta \, \mathbf{u} + \sin \beta \, \mathbf{v} \\
&= \langle \mathbf{n}, \mathbf{p} \rangle \mathbf{n} + \cos \beta (\mathbf{p} - \langle \mathbf{p}, \mathbf{n} \rangle \mathbf{n}) + \sin \beta \, \mathbf{n} \wedge \mathbf{p} \\
&= \cos \beta \, \mathbf{p} + (1 - \cos \beta) \langle \mathbf{n}, \mathbf{p} \rangle \mathbf{n} + \sin \beta \, \mathbf{n} \wedge \mathbf{p}.
\end{aligned}
\tag{4.9}
$$

We will now calculate $R(P)$ using the rotation $\varphi_{\mathbf{q}}$ in (4.8). In this case, the point P is represented by the pure quaternion $P = \hat{\mathbf{p}}$, and $\mathbf{q}$ is the unit quaternion $\mathbf{q} = \cos \theta + \hat{\mathbf{n}} \sin \theta$. To simplify the calculations, let us take $c = \cos \theta$ and $\mathbf{t} = \hat{\mathbf{n}} \sin \theta$, therefore $\mathbf{q} = c + \hat{\mathbf{t}}$.

We then have

$$R(P) = \varphi_{\mathbf{q}}(\mathbf{p}) = \mathbf{q}\hat{\mathbf{p}}\mathbf{q}^{\star}$$
$$= (c + \hat{\mathbf{t}})(\hat{\mathbf{p}})(c - \hat{\mathbf{t}})$$
$$= (c + \hat{\mathbf{t}})(\langle \hat{\mathbf{p}}, \hat{\mathbf{t}}\rangle - \hat{\mathbf{p}} \wedge \hat{\mathbf{t}} + c\hat{\mathbf{p}})$$
$$= c\langle \hat{\mathbf{p}}, \hat{\mathbf{t}}\rangle - \langle \hat{\mathbf{t}}, -\hat{\mathbf{p}} \wedge \hat{\mathbf{t}}\rangle - c\langle \hat{\mathbf{t}}, \hat{\mathbf{p}}\rangle + \tag{4.10}$$
$$+ \hat{\mathbf{t}} \wedge (-\hat{\mathbf{p}} \wedge \hat{\mathbf{t}}) + c(\hat{\mathbf{t}} \wedge \hat{\mathbf{p}}) \quad + c(-\hat{\mathbf{p}} \wedge \hat{\mathbf{t}} + c\hat{\mathbf{p}}) + \langle \hat{\mathbf{p}}, \hat{\mathbf{t}}\rangle\hat{\mathbf{t}})$$
$$= \langle \hat{\mathbf{t}}, \hat{\mathbf{p}}\rangle\hat{\mathbf{t}} - \langle \hat{\mathbf{t}}, \hat{\mathbf{t}}\rangle\hat{\mathbf{p}} + 2c(\hat{\mathbf{t}} \wedge \hat{\mathbf{p}}) + c^2\hat{\mathbf{p}} + \langle \hat{\mathbf{p}}, \hat{\mathbf{t}}\rangle\hat{\mathbf{t}}$$
$$= (c^2 - \langle \hat{\mathbf{t}}, \hat{\mathbf{t}}\rangle)\hat{\mathbf{p}} + 2\langle \hat{\mathbf{p}}, \hat{\mathbf{t}}\rangle\hat{\mathbf{t}} + 2c(\hat{\mathbf{t}} \wedge \hat{\mathbf{p}}).$$

Substituting the values of $c = \cos\theta$ and $\mathbf{t} = \hat{\mathbf{n}}\sin\theta$, in the expression of $\varphi_{\mathbf{q}}(\mathbf{p})$ obtained in Equation (4.10), we have

$$R(P) = (\cos^2\theta - \sin^2\theta)\hat{\mathbf{p}} + 2\sin^2\theta\langle \hat{\mathbf{p}}, \hat{\mathbf{n}}\rangle\hat{\mathbf{n}} + 2\sin\theta\cos\theta(\hat{\mathbf{n}} \wedge \hat{\mathbf{p}})$$
$$= \cos 2\theta\hat{\mathbf{p}} + (1 - \cos 2\theta)\langle \hat{\mathbf{n}}, \hat{\mathbf{p}}\rangle\mathbf{n} + \sin 2\theta(\mathbf{n} \wedge \hat{\mathbf{p}})).$$

Comparing this equation with Equation (4.9), we see that $\varphi_{\mathbf{q}}(\mathbf{p})$ is a rotation of an angle 2θ about the axis defined by the unit vector $\mathbf{n}$.

The results demonstrated above can be summarized in

Theorem 4.3 (Representation of rotations by quaternions). *A rotation of an angle 2θ about an axis defined by a unit vector $\mathbf{n}$ is represented by $\varphi_{\mathbf{q}}(\mathbf{v}) = \mathbf{q}\hat{\mathbf{v}}\mathbf{q}^{-1}$, where $\mathbf{q} = \cos\theta + \hat{\mathbf{n}}\sin\theta$.* ❏

Therefore, we have a transformation $\varphi\colon S^3 \to \mathrm{SO}(3)$, $\varphi(\mathbf{q}) = \varphi_{\mathbf{q}}$, representing the rotations in space by points of the unit sphere in $\mathbb{R}^4$. Following up, from the definition of $\varphi_{\mathbf{q}}$, we have $\varphi_{\mathbf{q}} = \varphi_{-\mathbf{q}}$. Geometrically, this fact is obvious, because a rotation of an angle θ about the $\mathbf{n}$-axis is the same to a rotation of an angle $2\pi - \theta$ about the $-\mathbf{n}$-axis. Therefore, the transformation $\varphi_{\mathbf{q}}$ is not injective. However, it can be shown, without a lot of difficulty, that

$$\varphi_{\mathbf{q}_1} = \varphi_{\mathbf{q}_2} \Leftrightarrow \mathbf{q}_1 = \pm\mathbf{q}_2.$$

We left this demonstration for the exercises at the end of this chapter.

The above result shows that only the antipode points of the sphere define the same rotation. Therefore, the space $\mathrm{SO}(3)$ is naturally identified with the unit sphere S^3 in $\mathbb{R}^4$, with the antipode points identified. We know this is the real projective space of dimension 3, $\mathbb{RP}^3$. However, we will not use this association between $\mathrm{SO}(3)$ and $\mathbb{RP}^3$. The important aspect of the above result is that the rotation space $\mathrm{SO}(3)$ can be represented by unit quaternions. Care should only be taken with antipode points of the sphere representing the same rotation.

An important fact in relation to the representation of rotations in $\mathrm{SO}(3)$ by quaternions, is that the product operation is preserved, that is

$$\varphi_{\mathbf{u}_1\mathbf{u}_2} = \varphi_{\mathbf{u}_1}\varphi_{\mathbf{u}_2}.$$

Checking this is immediate: for every vector $\mathbf{v} \in \mathbb{R}^3$, we have

$$\varphi_{\mathbf{u}_1 \mathbf{u}_2}(\mathbf{v}) = \mathbf{u}_1 \mathbf{u}_2 \hat{\mathbf{v}}(\mathbf{u}_1 \mathbf{u}_2)^{-1} = \mathbf{u}_1 \mathbf{u}_2 \hat{\mathbf{v}} \mathbf{u}_2^{-1} \mathbf{u}_1^{-1} = \mathbf{u}_1 \varphi_{\mathbf{u}_2} \hat{\mathbf{v}} \mathbf{u}_1^{-1} = \varphi_{\mathbf{u}_1} \varphi_{\mathbf{u}_2}(\mathbf{v}).$$

This fact shows that we can substitute the product of two rotation matrices by the product of the quaternions representing them. In particular, notice that this result answers a question we had in the beginning of the chapter: if the rotations R_1 and R_2 have axes defined by the unit vectors $\mathbf{u}_1$ and $\mathbf{u}_2$, respectively, then the axis of the rotation product $R_1 R_2$ is the product of quaternions $\hat{\mathbf{u}}_1 \hat{\mathbf{u}}_2$.

4.7.2 Exponential and Logarithm

Consider a vector $\mathbf{v} \in \mathbb{R}^3$, $\mathbf{v} \neq 0$. and let $\hat{\mathbf{v}} = \mathbf{v}/|\mathbf{v}|$ be the pure unit quaternion obtained from $\mathbf{v}$ through normalization. Using the notation $\theta = |\mathbf{v}|$, we have $\mathbf{v} = \theta \hat{\mathbf{v}}$. Substituting the expression $\mathbf{v} = \theta \hat{\mathbf{v}}$ in the Taylor series of the exponential function, we obtain,

$$e^{\mathbf{v}} = e^{\theta \hat{\mathbf{v}}} = \sum_{n=0}^{\infty} \frac{(\theta \hat{\mathbf{v}})^n}{n!}.$$

An immediate calculation, taking into account $\hat{\mathbf{v}}^2 = -1$, gives us

$$e^{\mathbf{v}} = e^{\theta \mathbf{v}} = \cos \theta + \hat{\mathbf{v}} \sin \theta = \cos |\mathbf{v}| + \hat{\mathbf{v}} \sin |\mathbf{v}|.$$

We can define the exponential $\exp \colon \mathbb{R}^3 \to S^3$ having

$$\exp(\mathbf{v}) = \begin{cases} \cos |\mathbf{v}| + \sin |\mathbf{v}| \hat{\mathbf{v}}, & \text{if } \mathbf{v} \neq 0; \\ \mathbf{1} = (1, 0, 0, 0) & \text{if } \mathbf{v} = 0, \end{cases}$$

where $\hat{\mathbf{v}} = |\mathbf{v}|/|\mathbf{v}|$. We will also use the notation $\exp(\mathbf{v}) = e^{\mathbf{v}}$. Notice that if $\mathbf{v}$ is a unit quaternion, then $\mathbf{v} = \cos \theta + \sin \theta \hat{\mathbf{u}}$, with $|\mathbf{u}| = 1$. As $|\theta \hat{\mathbf{u}}| = \theta$, we have

$$e^{\theta \mathbf{u}} = \cos \theta + \sin \theta \hat{\mathbf{u}},$$

which is a similar expression to $e^{i\theta} = \cos \theta + i \sin \theta$ for complex numbers.

Now we can define the logarithm function $\log \colon S^3 \to \mathbb{R}^3$, as being the inverse of the exponential. Given $\mathbf{q} \in S^3$, $\mathbf{q} = \cos \theta + \sin \theta \hat{\mathbf{u}}$, $|\hat{\mathbf{u}}| = 1$, we have

$$\log(\mathbf{q}) = \theta \mathbf{u} \in \mathbb{R}^3.$$

It is immediate to verify that $e^{\log(\mathbf{q})} = \mathbf{q}$.

It is sometimes useful to have the definition of the exponential and of the logarithm explicitly showing the coordinates: if $\mathbf{u} = w + x\mathbf{i} + y\mathbf{j} + z\mathbf{k} \in S^3$, we can write

$$\log(w, x, y, z) = \arccos(w) \frac{(x, y, z)}{\sqrt{x^2 + y^2 + z^2}} = \frac{\arccos(w)}{\sqrt{x^2 + y^2 + z^2}} (x, y, z).$$

Similarly, if $\mathbf{v} = (x, y, z) \in \mathbb{R}^3$, we have

$$\exp(x, y, z) = \begin{cases} \cos(\sqrt{x^2 + y^2 + z^2}) + \frac{\sin(\sqrt{x^2+y^2+z^2})}{\sqrt{x^2+y^2+z^2}}(x, y, z), & \text{if } (x, y, z) \neq 0; \\ (1, 0, 0, 0) & \text{if } (x, y, z) = 0. \end{cases}$$

Now we can define arbitrary powers of unit quaternions extending, in a natural way, the result for complex numbers: if $\mathbf{q}$ is a unit quaternion and $t \in \mathbb{R}$, we have

$$q^t = e^{t \log \mathbf{q}} = e^{t\theta\hat{\mathbf{u}}} = \cos(\theta t) + \sin(\theta t)\hat{\mathbf{u}}. \tag{4.11}$$

Observe that the expression $c(t) = \mathbf{q}^t$ (with t varying in the interval $[0, 1]$) defines a curve $c: [0, 1] \to S^3$ connecting the point $c(0) = \mathbf{1} = (1, 0, 0, 0)$ (north pole of the sphere S^3) to the point $c(1) = \mathbf{q}$. We will give a geometric interpretation of this curve, and at the same time, of the exponential function. For this, we need some definitions.

A *maximum circle* of the sphere S^3 is a circle of radius 1, in other words, a circle contained in S^3 that has maximum radius. A maximum circle is obtained by the intersection of S^3 with a subspace of dimension 2 in $\mathbb{R}^4$. A *geodesic* of S^3 is a maximum circle parameterized by a curve $c: [0, 1] \to S^3$, such that the speed $|c'(t)|$ is constant. This implies that the image of a uniform partition of the parameter space $[0, 1]$ results in a uniform partition of the maximum circle. (The geodesics are the "straight lines" of the sphere S^3, when tracing a path between two points on the sphere; The smallest path is always along a geodesic).

Let us return to Equation (4.11), from which the curve $c(t) = \mathbf{q}^t$ geometrically represents a family of rotations about the same $\mathbf{u}$-axis with angles given by $2\theta t$. In S^3, these rotations represent an arch of maximum circle connecting the quaternion $c(0) = \mathbf{1}$ to the quaternion $c(1) = \mathbf{q}$. We left as an exercise at the end of the chapter, the demonstration that

$$c'(t) = \frac{d}{dt}\mathbf{q}^t = \mathbf{q}^t \log \mathbf{q}.$$

Then it is easy to verify that the curve $c(t)$ holds the following properties:

1. $c'(0) = \theta\mathbf{u}$;

2. $|c'(t)| = |\theta|$;

3. $c''(t) = kc(t)$, where $k < 0$ is a constant.

Condition 3 above shows that $c(t)$, in fact, it describes an arch of maximum circle, and condition 2 guarantees us that $c(t)$ is a geodesic. In short, the curve $c(t)$ is a geodesic in S^3, beginning at point $\mathbf{1}$ and ending at point $\mathbf{q}$. (Of course, the maximum circle described by $c(t)$ is determined by the intersection of S^3 with the 2D subspace generated by the vectors $\mathbf{1}$ and $\mathbf{q}$.)

The previous result provides us with a geometric description of the following exponential application $\exp: \mathbb{R}^3 \to S^3$ (see Figure 4.7): by identifying the space $\mathbb{R}^3$ with the

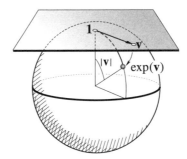

Figure 4.7. Exponential application.

tangent plane to the sphere S^3 at point $\mathbf{1}$ (north pole), we have that $\exp(\mathbf{v})$ is the extreme point of the arch that begins at $\mathbf{1}$, along the direction $\mathbf{v}$ and has length $|\mathbf{v}|$.

In particular, notice that the exponential defined here coincides with the exponential defined in differential geometry for any surface of $\mathbb{R}^n$ with the induced metric.

Exponential parameterization of SO(3). The result of the previous section shows that the exponential defined here naturally extends the exponential $\exp\colon \mathbb{R} \to S^1$ we defined in the beginning of the chapter. In this case, the exponential allowed a parameterization of the rotation space on the plane for a subset of $\mathbb{R}$. Also, in our case, the exponential $\exp\colon \mathbb{R}^3 \to S^3$ naturally defines a parameterization of the rotation space $\mathrm{SO}(3)$.

Given a vector $\mathbf{v} \in \mathbb{R}^3$, we have $\exp(\mathbf{v}) = \cos|\mathbf{v}| + \sin|\mathbf{v}|\hat{\mathbf{u}}$, where $\mathbf{u} = \mathbf{u}/|\mathbf{u}|$. Therefore, $\exp(\mathbf{v})$ is a unit quaternion representing a rotation about the $\mathbf{u}$-axis (this is the same axis of $\mathbf{v}$), of an angle $2|\mathbf{v}|$. That is, the exponential is a parameterization capturing the essence of parameterizing the rotation space in the $\{\text{axis}+\text{angle}\}$ space, as we previously promised: the axis is modeled by a vector, and the angle is its length (less of factor 2).

Of course, the exponential holds singularities as we are parameterizing $\mathrm{SO}(3)$ by $\mathbb{R}^3$. By using the geometric interpretation of the exponential, it is easy to conclude that the spheres of radius $k\pi$, $k = 1, 2, \ldots$ are singularity regions of this parameterization. They are the regions of $\mathbb{R}^3$ either mapped in the north pole or south pole of the sphere S^3. These singularities can be avoided in practice: when $\mathbf{v}$ gets closer to the sphere of radius π, we have a rotation of an angle $\theta = 2|\mathbf{v}|$ close to 2π; we then change for a rotation of an angle $2\pi - \theta$ about the $-\mathbf{v}$-axis that is equivalent and far away from the singularity.

Observing the definition of exponential parameterization attentively, we notice that we can have a problem of numerical instability in the calculation of the unit vector $\mathbf{u} = \mathbf{v}/|\mathbf{v}|$, when $v \to 0$. This problem can be avoided. In fact, we have

$$e^{\mathbf{v}} = \cos|\mathbf{v}| + \sin|\mathbf{v}|\hat{\mathbf{u}}$$
$$= \cos|\mathbf{v}| + \sin|\mathbf{v}|\frac{\mathbf{v}}{|\mathbf{v}|}$$
$$= \cos|\mathbf{v}| + \frac{\sin|\mathbf{v}|}{|\mathbf{v}|}\mathbf{v}.$$

Now we only need to observe that $\frac{\sin |\mathbf{v}|}{|\mathbf{v}|} \to 1$ when $|\mathbf{v}| \to 0$. In reality, for implementation purposes, we can substitute this function by an approximation given by its Taylor series:

$$\frac{\sin t}{t} = 1 + \frac{t^2}{3!} + \frac{t^4}{5!} + \cdots + \frac{t^{2(n-1)}}{(2n-1)!} + \cdots$$

4.7.3 Interpolating Quaternions

In this section, we will cover the topic of interpolating rotations using the representation by quaternions. We will only study the case for two rotations. Let us consider two rotations $\varphi_{\mathbf{u}}$ and $\varphi_{\mathbf{v}}$, of angles $\theta_{\mathbf{u}}$ and $\theta_{\mathbf{v}}$, about unit axes $\mathbf{u}$ and $\mathbf{v}$, respectively. These two rotations are represented by unit quaternions, as indicated below:

$$\varphi_{\mathbf{u}} \longleftrightarrow \mathbf{p} = e^{\theta_u \mathbf{u}} = \cos \theta_{\mathbf{u}} + \sin \theta_{\mathbf{u}} \hat{\mathbf{u}}$$
$$\varphi_{\mathbf{v}} \longleftrightarrow \mathbf{q} = e^{\theta_v \mathbf{v}} = \cos \theta_{\mathbf{v}} + \sin \theta_{\mathbf{v}} \hat{\mathbf{v}}$$

We then have

$$\varphi_u(\mathbf{x}) = \mathbf{p}\hat{\mathbf{x}}\mathbf{p}^* \qquad \text{and} \qquad \varphi_v(\mathbf{x}) = \mathbf{q}\hat{\mathbf{x}}\mathbf{q}^*.$$

Quaternions $\mathbf{p}$ and $\mathbf{q}$ define an arch of maximum circle on the sphere S^3 (see Figure 4.8(a)). Due to the representation of rotations by unit quaternions, an interpolation between the rotations φ_u and φ_v is simply a curve $g \colon [0,1] \to S^3$ on the sphere S^3, such that $g(0) = \mathbf{p}$ and $g(1) = \mathbf{q}$. That is, the interpolation of rotations is reduced to a problem of interpolating points in S^3.

Linear interpolation. As quaternions hold the usual linear structure of $\mathbb{R}^4$, we can perform a linear interpolation between $\mathbf{p}$ and $\mathbf{q}$. More precisely, we define the quaternion

$$\mathbf{a}(t) = (1-t)\mathbf{p} + t\mathbf{q},$$

and the interpolated rotation is given by $\varphi_{\mathbf{a}(t)}$, where

$$\varphi_{\mathbf{a}(t)}(\mathbf{x}) = \mathbf{a}(t)\hat{\mathbf{x}}\mathbf{a}(t)^*.$$

The problem of this method is that $\mathbf{a}(t)$ are not unit quaternions. In fact, geometrically, they form a "cord" of the maximum circle of the unit sphere $S^3 \subset \mathbb{R}^4$, containing the quaternions $\mathbf{p}$ and $\mathbf{q}$ (see Figure 4.8(b)).

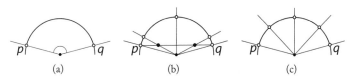

(a)　　　　　(b)　　　　　(c)

Figure 4.8. Quaternion interpolation.

We can bypass this problem by performing a *spherical radial interpolation*, consisting of taking the radial projection of $\mathbf{a}(t)$ on the unit sphere, as shown in Figure 4.8(b). That is,

$$\overline{\mathbf{a}}(t) = \frac{\mathbf{a}(t)}{|\mathbf{a}(t)|} = \frac{(1-t)\mathbf{p} + t\mathbf{q}}{|(1-t)\mathbf{p} + t\mathbf{q}|}.$$

While this method works, it does not provide a uniform sampling of the angles, even when the sampling is uniform in time. (That is, the projection is a maximum circle but it is not a geodesic.) This fact can be observed in Figure 4.8(c). In other words, the angular speed is variable (it accelerates and decelerates).

Geodesic interpolation. Our goal now is to correct the problem of spherical radial interpolation and to obtain an interpolation between two quaternions in which the interpolating curve has constant angular speed, that is, a uniform sampling in time generates a uniform sampling in the angles. As we know already, it is enough to perform the interpolation along a geodesic of the sphere S^3, connecting quaternions $\mathbf{p}$ to $\mathbf{q}$. This method is called *geodesic interpolation*.

Previously, we saw that curve $c(t) = \mathbf{b}^t$ is a geodesic connecting the north pole $\mathbf{1}$ of the sphere to the quaternion $\mathbf{b}$. This provides a geodesic interpolation between rotations $\varphi_{\mathbf{1}}$ and $\varphi_{\mathbf{b}}$.

How can we obtain a geodesic connecting $\mathbf{p}$ to $\mathbf{q}$? A simple calculation shows that the curve

$$g(t) = (\mathbf{qp})^t \mathbf{p}, \quad t \in [0, 1], \tag{4.12}$$

satisfies $g(0) = \mathbf{p}$ and $g(1) = \mathbf{q}$. It remains to show $g(t)$ is a geodesic. For this, observe that $g(t) = R_{\mathbf{p}}(c(t))$, where $c(t) = (\mathbf{qp})^t$ and $R_{\mathbf{p}}$ is the transformation of S^3 defined by $R_{\mathbf{p}}(\hat{\mathbf{v}}) = \hat{\mathbf{v}}\mathbf{p}$. Now we know $c(t)$ is a geodesic; On the other hand, the transformation $R_{\mathbf{p}}$ is orthogonal, therefore it preserves geodesics given it is an isometry of the sphere. (In more geometric terms: a geodesic is a maximum circle and sphere rotations preserve maximum circles.)

In conclusion, Equation (4.12) defines a geodesic interpolation between quaternions $\mathbf{p}$ and $\mathbf{q}$. This interpolation method is called *spherical linear interpolation* or, briefly, slerp.

Next, we will obtain an expression of geodesic interpolation which is very common in the literature; It expresses the interpolation parameter in terms of angles in the maximum circle connecting $\mathbf{p}$ to $\mathbf{q}$. Let θ be the angle between quaternions $\mathbf{p}$ and $\mathbf{q}$, that is, $\cos\theta = \langle \mathbf{p}, \mathbf{q} \rangle$ (Figure 4.9(a)).

As we will work with angles of the maximum circle, our problem is reduced to a planar one. That is, we work on the plane of the maximum circle, defined by $\mathbf{p}, \mathbf{q}$ and the origin. Observing Figure 4.9(b), we have:

$$\mathbf{p} = (\cos(\theta_0), \sin(\theta_0)),$$
$$\mathbf{q} = (\cos(\theta_0 + \theta), \sin(\theta_0 + \theta)).$$

Therefore, in terms of angles, the interpolation between $\mathbf{p}$ and $\mathbf{q}$ is given by

$$\mathbf{g}(t) = (\cos(\theta_0 + t\theta), \sin(\theta_0 + t\theta)).$$

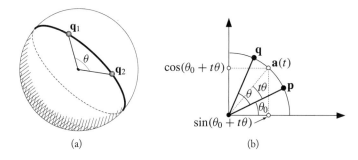

Figure 4.9. Spherical linear interpolation.

Now we should write this expression as a function of quaternions p and q. That is achieved with a simple calculation using trigonometry (due to space constraints, we will indicate a vector (x, y) on the plane by a column matrix $\left(\begin{smallmatrix} x \\ y \end{smallmatrix} \right)$):

$$
\begin{aligned}
\mathbf{g}(t) &= \begin{pmatrix} \cos(\theta_0 + t\theta) \\ \sin(\theta_0 + t\theta) \end{pmatrix} \\
&= \begin{pmatrix} \cos(\theta_0)\cos(t\theta) - \sin(\theta_0)\sin(t\theta) \\ \sin(\theta_0)\cos(t\theta) + \cos(\theta_0)\sin(t\theta) \end{pmatrix} \\
&= \begin{pmatrix} \frac{\cos(\theta_0)[\sin(\theta)\cos(t\theta) - \cos(\theta)\sin(t\theta)] + [\cos(\theta_0)\cos(\theta) - \sin(\theta_0)\sin(\theta)]\sin(t\theta)}{\sin(\theta)} \\ \frac{\sin(\theta_0)[\sin(\theta)\cos(t\theta) - \cos(\theta)\sin(t\theta)] + [\sin(\theta_0)\cos(t\theta) + \cos(\theta_0)\sin(\theta)]\sin(t\theta)}{\sin(\theta)} \end{pmatrix} \\
&= \begin{pmatrix} \frac{\cos(\theta_0)\sin((1-t)\theta) + \cos(\theta_0+\theta)\sin(t\theta)}{\sin(\theta)} \\ \frac{\sin(\theta_0)\sin((1-t)\theta) + \sin(\theta_0+\theta)\sin(t\theta)}{\sin(\theta)} \end{pmatrix} \\
&= \begin{pmatrix} \cos(\theta_0) \\ \sin(\theta_0) \end{pmatrix} \frac{\sin((1-t)\theta)}{\sin(\theta)} + \begin{pmatrix} \cos(\theta_0+\theta) \\ \sin(\theta_0+\theta) \end{pmatrix} \frac{\sin t\theta}{\sin(\theta)} \\
&= \mathbf{p}\frac{\sin((1-t)\theta)}{\sin\theta} + \mathbf{q}\frac{\sin(t\theta)}{\sin\theta}.
\end{aligned}
$$

In short, we have

$$
\operatorname{slerp}(\mathbf{p}, \mathbf{q}, t) = \mathbf{p}\frac{\sin((1-t)\theta)}{\sin\theta} + \mathbf{q}\frac{\sin(t\theta)}{\sin\theta}. \tag{4.13}
$$

4.7.4   Some Afterthoughts

The representation of rotations by quaternions brings many advantages. Among them, we can list:

- Smaller storage space (a quaternion has four components and a matrix holds nine, or 16 in homogeneous coordinates).

- We have a global parameterization, thus facilitating the interpolation of rotations (we interpolate in the unit sphere $S^3 \subset \mathbb{R}^4$).

- The interpolation of rotations is reduced to a problem of interpolating points in the unit sphere (unit quaternions).

- The exponential parameterization naturally translates to the way we think about rotation: axis+angle.

- The representation is intrinsic (i.e. it does not depend on coordinates)

- Quaternion product can be implemented in a more efficient way than matrix product.

On this last item, notice that multiplication of two quaternions has 16 products, while the multiplication of matrices has 27 products. In reality, even more surprisingly, the multiplication of two quaternions can be obtained with only 8 products (see exercises).

A problem that arises from using quaternions is that several existing geometry libraries are, in general, designed to work with matrices, and in particular with Euler's angles. Therefore, it is convenient to have the expressions to convert between the several representations (i.e. matrix, Euler's angles, quaternions, exponential parameterization). We will cover this subject in the next section.

4.8   Converting between Representations

In this section, we will study two types of conversions between representations: from quaternions to matrix, and from Euler's angles to quaternions. Previously, we already saw the conversion from Euler's angles to matrices.

4.8.1   Quaternions and Matrices

In this section, we will find the rotation matrix represented by

$$\varphi_{\mathbf{q}}(\mathbf{v}) = \mathbf{q}\hat{\mathbf{v}}\mathbf{q}^{-1},$$

where $\mathbf{q}$ is a unit quaternion.

Before, let us draw a parallel with complex numbers. There exists a relation between the product of complex numbers and linear transformations. Let us associate, to each complex number $z = a + bi$, a linear transformation $L_z \colon \mathbb{R}^2 \to \mathbb{R}^2$, defined by

$$L_z(x, y) = z(x + iy) = (a + bi)(x + iy) = (ax - by, bx + ay).$$

The transformation matrix of L_z is given by

$$\begin{pmatrix} a & -b \\ b & a \end{pmatrix},$$

which can be verified with a direct calculation. We will extend this result for quaternions.

Initially, we observe that, when a quaternion $\mathbf{q} = w\mathbf{1} + x\mathbf{i} + y\mathbf{j} + z\mathbf{k}$ represents a vector of $\mathbb{R}^3$ in homogeneous coordinates, the infinite coordinate is given by variable w. In Chapter 2 (Geometry), we took the coordinate of a point in the infinite as being the last coordinate. In this case, the vector corresponding to quaternion $\mathbf{q}$ is given by (x, y, z, w).

Let $\mathbf{q}_1 = (x_1, y_1, z_1, w_1)$ and $\mathbf{q}_2 = (x_2, y_2, z_2, w_2)$ be two quaternions. From the bilinearity of quaternion product, we know that the product $\mathbf{q}_1 \mathbf{q}_2$ is linear in $\mathbf{q}_1$ and $\mathbf{q}_2$. Therefore, we have two linear transformations: $L_{\mathbf{q}_1} \colon \mathbb{R}^4 \to \mathbb{R}^4$, and $R_{\mathbf{q}_1} \colon \mathbb{R}^4 \to \mathbb{R}^4$ defined, respectively, by the product to the left and right of quaternion $\mathbf{q}_1$:

$$L_{\mathbf{q}_1}(\mathbf{q}) = \mathbf{q}_1 \mathbf{q}, \quad \text{and} \quad R_{\mathbf{q}_1}(\mathbf{q}) = \mathbf{q}\mathbf{q}_1.$$

A direct calculation shows that the matrices of these linear transformations are given by

$$L_{\mathbf{q}_1}(\mathbf{q}) = \begin{pmatrix} w_1 & -z_1 & y_1 & x_1 \\ z_1 & w_1 & -x_1 & y_1 \\ -y_1 & x_1 & w_1 & z_1 \\ -x_1 & -y_1 & -z_1 & w_1 \end{pmatrix} \begin{pmatrix} x \\ y \\ z \\ w \end{pmatrix} \tag{4.14}$$

and

$$R_{\mathbf{q}_1}(q) = \begin{pmatrix} w_1 & z_1 & -y_1 & x_1 \\ -z_1 & w_1 & x_1 & y_1 \\ y_1 & -x_1 & w_1 & z_1 \\ -x_1 & -y_1 & -z_1 & w_1 \end{pmatrix} \begin{pmatrix} x \\ y \\ z \\ w \end{pmatrix}. \tag{4.15}$$

We can use any one of these two transformations to represent the product of two quaternions using linear transformations. We observe that this association is analog to the association between complex numbers and matrices we presented in the beginning of this section. Here we have two possibilities, given that the product of two quaternions is not commutative. (Notice that, if $\mathbf{q}_1$ is unitary, then transformations $L_{\mathbf{q}_1}$ and $R_{\mathbf{q}_1}$ are orthogonal.)

It is important to highlight that this relation between quaternions and linear transformations preserves the product and sum operations in each of these spaces. More precisely, the quaternion $\mathbf{q}_1(\mathbf{q}_2 + \mathbf{q}_3)$ corresponds to the matrix $L_{\mathbf{q}_1}(L_{\mathbf{q}_2} + L_{\mathbf{q}_3})$, or $R_{\mathbf{q}_1}(R_{\mathbf{q}_2} + R_{\mathbf{q}_3})$.

As quaternion product is associative, we have

$$(\mathbf{q}_1\mathbf{p})\mathbf{q}_2 = \mathbf{q}_1(\mathbf{p}\mathbf{q}_2).$$

By using transformations L and R, this expression can be written in the form

$$R_{\mathbf{q}_2}L_{\mathbf{q}_1}p = L_{\mathbf{q}_1}R_{\mathbf{q}_2}p.$$

This relation is very important from the computational point of view. In fact, in a quaternion product

$$\mathbf{q}_1\mathbf{q}_2\cdots\mathbf{q}_{i-1}\mathbf{q}_i\mathbf{q}_{i+1}\cdots\mathbf{q}_N, \tag{4.16}$$

if only quaternion q_i varies, we can combine the products $\mathbf{q}_1\mathbf{q}_2\cdots\mathbf{q}_{i-1}$ in a matrix $L_{\mathbf{q}_1\mathbf{q}_2\cdots\mathbf{q}_{i-1}}$, and the product $\mathbf{q}_{i+1}\cdots\mathbf{q}_N$ in a matrix $R_{\mathbf{q}_{i+1}\cdots\mathbf{q}_N}$, and then write the product of (4.16) in the form

$$L_{\mathbf{q}_1\mathbf{q}_2\cdots\mathbf{q}_{i-1}}R_{\mathbf{q}_{i+1}\cdots\mathbf{q}_N}\mathbf{q}_i.$$

We will now find the matrix in SO(3) corresponding to the rotation defined by $\varphi_\mathbf{q}(\mathbf{v}) = \mathbf{q}\hat{\mathbf{v}}\mathbf{q}^{-1}$. As $\mathbf{q}$ is unitary, we have $\mathbf{q}^{-1} = \mathbf{q}^\star$, and therefore it is proceeded that

$$\varphi_\mathbf{q}(\mathbf{v}) = L(\mathbf{q})R(\mathbf{q}^\star)(\hat{\mathbf{v}}). \tag{4.17}$$

If $\mathbf{q} = (x, y, z, w)$, then $\mathbf{q}^\star = (-x, -y, -z, w)$. Using the matrix of $L_{\mathbf{q}_1}$ in (4.14), with $\mathbf{q}_1 = \mathbf{q}$ and the matrix $R_{\mathbf{q}_1}$ in (4.15) with $\mathbf{q}_1 = \mathbf{q}^\star$, and substituting in (4.17), we obtain

$$L(\mathbf{q})R(\mathbf{q}^\star) = \begin{pmatrix} w & -z & y & x \\ z & w & -x & y \\ -y & x & w & z \\ -x & -y & -z & w \end{pmatrix} \begin{pmatrix} w & -z & y & -x \\ z & w & -x & -y \\ -y & x & w & -z \\ x & y & z & w \end{pmatrix}.$$

Performing the product, we obtain the matrix

$$\begin{pmatrix} w^2 + x^2 - y^2 - z^2 & 2xy - 2wz & 2xz + 2wy & 0 \\ 2xy + 2wz & w^2 - x^2 + y^2 - z^2 & 2yz - 2wx & 0 \\ 2xz - 2wy & 2yz + 2wx & w^2 - x^2 - y^2 + z^2 & 0 \\ 0 & 0 & 0 & |\mathbf{q}|^2 \end{pmatrix}. \tag{4.18}$$

(See Exercise 3 for another form of writing this matrix.) As the quaternion $\mathbf{q}$ is unitary, we have

$$|\mathbf{q}|^2 = w^2 + x^2 + y^2 + z^2 = 1.$$

In Cartesian coordinates, the above matrix can be written in the form

$$\begin{pmatrix} w^2 + x^2 - y^2 - z^2 & 2xy + 2wz & 2xz - 2wy \\ 2xy - 2wz & w^2 - x^2 + y^2 - z^2 & 2yz + 2wx \\ 2xz + 2wy & 2yz - 2wx & w^2 - x^2 - y^2 + z^2 \end{pmatrix}$$

as well as,

$$2 \begin{pmatrix} \frac{1}{2} - y^2 - z^2 & xy + wz & xz - wy \\ xy - wz & \frac{1}{2} - x^2 - z^2 & yz + wx \\ xz + wy & yz - wx & \frac{1}{2} - x^2 - y^2 \end{pmatrix}.$$

It is also important to solve the problem of finding the quaternion associated with the matrix L or R. We left this problem for the exercises at the end of the chapter.

4.8.2   Quaternions and Euler's Angles

Another important relation is to obtain the quaternion associated with a rotation using Euler's angles α β and γ. We have three rotations $R_x(\alpha)$, $R_y(\beta)$, $R_z(\gamma)$ of angles α, β, γ about axes $(1,0,0)$, $(0,1,0)$ and $(0,0,1)$, respectively. Each one of these rotations is represented by $\varphi_{\mathbf{q}_x}$, $\varphi_{\mathbf{q}_y}$, and $\varphi_{\mathbf{q}_z}$, where

$$\mathbf{q}_x = \cos\frac{\alpha}{2} + \sin\frac{\alpha}{2}(1,0,0),$$

$$\mathbf{q}_y = \cos\frac{\beta}{2} + \sin\frac{\beta}{2}(0,1,0),$$

$$\mathbf{q}_z = \cos\frac{\gamma}{2} + \sin\frac{\gamma}{2}(0,0,1).$$

The rotation in space is given by $R_z(\gamma)R_y(\beta)R_x(\alpha)$, and it will be represented by the quaternion $\mathbf{q} = \mathbf{q}_z\mathbf{q}_y\mathbf{q}_x$, obtained from the product of the three quaternions. Assuming

$$\mathbf{q} = w + \mathbf{i}x + \mathbf{j}y + \mathbf{k}z,$$

and performing the calculations, we obtain

$$w = \cos\frac{\alpha}{2}\cos\frac{\beta}{2}\sin\frac{\gamma}{2} + \sin\frac{\alpha}{2}\sin\frac{\beta}{2}\sin\frac{\gamma}{2},$$

$$x = \sin\frac{\alpha}{2}\cos\frac{\beta}{2}\cos\frac{\gamma}{2} - \cos\frac{\alpha}{2}\sin\frac{\beta}{2}\sin\frac{\gamma}{2},$$

$$y = \cos\frac{\alpha}{2}\sin\frac{\beta}{2}\cos\frac{\gamma}{2} + \sin\frac{\alpha}{2}\cos\frac{\beta}{2}\sin\frac{\gamma}{2},$$

$$z = \cos\frac{\alpha}{2}\cos\frac{\beta}{2}\sin\frac{\gamma}{2} + \sin\frac{\alpha}{2}\sin\frac{\beta}{2}\cos\frac{\gamma}{2}.$$

4.9   Comments and References

Hamilton's goal on searching for quaternions, consisted on discovering a multiplication structure, similar to the one of complex numbers, for the $\mathbb{R}^3$ and that could be used to represent rotations in space. He ended up discovering such structure for the $\mathbb{R}^4$. Today it is known that this type of structure exists only in $\mathbb{R}^2$, $\mathbb{R}^4$ and $\mathbb{R}^8$.

Quaternions were introduced in computer graphics in the article [Shoemake 85]. They were already used in the area of robotics.

For more information about orthogonal transformations of $\mathbb{R}^n$, and in particular, the demonstration of Euler's Theorem in its general form, we suggest [Lima 99].

4.9.1 Additional Topics

In this chapter, we studied the special group SO(3) of rotations in space, which consists in just a part of the representation of rigid motions in space. The complete group has dimension 6 (three degrees of freedom for rotations and three for translations), and is called *special Euclidean group*, denoted by SE(3). An appropriate representation of this group involves Screw Theory and the study of the Lie Groups of matrices. Complete material on these topics can be found in a good textbook in the area of robotics.

Quaternions represent only the tip of the iceberg in the study of intrinsic operations between geometric objects; see [Dorst et al. 07] and [Perwass 09].

An important topic we did not cover is the problem of interpolating the n rotations represented by quaternions. From what we covered, this problem is equivalent to the study of interpolation curves (e.g., Bezier, splines) in the sphere S^3; several works have been written on the subject. This subject is also very important in the area of GIS (geographic information systems). The interpolation of rotations using the matrix Lie Groups structure is an interesting topic. In [Alexa 02], for instance, there is a description on the interpolation of linear transformations using results from matrix Lie Groups.

Exercises

1. Demonstrate Chasles' Theorem: "Every positive rigid motion in $\mathbb{R}^3$ can be obtained by a rotation about a r-axis, followed by a translation along r" (this type of motion is called *screw*).

2. Describe a method to obtain the quaternion associated with the matrix L or R (as defined in this chapter), in homogeneous coordinates.

3. If $R \colon \mathbb{R}^3 \to \mathbb{R}^3$ is a rotation, show that $R(u \wedge v) = R(u) \wedge R(v)$.

4. Show that there exists no multiplicative structure in $\mathbb{R}^3$ among vectors, similarly to quaternions (or to complex numbers). (Hint: use the transformation $L_{\mathbf{q}}$ or $R_{\mathbf{q}}$ defined in the chapter, including the fact that a linear transformation in $\mathbb{R}^3$ has 1 real eigenvalue.)

5. If (r_{ij}) is a matrix of order 3, representing a rotation in $\mathbb{R}^3$, show that the Euler's angles of the parameterization given by $R_z(\alpha)R_y(\beta)R_z(\gamma)$ are determined by:

$$\alpha = \arctan2(\sqrt{r_{31}^2 + r_{32}^2}, r_{33});$$
$$\beta = \arctan2(\frac{r_{23}}{\sin \beta}, \frac{r_{13}}{\sin \beta});$$
$$\gamma = \arctan2(\frac{r_{32}}{\sin \beta}, -\frac{r_{31}}{\sin \beta}),$$

where $\arctan2(x, y)$ is the function $\arctan(x/y)$, using the sign of x and y to determine the quadrant of the resulting angle.

6. Consider the following problem: "to implement an interactive interface to rotate an object of $\mathbb{R}^3$ about an axis, by only using the mouse as input device." Two solutions exist in the literature for this problem: the *Metaball* and the *virtual sphere*.

    (a) Describe the model of these two solutions;

    (b) Compare the solutions and discuss the advantages and disadvantages of each one.

7. Show that, if a quaternion $\mathbf{q}$ commutes with every pure quaternion, then $\mathbf{q}$ is real. Use this fact to show that

    $$\varphi_{\mathbf{q}_1} = \varphi_{\mathbf{q}_2} \Leftrightarrow \mathbf{q}_1 = \pm\mathbf{q}_2,$$

    where $\varphi_{\mathbf{q}}$ is the representation of a rotation defined in (4.8).

8. If $\mathbf{p} = \cos\theta + \hat{\mathbf{v}}\sin\theta$ is a unit quaternion, and $t \in \mathbb{R}$, show that $\mathbf{q}\mathbf{p}^t\mathbf{q}^* = (\mathbf{q}\mathbf{p}\mathbf{q}^*)^t$.

9. If $\mathbf{q}$ is a unit quaternion, and $a, b \in \mathbb{R}$, show that:

    $$\mathbf{q}^a\mathbf{q}^b = \mathbf{q}^{a+b} \qquad \text{and} \qquad (\mathbf{q}^a)^b = \mathbf{q}^{ab}.$$

10. The calculations below show that the product of quaternions is commutative:

    $$\begin{aligned}
    pq &= \exp(\log(pq)) \\
    &= \exp[\log(p) + \log(q)] \\
    &= \exp[\log(q) + \log(p)] \\
    &= \exp(\log(q))\,\exp(\log(p)) = qp.
    \end{aligned}$$

    Where is the error?

11. Show that, in the case in which quaternion $\mathbf{q}$ is not unitary, the matrix (4.18) of the transformation $\varphi_{\mathbf{q}}$ is given by

    $$\frac{2}{|\mathbf{q}|^2}\begin{pmatrix}
    \frac{|\mathbf{q}|^2}{2} - y^2 - z^2 & xy + wz & xz - wy & 0 \\
    xy - wz & \frac{|\mathbf{q}|^2}{2} - x^2 - z^2 & yz + wx & 0 \\
    xz + wy & yz - wx & \frac{|\mathbf{q}|^2}{2} - x^2 - y^2 & 0 \\
    0 & 0 & 0 & \frac{|\mathbf{q}|^2}{2}
    \end{pmatrix}.$$

12. Show that

    $$e^{\theta\hat{\mathbf{u}}} = \cos\theta + \hat{\mathbf{u}}\sin\theta,$$

    substituting the expression of $\theta\hat{\mathbf{u}}$ in the series of potencies of the exponential function

    $$e^x = \sum_{n=0}^{\infty}\frac{x^n}{n}.$$

13. Let

$$R = \begin{pmatrix} r_{11} & r_{12} & r_{13} \\ r_{21} & r_{22} & r_{23} \\ r_{31} & r_{32} & r_{33} \end{pmatrix}$$

be a rotation matrix in $\mathbb{R}^3$. Show that the rotation axis is the vector

$$v = \frac{1}{\sin\theta}(r_{32} - r_{23}.r_{13} - r_{31}, r_{21} - r_{12}).$$

If the rotation angle θ is different from π and 0, then

$$\theta = \arcos\left(\frac{r_{11} + r_{22} + r_{33} - 1}{2}\right).$$

What happens if the rotation angle is very small? Describe a robust method to find the rotation axis (your method should include the cases for $\theta = 0$ and $\theta = \pi$).

14. Using the previous exercise, show that the matrix

$$R_y(90°)R_z(90°) = \begin{pmatrix} 0 & 0 & 1 \\ 1 & 0 & 0 \\ 0 & 1 & 0 \end{pmatrix}$$

represents a rotation of $120°$ about the axis of the $\mathbb{R}^3$ space defined by the unit vector $(1/\sqrt{3}, 1/\sqrt{3}, 1/\sqrt{3})$.

15. If $\mathbf{q}(t)$ is a curve in the unit sphere S^3, then $\mathbf{q}'(t) = \hat{v}(t)\mathbf{q}(t)$, where $\hat{v}$ represents a pure quaternion associated with the vector $\mathbf{v} \in \mathbb{R}^3$. (Hint: do $\mathbf{q}(t+s) = \mathbf{q}(t+s)\mathbf{q}(t)\mathbf{q}(t)^{-1}$, and use $\mathbf{q}'(t) = d\mathbf{q}(t+s)/ds$ at the point $s = 0$.)

16. If $\mathbf{q}$ is a unit quaternion, show that

$$\frac{d}{dt}\mathbf{q}^t = \mathbf{q}^t \log(\mathbf{q}).$$

17. Show the veracity of the equalities below:

    (a) $\mathrm{slerp}(\mathbf{p}, \mathbf{q}, t) = \mathbf{p}(\mathbf{p}^*\mathbf{q})^t$;
    (b) $\mathrm{slerp}(\mathbf{p}, \mathbf{q}, t) = (\mathbf{p}\mathbf{q}^*)^{1-t}\mathbf{q}$;
    (c) $\mathrm{slerp}(\mathbf{p}, \mathbf{q}, t) = (\mathbf{q}\mathbf{p}^*)^t\mathbf{p}$;
    (d) $\mathrm{slerp}(\mathbf{p}, \mathbf{q}, t) = \mathbf{q}(\mathbf{q}^*\mathbf{p})^{1-t}\mathbf{q}$.

    Then conclude that $\mathrm{slerp}(\mathbf{p}, \mathbf{q}, t) = \mathrm{slerp}(\mathbf{q}, \mathbf{p}, 1 - t)$.

18. If $\mathbf{v} = (x, y, z) \in \mathbb{R}^3$, $\mathbf{v} \neq 0$, show that the Jacobian matrix $d\exp_{\mathbf{v}}$ of the exponential function at the point $\mathbf{v}$ is given by

$$\begin{pmatrix} \left(\frac{c}{|v|^2} - \frac{s}{|v|^3}\right)x^2 + \frac{s}{|v|} & \left(\frac{c}{|v|^2} - \frac{s}{|v|^3}\right)xy & \left(\frac{c}{|v|^2} - \frac{s}{|v|^3}\right)xz \\ \left(\frac{c}{|v|^2} - \frac{s}{|v|^3}\right)xy & \left(\frac{c}{|v|^2} - \frac{s}{|v|^3}\right)y^2 + \frac{s}{|v|} & \left(\frac{c}{|v|^2} - \frac{s}{|v|^3}\right)yz \\ \left(\frac{c}{|v|^2} - \frac{s}{|v|^3}\right)xz & \left(\frac{c}{|v|^2} - \frac{s}{|v|^3}\right)yz & \left(\frac{c}{|v|^2} - \frac{s}{|v|^3}\right)z^2 + \frac{s}{|v|} \end{pmatrix},$$

where $s = \sin|\mathbf{v}|$ and $c = \cos|\mathbf{v}|$.

19. Show that the Jacobian matrix of the exponential function at the origin $\mathbf{v} = (0, 0, 0) \in \mathbb{R}^3$, is given by the matrix

$$\begin{pmatrix} 0 & 0 & 0 \\ 1 & 0 & 0 \\ 0 & 1 & 0 \\ 0 & 0 & 1 \end{pmatrix}.$$

(Hint: use the expression from the previous exercise for $\mathbf{v} \neq 0$, together with the definition of the derivative as a limit.)

20. This exercise shows that the multiplication of two quaternions can be performed with only 8 products, instead of 16 products, as would be expected. Given two quaternions $q_1 = (x_1, y_1, z_1, w_1)$ and $q_2 = (x_2, y_2, z_2, w_2)$, do

$$\begin{aligned}
a_1 &= (x_1 - y_1)(y_2 - z_2) & a_5 &= (z_1 - x_1)(x_2 - y_2) \\
a_2 &= (w_1 + x_1)(w_2 + x_2) & a_6 &= (z_1 + x_1)(x_2 + y_2) \\
a_3 &= (w_1 - x_1)(y_2 + z_2) & a_7 &= (w_1 + y_1)(w_2 - z_2) \\
a_4 &= (z_1 + y_1)(w_2 - x_2) & a_8 &= (w_1 - y_1)(w_2 + z_2).
\end{aligned}$$

Define $s = a_6 + a_7 + a_8$ and $t = (a_5 + s)/2$. Show that

$$q_1 q_2 = (a_2 + t - s)\mathbf{1} + (a_2 + t - s)\mathbf{i} + (a_3 + t - a_8)\mathbf{j} + (a_4 + t - a_7)\mathbf{k}.$$

5 | Color

Color plays an important role in the study of images, and more generally in computer graphics. It will therefore occupy us for more than one chapter. Here we start to approach the subject by asking, What is color? What are the mathematical models for color? How should colors be discretized? As usual, our approach will be based on the four universes paradigm, as follows:

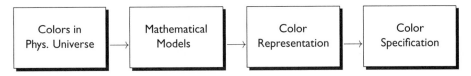

5.1   Color in the Physical Universe

Color arises from electromagnetic radiation within a range of wavelengths that affect the human eye. The visible electromagnetic spectrum lies between the wavelengths of

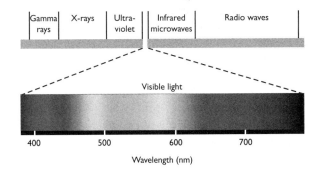

Figure 5.1. Electromagnetic spectrum; top diagram has a logarithmic scale. (See Color Plate I.)

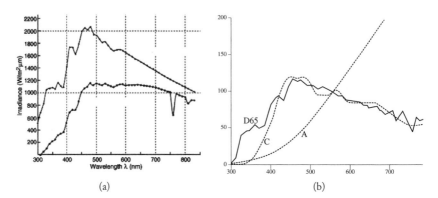

Figure 5.2. (a) Spectral distribution of sunlight and (b) of some standard illuminants.

$\lambda_a = 400$ nm and $\lambda_b = 800$ nm, approximately (the nanometer, abbreviated nm, is one billionth of a meter). Figure 5.1 shows the visible spectrum as a part of the much broader spectrum of all electromagnetic radiation, most of which is not visible. The energy associated with an electromagnetic wave is called *radiant energy*.

Color is a *psychophysical* phenomenon, that is, it has an important perceptual component, in addition to its physical aspects. There are three important areas in the study of color: *colorimetry*, *radiometry*, and *photometry*. Colorimetry deals with the representation and specification of color, ignoring its physical nature and propagation properties. Radiometry deals with physical measures associated with radiant energy, while photometry deals with perceptual measurements of radiant energy as illumination. In this chapter we discuss colorimetry; the other two areas are briefly covered in Chapter 18. We recommend that the reader consult Chapter 18 in parallel with this chapter.

When we perceive a color, our eyes are in fact being struck by light of different wavelengths, whose combined energies produce the color. For instance, white is typically a mixture of light from all wavelengths of the visible spectrum in roughly equal proportions.

A color can be completely specified by its *spectral distribution function* (or *spectral density function*), which associates to each wavelength λ the amount of radiant energy, or *irradiance*, at that wavelength, denoted by $E(\lambda)$. Figure 5.2(a) shows the spectral distribution of sunlight, measured above the atmosphere, and also the spectral distribution measured on the surface of the earth. Which one is which?

5.1.1   Color Temperature

We see most objects because they radiate light received from elsewhere. But all objects also emit *thermal radiation*, that is, electromagnetic radiation that is not a response to external stimulation, and is characterized by temperature. For objects at room temperature, thermal radiation is infrared: it has longer wavelengths than visible light. As the temperature increases, shorter wavelengths are produced in increasing proportion, even as the total

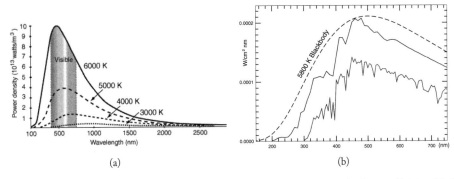

Figure 5.3. (a) Spectral distribution of black bodies at various temperatures, in degrees Kelvin; (b) the spectrum of sunlight is approximately that of a black body at around 5,800 K. (See Color Plate II.)

amount of thermal radiation emitted also grows. As a result, any object heated above 500°C or so emits visible light. The spectral distribution of this light, and hence its color, depends on the temperature of the object.

A *black body* or *perfect radiator* is an idealized object that *only* emits thermal radiation, and has no intrinsic preference for one wavelength over another. The German physicist Max Planck (1858–1947) was the first to find a very accurate formula for the spectral distribution function of a black body at any given temperature; he also explained that function in terms of quantization of radiant energy. Figure 5.3(a) shows those distributions for a number of different temperatures; Figure 5.3(b) compares the black body distribution with the spectral distribution of sunlight, shown in the previous figure.

Other processes in nature, besides heat, that cause matter to emit visible light. For instance, a fluorescent lamp or a computer screen work at a relatively low temperature but emit visible light. Their spectrum is dramatically different from that of the sun or other thermal sources: it is a combination of a few narrow peaks. However, because of the way color perception works (more on this later), the color of such a light source may be *perceived* as being the same as that of a black body at a certain temperature. It is convenient to identify such sources with the corresponding temperature, usually given in degrees Kelvin.

This method is used to specify the different types of white used in illumination standards, and for describing artificial light sources. Figure 5.2(b) shows the spectral distribution of three white light sources used as standards by the CIE (see Section 5.6): illuminant D65 (6504 K) corresponds to average daylight; illuminant C (6774 K) corresponds to average daylight after the exclusion of ultraviolet radiation; and illuminant A (2856 K) represents a standard incandescent lamp.

5.2   Spectral Color Space

As already stated, a color can be completely specified by its spectral distribution function. We will use this property to define a first mathematical model for color: a color is

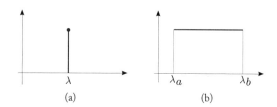

Figure 5.4. Spectral distribution of (a) a pure color and (b) equal-energy white.

represented by a function $C \colon \mathbb{R}_+ \to \mathbb{R}$, its spectral distribution function. Here $\mathbb{R}_+$ is the set of strictly positive real numbers, regarded as wavelengths.

We could have chosen to define spectral distribution functions only on the interval $[\lambda_a, \lambda_b]$ of visible wavelengths, but mathematically it is just as easy to work with functions on $\mathbb{R}_+$. This choice amounts to saying that points in spectral color space also carry information about invisible wavelengths (for example in the infrared and ultraviolet ranges).

We could also have restricted our range to the nonnegative real numbers, since a physical color cannot have a negative energy density at a certain wavelength; but it will be useful in calculations to assume it can.

The space of colors thus defined—that is, the space of spectral distribution functions $C \colon \mathbb{R}_+ \to \mathbb{R}$—is called *spectral color space* and is denoted by $\mathcal{E}$. In this sense "color" subsumes the notion of intensity or brightness: two points in $\mathcal{E}$ that are multiples of one another, $C'(\lambda) = r\, C(\lambda)$, are distinct, though they represent the same shade of color, with the same *relative* spectral distribution. This is further discussed in Section 5.7.

A color having energy at just one wavelength is called a *pure spectral color*; its spectrum is zero everywhere except at a single peak, as shown in Figure 5.4(a). In practice, any spectral peak has a finite width, but pure spectral colors are a useful idealization. By contrast, the spectrum shown in Figure 5.4(b) has the same density throughout the visible range; it defines *equal-energy white*.

5.3 Color Representation and Reconstruction

The spectral color space $\mathcal{E}$ is infinite-dimensional. In practice, therefore, we need to represent $\mathcal{E}$ by a finite-dimensional space. A simple method for doing this is *point sampling* of spectral distribution functions. Choose n wavelengths $\lambda_a \leq \lambda_1 < \lambda_2 < \cdots < \lambda_n \leq \lambda_b$, and represent a color $C \in \mathcal{E}$ by the vector $(C(\lambda_1), \ldots, C(\lambda_n)) \in \mathbb{R}^n$ (see Figure 5.5).

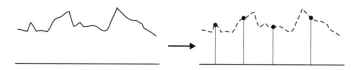

Figure 5.5. Color sampling.

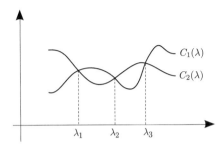

Figure 5.6. Two different colors $C_1(\lambda)$ and $C_2(\lambda)$ have the same sampling representation if their distributions happen to coincide at the sampling wavelengths—in this case at λ_1, λ_2 and λ_3.

Clearly some information is lost in this process; two colors may end up being represented by the same vector, as in Figure 5.6. In other words, the representation of $\mathcal{E}$ in this finite-dimensional space is not invertible. This is to be expected (see Chapter 1).

As we shall see, the fact that exact reconstruction is not possible—we cannot obtain the exact function $C(\lambda)$ from the samples $C(\lambda_i)$—is less of a problem in practice than it might seem, due to the limitations of our color perception apparatus. But it raises important practical questions: how many samples should we take to represent the spectral color space $\mathcal{E}$, and where should the samples be taken? (For computational economy, the fewer the better.) How can we reconstruct a color $C(\lambda)$ as well as possible, given its representation vector in $\mathbb{R}^n$?

5.3.1 The Perceptual Foundation for Trichromatic Models

The human retina has a few million photosensitive cells, called *cones*, responsible for color perception. Each cone is equipped with one of three distinct chemical photoreceptors, conventionally labeled L, M, S as they respond predominantly to long, medium and short wavelengths, respectively. The response of each kind of photoreceptor is then sent to the brain as an electrical impulse, and the brain fuses these responses into the sensation of color. In some sense, therefore, three numbers are sufficient to describe a color as perceived by the human eye.

A color perceived as white can have a more or less flat spectrum such as the one in Figure 5.4(b), but it does not have to. A combination of three pure spectral colors of particular, sufficiently separated wavelengths—one red, one blue and one green, say—will also be perceived as white if the proportions are right. This is the principle behind color video and computer screens, with their three types of light-emitting elements.

The *RGB color model* (or *Young–Helmholtz model* of color[1]), then, represents elements of spectral color space $\mathcal{E}$ by triples of numbers (vectors in $\mathbb{R}^3$), whose coordinates are the red, green and blue components of the color in some sense. We will have more to say about

---

[1]The trichromatic model of color vision was postulated as early as 1802 by the English scientist Thomas Young (1773–1829) and later refined by the German physicist Hermann von Helmholtz (1821–1894).

the choices involved in such a model soon; but for now, let's just assume that the mixture of equal parts red, green and blue appears white. Then the triples (t, t, t), where $t > 0$, represent different intensities of white, the triples $(t, 0, 0)$ represent different intensities of the same red, and so on.

Colors represented by triples of the form (t, t, t) are called *achromatic* (literally, color-less). If two such colors of different brightness (different values of t) are placed in close juxtaposition—say, areas on a computer screen—the less bright one is perceived as gray, or a darker gray than the other. Thus achromatic colors correspond to shades of gray, but the correspondence is subjective and context-dependent (Section 5.7.5). In any case, the color represented by $(0, 0, 0)$ is black—its intensity in all three wavelengths is zero—and the line in RGB space consisting of triples (t, t, t) is therefore called the *black-white line*.

5.3.2 Perceptual Reconstruction

Suppose, as an approximation, that each type of retinal photoreceptor is sensitive only to a single wavelength. Then any two colors that have the same intensities as each other at each of these three wavelengths—call them $\lambda_1, \lambda_2, \lambda_3$—will be perceived as identical. To reconstruct a color $C \in \mathcal{E}$, we just need the values $C(\lambda_1)$, $C(\lambda_2)$, $C(\lambda_3)$: the reconstruction $\widetilde{C}(\lambda)$ will be perceptually faithful as long as $\widetilde{C}(\lambda_i) = C(\lambda_i)$, for $i = 1, 2, 3$. See Figure 5.7. The colors C and $\widetilde{C}$ do not need to have the same spectral distribution

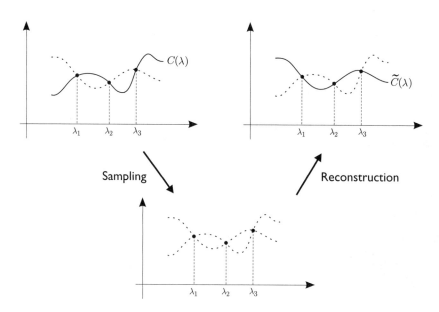

Figure 5.7. Perceptually equivalent reconstruction.

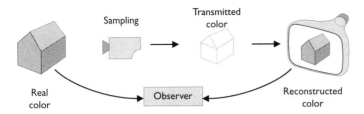

Figure 5.8. Sampling and reconstruction.

in order to be seen as identical; they just need to be represented by the same triples of numbers![2]

As mentioned, this property is what underlies the technology of color electronics; see the schematic diagram in Figure 5.8. A color from the real world is captured by video camera sensors and is then reconstructed on the television or computer screen. The reconstructed and original colors may seem the same to the viewer, but they do not have the same spectral distribution function.

Two perceptually equal colors with different spectral distributions are called *metameric*. Much of this chapter will be devoted to elaborating on the phenomenon of metamerism, removing some of the simplifying assumptions made in this section.

5.4 Physical Color Systems

Physical systems never produce or respond to a single wavelength; they are characterized by a spectral emission or response curve, like the one in Figure 5.9. The peak can be broad or narrow, or there can be more than one, but the fact is that the monochromaticity assumption made in the previous section is generally unrealistic. Our goal is to develop a theory that can handle physically realistic spectral curves.

5.4.1 Color Sampling Systems

Mathematically, a *color sampling system* is a finite collection of functions $s_1, s_2, \ldots, s_n$ from $\mathbb{R}$ to $\mathbb{R}$, each of which represents the *response curve* of a sensor in the system. The response curve of a sensor is a weighting function; it defines how the different wavelengths of an incoming color contribute to the number reported by the sensor when it sees that color. In symbols, a color C is represented by the vector of numbers $(c_1, c_2, \ldots, c_n)$, where

$$c_i = \int_0^\infty C(\lambda) s_i(\lambda) \, d\lambda. \tag{5.1}$$

---

[2]This is an oversimplification. Color perception also depends on texture, level of lighting, and other features. Not to mention an increasing body of evidence that certain individuals have *four* distinct photoreceptors in their cone cells, and so can discriminate colors regarded as the same by most people.

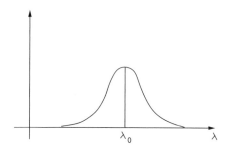

Figure 5.9. Typical spectral response of a color sensor.

The human eye is the prototypical color sampling system. Its color apparatus, as we have seen, has three types of sensors, each with a response curve similar to that of Figure 5.9, peaking at around $445, 540,$ and $570\,\text{nm}$, respectively. Each type of sensor samples the spectral distribution of a color according to its own response curve. Scanners and video cameras work on a similar principle.

5.4.2   Color Reconstruction Systems

Reconstruction systems play the reverse role as sampling systems: given a signal, they emit the color encoded by the signal. A computer screen is a familiar example; the color of each pixel is controlled by three numbers, each indicating the intensity of the corresponding color emitter (red, green, blue).

Mathematically, a color reconstruction system is defined by finitely many spectral distribution functions $P_1, \ldots, P_n : \mathbb{R} \to \mathbb{R}$, called the system's *primary colors*. The color reconstructed from a sample $(c_1, \ldots, c_n)$ is given by the linear combinations of the primary colors, with the given coefficients:

$$\widetilde{C}(\lambda) = \sum_{k=1}^{n} c_k P_k(\lambda). \tag{5.2}$$

The set of all colors that can be constructed by combining primary colors with *nonnegative* coefficients c_i is called the *gamut* of the system. In other words, the gamut is the cone over the convex hull of the set of primary colors. Colors metameric to those in the gamut are said to be (metamerically) *reconstructible*.

5.5   Tristimulus Values and Metameric Reconstruction

Consider a color reconstruction system whose primary colors are $P_1(\lambda), P_2(\lambda), P_3(\lambda)$. How can colors be sampled so that, upon reconstruction, they will be perceived as the same as the original? A solution to this problem is obviously desirable in situations such as the sampling/reconstruction scheme for television (see Figure 5.8).

To spell things out, given a color with spectral distribution function $C(\lambda)$, we wish to find a triple (c_1, c_2, c_3) such that the reconstructed color

$$\widetilde{C} = c_1 P_1 + c_2 P_2 + c_3 P_3 \tag{5.3}$$

is metameric to C. Such a triple is called the *tristimulus value* associated with C.

Since it involves perception, the problem will ultimately require the use of experimental data in its solution. But a bit of mathematical thinking helps figure out what sort of data is needed.

We first observe that the sampling/reconstruction procedure is *additive*: for a fixed sampling system, the vector representation of the sum of two colors (spectral distribution functions) C and C' is the sum of the vector representations of C and C', thanks to (5.1); and conversely, once we have fixed the primary colors P_1, P_2, P_3 of our reconstruction, the sum of two samples yields upon reconstruction the sum of the colors reconstructed from each sample.

For this reason it is possible (conceptually at least) to narrow down the problem to pure spectral colors. The spectral density function of a pure color of wavelength λ is indicated by δ_λ. Suppose we can find out the tristimulus value for δ_λ, that is, the numbers $q_1(\lambda)$, $q_2(\lambda)$ and $q_3(\lambda)$ such that the reconstructed color

$$q_1(\lambda)P_1(\lambda) + q_2(\lambda)P_2(\lambda) + q_3(\lambda)P_3(\lambda)$$

is metameric to δ_λ; and suppose moreover that we can do so for every wavelength λ in the visible spectrum. We obtain in this way three functions q_1, q_2, q_3, called the (*spectral*) *color matching functions* of the reconstruction system $\{P_1, P_2, P_3\}$. Figure 5.10 shows a possible graph for one of them.

One can also think of the three spectral matching functions as being the coordinate functions of a single curve in space,

$$\varphi \colon [\lambda_a, \lambda_b] \to \mathbb{R}^3, \qquad \varphi(\lambda) = (q_1(\lambda), q_2(\lambda), q_3(\lambda)).$$

This curve is called the reconstruction system's *spectral color map*. By definition, each of its points is the tristimulus value of a spectral color in $\mathbb{R}^3$.

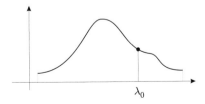

Figure 5.10. Graph of one of the spectral matching functions of an arbitrary reconstruction system.

Then, to represent an arbitrary color C, we write it as a sum[3] of pure colors δ_λ, with weights given by the spectral distribution $C(\lambda)$. We then take the weighted average of the representations of these δ_λ; these representations, as we recall, are the triples $(q_1(\lambda), q_2(\lambda), q_3(\lambda))$. We thus have justified the following result:

Theorem 5.1 (Theorem of metameric reconstruction). *Color matching functions solve the problem of metameric reconstruction stated at the beginning of this section. More precisely, if a spectral distribution $C(\lambda)$ is assigned the representation, or tristimulus value, (c_1, c_2, c_3), where*

$$c_i = \int_{\mathbb{R}} C(\lambda)\, q_i(\lambda)\, d\lambda \tag{5.4}$$

(the q_i being the color matching functions), *then the reconstruction of C* (5.3) *is metameric to C.* ❑

5.5.1   Acquisition of Color Matching Functions

How can one obtain the color matching functions for a given physical color reconstruction system? We will briefly describe the experimental procedure that lies at the basis of this determination.

The basic setup is shown in Figure 5.11. An observation panel that qualifies as a *diffuse reflector* (a surface that reflects light as evenly as possible in all directions and wavelengths—in essence, a matte white wall), is divided into two fields, one of which will be lit by a source with spectral distribution C, called a *test light*.

The other field is lit by the three primary light sources, with spectral distributions P_1, P_2 and P_3; their intensity is controlled so they can be combined in various proportions. The goal is to find the exact mixture of P_1, P_2 and P_3 that looks the same as C; this gives the components c_1, c_2, c_3 of C with respect to the primary colors P_1, P_2 and P_3.

In a properly normalized (calibrated) reconstruction system, the most intense white obtainable is given the coordinates $(1, 1, 1)$, as we assumed on page 113. To achieve this, we start by using a white reference light W as the test color (see also Section 5.7.5). That is, the three primary lights are shone on one side of the observation panel and adjusted so the resulting color looks the same as W on the other side. The intensity values w_1, w_2 and w_3 of the three primaries are then recorded.

The next step is to obtain the coordinates of the test color C of interest. Again, the intensities of the primary lights are adjusted so their color combination looks the same as the test color. Let these three intensities be denoted by β_1, β_2 and β_3. The normalized color coordinates of the test color C are then

$$c_i = \frac{\beta_i}{w_i}, \quad i = 1, 2, 3. \tag{5.5}$$

---

[3]More precisely, an integral. Mathematically advanced readers will have noticed that the δ_λ are not actual functions but Dirac deltas, in terms of which C can be written as $C(\lambda') = \int_{\mathbb{R}} C(\lambda)\delta_\lambda(\lambda')\, d\lambda$. Our justification of Theorem 5.1 is therefore honest, though informal. The reader interested in details can consult [Gomes and Velho 02] or [Gomes and Velho 97].

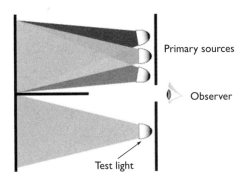

Figure 5.11. Experimental acquisition of color matching functions. (See Color Plate III.)

The most direct measurement of the matching functions $q_i(\lambda)$ would be obtained by taking for C a succession of monochromatic light sources of varying wavelength λ. In reality things are more complicated, because some pure colors are actually outside the gamut of P_1, P_2, P_3, and also because pure color sources with adjustable wavelength and sufficient intensity are difficult to come by. These problems can be circumvented by working with a number of *almost* monochromatic test colors and using the additivity property to work backwards toward the values that would be obtained for pure colors.

5.6   The Standard CIE-RGB System

Due to the great importance of color in industrial processes, it is necessary to establish standards for color specification. The International Commission on Illumination, abbreviated CIE for its French name (Comission internationale de l'éclairage) is the international body that defines standards in this area. In 1931 it established a standard trichromatic representation system based on the following set of primary colors, the last two of which were chosen for their ease of reproduction as lines in a mercury vapor discharge:

$$\begin{aligned}
P_1 &= R = \delta_{700} && \text{(a pure spectral red of wavelength 700 nm),} \\
P_2 &= G = \delta_{546.1} && \text{(a pure spectral green of wavelength 546,1 nm),} && (5.6) \\
P_3 &= B = \delta_{435.8} && \text{(a pure spectral blue of wavelength 435.8 nm).}
\end{aligned}$$

The color matching functions of this system have been acquired by careful experiments similar to those just described, refined over the decades. Their graphs are shown in Figure 5.12. Note that the color matching function for the first component, traditionally denoted by $\bar{r}$, takes on negative values for some values of λ; that is, a pure spectral color of wavelength 490 nm (for example) is metameric to a combination of standard blue (P_3) and standard green (P_2) with some standard red (P_1) subtracted! The physical meaning of this is that this pure color *cannot* be emulated by combining the three chosen primary colors in positive amounts; it lies outside the gamut of the system $\{P_1, P_2, P_3\}$. Instead, if one

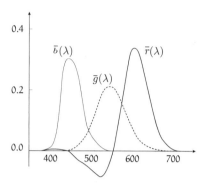

Figure 5.12. Color matching functions $\bar{r} = q_1$, $\bar{g} = q_2$, $\bar{b} = q_3$ of the CIE-RGB color system. They allow the calculation of tristimulus values (c_1, c_2, c_3) for any spectral distribution C via (5.4), and the metameric reconstruction of C via (5.3).

adds a small amount of red to it, the result will match a certain mixture of blue and green (see also Exercise 1). This is a drawback of the CIE-RGB color space, and has led to the introduction of alternatives, notably the CIE-XYZ color space discussed in Section 5.8.

Here a note about terminology is in order: the expression *color space* is used with several related meanings. It can refer to:

❑ the vector space where colors are represented—in this case $\mathbb{R}^3$;

❑ the representation itself, that is, a rule associating to a spectral distribution function its tristimulus value (c_1, c_2, c_3), or an n-tuple in a higher-dimensional representation space;

❑ the representatives whose coordinates range from 0 to 1, or the spectral distribution functions they represent (this color is outside our mRGB color space, where m stands for "monitor"); or

❑ the reconstruction rule, assigning to a tristimulus value or n-tuple a certain spectral distribution function (CIE-RGB color space is defined by (5.3) and (5.6)).

The *color solid* is a subset of color space in the first sense. It is defined as the set of tristimulus values corresponding to physically meaningful spectral distribution functions; we will revisit it in Section 5.7.4.

5.7   The Geometry of Color Space

We now explore further the way colors are distributed in a given 3D color space.

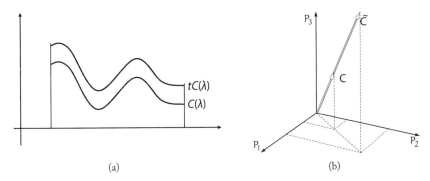

Figure 5.13. (a) Proportional spectral distributions are represented by (b) proportional vectors in color space.

5.7.1 Luminance and Chromaticity

Consider a spectral distribution function $C(\lambda)$ and a real number $t > 0$. On page 112 we discussed briefly how the spectral distribution function $C' = tC$, proportional to C, represents in a sense the same color as C; they differ only in intensity or brightness, or, to use the technical term, *luminance*. A pair of such functions is shown in Figure 5.13(a); in that example $C' = tC$ is brighter than C since $t > 1$. The luminance of C' is t times the luminance of C, and the same is true of other photometric measurements for the two. (For more on luminance and photometric measurements in general, see Chapter 18.)

Since the procedures involved are linear, the representation of C' in a color space is t times that of C (Figure 5.13(b)). It follows that all points on a ray going through the origin (minus the origin itself) have the same color information. This information is called *chromaticity*, or *chroma*. Chromaticity space is the set of straight lines in $\mathbb{R}^3$ passing through the origin—that is, the projective plane of Section 2.6.

5.7.2 Invisible, Nonphysical, and Nonreconstructible Colors

Because we defined the spectral color space $\mathcal{E}$ as the space of functions $\mathbb{R}_+ \to \mathbb{R}$, there exist spectral distribution functions that vanish through the visible band of the spectrum; they are called *invisible*. All invisible colors are represented by the zero tristimulus value. We will not consider them further.

There exist also spectral distribution functions $C(\lambda)$ that take negative values at some or all λ. These are nonphysical (see page 112) but can be useful in computations. Of the visible, physical colors—those are nonnegative everywhere and positive somewhere in the visible range—we have called those that are in the gamut of the primary colors P_1, P_2, P_3 reconstructible (page 116). A color may be visible and yet not be reconstructible from a given set of primaries: this happens when its reconstruction would require a linear combination involving one or more negative coefficients, like the pure spectral color of wavelength 490 nm in CIE-RGB color space (see page 119).

The *color solid* is the set of tristimulus values in $\mathbb{R}^3$ that correspond to visible physical colors, together with the origin $(0,0,0)$. Points in the color solid are also called *visible colors*; the context should make it clear whether we are referring to a spectral density function or a tristimulus value.

5.7.3 The Maxwell Triangle and Chromaticity Coordinates

Recall that chromaticity space is the projectivization of the color space $\mathbb{R}^3$; a vector $c \in \mathbb{R}^3$ has the same chromaticity as any nonzero multiple of it (while their *luminance* differs).

In our study of the projective plane, we represented affine points by the plane $z = 1$ (this is the model $\mathcal{P}$ of affine space; see Figure 2.7). We seek an analogous construction for the color space, looking for a subset of $\mathbb{R}^3$ where each point represents a different chromaticity. Every reconstructible chromaticity in the given system of primaries P_1, P_2 and P_3 should be representable on that plane. A good choice is the plane with equation

$$x + y + z = 1,$$

called the *Maxwell plane*, after the Scottish mathematician and physicist James Clerk Maxwell (1831–1879). In this plane, the triangle with vertices $(1,0,0)$, $(0,1,0)$ and $(0,0,1)$ is the *Maxwell triangle*, shown in Figure 5.14.

Colors C in the gamut of the system have a representative in the Maxwell triangle (that is, the line going through the origin and C intersects the triangle). Colors outside the Maxwell triangle have some negative coordinate; they represent either nonphysical colors, or those outside the gamut.

The projection coordinates of a color on the Maxwell plane are known as the *chromaticity coordinates* of the color. Their calculation is straightforward. Let (c_1, c_2, c_3) be a color's tristimulus value; we must find a vector proportional to it lying on the Maxwell plane. Since the Maxwell plane is the plane of points whose coordinates add up to 1, we just need to divide (c_1, c_2, c_3) by the sum of its coordinates:

$$c_i^{\star} = \frac{c_i}{c_1 + c_2 + c_3}, \quad i = 1, 2, 3. \tag{5.7}$$

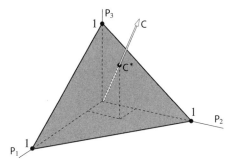

Figure 5.14. The Maxwell triangle (in gray) lies on the Maxwell plane $x + y + z = 1$.

By construction, the sum of the coordinates c_i^* is unity. We have effectively proved this:

Theorem 5.2. *The chromaticity coordinates of a color are the barycentric coordinates of the corresponding point on the Maxwell plane, relative to the vertices of the Maxwell triangle.* ❏

5.7.4   The Chromaticity Diagram

Our next goal is to visualize the set of visible colors in $\mathbb{R}^3$, which we have called the *color solid* (page 123; recall that black is included). Clearly, if (r, g, b) is a visible color and $t \geq 0$ is a real number, the color (tr, tg, tb) is also visible. A set with this property is called a *cone*.

It is therefore possible, and convenient, to describe the color solid just by its intersection with the Maxwell plane. This intersection is called the *chromaticity diagram*; it is shown in Figure 5.15 for the case of the CIE-RGB color space. Conventionally, the Maxwell plane and the chromaticity diagram inside it are depicted using the coordinates corresponding to the first two primary colors, P_1 and P_2; that is, in Figure 5.14 we view the Maxwell plane from above, rather than orthogonally along the axis of the white arrow. That is why the axes in Figure 5.15 are labeled "red" and "green." The origin is labeled "blue" since it corresponds to the P_3 direction.

The color solid is also easily seen to be *convex*: if (r_1, g_1, b_1) and (r_2, g_2, b_2) are visible colors, any point along the line segment joining them is also a visible color (see Exercise 10).

Another property of the color solid, this time an experimental one, is that pure spectral colors are located on its boundary; it follows that their projections on the Maxwell plane also appear on the boundary of the chromaticity diagram. See Figure 5.15, where a few such colors are marked, with their corresponding wavelengths (see also Exercise 11).

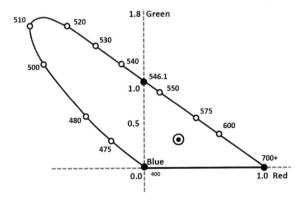

Figure 5.15. Chromaticity diagram on the Maxwell plane for the CIE-RGB color space. The horizontal axis is the red coordinate, corresponding to P_1, and the vertical axis is the green coordinate, corresponding to P_2. Wavelengths are indicated around the curved edge.

Together with the U-shaped arc formed by the image of the spectral color map, the boundary of the chromaticity diagram is completed by a straight line joining the two extremes of the spectral color map, the longest-wavelength red and the shortest-wavelength purple-blue. This segment is called the *line of purples*.

5.7.5   The White Point

Since the primary colors of a reconstruction system are arbitrary, there is no absolute requirement that they sum up to white. However, it is intuitive and useful that they should. As long as white is in the gamut of the primaries, there is no obstacle to achieving this; if it is not so already, one simply replaces the primaries P_1, P_2, P_3 by appropriate multiples of themselves (which, as we recall, does not change their chromaticity), so that after this normalization, white is given by $P_1 + P_2 + P_3$.

The corresponding point on the Maxwell plane, marked with a circled dot in Figure 5.15, is called the *white point*; clearly it has chromaticity coordinates $\left(\frac{1}{3}, \frac{1}{3}, \frac{1}{3}\right)$, since it comes from a spectral distribution with tristimulus value $(1, 1, 1)$.

But what is white? Snow is white because its myriad surfaces reflect and diffuse all wavelengths equally, but ultimately its color depends on what it is lit by: even the purest snow will look a different hue under a blue sky than under an overcast one, and light sources lack an intrinsic definition of whiteness (compare Section 5.1.1). Therefore, establishing a color space requires fixing a standard white spectral distribution. The CIE defines the standard white for its RGB color space (and also for the CIE-XYZ space to be discussed next) as the equal-energy white of Figure 5.4(b), also known as Illuminant E. It is that spectral distribution that, by definition, has chromaticity coordinates $\left(\frac{1}{3}, \frac{1}{3}, \frac{1}{3}\right)$.

This is a natural choice, but it should be understood that it is nonetheless an arbitrary one, like the choice of wavelengths for the primaries. For example, the equal-energy spectrum in the frequency domain is arguably just as natural, and it is not the same.[4]

Another role of white is in contrast to gray. No physical system can support arbitrarily large color intensities: its light sensors or emitters would be destroyed. In practice, then, a further choice is made: one scales the primaries—this time, all by the same factor—in such a way that now it is the *maximum intensity white* desired for the system that is given by $P_1 + P_2 + P_3$. This point in the color cone, with tristimulus value $(1, 1, 1)$, is then called simply "white." Other points on the segment joining this point to the origin—the black-white line (see page 114)—are grays of varying darkness, while points on the continuation of this line beyond $(1, 1, 1)$ are assumed to be of no interest.

Note again the arbitrariness of this choice: a pixel on your computer screen that is currently showing "white" will still be showing "white" if you dim the screen slightly, even though it is now emitting exactly the same energy that its "gray" neighbor with tristimulus value $(0.8, 0.8, 0.8)$ was emitting before. White is a relative notion.

---

[4]Since $\lambda \propto 1/\nu$, we have $dE/d\lambda \propto \nu^2 \, dE/d\nu$: the spectral distribution having the same energy for equal wavelength increments has a quadratically decreasing energy density when expressed in the frequency domain.

5.8 The CIE-XYZ Color System

We have seen that the CIE-RGB system does not reconstruct all visible colors; those colors that do not lie in the first quadrant of the chromaticity diagram in Figure 5.15 are not reconstructible. A natural question follows: is there a color system in which all visible colors are reconstructible?

The answer is yes, but at the cost of introducing primary "colors" that are nonphysical. Indeed, consider again Figure 5.15: we can certainly find a triangle on the Maxwell plane big enough to contain all of the chromaticity diagram. We can even do this while insisting that the white point remain at the barycenter of the big triangle. If we now choose the vertices of this triangle as the primary colors of a new color space, the Maxwell plane of the new space will be the same as the old, but the new Maxwell triangle will be the big triangle we chose. In other words, the new Maxwell triangle will contain all of the chromaticity diagram, and all visible colors will be reconstructible from these primaries.

Why must at least one of the new primaries be nonphysical? Because if all three vertices of the new Maxwell triangle were inside the chromaticity diagram, the triangle they form would also be contained in the chromaticity diagram (it being a convex set). So each of these two sets would be contained in the other, meaning they would be equal; but we know that the chromaticity diagram is not a triangle.

At the same time the CIE introduced its RGB system, in 1931, it defined a derived system, called XYZ, in which all visible colors have a representation with positive coordinates. The criteria that led to the selection of the new system's primaries need not concern us; suffice it to say that these primaries, called X, Y, Z, were defined as linear combinations

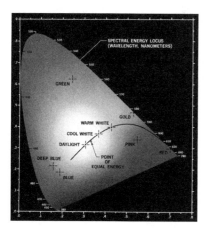
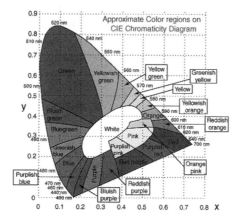

Figure 5.16. Chromaticity diagram in the CIE-XYZ system, showing in black the *Planck curve*, which represents the colors emitted by a black body at each temperature (left). (The colors shown are approximations; many colors in the diagram do not lie in the gamut of a printer or computer screen.) Common color names and the regions they signify (right). (See Color Plate IV.)

of the primaries R, G, B listed in (5.6), so the conversion between the two color spaces is simply a matter of a linear change of coordinates. Specifically, it is given by

$$\begin{pmatrix} x \\ y \\ z \end{pmatrix} = \begin{pmatrix} 0.49 & 0.31 & 0.20 \\ 0.18 & 0.81 & 0.01 \\ 0.00 & 0.01 & 0.99 \end{pmatrix} \begin{pmatrix} r \\ g \\ b \end{pmatrix},$$

where of course (r, g, b) is the tristimulus value in the RGB system and (x, y, z) the one in the XYZ system. Using this change of basis transformation, we can calculate all the colorimetric quantities of CIE-XYZ space from the corresponding ones in CIE-RGB space: color matching functions, chromaticity diagram, and so on. Figure 5.16 shows the chromaticity diagram of the CIE-XYZ system. Notice that all visible colors lie in the first quadrant.

5.9 Dominant Wavelength and Complementary Colors

Using the chromaticity diagram, we can define the *dominant wavelength* of a nonwhite visible color c. To this end, we trace a ray from the white point through the point in the diagram representing c. The point where this ray intersects the curved boundary of the chromaticity diagram (Figure 5.17(a)) marks the unique spectral color that can be combined with white to give c. Its wavelength is called the *dominant wavelength* of c. If the ray intersects the boundary at the line of purples, there is no dominant wavelength.

The notion of complementary colors is defined similarly. Given a spectral color c, consider the line going through c and the white point. If this line intersects the curved boundary of the chromaticity diagram again, the pure spectral color defined by the second intersection is *complementary* to c, and is denoted by c'; an example appears in Figure 5.17(b). The spectral colors c and c' can be combined in appropriate proportions to

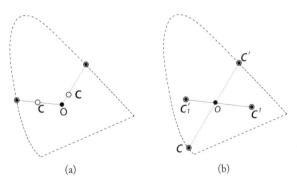

Figure 5.17. Definition of the (a) dominant wavelength and (b) complementary color.

obtain white. More generally, two colors c_1 and c_1' are called complementary if the white point falls inside the segment containing them; this means that white can be obtained as a combination of the two, not necessarily in equal amounts.

From Figure 5.16 we see that the color complementary to the CIE primary green is a reddish purple, known as *magenta*. The color complementary to red is a blue-green called *cyan*. The CIE primary blue, which has a rather short wavelength, has yellow as it complementary color, but the name "blue" more commonly applies to colors with dominant wavelength from 440 to 490 nm, which are complementary to various shades of orange.

5.10 Color Systems and Computer Graphics

We now consider in more detail the applications of colorimetry to computer graphics. We discuss three types of color system of interest: device color systems, color standard systems, and color interface systems.

5.10.1 Device Systems

Color output devices reconstruct colors from their mathematical representation. Color handling in the context of image generation is a complex problem that will be examined in the next chapter; here we just discuss how color systems relate to graphics output devices. Similar considerations apply to color input devices, such as scanners, based on color sampling.

A reconstruction system, as we saw in Section 5.4.2, is defined by a collection of primary colors. If an output device uses three primary colors, its gamut, or set of achievable colors, is a triangle in the chromaticity diagram.

This is the case with the most common type of device color systems, namely, RGB monitor color systems. Although there are many other color device systems in use in computer graphics, we will concentrate on those.

The RGB system for color monitors. Both the older CRT (cathode ray tube) and the more modern LCD (liquid crystal display) screens have red, green, and blue subpixel light emitters that can be made to shine with different intensities. That is, these monitors use particular red, green and blue primary colors, depending on the model; the resulting color space is generically called mRGB ("m" for monitor).

The unit along each primary axis is chosen so that $(1, 1, 1)$ corresponds to the brightest white obtainable, and the coordinates take values in the interval $[0, 1]$. Therefore, the mRGB color solid is a cube. We show it in Figure 5.18, together with the corresponding Maxwell triangle and the complementary colors cyan, magenta, and yellow. As already mentioned, the corner $(0, 0, 0)$ represents black and the corner $(1, 1, 1)$ is the maximum-brightness white in the monitor in question.

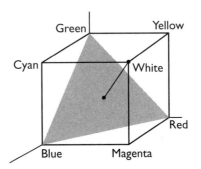

Figure 5.18. The mRGB color cube.

The three primary colors of an mRGB system depend on the particular phosphors or LCD technology used. When plotted on the chromaticity diagram of the CIE-XYZ system, as shown on the left in Figure 5.19, these colors form the vertices of a triangle, which is the image of the color cube on the Maxwell plane—in other words, the gamut of the system.

Different monitor models have different gamuts. On the right in Figure 5.19 we show the gamuts of two different mRGB systems. Only the colors in the hexagonal intersection, shaded light gray, can be represented in the color space of both monitors. It is important to keep this in mind when converting between the color systems of different devices.

A very common problem consists of performing a color transformation between an mRGB space and the CIE-XYZ or CIE-RGB models; this is the subject of Exercise 15.

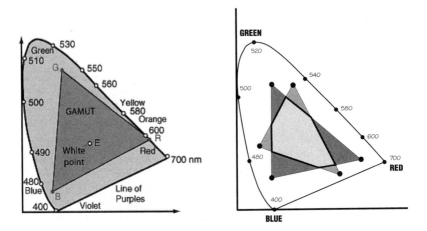

Figure 5.19. Chromaticity triangle of the mRGB space.

Generally, the chromaticity coordinates of the primary colors of the device are provided by the manufacturer.

5.10.2 Standard Color Systems

Standard color systems are independent of any particular physical device, and serve as a common language in contexts such as image storage or conversion between device systems.

Color standards are generally set up or ratified by some normative institution. We have already discussed two CIE standards (RGB and XYZ). There are also many industry-developed standards and others controlled by organizations such as the Society for Motion Picture and Television Engineers (SMPTE) and the National Television System Committee (NTSC).

An example of the latter is the *NTSC chrominance-luminance system* established in 1953 for color television in the United States, also known as the *video component system* Because the system had to be compatible with existing black-and-white TV sets, one of the broadcast signals was chosen to encode the luminance, denoted by Y, while the other two were based on RGB color components, and are called the *chrominance* signal. However, instead of sending two of the RGB components, it was more economical to choose linear combinations of the three components that represented in some sense departures from black-and-whiteness, namely $B-Y$ and $R-Y$, so that a broadcast in black and white simply sets these two signals to zero.

This system was later revised to the so-called SMPTE 170M standard, whose primaries are based on a particular set of phosphors widely used in the video and television industries. The NTSC luminance was empirically determined[5] to be

$$Y = 0.299\,R + 0.587\,G + 0.114\,B, \tag{5.8}$$

where R, G, B are the tristimulus values relative to the standard primaries. We see that the largest contribution to luminance comes from the G component, located in the middle of the visible spectrum, while the R and B components contribute relatively less.

Using Equation (5.8), we see the change of coordinates giving the $(Y, R-Y, B-Y)$ values of a color is

$$\begin{pmatrix} Y \\ R-Y \\ B-Y \end{pmatrix} = \begin{pmatrix} 0.299 & 0.587 & 0.114 \\ 0.711 & -0.587 & -0.114 \\ -0.299 & -0.587 & 0.886 \end{pmatrix} \begin{pmatrix} R \\ G \\ B \end{pmatrix}. \tag{5.9}$$

Figure 5.20 shows the color solid of the (Y, R–Y, B–Y) space, which is the parallelogram obtained as the image under the transformation (5.9) of the unit color cube of RGB space (Figure 5.18).

---

[5] See the discussion about luminous efficiency function in Chapter 18.

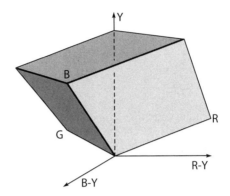

Figure 5.20. Color solid of the (Y, R−Y, B−Y) space.

The chromaticity set of the system can be obtained simply by disregarding the luminance information, that is, by taking the orthogonal projection of the color solid in Figure 5.20 onto the (R–Y, B–Y) plane, as shown on the left in Figure 5.21. The result is the *color hexagon* of video component system, also shown in Figure 5.21.

Hundreds of variations on the basic luminance-chrominance system just described have been proposed, and many are in current use. These systems differ amongst themselves by scale settings or rotations on the chrominance plane. For example, the YIQ system, a television standard, has axes along the blue-orange and green-magenta directions (roughly halfway between the axes as shown on the right in Figure 5.21); the information along the green-magenta axis can be encoded at low resolution without perceptible loss in quality, saving some bandwidth. Other systems, such as that underlying Kodak's PhotoCD image format, have been adapted specifically for use in computer graphics.

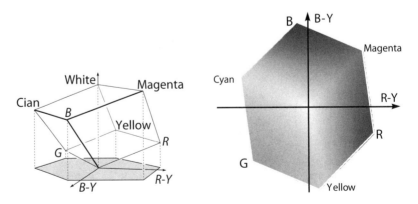

Figure 5.21. Chrominance hexagon of the Y, R–Y, B–Y system. (See Color Plate V.)

5.10.3   Interface Systems

Interface systems are designed to allow users to specify colors easily and intuitively, and constitute a significant contribution of computer graphics to colorimetry. The fundamental example is the HSV system, where H stands for hue, S for saturation and V for value, the latter being defined as the maximum of the R, G, and B components.

In most interface systems, the user specifies separately the brightness (luminance) and the chromaticity of the desired color, typically via a graphical interface.

A color on the boundary in the chromaticity diagram (or of the gamut, depending on context) is called *saturated* or *pure*. Mixing with white—which corresponds to taking a point on the line segment drawn from the white point to the saturated color—reduces the *saturation*. Informally speaking, the saturation indicates how far along a color lies along a ray drawn from the white point through the color.

In most interface systems the chromaticity is specified separately by the saturation and the *hue* (also called the *shade*), which measures the direction in the chromaticity diagram in which the color lies relative to the white point. (Compare with the notion of the dominant wavelength in Section 5.9.)

So to select a color, a user might first select the hue by clicking on the edge of an appropriate diagram; this has the advantage that saturated colors are easier to tell apart from one another than corresponding unsaturated ones. Then the chosen color is mixed with white to reach the desired saturation.

The HSV system was introduced by Alvy Ray Smith in [Smith 78], as a way to specify colors interactively on a monitor, and more specifically in painting software. As we have seen, the V coordinate of a color is chosen as $\max(R, G, B)$; the simplicity of this choice recommended it over other possible choices (such as the luminance) for implementation in the relatively slow computers of the 1970s.

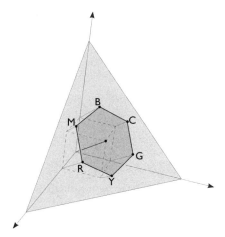

Figure 5.22. Projection of the mRGB cube.

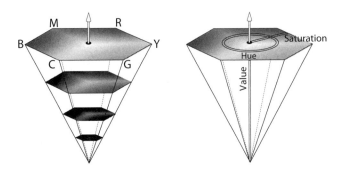

Figure 5.23. Color solid of the HSV system. (See Color Plate VI.)

The set of points in the mRGB color cube that has a constant V, say t, is formed by three sides of a cube of side t (the three sides away from the origin or black point; see the dotted lines in Figure 5.22). The orthogonal projection of these three sides onto the Maxwell plane forms a regular hexagon, each of whose vertices corresponds to a saturation-t version of either a primary (RGB) or complementary (CMY) color. As t increases, so does the size of the hexagon.

The hexagonal projections of all cubes, for all values of t ranging from 0 to 1, can be imagined to be stacked together forming a hexagonal pyramid, as in Figure 5.22. The topmost hexagon corresponds to the sides of the unit cube, whose colors have value 1.

Each hexagonal cross section, parallel to the basis of the pyramid, represents a set of colors of the unit cube with the following characteristics: the hue of each color appears on the edges of the hexagon; the saturation of the color decreases as we approach the center of the hexagon along the radial direction; the value of the colors in each hexagon is constant and is proportional to the distance from the plane of the hexagon to the vertex of the pyramid. The axis of the pyramid, formed by the centers of the hexagons, corresponds to the diagonal of the cube: the black-white line. On the right, Figure 5.23 shows the variation of these parameters in the HSV pyramid of the system.

5.11   Comments and References

The encoding of color in the computer requires the use of a finite number of bits. The discretization of color spaces gives rise to a number of interesting problems in computer graphics. We will take a look at some of them in the next chapter.

This chapter only scratched the surface of colorimetry and color systems. We have not discussed subtractive color systems, which are essential for color reproduction in print (see Exercise 6). Other important topics include pigment systems, departures from Grassmann's Law (perceptual nonlinearity), and nonchromaticity features of color, a vast subject which includes the simulation of painting effects with different types of paints, such as watercolor. A more extensive treatment of color along the conceptual lines of this chapter can be found in [Gomes and Velho 02, Chapters 3 and 4] and in [Gomes and Velho 97].

Exercises

1. Explain exactly how negative values of the color matching function $\bar{r}$ in the CIE-RGB system can be obtained experimentally (compare Section 5.6).

2. Explain the relation between the chromaticity diagram of Figure 5.15 and the functions of Figure 5.12.

3. The following text was extracted from an article in a technical magazine:

> The IHS system is the same as the HSV color system. IHS transformation consists of the spectral rotation of images. The goal is to perform linear transformations that involve the combined processing of several image bands to generate new chan- nels. IHS transforms the original RGB channels into the IHS channels and vice versa. The attributes that permit distinguishing one color from another color are: intensity (I), shade (H) and saturation (S). Intensity is associated with the bright- ness of the point, shade is related to the dominant wavelength and saturation has to do with the purity of the color.

   The IHS system is the same as the HSV color system, and the text quoted attempts to describe the relation between it and another color space. But it is not very clear and even has incorrect statements. Interpret and rewrite the text clearly and precisely.

4. Consider the plane $ax + by + cz = 1$ in a color space $\mathbb{R}^3$.

   (a) What conditions must a, b, c satisfy to ensure that this plane has a representative of every color that can be reconstructed by the system? Interpret this criterion geometrically.

   (b) Define "generalized chromaticity coordinates" of a color $C = (c_1, c_2, c_3)$ using the plane of item (a) instead of the Maxwell plane.

5. Let $A = (a_{ij})$ be a matrix that transforms from a vector $\mathcal{P} = (\mathcal{P}_\infty, \mathcal{P}_\in, \mathcal{P}_\ni)$ of primary colors to another such vector $\mathcal{Q} = (\mathcal{Q}_\infty, \mathcal{Q}_\in, \mathcal{Q}_\ni)$; in symbols, $\mathcal{Q} = \mathcal{A}\mathcal{P}$. Show how A can be used to find the color matching functions of the reconstruction system $\mathcal{Q}$ from those of $\mathcal{P}$.

6. In this chapter, we studied additive color systems, based on the primary colors RGB; color combinations in this system are shown on the left in Figure 5.24. The image on the right illustrates the color combination in another type of system, called *subtractive*.

   (a) Explain a subtractive system in terms of a basis in color space.

   (b) Why are subtractive systems the chosen reconstruction systems for color printers, instead of an RGB system?

   (c) Give other examples of applications of subtractive systems.

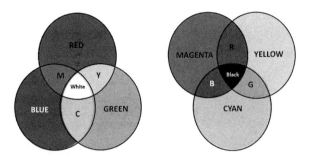

Figure 5.24. Additive and subtractive color systems (Exercise 6; see Color Plate VII).

7. Explain in detail how to obtain the CIE-XYZ chromaticity diagram (Figure 5.16), starting from information on the CIE-RGB system.

8. Show that the chromaticity diagram of a color space generated by a finite number of colors $C_1, C_2, \ldots, C_n$ is the convex hull of these colors on the Maxwell plane.

9. Explain why spectral color space has infinite dimension.

10. Justify in detail why the (visible) color solid is a convex cone in $\mathbb{R}^3$.

11. Suppose the spectral color map for the visual apparatus of the highly evolved animal species *Ludovicus vetustus* is given on the Maxwell plane of some 3D color space by the parameterization $q_1+q_2 = 3t^4 - 2t^2$, $q_1-q_2 = t$, where $t \in [-1, 1]$ parameterizes the visible range of wavelengths for this organism (that is, the range is $[\lambda_0 - \kappa t, \lambda_0 + \kappa t]$ for some λ_0 and κ).

    (a) Draw the spectral color map on the Maxwell plane. (Hint: plot it first in the coordinates $t = q_1 - q_2$ and $u = q_1 + q_2$, then transform from (t, u) to the Maxwell plane coordinates (q_1, q_2) by rotating and scaling.)

    (b) Complete the chromaticity diagram. Do all pure spectral colors lie on its boundary?

    (c) Plot or write formulas for possible color matching functions for the three primary colors.

    (d) Can the three primaries P_1, P_2, P_3 of the color space we are using be the pure spectral colors corresponding to $t = 1, -1, 0$, in this order? How about $t = 0.5, -0.5, 0$? Are there other possibilities for the primary colors?

    (e) From the functions in part (c), deduce possible graphs or possible formulas for the spectral response of the three types of photosensors, called A, B, C, present in the eyes of *L. vetustus*. (Hint: assign specific numbers to the responses of A, B, C to the three primaries. Do you get a physically meaningful answer if the values you choose are $(1, 0.5, 0)$ for A, $(0.5, 1, 0)$ for B, and $(0, 0, 1)$ for C? What if you replace each 0.5 by 0.1?)

12. Figure 5.25 shows the color reconstruction functions $r(\lambda)$, $g(\lambda)$ and $b(\lambda)$ of a system.

    (a) Sketch the chromaticity diagram of the system.

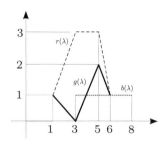

Figure 5.25. Figure for Exercise 12.

(b) Check whether the color of the plane whose (r, g) chromaticity coordinates are given by $(2, 2)$ can be represented in this system.

13. Why does the boundary of the chromaticity diagram include a straight line segment, the line of purples?

14. Consider a monitor whose three primary colors have chromaticity coordinates (0.628, 0.346), (0.268, 0.588) and (0.150, 0.070) in the CIE-XYZ space. Show that the color with chromaticity coordinates (0.274, 0.717) cannot be displayed faithfully on this monitor.

15. Given the chromaticity coordinates of the primary colors of a monitor in CIE-XYZ space, explain how to find the CIE-XYZ chromaticity coordinates of an arbitrary color (r, g, b) in the color cube of the monitor.

16. Consider the color solid of a graphics device and a color that cannot be represented in this device.

    (a) What are some situations that can cause this problem?

    (b) Propose at least two methods for approximating the color with an existing color in the system of this device.

    (c) Discuss the pros and cons of each method proposed.

17. Draw a geometric depiction of the *complementary color* concept using the chromaticity diagram.

18. Describe a method to change the coordinates between the RGB system of the monitor, mRGB, and the HSV system.

19. Two white light sources appear identical. However, when each is used to illuminate the same object, the object's color appears to change. Explain.

6 Image

A digital image is the materialization of a number of processes in computer graphics. Images are present in all areas of computer graphics, either as a final product, as in the case of rendering, or as an essential part of the interaction process, as in the case of modeling. In this chapter we will develop a conceptualization of the digital image, including mathematical models and representation techniques on the computer. Our emphasis will be the conceptual aspects of displaying images, a problem that is directly related to displaying colors.

6.1 Image Abstraction Paradigms

We begin by defining the proper mathematical models to represent and manipulate images on the computer. The four universes paradigm provides an appropriate framework for the image models we will study.

6.1.1 Mathematical Model of an Image

When we look at a photograph or a real scene we receive, from each point in space, a luminous pulse associating color information to that point. A natural mathematical model describing an image is a function that is defined on a 2D surface and includes values in a color space.

A *continuous image*[1] is an application $f \colon U \subset \mathbb{R}^2 \to C$, where C is a color space. In general, we have $C = \mathbb{R}^n$. The function f is called the *image function*. The set U is called the *image support*, and the set $f(U)$ of the values of f, a subset of C, is called the set of *image colors* or *color gamut* of the image. Most frequently, $n = 3$ or $n = 1$. For $n = 3$ we have a space of trichromatic color representation, a space with a basis of primaries R, G, B, and therefore a color image. When $n = 1$, we say the image is *monochrome*.

---

[1] Here *continuous* means *nondiscrete*; it does not take its usual meaning in topology, where the application f is continuous.

A monochrome image can be geometrically visualized as the graph $G(f)$ of the image function f,

$$G(f) = \{(x, y, z); (x, y) \in U \text{ and } z = f(x, y)\},$$

considering the intensity values as the heights $z = f(x, y)$ at each point (x, y) of the domain. This geometric interpretation provides a more intuitive view of certain aspects of the image. In the graph of Figure 1.5 (illustrating a monochrome image and the graph of its function) for instance, it is easy to identify discontinuity regions of the function, which correspond to abrupt variations in the intensity of the image points.

When $f \colon U \to \mathbb{R}^3$ is a color image, we can write $f(x, y) = (f_1(x, y), f_2(x, y), f_3(x, y))$, where $f_i \colon U \to \mathbb{R}$. This way, each f_i is a monochrome image. A color image f is therefore formed by three monochrome images, called the *color components* of f.

6.2   Image Representation

In the representation of an image $f \colon U \to \mathbb{R}$, we should take into account two aspects: the *spatial representation*, the representation of the support set U, and the *color representation*, the representation of the color space f.

6.2.1   Spatial Representation

Uniform sampling is the method most used for space discretization of an image. In this method, we consider the support set of the image as being the rectangle

$$U = [a, b] \times [c, d] = \{(x, y) \in \mathbb{R}^2 \,;\, a \le x \le b \text{ and } c \le y \le d\}$$

and discretize this rectangle using the points of a 2D grid. More precisely, we can assume, without loss of generality, that $a = c = 0$, and the discretization grid P_Δ is the set

$$P_\Delta = \{(x_j, y_k) \in \mathbb{R}^2\},$$

where

$$x_j = j \cdot \Delta x, \quad j = 0, 1, \ldots, m - 1, \ \Delta x = b/m,$$
$$y_k = k \cdot \Delta y, \ k = 0, 1, \ldots, n - 1, \ \Delta y = d/n.$$

This grid is shown in Figure 6.1. Notice that the above reconstruction generalizes itself immediately for $\mathbb{R}^n$.

The grid P_Δ is formed by a set mn of cells

$$c_{jk} = [j\Delta x, (j + 1)\Delta x] \times [k\Delta y, (k + 1)\Delta y],$$

$j = 0, \ldots, m - 1, k = 0, \ldots, n - 1$. The representation of the image function f is reduced to obtaining a color value for the image function f in each of those cells. Two simple and broadly used methods to obtain the representation in the cell are point and area sampling.

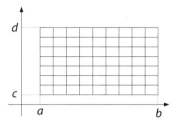

Figure 6.1. Uniform grid of the matrix representation of the image.

In *point sampling*, we choose a point (x_j, y_k) of the cell c_{jk}, and we represent f by its value $f(x_j, y_k)$ at that point. The point (x_j, y_k) can be, for instance, the center of the cell, as shown in Figure 6.2(a) for the unidimensional case.

In *area sampling*, we represent the function in the cell c_{jk} by its average value of f

$$f_{jk} = \frac{1}{\text{area}(C_{jk})} \int_{C_{jk}} f(x, y) dx dy$$

in the cell. Figure 6.2(b) illustrates area sampling for the unidimensional case.

There are advantages and disadvantages to each of these two methods in each cell of the matrix representation. A complete comparative analysis of these two methods requires more advanced mathematical methods and is outside the scope of this book. However, we can affirm intuitively that if the function f has large variations within the cell region, then area sampling is certainly a more reasonable choice for representing the average function variation.

Whether one uses point or area sampling in each cell, the final representation of the image f is given by a matrix A of order $m \times n$, $A = (a_{jk})$, where the value of each element a_{jk} is given by the representation of f in each cell c_{jk}. Each one of the cells c_{jk} is called a *pixel* (an abbreviation of *picture element*). This image representation is called a *matrix representation*.

Each element a_{jk}, $j = 0, \ldots, m - 1$ and $k = 0, \ldots, n - 1$ of the matrix represents the function in the cell c_{jk}. This representation is a vector of the color space, indicating

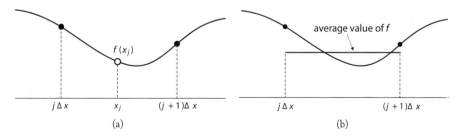

Figure 6.2. (a) Point and (b) area sampling.

the color of the pixel corresponding to cell c_{jk}. If the image is monochrome, (a_{jk}) is a real matrix where each element is a real number representing the luminance value of the pixel.

The number of lines m of the pixel matrix A is called the *vertical resolution* of the image, and the number of columns n is called the *horizontal resolution*. We call the product $m \times n$ of the vertical by the horizontal resolution the *spatial resolution*, or *geometric resolution*, of the matrix representation.

Each element a_{jk} of the matrix representation is generically called a *sample* of the image. It is common to consider a_{jk} as representing the value of f at the pixel with coordinates (j, k). We know conceptually that this interpretation might not always be correct because the image may not have been represented using point sampling. But if one is interested only in the discretized image, this interpretation can be appropriate. Meanwhile, we should stress that if we need to reconstruct image f, it is necessary to know the lengths Δx and Δy. In other words, the spatial resolution mn, given in absolute terms, does not provide much information about the actual resolution of the image after it is reconstructed. An interesting case happens in graphics devices where the values of Δx and Δy appear in the *resolution density* of the devices. This density provides the number of pixels per linear unit of measure, generally given as pixels per inch (ppi), also called dots per inch (dpi). In some graphics devices the resolution density governs the pixel dimensions in the reproduction of the image.

In Figure 6.3, we show an image in two different spatial resolutions. Notice that to maintain the dimensions of the low resolution image equal to the dimensions of the other image, we increased the size of the cell at each pixel.

The problem of knowing the values of Δx and Δy is directly connected to the problem of the scale used in the image representation. In fact, taking $\Delta x = \Delta y = 1$ is equivalent to considering the samples a_{jk} of each pixel as the values of the pixel in the coordinates (j, k) of the pixel. This assumption can always be made in problems whose solution does not depend on the scale used; for instance, the solution, after solving the problem for f, can be applied to the function $g(x) = sf(\frac{x}{s})$, obtained from f by a dilation by the

Figure 6.3. Different spatial resolutions of an image. (*Original photo from Kodak Photo CD ©Eastman Kodak Company.*)

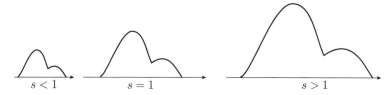

Figure 6.4. Dilation of a function.

factor $s > 0$ (see Figure 6.4). In the case of the matrix representation, a change of scale provides a corresponding change in the cell dimensions $\Delta x \Delta y$ of the representation (i.e., sampling) grid.

6.2.2 Color Representation

The color space is represented by the space $\mathbb{R}^3$; therefore, the problem of color representation is the problem of representing real numbers. The question is, how many bits should we use to represent color? The number of bits is called the *color resolution* of the image. This process of discretizing the color space of an image is called *quantization*.

From a computational point of view, color discretization is directly related to the problem of discretizing the space $\mathbb{R}^3$. We have the option of using floating-point arithmetic with either 32 or 64 bits. However, from an image point of view, the issue is more delicate, requiring us to consider issues of human perception. For example, the gradation of intensities shown in Figure 6.5 has only 256 intensity levels (eight bits), yet most viewers see a continuous gradient rather than distinct levels. We will study the problem of color quantization further on in Section 6.5.

Figure 6.5. Gradation with 256 intensity levels.

6.3 Matrix Representation and Reconstruction

The reconstruction of the image function, starting from its matrix representation $f(x_j, y_k)$, is an interpolation problem. Given a grid P_Δ in $\mathbb{R}^n$, with cells C_m, $m = 0, 1, \ldots, k - 1$, let us take a point $p_k \in C_k$ in each of the cells C_k. An n-dimensional *reconstruction kernel* associated to P_Δ is a function $\phi \colon \mathbb{R}^n \to \mathbb{R}$ such that $\phi(0) = 1$ and the family of functions $\{\phi(x - p_i); i = 0, \ldots, k - 1\}$ is linearly independent. If f_j are samples of a matrix representation of function f in grid P_Δ, a *reconstruction* of f using kernel ϕ is given by

$$f_r = \sum_j f_j \phi(x - p_j). \tag{6.1}$$

Notice that the kernel is translated to the position p_j of each sample, it is multiplied by the value of the sample, and then the results are added. Equation (6.1) defines the function f_r as a linear combination in the basis $\phi(x - p_j)$. Observe from (6.1) that $f_r(p_j) = f_j$; therefore, the point sampling of the reconstruction function f_r has the same matrix representation of function f.

There are numerous possibilities for interpolating the samples of f in the matrix representation. An ideal interpolation method would obtain the exact reconstruction of the original image function. Finding this ideal method is related to the problem described in Section 6.2.1. The three reconstruction kernels most broadly used in image processing are constant, triangular, and cubic kernels.

In the unidimensional case, the *constant kernel* (also called *box*, or *Haar kernel*) is given by the function

$$h_0(x) = \begin{cases} 1 & \text{if } 0.5 \leq x \leq 0.5 \\ 0 & \text{otherwise.} \end{cases} \tag{6.2}$$

The graph of h_0 is shown in Figure 6.6(a). The *triangular kernel* is defined by

$$h_1(x) = \begin{cases} 1 - |x| & \text{if } |x| \leq 1 \\ 0 & \text{otherwise.} \end{cases} \tag{6.3}$$

The graph of h_1 is shown in Figure 6.6(b). The *cubic kernel* is defined by

$$h_3(x) = \begin{cases} 1 - 2|x|^2 + |x|^3 & \text{if } 0 \leq |x| \leq 1 \\ 4 - 8|x| + 5|x|^2 - |x|^3 & \text{if } 1 \leq |x| \leq 2 \\ 0 & \text{otherwise.} \end{cases} \tag{6.4}$$

The graph of h_3 is shown in Figure 6.6(c).

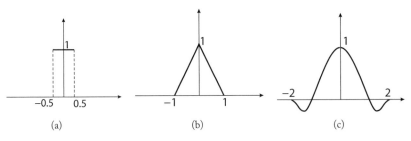

Figure 6.6. Reconstruction kernels: (a) constant, (b) triangular, and (c) cubic.

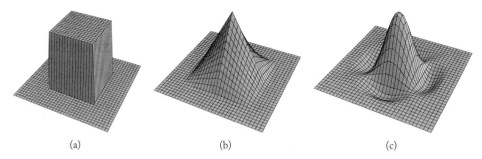

(a) (b) (c)

Figure 6.7. 2D reconstruction kernels corresponding to the (a) constant, (b) triangular, and (c) cubic kernels. (*Courtesy of Moacyr Silva.*)

Notice that the reconstruction kernels h_0, h_1, and h_3 are piecewise polynomial functions, formed by polynomials of degrees 0, 1, and 3, respectively.

We can construct 2D kernels $h_i^2(x, y)$ from those unidimensional reconstruction kernels by placing $h_i^2(x, y) = h_i(x)h_i(y)$. The 2D kernel $h_1^2(x, y)$ obtained from the triangular kernel is called the *Bartlett kernel*. Figure 6.7 shows the graph of the 2D kernels derived from the constant, triangular, and cubic kernels.

We will now analyze the reconstruction process of each of these kernels. For this, without loss of generality, we will assume that the function is sampled in an integer grid. That is, in the unidimensional case, we have $f_j = f(j)$, $j \in \mathbb{N}$, and in the case of images we have $f_{j,k} = f(j, k)$, $j, k \in \mathbb{N}$.

6.3.1 Reconstruction with the Constant Kernel

Because of the procedure used, reconstruction with the constant kernel is known as *nearest neighbor* reconstruction. From Equation (6.1), the reconstruction function of the constant kernel is given by

$$f_r = \sum_j f_j h_0(x - j).$$

We will analyze f_r in the interval $[j - 1, j]$. From the above equation we have

$$f_r(x) = f_{j-1}h_0(x - j + 1) + f_j h_0(x - j).$$

Taking into account the definition of kernel h_0 in (6.2), we have

$$f_r(x) = \begin{cases} f_{j-1} & \text{if } j - 1 < x \leq \frac{2j-1}{2} \\ f_j & \text{if } \frac{2j-1}{2} < x < j, \end{cases}$$

in other words, at each point $x \in [j - 1, j]$, the value of $f_r(x)$ is given by the value of the sample f_{j-1}, or f_j in the extremes of the closest interval to point x. Figure 6.8(a) shows the reconstruction process starting from three samples.

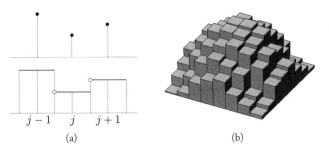

$$j-1 \quad j \quad j+1$$

(a)   (b)

Figure 6.8. Reconstruction with (a) unidimensional and (b) 2D constant kernels.

The reconstruction function is a piecewise constant function (step or staircase function). Figure 6.8(b) illustrates the reconstruction in the 2D case. Notice that the reconstruction process with the constant kernel introduces several nonexistent discontinuities in the original image.

6.3.2 Reconstruction with the Triangular Kernel

Reconstruction with the triangular kernel is called *reconstruction by linear interpolation*. From the reconstruction Equation (6.1), we have the reconstruction function given by

$$f_r(x) = \sum_j f_j h_1(x - j).$$

We will analyze f_r in the interval $[j-1, j]$. Taking into account the definition of kernel h_1 in (6.3), we have

$$
\begin{aligned}
f_r(x) &= f_{j-1} h_1(x - (j-1)) + f_j h_1(x - j) \\
&= f_{j-1}(1 - |x - (j-1)|) + f_j(1 - |x - j|) \\
&= f_{j-1}(1 - (x - j + 1))) + f_j(1 - (j - x)).
\end{aligned}
$$

Changing variables $s = x - j + 1$, where $j - x = 1 - s$, from the above equation, it follows that

$$f_r(s) = (1 - s)f_{j-1} + s f_j.$$

Therefore, $f_r(s)$, $0 \leq s \leq 1$ is obtained by the linear interpolation of the samples f_{j-1} and f_j in the extremes of interval $[j-1, j]$. The process is illustrated in Figure 6.9(a). Notice that the reconstruction function is continuous and therefore does not introduce discontinuities as reconstruction with the constant kernel does.

Reconstruction with the Bartlett kernel. The Bartlett kernel $h_1^2(x, y)$ is given by $h_1^2(x, y) = h_1(x)h_1(y)$. Following up from Equation (6.1), the reconstruction function is given by

$$f_r(x, y) = \sum_{j,k} f_{j,k} h_1^2(x - j, y - k).$$

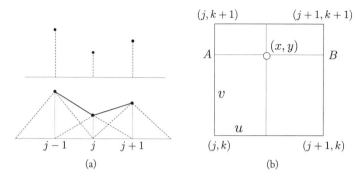

(a)   (b)

Figure 6.9. Reconstruction with (a) unidimensional and (b) 2D triangular kernels.

We will analyze f_r in the cell $[j, j + 1] \times [k, k + 1]$ of the matrix representation (see Figure 6.9(b)). In this cell, the function $f_r(x, y)$ is given by

$$f_r(x, y) = f_{j,k}h_1^2(x - j, y - k) + f_{j+1,k}h_1^2(x - j - 1, y - k)$$
$$+ f_{j+1,k+1}h_1^2(x - j - 1, y - k - 1) + f_{j,k+1}h_1^2(x - j, y - k - 1).$$

Taking into account the definition of h_1^2, and changing variables $u = x - j$ and $v = y - k$, we obtain

$$f_r(u, v) = (1 - u)(1 - v)f_{j,k} + (1 - u)vf_{j,k+1} + u(1 - v)f_{j+1,k} + uvf_{j+1,k+1}.$$

This equation can still be rewritten in the form

$$f_r(u, v) = (1 - u)[(1 - v)f_{j,k} + vf_{j,k+1}] + u[(1 - v)f_{j+1,k} + vf_{j+1,k+1}]. \qquad (6.5)$$

Equation (6.5) has important geometric implications (see Figure 6.9(b)): the term $(1 - v)f_{j,k} + vf_{j,k+1}$ in Equations (6.5) linearly interpolates the samples $f_{j,k}$ and $f_{j,k+1}$, obtaining the value of the reconstruction function f_r at point A with coordinates

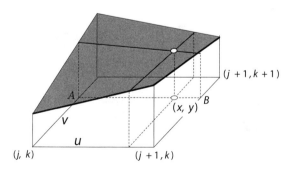

Figure 6.10. Bilinear interpolation.

$(j, (1-v)k + v(k+1))$. Similarly, the term $(1-v)f_{j+1,k} + vf_{j+1,k+1}$ in Equation (6.5) linearly interpolates samples $f_{j+1,k}$ and $f_{j+1,k+1}$, obtaining the value of the reconstruction function f_r at the point B with coordinates $(j+1, (1-v)k + v(k+1))$. Finally, Equation (6.5) obtains the value of the reconstruction function at point (x, y) of the cell by interpolating the values at the points A and B: $f_r(x, y) = (1-u)f_r(A) + uf_r(B)$.

Reconstruction with the Bartlett kernel is called *reconstruction by bilinear interpolation*. Figure 6.10 illustrates the reconstruction process in space.

Examples and final considerations. We have observed that the three interpolation kernels h_0, h_1, and h_3 are piecewise polynomials functions, with degrees 0, 1, and 3, respectively. As we increase the degree, we obtain better properties from those kernels: h_0 is not continuous; h_1 is continuous; and the reader can verify that kernel h_2 is of class C^2.

Choosing the most appropriate reconstruction kernel is a difficult task. Images have discontinuity points forming *edges*, which are the points at the boundary of the objects in the image. Those edges have great importance in the process of image perception by the human eye. In reality, in the process of primary vision, the eye processes the edges of the image. This fact led David Marr (1945–1980), a British neuroscientist and psychologist pioneer in the mathematical study of vision, to formulate what is now known as *Marr's conjecture*: an image is determined uniquely by its edge points.

Edges have a major influence on the image as a whole, and there are many challenges to accurately reconstructing them. On the one hand, the discontinuities of a reconstruction kernel introduce new edges in the reconstructed image, which would be quickly noticed

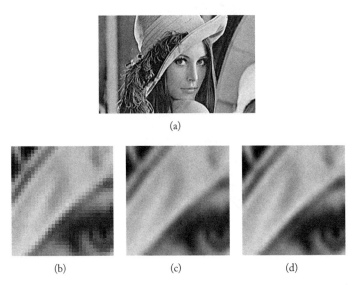

(a)

(b)   (c)   (d)

Figure 6.11. Reconstruction of (a) Lena's photograph using (b) constant, (c) Bartlett (bilinear), and (d) cubic kernels.

and interpreted by the human eye as new information about the image. On the other hand, when we use a smoother reconstruction kernel (class C^k with $k \geq 2$, for instance) we take the risk of destroying the edges in the reconstruction. In this case, the reconstructed image is seen as a little blurry. A better method is reconstruction with the cubic interpolation kernel. Although it is class C^2, its two lateral lobes, where the cubic interpolation kernel assumes negative values, help prevent the reconstructed image from becoming too soft.

The images in Figure 6.11 show a detail of the eye from Lena's image.[2] This detail was reconstructed to a larger scale using the constant, Bartlett, and cubic kernels. The higher reconstruction quality of the cubic kernel is clear.

6.4   Elements of a Digital Image

A *digital image* is an image $f: U \rightarrow \mathbb{R}^3$, where the support U and the color space are both discretized. This image is characterized by three factors: spatial resolution (number of pixels), number of color components, and color resolution.

As we previously saw, the *gamut* of f is the finite set of colors $f(U)$; f is a *binary image* when its gamut has only two colors. An image $f: U \rightarrow \mathbb{R}^n$ has *continuous support* when f can be calculated at any point of U. The color space of f is continuous when its color is represented using floating-point arithmetic (in simple or double precision).

From the various methods and techniques for image processing and manipulation on the computer, we can idealize an image in four different representations: continuous-continuous, continuous-quantized, discrete-continuous, or discrete-quantized. In practice, the continuous-continuous image serves as a concept used in the development of the mathematical methods for image processing. The discrete-quantized image is the representation used by several graphics devices. The discrete-continuous image is a convenient representation for a large portion of image operations because the image function assumes floating point values which, although represented by a finite number of bits, provide a good approximation for real numbers in calculations involving color. An image of the type discrete-quantized is what we call a *digital image*.

6.4.1   Frequency Histogram

We can create an interesting mathematical model for an image f by considering it as being a randomly defined variable in its representation grid (this 2D random variable is called a *random field*). This means that we have a probability distribution associated to the occurrence of the various colors at each pixel. The parameters of this distribution vary from one pixel to another, and we do not have a priori knowledge of this distribution. Several applications involve researching methods for estimating this distribution.

An approximation for this distribution, assuming cases that do not depend on pixel positions, is given by the frequency histogram (familiar from statistics). In the case of an

---

[2]Lena was a model photographed for *Playboy* magazine in 1972. Her picture was scanned and has been used as a test image by the image processing community.

 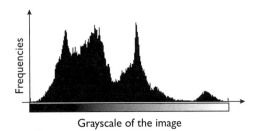

Figure 6.12. Frequency histogram of an image. (*Original photo from Kodak Photo CD ©Eastman Kodak Company.*)

image we have a *color histogram* associating, for each color intensity c in the image, its occurrence frequency, or the number of pixels in the image with color c.

Figure 6.12 shows an intensity histogram of an image of a house with 256 intensity levels (eight bits). This histogram indicates the image does not have a good balance of intensity levels: rather than having predominantly average-intensity regions, there are large regions with high intensity values and many regions with low intensity values.

For color images, we can compute the histogram of each of the color components, or instead compute a 3D histogram. A graphics representation of this histogram can be obtained by associating to each color $c = (c_x, c_y, c_z)$ a sphere holding the given color, whose radius is proportional to the number of occurrences of the color in the image.

6.5   Color and Image Quantization

We previously saw that the color discretization process is known as *quantization*. This process converts an image with a continuous set of colors into an image with a discrete set of colors. In this section, we will review several of the current issues in color discretization.

First, we must define the concept of quantization. Consider a finite subset $R_k = \{p_1, p_2, \ldots, p_k\}$ of $\mathbb{R}^n$. We call *quantization* of k levels, a surjective transformation $q: \mathbb{R}^n \to R_k$. If $k = 2^m$, the set R_k can be encoded on the computer using m bits. In this case, it is common to affirm that the transformation q is a quantization of m bits. Each element p_i of the set R_k is called a *quantization level*. The set R_k is called the *codebook* of the quantization transformation.

It is very common to have a quantization transformation between finite subsets of $\mathbb{R}^n$. In other words, given a subset $R_j = \{q_1, \ldots, q_j\}$, with $j > k$, we have a quantization transformation $q: R_j \to R_k$. If $j = 2^n$ and $k = 2^m$, we have a quantization of n for m bits.

The quantization of a digital image consists of quantizing the color gamut of the image, implying in the color quantization information of each image pixel. More precisely, if $f: U \to \mathbb{R}^3$ is a discrete-continuous or discrete-discrete image, the result of the quantization of $f(x, y)$ is a discrete-discrete image

$$f' : U \to R_k, \quad \text{such that} \quad f'(x, y) = q(f(x, y)),$$

where q is the quantization transformation. Consequently, the quantization changes the color resolution of the image.

There are two main purposes for quantizing an image: displaying and compressing. To display an image in some graphics device, the color gamut of the image cannot be larger than the available colors in the physical color space of the equipment. In this case, the quantization space R_k is directly linked to the color space of the graphics display device.

Quantization is also useful because the quantization of an image reduces the number of bits used to store its color gamut, which reduces the required storage space for the image as well as the volume of data. This compression is helpful if, for example, the image must be transmitted through some communication channel.

6.5.1 Cells and Quantization Levels

Let us consider a quantization transformation $q \colon \mathbb{R}^n \to R_k$. Each quantization level $p_i \in R_k$, corresponds to a subset of colors $C_i \subset \mathbb{R}^n$, determining the quantized colors for a color p_i; that is

$$C_i = q^{-1}(p_i) = \{c \in C \,;\, q(c) = p_i\}.$$

The finite family of sets C_i, constitutes a partition of the color space $\mathbb{R}^n$; that is, $C_i \cap C_j \neq \varnothing$ if $i \neq j$. Each one of the partition sets C_i is called a *quantization cell*. For each of these cells, the quantization function assumes a constant value given by the quantization level p_i.

If c is a color in a cell c_i, we define the *quantization error* of color c, e_q, as $e_q = |c - q(c)| = |c - p_i|$. The smaller the cell diameter, the less the quantization error will be.

Unidimensional and multidimensional quantization. Let us consider the case for $n = 1$, known as *unidimensional quantization*. Let q_i, $1 \leq i \leq L$ be the quantization levels assumed by the quantization transformation q. In this case, the quantization cells are the intervals

$$c_{i-1} < c \leq c_i, \ 1 \leq i \leq L.$$

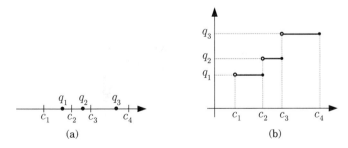

(a)   (b)

Figure 6.13. Quantization levels and graph of the quantization function.

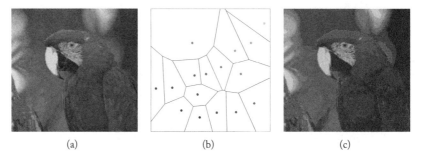

(a)   (b)   (c)

Figure 6.14. 2D quantization. (*Original photo from Kodak Photo CD ©Eastman Kodak Company.* See Color Plate VIII.)

In Figure 6.13(a), we show three quantization intervals $(c_{i-1}, c_i]$, and in each interval we have the associated quantization level q_i. In Figure 6.13(b), we show the graph of the quantization function q, which is constant at each quantization interval.

In the unidimensional quantization, the quantization cell is always an interval and we can vary only its length. In *multidimensional quantization*, this is, $n \geq 2$, the quantization cells are regions of the color space that represent a more complex geometry. In Figure 6.14(a), we show an image whose color space is 2D (a space with components RG). In Figure 6.14(b), we show the cell subdivision of the color space of the image in 16 colors. In Figure 6.14(c), we show the quantized image.

Scalar and vector quantization. Consider a unidimensional quantization $q_1 \colon \mathbb{R} \to R_k$, and let us define a quantization $q \colon \mathbb{R}^n \to R_k \times \cdots \times R_k$, placing $q(x_1, \ldots, x_n) = (q_1(x_1), \ldots, q_1(x_n))$. That is, we quantize each component of the vector $(x_1, \ldots, x_n)$ separately. This quantization is called *scalar quantization*. When the multidimensional quantization is not computed component by component, it is called *vector quantization*.

6.5.2 Two-Level Quantization

Consider the quantization of a grayscale image represented by 8 bits (256 gray levels) for 1 bit. In this case, we have two quantization levels that we will indicate by 0 and 1 (black

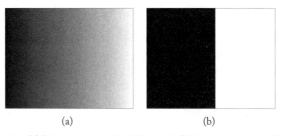

(a)   (b)

Figure 6.15. (a) Linear ramp with 8 bits, and (b) its quantization for 1 bit.

Figure 6.16. Periodic pattern of a dithering method.

and white). A simple method of quantization consists of taking the threshold $L = 128$ (the middle value of the image intensities) and quantizing by 1 all intensities $c \geq 128$ and by 0 all intensities $c < 128$. In this case, the quantization of the linear *ramp* displayed in Figure 6.15(a) is given by the image shown in (b).

Now consider Figure 6.16. The gray level in Figure 6.16(a) is above 50%; therefore its quantization by the previous method should result in a completely black image. Of course, the image shown in Figure 6.16(b) is not solid black: it was obtained by processing the quantized image, instead of just quantizing the color gamut of the image. Observe this image by moving your eyes away from the book or with your eyes half-opened; you will not see a great difference between the continuous tones of the two images.

As Figure 6.16 demonstrates, better quantization results can be obtained by taking into account the spatial color distribution in the image. This fact is crucial when quantizing for a very small number of bits. In Section 6.9 we will discuss the dithering technique, which is a filtering, used together with quantization, aimed at obtaining better spatial distribution of the colors in the image domain.

6.5.3   Perception and Quantization

A central problem of quantization is *quantization boundaries*, or *quantization contours*, in which the boundaries between quantization levels become visible.

Consider a monochrome image function $f \colon U \to C$ whose color space is quantized in L levels. This quantization determines a partition of the image domain U in subsets (quantization cells) C_i defined by

$$C_i = f^{-1}(c_i) = \{(x, y) \in U \, ; \, f(x, y) = c_i\}.$$

In other words, each subset C_i of the partition is constituted by the image pixels, whose intensity assumes the quantization level c_i. If we have a small number of quantization levels, we will have a small number of regions C_i in the partition. If the image function behaves well (e.g., class C^1), the boundary unbundling those regions will be defined by a regular curve. Depending on the value difference between the quantization levels of two neighboring cells, the curve of the boundary will be perceptible to the human eye.

(a)   (b)

(c)   (d)

Figure 6.17. Quantization outline. (*Original photo from Kodak Photo CD ©Eastman Kodak Company.*)

Figure 6.17(a)–(d) illustrates this problem. We have 256, 18, 8, and 2 quantization levels in images (a), (b), (c), and (d), respectively. As we reduce the number of quantization levels, the quantization boundary becomes more perceptible. Much of the research in quantization methods aims to obtain a quantization in which the boundary is not perceptible.

6.6 Quantization and Cell Geometry

Quantizing an image begins with dividing the image into cells and then choosing a quantization level within each cell. How can we set up the distribution quantization cells? There are two approaches: uniform and nonuniform.

6.6.1 Uniform Quantization

One option for determining quantization cells is to divide the color space into congruent cells and, in each cell, take its center as the cell's quantization level. This method is called *uniform quantization*. In the case of scalar quantization with L levels, the quantization cells are the intervals $(c_{i-1}, c_i]$ of equal length; that is, $c_i - c_{i-1} = $ constant, and in each cell the quantization value is given by the average

$$q_i = \frac{c_i + c_{i-1}}{2}, \quad 1 \leq i \leq L. \tag{6.6}$$

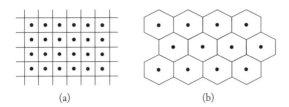

Figure 6.18. Cells of 2D uniform quantization.

Figure 6.18(a) shows a scalar quantization of $\mathbb{R}^2$, starting from uniform quantization in each of the coordinate axes. Figure 6.18(b) shows another geometry of 2D uniform vector quantization cells.

6.6.2 Nonuniform Quantization

Although uniform quantization is easy to obtain, it is not necessarily the best option. Usually color values in an image are not evenly distributed: in each region of the image certain colors occur with higher frequency than other colors. If we subdivide each region into a larger number of quantization cells, we will be reducing the size of the cells, and therefore quantization error will decrease for the image elements containing those colors. Essentially, we will minimize the difference between the original and the quantized image.

This method of partitioning the color space into incongruent cells is called a *nonuniform quantization* and is said to be *adaptive* when the geometry of the cells is chosen according to the specific characteristics of the color distribution in the image.

Figure 6.19 shows an image we will use to compare different quantization methods. This Figure is quantized in 24 bits (8 bits per R, G, and B channel) and does not present perceptible quantization contours. For the sake of comparison, Figure 6.19(b) and (c) show uniform quantization of the image (a) using 8 and 4 bits. Notice that the quantization contours are perceptible in those two images.

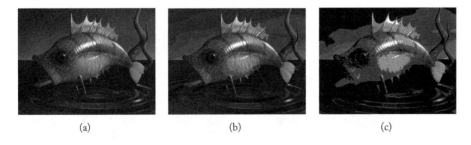

Figure 6.19. (a) Color image with 24 bits. (b) Uniform quantization of (a) with 8 bits. (c) Uniform quantization of (a) with 4 bits. (See Color Plate IX.)

6.7   Adaptive Quantization Methods

In this section we will examine several methods of adaptive quantization. They can be divided into three distinct categories depending on whether we begin by determining levels or cells.

In *direct selection methods* the quantization levels are determined first and the corresponding cells are then calculated by mapping each color in space to the closest level:

$$q(c) = c'_i \quad \Longleftrightarrow \quad d(c, c'_i) \leq d(c, c'_j),$$

for every $1 \leq j \leq N$, with $j \neq i$. Here, d represents a metric in the color space $\mathbb{R}^n$ (the Euclidean metric, for instance).

Spatial subdivision methods instead start by determining the quantization cells C_i, $i = 1, \ldots, N$; then the quantization transformation is determined by choosing a quantization level c'_i in each cell C_i. In addition to these two approaches, *hybrid methods* simultaneously determine the cells and quantization levels in an interdependent way.

6.7.1   Quantization by Direct Selection

An example of the quantization by direct selection method is the *populosity algorithm*.

Populosity algorithm. This method initially constructs the frequency histogram of the image and then chooses, for the K quantization levels, the K colors appearing more frequently in the image gamut (i.e., the most populous colors). The quantization function can be defined by taking, for each color c of the image gamut, $q(c)$ as being the closest quantization level by using, for instance, the square of the Euclidean metric. Of course if there is more than one quantization level conforming to the minimality condition, then we should make a decision about the value of $q(c)$. A possible solution is to randomly choose one of the possible levels. A better method would take into account the quantization values of the neighboring pixels.

The problem with the populosity algorithm is that it totally ignores colors in low density regions of the color space. A highlight in one image can disappear completely in this

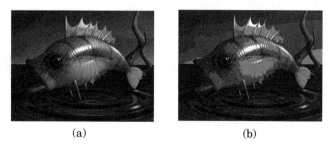

(a)   (b)

Figure 6.20. Populosity algorithm: quantization with (a) 8 and (b) 4 bits. (See Color Plate X.)

quantization process since it only occupies a small number of image pixels. But the algorithm can be used with satisfactory results for images with a uniform color distribution.

Figure 6.20(a) shows a reproduction of Figure 6.19 quantized to 8 bits with the populosity algorithm. Figure 6.20(b) shows the same image quantized to 4 bits with the same algorithm. Compare these results to the uniform quantization of this image in Figure 6.19(b) and (b).

6.7.2 Quantization by Spatial Subdivision

Quantization by spatial subdivision, in which the quantization cells are first determined and then the quantization level in each cell calculated, is illustrated here by the median cut algorithm, which uses a recursive subdivision process of the color space. This is a simple and effective method.

Histogram equalization and the median cut algorithm. We can quantize an image by choosing quantization levels using the same number of pixels. This quantization transformation performs a histogram equalization of the image; in other words, it replaces the original histogram with a histogram corresponding to a uniform distribution of the pixel intensities, as illustrated in Figure 6.21.

The histogram is equalized according to the *median*. Given a finite and sorted set of points in space

$$C = \{c_1 \leq c_2 \leq \cdots \leq c_{n-1} \leq c_n\},$$

the *median* m_C of this set is defined by the middle element $c_{(n+1)/2}$ if n is odd, and by the average of the two intermediate elements if n is even. The median is a statistical localization measure that divides the given set C in two parts with an equal number of elements. Observe that, unlike an average, the median is not influenced by the magnitude of the elements of the set. It is important that the median calculation take into consideration the frequency of each element c_i of the set C: the construction of a frequency histogram associated to the dataset is an important stage in the calculation of the median.

For monochrome images, the quantization process by histogram equalization consists of performing successive subdivisions of the interval of image intensities using the median of the intensity set at each subdivision.

For color images, the extension of the quantization algorithm by histogram equalization described above is known as the *median cut algorithm*. Simply put, the method consists

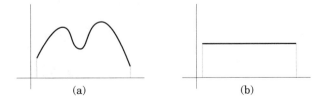

(a)   (b)

Figure 6.21. Histograms: (a) original; (b) equalized.

of using the quantization algorithm by histogram equalization in each of the color components of the image gamut. Because of its ease of implementation, computational efficiency, and good perceptual results from quantizing images from 24 to 8 bits, this is one of the most popular quantization algorithms in the computer graphics community.

We now give a detailed description of the median cut algorithm for the RGB color cube. We use K to denote the number of desired quantization levels. Let us take a parallelepiped

$$V = [r_0, r_1] \times [g_0, g_1] \times [b_0, b_1]$$

of minimum volume containing all the colors in the gamut of the image to be quantized. Next, we take the color space component, whose direction the parallelepiped V has the edge with largest length. We will assume it to be the green component g. We then sort the colors of the image gamut by the g component, and we calculate the median m_g of the color set based on this sorting. We therefore divide region V in two subregions:

$$V_1 = \{(r, g, b) \in C; g \leq m_g\}, \quad \text{and} \quad V_2 = \{(r, g, b) \in C; g \geq m_g\}.$$

We then apply the same subdivision method to each of the regions V_1 and V_2. We continue the subdivision process recursively until one of the two following conditions is satisfied: the two subregions V_1 and V_2 obtained do not contain colors of the image gamut or the desired number K of quantization cells is already obtained.

After subdividing the color space by the desired number of cells, we determine the quantization level of each cell. To obtain the quantization value of one pixel of the image, we locate the cell containing the color of that pixel and perform the quantization for the corresponding level of that cell. The algorithm can be efficiently implemented using an appropriate spatial data structure in the process of recursive subdivision of the color space (a natural structure is a kd-tree).

When quantization is performed by a subdivision of the color space based on the median color value in each region, the result is an image that has approximately the same number of pixels with the value of each corresponding quantization cell. This process is equivalent to a histogram equalization of the image.

(a)   (b)

Figure 6.22. Median cut algorithm: quantization with (a) 256 and (b) 16 colors. (See Color Plate XI.)

Figure 6.22(a) shows a reproduction of the image in Figure 6.19, quantized with 8 bits by the median cut algorithm. Figure 6.22(b) is a reproduction of a quantization in 4 bits by the median cut algorithm. Compare these results to those in Figures 6.19 and 6.20.

6.8   Optimization and Quantization

Hybrid methods of quantization aim to optimally partition the image. Given a fixed number of desired quantization levels, there are a many possible determinations of quantization cells and of quantization levels in each cell.

6.8.1   Quantization Error

To choose an optimal determination, we must first precisely define "optimal" by defining an objective function that should be minimized. If q is the quantization transformation and c a color to be quantized, then

$$c = q(c) + e_q, \tag{6.7}$$

where the vector e_q is a measure of the error introduced by the quantization process. The quantization error is measured by the distance $d(c, q(c))$ between the original and the quantized colors, c and $q(c)$, respectively.

Several metrics d in the color space can be chosen, aiming at measuring the quality of the quantization process. The choice of these metrics should take into account both computational efficiency and color perception. In reality, it is enough to have a function measuring the proximity between two colors, which is known as *distance function*. A widely used distance function is the square of the Euclidean distance; that is, $d(c_1, c_2) = \langle c_2 - c_1, c_2 - c_1 \rangle$, where $\langle \, , \rangle$ is an inner product in the color space.

One of the perceptual factors affecting quantization error is the color occurrence frequency in the image: if a color occurs with a high frequency, its quantization in the image will be more easily noticed than the quantization of a color that has low frequency. The equation below takes frequency into account to measure the error, while quantizing a color set in a region R of an image for a color c:

$$E = \int_R p(c) d(c, q(c)) dc, \tag{6.8}$$

where p is the occurrence probability of color c in the color space of the image. The use of the above equation for measuring the distortion introduced by the quantization is quite intuitive: on measuring the error, we should take into consideration the occurrence probability of color c in the color space to be quantized; the quantization error is then multiplied by the probability, resulting in a weighted average. As we already pointed out, in general, we do know in advance the probability distribution. It is common to use Equation (6.8), replacing the probability by the frequency of the color in the color histogram of the image.

If we want a quantization in N levels, we will have a partition of the color space in N cells $K_1, K_2, \ldots, K_N$. Indicating by q_j the quantization level of cell K_j, and applying Equation 6.8 to each cell, we have

$$E = \sum_{1 \leq j \leq N} \int_{K_j} p(c) d(c, q_j) dc. \tag{6.9}$$

When the color space to be quantized is finite, each cell is a cluster of colors. In this case, Equation (6.9) can be written in the form

$$E = \sum_{1 \leq j \leq N} \sum_{c \in K_j} p(c) d(c, q_j). \tag{6.10}$$

Notice that E depends on the space partition in cells K_j, and on the quantization levels q_j in each cell.

The quantization problem should ideally therefore be solved by minimizing the value of E, given by Equation (6.10), on every possible partition with N elements of the color space and on every possible choice of quantization level. This problem of combinatorial optimization is known as *cluster analysis*. The large variety of partitions with N elements of the color space makes this problem intractable from a computational point of view. This way, in general, the optimization methods used for solving the quantization problem use some type of heuristic. These heuristic methods solve only particular cases of the problem or find a solution without guaranteeing it to be an optimal one.

6.8.2 Color Cluster Quantization

In this section, we will examine a color quantization method using optimization techniques. As we previously saw, in practice we have a finite set of M colors and want to represent it using N colors, with $M > N$. In this context, a quantization cell has a finite number of colors and is called a *cluster* of colors. The theorem below provides the optimal quantization level of a cluster.

Theorem 6.1. *Let $R_M = \{c_1, \ldots, c_M\}$ be a cluster of M colors in $\mathbb{R}^n$ belonging to the image gamut. The optimal quantization level c of R_M is the color*

$$c = \frac{1}{\sum_i f_i} \sum_{j=1}^{M} f_j c_j, \tag{6.11}$$

where f_i is the frequency of color c_i in the image. The quantization error in the cluster is given by

$$E(R_M) = \frac{1}{(\sum_k f_k)^2} \sum_{j=1}^{M} f_j \| \sum_{i=1}^{M} f_i (c_i - c_j) \|^2. \tag{6.12}$$

Proof: Let us take the square of the Euclidean metric to define the distance function in the color space $d(c, c_j) = ||c - c_j||^2$. As we have only one cell, $j = 1$ in Equation (6.10). The quantization error in cell R_M depends only on the choice of the quantization level c. By placing the probability distribution for the frequency histogram in Equation (6.10), the quantization error in the cluster R_M is given by

$$E_M(c) = \sum_{c \in R_M} f_j d(c_j, c) = \sum_{j=1}^{M} f_j ||c_j - c||^2. \tag{6.13}$$

The gradient of this error function can be easily calculated, resulting in

$$\text{grad}(E_M)(c) = \sum_{j=1}^{M} 2 f_j (c - c_j).$$

Setting the equation equal to 0 and solving it, we obtain the critical point c given in Equation (6.11). As E_M is a convex function, c is, in reality, a minimum point. Replacing the value of c in Equation (6.11) in function E_M given in Equation (6.13) and performing the calculations, we obtain the quantization error given in Equation (6.12). This concludes the demonstration of the theorem. ❏

An interesting particular case is the error obtained for a two-color cluster.

Corollary 6.2. If $R_2 = \{c_i, c_j\}$ is a cluster of two colors in the image gamut, the optimal quantization level is given by

$$c = \frac{f_i}{f_i + f_j} c_i + \frac{f_j}{f_i + f_j} c_j. \tag{6.14}$$

The quantization error is given by

$$E(c_i, c_j) = \frac{f_i f_j}{f_i + f_j} ||c_i - c_j||^2. \tag{6.15}$$

❏

A geometric interpretation of the corollary provides a good visualization of the result. In fact, from Equation (6.11), in the case of two colors c_i and c_j, the quantization error is given by

$$E_2(c) = f_i(c - c_i)^2 + f_j(c - c_j)^2.$$

The graph of this function is shown in Figure 6.23, which shows a parabolic arc obtained as the sum of the two parabola arcs $f_i(c - c_i)^2$ and $f_j(c - c_j)^2$, shown in the figure as dashed lines. The quantization level is given by the minimum point of the parabola.

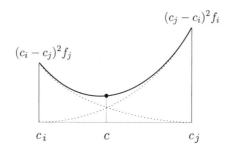

Figure 6.23. Quantization error in a two-color cluster.

Notice when two colors occur with the same frequency, that is $f_i = f_j$, then from Equation (6.14), we have

$$c = \frac{c_i + c_j}{2};$$

in other words, c is the median point of the segment $\overline{c_i c_j}$ in the color space. In the unidimensional case, the color set is naturally sorted and a quantization cell is an interval of the straight line. An optimal solution for the cluster analysis problem can be found. We left the deduction of this solution as an exercise (10).

6.8.3   Optimized Quantization by Binary Clustering

In this section we describe a quantization algorithm that obtains an optimal solution by a process of successive approximations over two-color clusters: at each stage of the process, two colors are replaced by an optimal quantization level given by the Corollary of Theorem 6.1.

The input to the algorithm is the color gamut of the image $C = \{c_1, \ldots, c_M\}$. Each color c_i has frequency f_i, and to each color c_i we associate an accumulated quantization error $E(c_i)$, initially assuming value 0. The quantization is obtained by the following procedure:

1. Calculate the histogram of the image.

2. Use Equation (6.15) to calculate the quantization error $E(c_i, c_j)$ between all pairs of colors $\{c_i, c_j\}$ in the image gamut C.

3. Choose the binary cluster $R_2^0 = \{c_i, c_j\}$, minimizing the quantization error $E(c_i, c_j)$ calculated in Step 2.

4. Use Equation (6.14) to calculate the quantization level c_{ij} of cluster $R_2^0 = \{c_i, c_j\}$ chosen in Step 3.

5. Replace, in the image gamut C, the cluster $R_2^0 = \{c_i, c_j\}$ with its quantization level c_{ij}. This results in a new image gamut C' with $M - 1$ colors. The frequency of

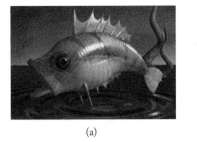

(a)   (b)

Figure 6.24. Optimized quantization for binary clustering. (See Color Plate XII.)

color c_{ij} is given by the sum $f(c_{ij}) = f_i + f_j$ of the frequencies of the two quantized colors. The accumulated quantization error of color c_{ij} is given by the sum

$$E(c_{ij}) = E(c_i, c_j) + E(c_i) + E(c_j).$$

6. Use the new quantized gamut C' as input for Step 2 of the algorithm, and repeat the Steps 2–6 until the desired number of colors in the image gamut is obtained.

The above process gives us the quantization levels. From them we can calculate the quantization cells as previously described.

Notice that the spatial correlation of the color in the image gamut is lost after the process of grouping colors in pairs in the quantization. This correlation loss can be minimized if we append a final step to the algorithm: after the calculation of the quantization cells, recalculate the quantization level in each cell using Equation (6.11).

Figure 6.24(a) and (b) shows a quantization of the image of a fish[3] for 256 and 16 colors, respectively. Compare the image of the fish in Figure 6.24 with the images of the quantized fish for the same number of colors using the median cut algorithm (Figure 6.22). We can clearly see the superiority of the binary clustering algorithm. However, it is more computationally expensive.

6.9 Dithering

There are special situations in which quantization triggers accentuated tonal discontinuities in an image, making it difficult to avoid the perception of quantization contours, even when using good quantization algorithms. This is the case for the two-level quantization, which is required for displaying images in bitmap-based output graphics devices. This problem is relevant because there are many output graphics devices within that category and, despite their limitations, we want to display monochrome images in those devices while keeping the halftone information. Examples of such devices include laser and inkjet printers.

---

[3]"Fish out of Water" by Mike Miller, created in PovRay.

The two-level quantization is based on a characteristic of human vision: the eye integrates the luminous stimuli received within a certain solid angle. Consequently, we may perceive intensities that do not necessarily exist in an image. Those colors result from the integration process given by the average of the color intensities within the neighborhoods of each image element contained in the solid angle. Therefore, for a certain resolution, our interest is in the average intensity within small regions of the image: what matters is the average tonal value in a region and not at a pixel.

Given a region $R_k(i,j)$, $i,j \in \mathbb{Z}$ in the domain of an image, the average intensity I_m of the image in this region is defined by the weighted average

$$I_m = \frac{1}{|R_k|} \sum_i \sum_j f(i,j), \qquad (6.16)$$

where $|R_k|$ indicates the number of pixels in the region. We can redistribute the values $f(i,j)$ in this region given that the average I_m remains equal.

6.9.1 Resolution and Perception

Physically, perceptual resolution is measured by the *visual sharpness*, which is the ability of the human eye to detect details when observing a scene. The eye has viewing angles of $150°$ and $120°$ in the horizontal and vertical directions, respectively. However, the eye does not distinguish separate details for an angle smaller than 1 minute, that is, $1/60°$. This angle is called the *visual sharpness angle*. This limitation depends on the wavelength of the visible light, on the geometry of the eye's optic system, and, most importantly, on the dimension and distribution of the eye's photosensory cells. This is because the eye distinguishes two objects when the light emitted by them touches different cells.

The perception of image details depends on three parameters: the distance of the image to the eye, the resolution density, and the eye opening space.

The distance of the image to the eye is measured along the optic axis. As we increase the distance from the image to the eye, we notice fewer image details.

Resolution density measures the relation between the pixel area and the distance to adjacent pixels. The smaller the pixel area and the distance between the adjacent pixels, the less the eye is capable of distinguishing between two neighboring pixels. The relation between the pixel area and the distance between pixels is expressed by the *resolution density* of the image, indicating the number of pixels per linear unit in the image in both vertical and horizontal directions. This way, the larger the resolution density, the smaller the possibility the eye has of distinguishing neighboring pixels.

The eye opening space also affects our ability to perceive details: observing an image with eyes a bit closed further decreases the eye's field of view, resulting in an increase of the visual sharpness angle and decreasing our perception of image details.

When displaying a digital image, we should look for the means of increasing its perceptual resolution while preventing the eye from noticing image artifacts introduced by the color quantization or by its spatial discretization. We can manipulate the three parameters

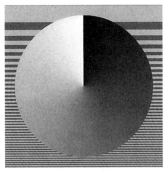

Figure 6.25. Half-tone images for dithering comparison.

given above in order to increase perceptual resolution: we can display the image at a greater distance, display it in a device with larger resolution density, or invite viewers to squint.

For over a century, the printing industry (e.g., newspapers, magazines) has been using analog methods to perceptually produce the halftone for an image reproduction process holding only two tones. For this reason, these methods are known among professionals as *digital halftone algorithms* and more commonly as *dithering* (meaning hesitation). Some dithering algorithms can implement digital versions of the analog process of halftone generation.

Test images. To demonstrate several dithering methods, we will use the images in Figure 6.25. The face is an illustration by Candido Portinari[4] rendered with chalk, and it was selected for presenting subtle halftone variations in the face as well as detailed information (high frequencies) in several hair threads. The other image is a synthetic one generated in a way to contain high frequency information with a variable level of detail, together with soft tone gradations.

You should not notice any difference between the existing gray tones in Figure 6.25 and the ones in a photograph—both images were reproduced using dithering methods.

6.9.2 Quantization for Two Colors

As we saw previously, a quantization transformation $q\colon \mathbb{R}^n \to R_k$ is defined in a color space $\mathbb{R}^n$, taking values in a subset R_k of that space. If $f\colon U \subset \mathbb{R}^2 \to \mathbb{R}^n$ is an image, its quantization is obtained by performing the composed operation $q \circ f\colon U \subset \mathbb{R}^2 \to R_k$. This method for quantizing an image in two stages is not appropriate for some types of quantization. Consider, for instance, the quantization for two colors (0 and 1) of an image with a constant gray tone. In this case, the quantization transformation has two cells and therefore, for the above method, every pixel in the image will be quantized for either 0 or 1,

---

[4]We thank the Portinari Project for granting permission to use this image, digitized from a drawing by Brazilian artist Cândido Portinari (1938). Courtesy of João Cândido Portinari.

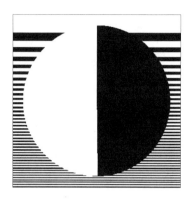

Figure 6.26. Quantization with constant threshold.

resulting in an image either totally white or totally black—obviously an unacceptable result in perceptual terms. In this section we will study quantization operations acting directly over the color distribution in the image support.

Quantization with constant threshold. Let us analyze again the problem of two-color quantization. The simplest method of quantizing an image f consists of establishing an intensity threshold L_0 and using the following rule: if $f(x, y) \geq L_0$, then quantize $f(x, y)$ for 1; otherwise, quantize $f(x, y)$ for 0.

Figure 6.26 shows the test images of Figure 6.25 quantized for one bit, using the constant threshold method above with a threshold of 50%. This method results in highly visible quantization boundaries, making it an undesirable method.

Quantization with random threshold. A naive improvement of the quantization with constant threshold method is to add a random value to the threshold L_0 before quantizing

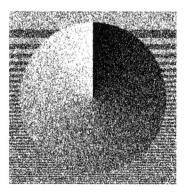

Figure 6.27. Dithering by random threshold.

each pixel. In this case, the quantization boundary from the previous method disappears. Figure 6.27 shows the images of Figure 6.25 quantized in two levels with a threshold of 50% with a random perturbation of the threshold. The quantization contours, which were extremely visible in Figure 6.26, are no longer present in Figure 6.27.

This algorithm is known as *quantization by random threshold*. In the quantization of each pixel, the random perturbation of the intensity threshold separates a pixel's intensity from the intensity of its neighbors. The quantization outline thus becomes a nonconnected curve, hindering the boundary perception separating two different quantization regions. However, the random perturbation also destroys high frequency information of the image (boundary information of the objects in the image) and furthermore introduces a lot of noise, as is apparent in Figure 6.27.

Quantization with a threshold function. Rather than use either of the extremes described above—using a fixed threshold or randomly perturbing the threshold—we recommend varying the quantization threshold in a deterministic way. To do so, we introduce a threshold function. We define a function $L \colon U \subset \mathbb{R}^2 \to \mathbb{R}$ in the same domain as image f and we use the following quantization algorithm: if $f(x, y) \geq L(x, y)$, then quantize $f(x, y)$ for 1; otherwise, quantize $f(x, y)$ for 0. Function L, is called a *threshold function*.

We can consider the quantization method with threshold function L to be a operation L in the image space associating, to each image f, the quantized image $\tilde{f} = L(f)$.

We previously showed that a two-color quantization method can be defined by a threshold function. Conversely, if $f \colon U \subset \mathbb{R}^2 \to \mathbb{R}$ is an image and $\tilde{f}$ is a quantization of f for two tones (0 and 1), then it is easy to see that a threshold function L exists such that $L(f) = \tilde{f}$ (however, L is not unique). This way, a two-color quantization operation is completely characterized by a threshold function.

6.9.3   Classification of Dithering Methods

Dithering methods can be classified in two ways: according to the arrangement of the cells and according to the type threshold function used.

Clustered and dispersed dithering. We say a dithering method is *clustered* if a partition $U = \bigcup_{i=1}^{M} U_i$ of the image domain U exists in connected regions U_i so that, by taking the average intensity c_i of each region U_i, the set

$$\overline{U}_i = \{(x, y) \in U_i \; ; \; L(x, y) \geq c_i\}$$

is a simple subset of U_i (that is, a connected set without holes). Geometrically, this means that the pixels in U_i that are quantized for the value 1 become clustered in each set U_i of the partition. This fact is illustrated in Figure 6.28, which shows the threshold function in the continuous domain, its discretization, and the distribution of the values 0 and 1 for a 50% tone. Clustered-dot methods give good results in devices that do not reproduce isolated points well, such as laser printers and high resolution photocomposites.

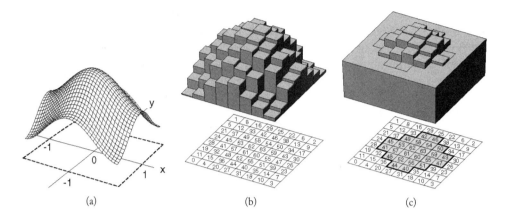

Figure 6.28. (a) Threshold function, (b) its discretization, and (c) the distribution of the values 0 and 1 for a 50% tone [Ostromoukhov and Hersch 95].

Non-clustered methods are called *dispersed dithering*. The quantization algorithm by random modulation previously studied is an example of a nonperiodic dispersed-dot dithering technique. Dispersed-dot methods are more appropriate to graphics devices that allow precise control of the pixel locations in the image, as in the case of video monitors.

Periodic and nonperiodic dithering. If the threshold function is periodic we say the dithering method is periodic; otherwise it is nonperiodic.

If a dithering method is periodic, there exist constants P_x and P_y, which are the periods along directions x and y, such that

$$L(x + P_x, y + P_y) = L(x, y).$$

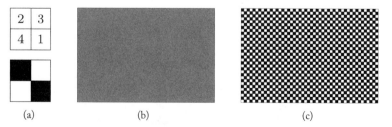

Figure 6.29. Periodic pattern of a dithering method. (a) The cell thresholds of a periodic dithering with period 2. (b) The pattern of that cell for an intensity value of 50%. (c) The quantization of the image in (b) using that dithering cell.

Geometrically, this means that the image is divided into blocks of order $P_x \times P_y$ and, in each block, the threshold function assumes the same values. Therefore, it is enough to define the threshold function in one block only. This block is called the *cell* or *dithering matrix*.

If c_i is the average of intensities at the dithering cell and f is an image with constant tone equal to c_i, then the dithering of f consists of replicating the pattern defined by the dithering cell (see Figure 6.29).

Using a quantization with a constant threshold function, we obtain a completely white or completely black image. The above result is a better compromise. If you hold the book at a distance or partially close your eyes, you will see that Figure 6.29(c) appears as a gray tone rather than as a pattern of black and white points.

6.10 Dithering Algorithms

In the following sections we will give examples of the periodic clustered, periodic dispersed, and nonperiodic dispersed dithering. These three classes of algorithms are illustrated in Figure 6.30. Notice there is also a fourth option: nonperiodic clustered-dot dithering. A first dithering method with these characteristics was first introduced by [Velho and Gomes 91], but a description of this dithering method is outside the scope of this book (for more information, see [Gomes and Velho 02] or [Gomes and Velho 97]).

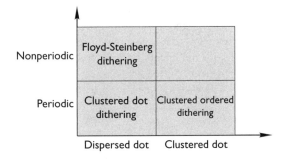

Figure 6.30. Classes of dithering algorithms.

6.10.1 Periodic Clustered Dithering

Periodic clustered dithering algorithms use a linear threshold periodic function responsible for the clustering (see Figure 6.28). Different choices of the threshold function obtain different pixel clustering geometries. Figure 6.31 shows examples of elliptic and circular clusters.

Notice that the algorithm tries a computer simulation of the traditional photographic method of obtaining halftone, which has broadly used by the graphic arts industry since

Figure 6.31. Different cluster geometries.

the early 1850s. Analog halftone is created with a special photographic camera. The image is re-photographed on high-contrast film, over which is superimposed a screen with a grid. This way the light intensity originating from the photograph is modulated by the grid before touching the film. Each small grid slot acts as a lens, focusing at a point the light originating from the image. The image luminance in each region determines the dimension of the point: areas of bright luminance produce small points; areas with average, gray luminance produce points with average dimension, and dark, low-luminance areas of the image produce large points, which generally bleed into each other. The film geometry depends on several factors, such as the kind of screen used and the exposure time.

The classic and broadly used example of periodic clustered dithering is ordered dithering. In Figure 6.32(a) we show an example of an ordered clustered-pixel dithering cell. This cell has order 6. We therefore have 36 different levels from the gray threshold defined by the matrix, which gives a total of 37 intensity levels in the cell. The numerical values of the lines and columns of the matrix indicate the order of the threshold values, but not the absolute gray value in the image. Before applying the ordered dithering algorithm, we normalize the gray intensities for the image in the interval $[0, 35]$, resulting in the average gray intensity of 17.5. In Figure 6.32(b) the image pixels with an intensity above the average 17.5 are shaded while the pixels with intensity below this average are shown in white. We create the quantized image by repeating this clustering pattern periodically across the image.

35	30	18	22	31	36
29	15	10	17	21	32
14	9	5	6	16	20
13	4	1	2	11	19
28	8	3	7	24	25
34	27	12	23	26	33

(a)

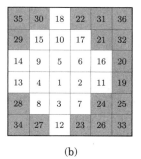

(b)

Figure 6.32. Ordered, clustered dithering cell.

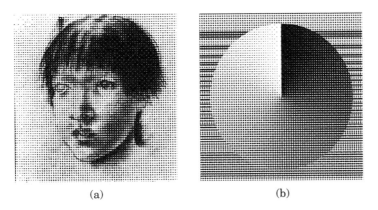

Figure 6.33. Clustered-pixel ordered dithering.

If we have an image whose gray intensity is constant but above the average, the bright region of Figure 6.32 will be reduced. If the gray intensity of the image is below the average, the opposite happens: the bright region of Figure 6.32(b) increases its area. For an image with a variable gray tone range, the pattern periodically repeats itself; however, the area with bright region varies according to the luminance of the image in each region. The final result is a cluster of pixels varying across the quantized image, sometimes increasing, other times reducing the clustering area. Figure 6.33 shows the images of Figure 6.25 quantized for 1 bit and processed with the ordered dithering method defined by the cell of Figure 6.32.

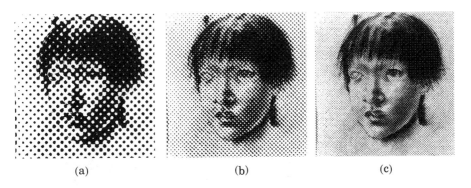

Figure 6.34. Variation of the cell size in ordered dithering: (a) 5 lpi, (b) 10 lpi, (c) 20 lpi. (The resolution of the original image was 75 dpi. These images are printed with a screen frequency of 150 lpi.)

Perceptually speaking, what matters with this algorithm is the clustering dimensions, not the image resolution of the display device. The clustering dimensions are determined by the order of the dithering cell, which in this context is also called the *dithering cell*, or *halftone cell*. The algorithm distributes the gray levels of the pixels from the original image through the pixel clustering in the halftone cell, meaning there is an exchange of tone by spatial resolution.

Therefore, the resolution that most matters to the best perceptual quality to the quantized image is the density of the dithering cells, not the pixel density of the device. (Of course, the pixel density indirectly affects the output as it imposes constraints on pixel clustering resolutions.) The density of dithering cells is called *screen frequency* and is measured in *lines per inch* (lpi). Screen frequencies with good image quality are usually between 120 and 150 lpi.

One of the secrets to getting good results with the ordered dithering algorithm is to achieve a good balance between the appropriate screen frequency and the dimension of each cell. In Figure 6.34 we show the image printed with different screen frequencies.

6.10.2 Periodic Dispersed Dithering

While clustered-dot ordered dithering aims at simulating the film obtained from the traditional halftone process, dispersed dithering tries to distribute the quantization thresholds in the cell to generate a texture whose frequency distribution is equal to the existing texture in the middle tones of the image. Dispersed-dot ordered dithering is also known as *Bayer dithering* after Dr. Bryce E. Bayer (of Eastman Kodak), who created a family of dithering cells that minimize the existing frequencies in texture patterns produced by the algorithm in constant intensity regions in the image.

The texture pattern obtained from dispersed ordered dithering is constructed by distributing the quantization thresholds as evenly as possible in the dithering cell. As with clustered-dot ordered dithering, the dithering cell is the crucial point of the algorithm.

The Bayer dithering cell of order 2 is given by

2	3
4	1

The intensity distribution using this dithering cell is shown in Figure 6.35. This way, in a region with constant average intensity ($= 1.5$), the texture consists of repeating the pattern defined by Figure 6.35(c), resulting in a checkerboard pattern.

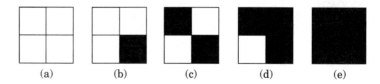

(a)   (b)   (c)   (d)   (e)

Figure 6.35. Intensity distribution with Bayer dithering of order 2.

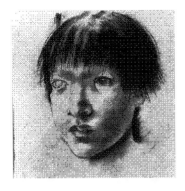 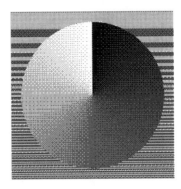

Figure 6.36. Bayer dithering of order 4.

The Bayer cell of order 4 is given by

2	16	3	13
10	6	11	7
4	14	1	15
12	8	9	5

Notice that the distribution of each grouping of four successive levels $1, 2, 3, 4; 5, 6, 7, 8$; etc., uses the same distribution combinatorial position as the four levels $1, 2, 3, 4$ in the dithering cell of order 2. This repetition is due to the recursive nature of the algorithm used to generate Bayer cells of any order.

Figure 6.36 shows the images in Figure 6.25 quantized for one bit and processed with a Bayer dithering.

The dispersed ordered dithering algorithm is useful for displaying images in devices with good precision for pixel positioning. This was the preferred method for displaying images in monitors at the time when graphics boards did not have enough color resolution.

6.10.3 Non-Periodic Dispersed Dithering

Several dithering methods produce nonperiodic patterns. The classic example, while of bad quality, is the dithering algorithm by random modulation previously studied. Compared to the dispersed ordered dithering algorithm, nonperiodic dispersed dithering has the same

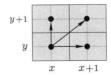

Figure 6.37. Error propagation in the Floyd-Steinberg algorithm.

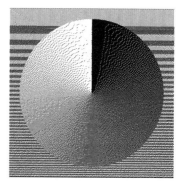

Figure 6.38. Dithering by the Floyd-Steinberg algorithm.

computational cost but much worse perceptual performance. For this reason, the dithering by random modulation algorithm is used only in historical and academic contexts.

Another method, also already classic and quite popular, is the Floyd-Steinberg algorithm [Floyd and Steinberg 75]. This algorithm calculates the effective error introduced in the quantization of each element and distributes it to its neighbors. In this way, global error tends to be minimized. For each pixel $I(x, y)$, the quantization error at that pixel is distributed among the pixel coordinates

$$(x + 1, y), (x, y + 1), \text{ and } (x + 1, y + 1), \tag{6.17}$$

with weights of 3/8 for pixels $(x + 1, y)$, $(x, y + 1)$, and of 2/8 for pixel $(x + 1, y + 1)$ (see Figure 6.37). In Figure 6.38, we show the two test images of Figure 6.25 quantized for 1 bit and filtered with the Floyd-Steinberg algorithm.

The problem with the Floyd-Steinberg method is that it uses a very simple error propagation strategy, biased toward the north, east, and northeast directions. This results in a certain directionality during the pixel dispersion and is clearly visible in Figure 6.38. There are several generalizations for this algorithm that seek greater uniformity in the pixel distribution.

6.11   Quantization and Dithering

Dithering techniques, besides being used for quantizing two-level (1 bit) images, can also be used to avoid or minimize the perception of quantization contours. The image in Figure 6.39(a) is quantized without dithering from 24 to 8 bits. In Figure 6.39(b) we show the same image quantized by the same number of bits, using instead the Floyd-Steinberg dithering algorithm. Notice that the use of dithering almost entirely eliminates the quantization contours in the quantized image.

(a) Without dithering						(b) With dithering

Figure 6.39. Quantizing from 24 to 8 bits. (See Color Plate XIII.)

6.12   Image Coding

So far we have studied the mathematical models and the representation methods of images according to the four universes paradigm. A summary of our study is given by the first two levels of the diagram shown in Figure 6.40.

Coding is the third level of an image abstraction, as presented in the model in the beginning of this chapter. In codification, the discrete representation of the image is quantized and the resulting digital image is transformed into a set of symbols organized according

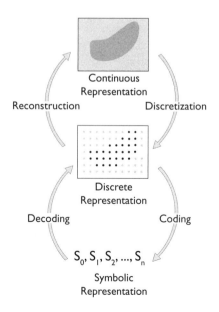

Figure 6.40. Abstraction levels in the representation of an image.

to a data structure. For example, consider the monochrome image of geometric resolution $m \times n$ and quantized in 256 color levels (8 bits). One possible codification of this image consists of using the matrix directly obtained, starting from the representation matrix previously introduced in this chapter. In this representation, the image could be encoded by a descriptor containing the geometric resolution of the image (order $m \times n$ of the representation matrix); by the number of pixel components (1 component for mono images), and the number of quantization bits per component (8 bits); or by a list with $m \times n$ elements of the pixel matrix.

Even for a very simple coding scheme such as this one, the corresponding data structure should allow for the stored information to be correctly retrieved. Details, such as matrix elements ordering by line or by column, should be stipulated in the specification of the representation.

There are several coding methods aimed at obtaining a compact image code. They are directly associated with several image compression techniques and image formats (e.g., TIFF, GIF, JPEG), but a study of these methods is outside the scope of this book.

6.13   Comments and References

This chapter has explored the challenge of displaying images. A classic work for the computer graphics community in the quantization area is [Heckbert 82]. This work introduced the median cut algorithm, which is certainly the algorithm most used by the computer graphics community due to its ease of implementation, perceptual quality, and computational efficiency for quantizing color images from 24 to 8 bits (see Section 6.7.2).

The image quantization algorithm using optimization presented in Section 6.8.2 was published in [Velho et al. 97]. In this work the reader finds a pseudocode of the algorithm, additional examples, and further comparisons with other methods.

The ordered dithering matrices described in Section 6.10 were obtained from [Ulichney 87]. In that book, a recursive algorithm is used for generating the dispersed-dot dithering matrices, based on planar subdivision. The matrices obtained with the algorithm coincide with the Bayer matrices.

6.13.1   Additional Topics

The study of images is extensive, and there were many topics we could not include in this chapter. One important problem is image compression, including sequence of images (digital video). This area of compression requires the use of techniques in the frequency domain, which leads to the study of Fourier and wavelets transforms. Image and video compression formats, such as JPEG and MPEG are, respectively, part of this area of study as well. (MPEG format is also used to compress films in DVD.) Students interested on dynamic systems should research fractal image coding. The study of multi-resolution image representation, in particular using wavelets, is also very important.

Additional topics that can be linked to this chapter include nonperiodic clustered dithering methods; blue dithering algorithms used by inkjet printers; artistic dithering effects, such as pen-and-ink; and image warping and morphing methods, which are broadly used in morphing special effects (see [Gomes et al. 98]). One can also study image analysis, which is very important to computational vision and includes topics such as linear and nonlinear filters and edge detection.

Exercises

1. Discuss possible methods of image representation besides point and area sampling. (Hint: use the reconstruction kernels studied in this chapter.)

2. Describe a method for representing and reconstructing an image using linear interpolation with barycentric coordinates.

3. Every introductory course in numerical analysis covers an interpolation topic beginning with Lagrange polynomials.

    (a) Define Lagrange polynomials and describe their associated interpolation method.

    (b) What is the importance of Lagrange polynomials?

    (c) How do you extend interpolation with Lagrange polynomials for dimension 2?

    (d) Discuss the disadvantages of using Lagrange polynomials to reconstruct images, taking into account both perceptual and computational aspects.

4. Consider the problem of scaling an image by changing its dimensions from $w \times h$ to $sw \times sh$, $s > 0$. If $s > 1$ or $s < 1$, we have either a amplification or a reduction of the dimensions, respectively. Analyze the difficulties of solving this problem for discrete images and describe a methodology to solve it.

5. Give an argument justifying why the Euclidean metric does not have good perceptual properties.

6. Consider a 2D set of 9 different colors, shown in Figure 6.41(a), and assume those colors constitute the image gamut whose frequencies are supplied by the table in Figure 6.41(b). Show that the quantization of this set by the median cut algorithm in four levels q_1, q_2, q_3, q_4, is given by Figure 6.41(c).

7. Taking into account the RGB luminance equation, covered in Chapter 5, define a metric in the color space having better perceptual qualities than the Euclidean metric.

8. If three colors c_1, c_2, c_3 are the vertices of an equilateral triangle, and these colors are present in an image with the same frequency, show that the optimal quantization level for them is given by the barycenter of the triangle. Generalize this result for dimension 3. How can this result be used to improve the quantization algorithm by binary clustering studied in this chapter?

9. Show that the cells of a scalar quantization are parallelepipeds in the $\mathbb{R}^3$ color space.

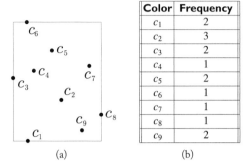

Color	Frequency
c_1	2
c_2	3
c_3	2
c_4	1
c_5	2
c_6	1
c_7	1
c_8	1
c_9	2

c (color)	$q(c)$
c_1, c_2	q_1
c_3, c_4, c_6	q_2
c_5, c_7	q_3
c_8, c_9	q_4

(a)   (b)   (c)

Figure 6.41. Figure for Exercise 6.

10. The problem of unidimensional quantization consists of quantizing the real function of a variable $f : [a, b] \to \mathbb{R}$ defined within an interval [a, b].

   (a) Define the concept of uniform quantization of f and illustrate it using the graph of the function.

   (b) Assuming that the color probability distribution in f is uniform, show that the optimal quantization is given by the uniform quantization of f and that the quantization level is the median point of the quantization interval.

   (Hint: use the quantization error given in Equation (6.10) to perform the calculations.)

11. Why would one use polynomial functions or piecewise polynomials for reconstruction? Discuss.

12. Describe a method to create a monochrome (grayscale) image from a RGB image.

13. The Adobe *Photoshop* program includes the three image reconstruction options discussed in this chapter. Under what circumstances does the program need to reconstruct images? Use the program to experiment with some images. Use images with different resolutions, and, using the three reconstruction methods, analyze the image reconstruction with the *zoom* tool of the program.

7 | Planar Graphics Objects

Graphics objects are at the heart of computer graphics: the whole purpose of the field is to synthesize, process, and analyze graphics objects. In this chapter we introduce the concept of a graphics object and study the specific group of planar graphics objects that includes curves and regions of the Euclidean plane.

7.1 Graphics Objects

We know that computer graphics transforms geometric models into images. From the point of view of the four universes paradigm, we characterize the elements manipulated by computer graphics processes as being elements of the mathematical universe and call such elements *graphics objects*:

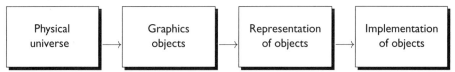

From the mathematical point of view, a *graphics object* is a subset $S \subset \mathbb{R}^m$ and a function $f\colon S \subset \mathbb{R}^m \to \mathbb{R}^n$. The set S is called the *geometric support* and f is called the *attribute function* of the graphics object. The dimension of the geometric support S of a graphics object is called its *dimension*.

Example 7.1 (Space subsets). Any subset of the Euclidean space $\mathbb{R}^m$ is a graphics object. In fact, given $S \subset \mathbb{R}^m$, we define an attribute function

$$f(p) = \begin{cases} 1 & \text{if } p \in S, \\ 0 & \text{if } p \notin S. \end{cases}$$

In general, the values of $f(p) = 1$ are associated to a certain color, called the *object color*. In this case, the attribute function simply characterizes the points of the set S and, for this reason, is called the *characteristic function* of the graphics object.

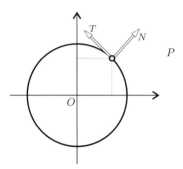

Figure 7.1. A circle and its tangent and normal vector fields.

The characteristic function completely defines the geometric support of the graphics object; in other words, if p is a point of the space $\mathbb{R}^m$, then $p \in S$ if and only if $f(p) = 1$. These two problems are therefore equivalent: to determine an algorithm to calculate $f(p)$ at any point $p \in \mathbb{R}^m$; and to determine whether a point $p \in \mathbb{R}^n$ belongs to the geometric support S of the graphics object.

The second problem is a *point membership classification problem*. Many of the challenges related to graphics objects rely on the solution of the point membership classification problem; therefore, we need efficient and robust algorithms to solve this problem. ❏

Example 7.2 (Image). In Chapter 6 we saw that an image is a function $f \colon U \subset \mathbb{R}^2 \to \mathbb{R}^n$, where $\mathbb{R}^n$ is a representation of the color space. This means that an image is a graphics object whose geometric support is the subset U on the plane (usually a rectangle) and the attribute function associates a color to each point on the plane. ❏

Example 7.3 (Circle and vector field). Consider the unit circle S^1 centered at the origin, whose equation is given by
$$x^2 + y^2 = 1.$$
The application of the plane $N \colon \mathbb{R}^2 \to \mathbb{R}^2$ given by $N(x, y) = (x, y)$ defines a unit vector field normal to S^1. The application $T \colon \mathbb{R}^2 \to \mathbb{R}^2$, given by $T(x, y) = (y, -x)$, defines a vector field tangent to the circle (Figure 7.1).

The circle is a unidimensional graphics object on the plane, and the two vector fields are attributes of the circle (they can represent, for instance, physical attributes such as tangential and radial accelerations). The attribute function is given by $f \colon S^1 \to \mathbb{R}^4 = \mathbb{R}^2 \times \mathbb{R}^2$, $F(p) = (T(p), N(p))$. ❏

7.2   Planar Graphics Objects

Specifying a graphics object means defining the geometry and the topology of the geometric support, as well as its attribute function. In general, this specification is done in the mathematical universe, and the object should be represented in such a way that it can be manipulated on the computer. The description, specification, and representation of the geometric support of graphics objects is called *modeling*.

There are several kinds of graphics objects, which can be classified by their dimensions in ambient space. When the dimension of the ambient space is 2 ($m = 2$) we have *planar graphics objects* (which we will discuss in this chapter) and when $m \geq 3$ we have *spatial objects*, which we will explore in Chapter 8.[1]

We begin with planar graphics objects partly because they are simple to study but are simultaneously a good way of introducing the inherent challenges associated with arbitrary graphics objects. Furthermore, graphics output devices, such as monitors and printers, and even some input devices are designed for representing planar objects. This allows objects to be reconstructed in those devices for their subsequent visualization. Ultimately, the visualization of any graphics object is executed through planar objects. For instance, a 3D scene is visualized through an image, which is a planar graphics object. Planar objects are of great importance in some applications. Electronic publishing systems, for instance, work exclusively with planar graphics objects. A detailed study of planar objects is the main thrust of 2D computer graphics.

Planar graphics objects can have dimensions 1 or 2. If the attribute function is constant, the objects correspond to subsets on the plane: intuitively, if the dimensions of the graphics object are 1 we obtain a *planar curve*; if they are 2 we obtain a *region on the plane*. In the following sections we will introduce a more formal definition of these two objects.

7.2.1   Planar Curves

A *planar curve*, also called a *planar topological curve*, is a planar object of dimension 1. A planar curve cannot have self-intersections. A subset $c \subset \mathbb{R}^2$ is a planar curve if, locally, c has the topology of an open interval $(0, 1)$ or of a half-open interval $(0, 1)$. This means, given an arbitrary point $p \in c$, there exists an open disk

$$D^2(\varepsilon, p) = \{(x, y) \in \mathbb{R}^2 \; ; \; ||x - p|| < \varepsilon\}$$

with center in p and radius $\varepsilon > 0$, such that $D^2(\varepsilon, p) \cap c$ has the topology of the interval $(0, 1)$ (arc AB of the curve in Figure 7.2 (left)), or $D^1 \cap c$ has the topology of the interval $(0, 1]$ (arc AP of the curve in Figure 7.2 (right)). The term *to have the topology of* means the sets are homeomorphic.

To illustrate the difference, Figure 7.3 shows a subset of dimension 1 on the plane that is *not* a planar curve. In fact, taking a disk $D^2(\varepsilon, p)$ with center at point p, the intersection $D^2(\varepsilon, p) \cap c$ does not have the topology of an interval (instead, it has the topology of the letter "X"). A topological curve is said to be *closed* if it has the topology of a circle.

The simplest method for describing a planar graphics object of dimension 1 is by an analytical equation characterizing the points of the object in a coordinate system. There are two ways of describing curves by equations: by *parametric* or *implicit* specification.

---

[1]Objects with fractional dimensions, called *fractal objects*, are outside the scope of this book.

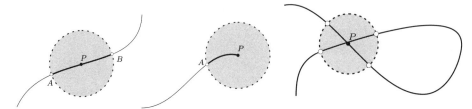

Figure 7.2. Local topology of a planar curve.

Figure 7.3. Subset on the plane which is not a planar curve.

Parametric description. In the parametric description, a curve is defined by a function $\gamma\colon I \subset \mathbb{R} \to \mathbb{R}^2, \gamma(t) = (x(t), y(t))$, where I is an interval of the real line. The parametric equation has an interesting physical interpretation: if we consider variable $t \in I$ to be time, then curve γ represents the path of a particle on the plane (Figure 7.4).

When a curve is given by a parametric equation $\gamma(t)$, the set $\gamma(I)$ is called a *line* of the curve. It is important to note that the line of a parametric equation does not always represents a topological curve. For instance, this line can have self-intersections, which cannot happen in a topological curve (see Figure 7.4). Also, a curve can have an infinite number of different parameterizations. When the line of a parametric curve is a topological curve, we can see the parameterization defining a coordinate system on the curve.

Example 7.4 (Parametric equation of a straight line). Consider a straight line r on the plane, a point $P \in r$, and a vector v determining its direction (see Figure 7.5).

The vector form of the parametric equation of a straight line r is given by

$$\gamma(t) = p + tv, \quad t \in \mathbb{R}.$$

If $p = (x_0, y_0)$, $v = (v_1, v_2)$, and $\gamma(t) = (x(t), y(t))$, then we can write the parametric equation of r with coordinates

$$(x(t), y(t)) = (x_0, y_0) + t(v_1, v_2);$$

in other words, $x(t) = x_0 + tv_1$ and $y(t) = y_0 + tv_2$. ❏

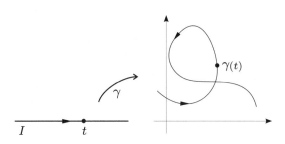

Figure 7.4. Parametric description of a planar curve.

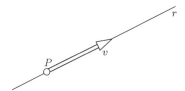

Figure 7.5. Straight line on the plane.

Example 7.5 (Graph of a function). Given a real function of a real variable $f : I \subset \mathbb{R} \to \mathbb{R}$, its graph is defined by the set

$$G(f) = \{(x, f(x)) ; \ x \in I\}.$$

The graph of f defines a topological curve on the plane, which can be easily parameterized by the equation

$$\gamma(t) = (t, f(t)).$$

If r is a nonvertical straight line on the Euclidean plane, then r is the graph of the linear function $f(x) = ax + b$, $a, b \in \mathbb{R}$. Therefore, r can be parameterized by placing $\gamma(t) = (t, at + b)$. Observe that this parameterization is very different from that of a straight line, obtained previously. ❑

Example 7.6 (Parametric equation of a circle). The parameterization of a unit circle centered at the origin is given by $\gamma(t) = (\cos(t), \sin(t))$. Geometrically, the parameter t represents the angle, in radians, between the segment OP and the x-axis of the coordinate system (see Figure 7.6). ❑

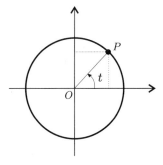

Figure 7.6. Parametric description of a unit circle.

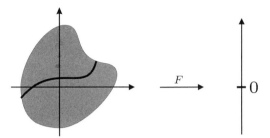

Figure 7.7. Implicit description of a curve.

Implicit description. The parametric equation defines a curve as being the trajectory of a point. The implicit description describes a curve as the set of the roots of an equation for the variables x and y. More specifically, we have a function $F \colon U \subset \mathbb{R}^2 \to \mathbb{R}$, and its geometric support is defined as the set of solutions of the equation $F(x, y) = 0$. Being an equation with two variables it has an infinite number of solutions.

The set of roots of the equation $F(x, y) = 0$ is called the *inverse image* of 0 by function F and is indicated by $F^{-1}(0)$. Therefore,

$$F^{-1}(0) = \{(x, y) \in \mathbb{R}^2 \; ; \; F(x, y) = 0\}.$$

Figure 7.7 illustrates this description. If the function $F(x, y)$ is a polynomial of degree g for the variables x and y, we say the curve is *algebraic* of degree g.

Example 7.7 (Implicit equation of a straight line). A straight line has an implicit equation given by $ax + by + c = 0$, where $ab \neq 0$. Therefore, we have an algebraic curve of degree 1. ❏

Example 7.8 (Implicit equation of a circle). A circle, centered at the origin $(0, 0)$ and with radius r, is defined by the second-degree algebraic equation

$$x^2 + y^2 - r^2 = 0.$$

In general, the conics (circle, ellipse, parabola, and hyperbola) constitute the family of (nondegenerated) algebraic curves of degree 2, defined by the implicit equation

$$ax^2 + by^2 + cxy + dx + ey + f = 0,$$

where $a^2 + b^2 + c^2 \neq 0$. ❏

The implicit equation $F(x, y) = 0$ does not always define a topological curve on the plane. An example is equation $x^2 - y^2 = 0$, whose solution is the set formed by the pair of straight lines $y = x$ or $y = -x$, intersecting at the origin. Depending on function F, the inverse image $F^{-1}(0)$ can have a quite complex topology. There exists a simple condition so the implicit object $F^{-1}(0)$ is a topological curve: *for every point $(x_0, y_0) \in F^{-1}(0)$, we should have*

$$\frac{\partial F}{\partial x}(x_0, y_0) \neq 0 \quad \text{or} \quad \frac{\partial F}{\partial y}(x_0, y_0) \neq 0.$$

In other words, the gradient vector

$$\text{grad}(F) = (\frac{\partial F}{\partial x}, \frac{\partial F}{\partial y})$$

does not cancel itself at the points in the graphics object.[2] In this case, we say that 0 is a *regular value* of function F.

The regular value condition is verified for the case of a circle, a straight line, and conics in general. In the previously mentioned example $F(x, y) = x^2 - y^2$, the condition is not verified at the origin $(x = 0, y = 0)$, which is exactly the point where the set does not have, locally, the topology of an interval. Despite its great theoretical importance, from a computational point of view, the regular value condition can be difficult to verify.

7.2.2 Planar Regions

In this section we will examine planar graphics objects of dimension 2 whose attribute function is constant. We already saw these objects corresponding to 2D subsets on the plane. The most common case is when they correspond to open and closed regions on the plane, which we will define next.

A subset S on the plane is an *open region* if, for every point $p \in S$, there exists an open disk $D^2(\varepsilon, p) = \{(x, y) \in \mathbb{R}^2 \; ; \; ||x - p|| < \varepsilon\}$ such that the intersection $D^2(\varepsilon, p) \subset S$ (Figure 7.8(a)). We say S is a *region with border* if, for every point $p \in S$, one of the following conditions is satisfied:

1. An open disk $D^2(\varepsilon, p)$ exists, such that $D^2(\varepsilon, p) \subset S$, or

2. $D^2(\varepsilon, p) \cap S$ has the topology of the semidisk

$$D_+^2 = \{(x, y) \in \mathbb{R}^2; x^2 + y^2 < \varepsilon \quad \text{and} \quad y \geq 0\}.$$

This concept is illustrated in Figure 7.8(b). The points of the region satisfying the second condition above, are called *boundary points*, or *points of the border*. If a region contains all of the boundary points and is limited, it is called a *planar region* or *2D solid*.

The simplest method of specifying a planar region is by using curves delimiting their border. In fact, a classic result in topology, called *Jordan's Curve Theorem*, affirms that a topological closed curve γ divides the plane in two open regions, one being limited and the other unlimited. Besides, the boundary of each of those regions is constituted by the curve γ.

---

[2]This result comes from the implicit function theorem studied in calculus courses.

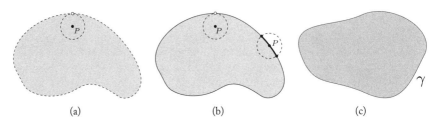

Figure 7.8. (a) Open planar region; (b) closed planar region; (c) planar region defined by a Jordan curve.

The closed topological curves are called *Jordan curves*. The region limited by γ is called *region defined by curve γ*, or *internal region to curve γ*. This result is illustrated by the region shown in Figure 7.8(c).

Of course, when a region is specified through its boundary curve, the representation of the region is reduced to the representation of the curve. However, the problem is not that simple because the specification of the region consists of two ingredients: specifying the boundary curve and determining an algorithm to solve the point membership classification problem.

In this case, the solution to the point membership classification problem consists of verifying whether a certain point on the plane belongs to either the internal or external region to the curve. When we know the boundary curve, we should determine a method to decide if a certain point p on the plane belongs to the region.

If a closed topological curve on the plane is defined by an implicit equation $F(x,y) = 0$, the point membership classification problem is easy to solve. In fact, given a point p on the plane, if $F(p) = 0$, p is a point on the curve, $F(p) > 0$, p is an exterior point, and if $F(p) < 0$, then p is a point in the region interior to the curve (or vice versa). This fact allows us to define regions on the plane through implicit inequations of the type $F(x,y) \leq 0$, $F(x,y) \geq 0$, $F(x,y) < 0$ or $F(x,y) > 0$.

In the Jordan Curve Theorem, it is important that the curve be closed and not have self-intersections. If the curve is closed but has self-intersections, it delimits more than one region on the plane (Figure 7.9).

Figure 7.9. Four regions limited by a nonsimple curve.

7.2.3   Implicit or Parametric?

Given that there are both implicit and parametric methods of describing graphics objects, which one should we use? The answer to this type of question is, invariably, "It depends on the situation."

We could give dozens of examples and analyze the advantages or disadvantages of using either an implicit or parametric specification in each of them. Instead, in this section, we want to demonstrate the basic difference between these two methods through two problems: one dealing with point sampling and the other with point membership classification.

For both these problems, let us consider a graphics object O_1 defined by a parameterization $\gamma\colon I \subset \mathbb{R} \to \mathbb{R}^2$ and a graphics object $O_2 = F^{-1}(0)$ defined by an implicit function $F\colon \mathbb{R}^2 \to \mathbb{R}$.

Point sampling. Consider this point sampling problem: given a graphics object with geometric support S, we want to determine a set of points $p_1, p_2, \ldots, p_n$ such that $p_i \in S$.

This problem can be easily solved for curve γ: we choose arbitrary points $t_1, t_2, \ldots, t_n$ in the interval I and we calculate the points $p_i = \gamma(t_i)$. In other words, we can easily obtain samples of the curve γ, by choosing samples $t_i \in I$ in the parameterization interval and then calculate the value $\gamma(t_i)$ in those points.

On the other hand, to do a sampling in object O_2, we should find solutions for the equation $F(x, y) = 0$. Depending on the function F, this can be an extremely complex problem.

Point membership classification. Consider the following point membership classification problem: given a point $p \in \mathbb{R}^2$ and a graphics object with geometric support S, we must determine if $p \in S$.

Given a point p on the plane, we can easily verify whether p is a point of the implicit object O_2 by calculating the value $F(p)$ of the implicit function in p. In fact, p belongs to the object if and only if $F(p) = 0$.

On the other hand, to verify if point $p = (p_1, p_2)$ belongs to curve γ, we should verify if equation $\gamma(t) = p$ has solutions. Assuming $\gamma(t) = (x(t), y(t))$, we obtain a system of equations

$$x(t) = p_1$$
$$y(t) = p_2.$$

Depending on the coordinate functions $x(t)$ and $y(t)$ of curve γ, the solution of this system can be quite difficult.

Conclusion. The point membership classification problem is easier to solve for objects defined implicitly than parametrically. The opposite is true of the sampling problem.

7.3   Polygonal Curves and Triangulation

Polygonal curves are widely used in computer graphics. Let $p_0, p_1, \ldots, p_n$ be different points on the plane. A polygonal curve is defined as being the union of the segments $p_0p_1, p_1p_2, p_2p_3, \ldots, p_{n-1}p_n$ (see Figure 7.10). The points p_i are called *vertices* of the curve. The segments p_ip_{i+1} are called *edges* of the polygonal curve. A polygonal curve can be easily parameterized provided each segment has a linear parameterization.

Figure 7.10. Polygonal curve.

Polygonal curves play an important role in computer graphics because they are easy to specify and represent (see Chapter 1) and because a great variety of planar curves can be approximated by a polygonal curve. Such curves are called *rectified*. The concept of "approximation" here means that, given an error $\varepsilon > 0$, the distance of any point of the polygonal curve to the original curve c is smaller than ε. For values sufficiently small of ε, this means the polygonal curve is contained in a "neighborhood" of radius ε of the curve, as illustrated in Figure 7.11(a).

From the Jordan Curve Theorem, a closed polygonal curve without self-intersections delimits a region on the plane. In this case, the region is called the *polygonal region*. As we saw previously, a polygonal region is represented by the polygonal curve defining its boundary.

In general, the border of a region is a curve that can be approximated by a polygonal curve. Therefore, a great variety of planar regions can be approximated by polygonal regions (see Figure 7.11(b)). In summary, it is common to represent topological or parametric curves by polygonal curves, and regions by polygonal regions. Of course, such representation is usually neither unique nor exact.

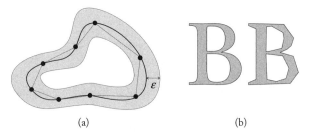

(a)   (b)

Figure 7.11. (a) Approximation of a curve by a polygonal one; (b) polygonal approach to a planar region.

7.3.1   Triangulation

A *triangulation* of a region on the plane is a collection $\mathcal{T} = \{T_i\}$ of triangles, in which, given two different triangles T_i, T_j in the triangulation $\mathcal{T}$ such that $T_i \cap T_j \neq \varnothing$, we have

$$T_i \cap T_j = \text{common vertex}, \quad \text{or} \quad T_i \cap T_j = \text{is a common edge}.$$

Figure 7.12(c) shows a triangulation, while (a) and (b) show collections of triangles that are not a triangulation.

The existence of a triangulation in a region allows us to define the region's attributes in a localized way in each triangle. Each triangle has a natural system of local coordinates, defined by the barycentric coordinates. This system allows, for instance, the attributes defined at the vertices of the triangulation to be linearly extended for the whole triangle and consequently to reconstruct the attributes in the entire region (we can also use barycentric coordinates to define nonlinear reconstruction functions). On the other hand, the existence of a triangulation allows the use of robust data structures for encoding the region on the computer. In summary, the triangulation of a region is a solution to the representation (i.e., discretization) and reconstruction problems of a 2D graphics object.

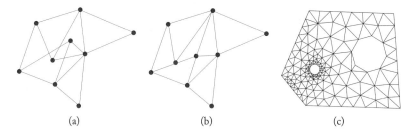

(a)   (b)   (c)

Figure 7.12. (a), (b) Nontriangulations; (c) triangulation.

7.4   Representation of Curves and Regions

Thus far we have examined planar graphics objects (curves and regions) from the mathematical universe point of view: the Euclidean plane is a "continuum" of points and the graphics objects are defined by points on the plane.

To manipulate objects on the computer we must obtain a representation, that is, a discrete specification of the elements defining the graphics object. As the attribute function of the objects we are studying is constant, we need to worry only about the geometry and topology representations of the geometric support.

In general, we use approximate representations of graphics objects. In other words, the graphics objects being represented are approximations of the object in the mathematical universe. These approximations are, in general, obtained in a way to preserve the topology of the represented object.

The representation methods we will study are based on the divide-and-conquer strategy: we divide the geometric support of the graphics object, or the space where the graphics object is embedded, in a way to obtain a simple representation in each element of the subdivision. There are two representation methods based on this principle: representation by intrinsic decomposition and representation by spatial decomposition.

7.4.1   Piecewise Linear Representation

In this section we will explore the intrinsic decomposition method of representation. We will cover only the case in which the elements of the representation are linear. This representation is called *piecewise linear representation*.

We previously saw that a curve (topological or parametric) can be represented by a polygonal curve, and a planar region can be represented by a polygonal region (i.e., a region whose boundary is a polygonal curve) or by a triangle. In all these cases, we are representing the object by its linear decomposition in "pieces."

When we represent a region by decomposing it into polygons, we are said to "polygonize a region." This decomposition can go beyond just polygonizing the border by transforming the region into a single polygon. That is, for instance, the case of a triangulation decomposing the region into a family of triangles.

Polygonization methods depend on the way the object is specified. Below, we will briefly describe two polygonization methods, one for curves defined parametrically and the other for curves defined implicitly. The general problem of region polygonization will be revisited in Chapter 8 as a particular case of surface polygonization in the space (a region is a planar surface).

Polygonization of parametric curves. Suppose the curve γ to be polygonized is defined in the interval $I = [a, b]$. The simplest method for approximating γ by a polygonal curve is

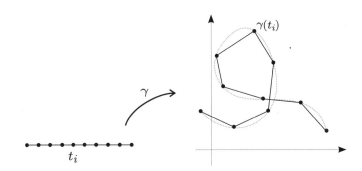

Figure 7.13. Polygonal approximation of a planar curve.

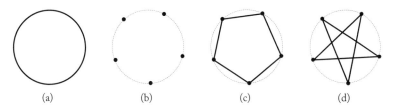

Figure 7.14. Point sampling and reconstruction: (a) circle; (b) representation by point sampling; (c) reconstruction with correct structuring; (d) reconstruction with incorrect structuring.

uniform polygonization. We obtain a partition of the interval I

$$a = t_0 < t_1 < t_2 < \cdots < t_n = b$$

and evaluate the curve γ at the partition points t_i, obtaining a sequence of points $p_0, p_1, \ldots, p_n$, with $p_i = \gamma(t_i)$. These points constitute the vertices of a polygonal curve approximating γ (Figure 7.13). When the partition points t_i satisfy $t_i = i\Delta t$, with $\Delta t = (b - a)/n$, we say the sampling is *uniform*.

An interesting way to interpret the above process consists of observing that we are representing the curve γ using a point sampling with the samples $\gamma(t_i)$; we are also reconstructing an approximation to the curve γ using linear interpolation starting from the samples.

However, to do the reconstruction, the sampling points should be ordered correctly. This fact emphasizes an important aspect: we should associate the sampling of a graphics object to a sampling structure to correctly obtain the reconstruction. In this case, the structuring consists of performing an ordering. Figure 7.14(b) shows a representation by point sampling and its reconstruction.

Polygonization of implicit curves. As we previously saw, we should take samples $p_1, \ldots, p_n$ on the curve $\gamma = F^{-1}(0)$. For this, we should find n roots of the equation $F(x, y) = 0$. Notice, however, these solutions should be structured in a way that obtains the correct polygonal curves in the reconstruction. We will provide an answer that simultaneously solves the sampling and structuring problems.

The strategy for obtaining a polygonal representation has the following stages:

1. Build a triangulation $\{T_i\}$ in the domain of function F;

2. Approximate, in each triangle T_i, function F by a linear function $\tilde{F}$;

3. Solve the problem $\tilde{F}(x, y) = 0$ in each triangle. In a generic way, the solution of this equation is a straight line segment approximating the curve $F(x, y) = 0$ in the triangle.

4. The structuring of the polygonization is inherited from the existent structuring in the triangulation.

We will provide more details about each of the stages above.

Construction of a triangulation. Consider a function $F\colon \mathbb{R}^2 \to \mathbb{R}$ and a curve $\gamma = F^{-1}(0)$ defined implicitly by F. Let $Q = [a, b] \times [a, b]$ be the square of the plane containing the curve γ. We take a uniform partition of the interval $[a, b]$

$$a = t_0 < t_1 < t_2 < \cdots < t_n = b,$$

where $t_{i+1} - t_i = \Delta t = (b - a)/n$. The Cartesian product of this partition defines a *uniform grid* of the square Q, as shown in Figure 7.15(a). Starting from the grid, we obtain a triangulation of the square, as indicated by Figure 7.15(b).

Linear approximation. We define a function $\tilde{F}\colon Q \to \mathbb{R}$ that is linear in each triangle from the decomposition and coincides with F in the triangle vertices. This function is obtained in the following way: let v_0, v_1 and v_2 be the triangle vertices; we define $\tilde{F}(v_i) = F(v_i)$ in a way that coincides with F in the vertices. We now extend $\tilde{F}$ linearly for other triangle points: given an arbitrary point p on the triangle, we can write p in barycentric coordinates:

$$p = \lambda_1 v_1 + \lambda_2 v_2 + \lambda_3 v_3,$$

with $\lambda_i \geq 0$, and $\lambda_1 + \lambda_2 + \lambda_3 = 1$. We then define

$$\tilde{F}(p) = \lambda_1 F(v_1) + \lambda_2 F(v_2) + \lambda_3 F(v_3),$$

Once the function $\tilde{F}$ is defined, the polygonal curve approximating γ is given by the inverse image $\tilde{F}^{-1}(0)$. By solving the linear equation $\tilde{F}(x, y) = 0$ in each triangle, we obtain the edges of the polygonal curve. Figure 7.15(c) shows the polygonization of a circle using this method.

While detailed mathematical analysis of the method and implicit polygonization described above is an interesting and delicate problem, it suffices to here state that if 0 is a regular value of function F, and the triangles are sufficiently small, then $\tilde{F}^{-1}(0)$ is a polygonal curve which approximates $\gamma = F^{-1}(0)$.

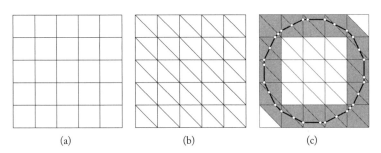

(a)   (b)   (c)

Figure 7.15. (a) Grid; (b) triangulation; (c) circle polygonization. (*Courtesy L. H. de Figueiredo.*)

Polygonization and attributes. We know when representing a graphics object, we should represent its geometry (geometric support) and attributes. In the case of a piecewise linear curve representation, we should represent each attribute in a way to obtain the attributes of the polygonal curve.

As an example, consider the case of the normal vector field to the curve. We can represent this field by sampling it at each vertex of the polygonal curve. Another method would be to calculate the normal of each edge (an immediate calculation) and to take the representation of the normal at each vertex v as being the average of the normal vectors in the incident edges in v.

7.4.2   Representation by Spatial Decomposition

The simplest and most broadly used representation by spatial decomposition is the matrix representation that we used in Chapter 6 to represent images. This representation is crucial for the process of visualizing planar graphics objects, as we will see further on.

The goal of the matrix representation is to discretize the graphics object as a union of rectangles on the plane. More precisely, we take the lengths Δx and Δy in the x and y axes, respectively, and define a uniform grid $\Delta = \Delta(\Delta x, \Delta y)$ on the plane by the set

$$P_\Delta = P_{\Delta x \Delta y} = \{(m\Delta x, n\Delta y) \; ; \; m, n \in \mathbb{Z}\},$$

as shown in Figure 7.16(a). The grid P_Δ is formed by a set mn of cells

$$c_{jk} = [j\Delta x, (j+1)\Delta x] \times [k\Delta y, (k+1)\Delta y],$$

$j = 0, \ldots, m - 1, \ k = 0, \ldots, n - 1.$ Representing a graphics object is reduced to obtaining a representation in each of those cells. In the case that the planar object is an image, as we previously saw, the cells are called *pixels* and the points $(j\Delta x, k\Delta y)$ are the *vertices* of the grid. The coding of a grid cell is quite simple. We have two options. We can either represent the cell by the coordinates of one of its vertices. More precisely, each cell is completely determined by its position $(i\Delta x, j\Delta y)$, where $i, j \in \mathbb{Z}$. Given that the position coordinates are integer numbers, this representation is quite attractive from a computational point of view. Our other option is to take the coordinates of a point on the

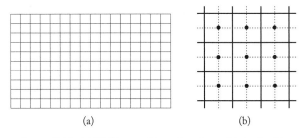

(a)   (b)

Figure 7.16. (a) Uniform grid on the plane; (b) dual grid.

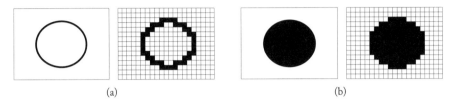

Figure 7.17. Matrix representation of (a) a circle and (b) a disk.

cell. A good choice is to take the centroid of the cell. The cell centroids determine another grid on the plane called the *dual grid*, shown in stippled lines in Figure 7.16(b).

The matrix representation of a graphics object consists of approximating its geometry by the union of rectangular blocks of the grid. Figure 7.17(a) and (b) show examples of representations by spatially decomposing a circle and a circular region on the plane, respectively.

Matrix representation and topology. As we already observed, generally a representation provides an exact topology and just an approximated description of the geometry of the object. For this to happen the dimensions of the grid cells should be small in relation to the dimensions of the object. Figure 7.18 shows a planar region whose topology is a disk with a hole in the center. Notice that in the grid representation of the illustration in the center the hole disappears, while the matrix representation correctly represents the topology of the region.

Geometry and attributes. In the matrix representation, each cell locally approximates the geometry of the graphics object. The representation of the object should take into account the values of the attribute function in each cell. We saw this problem when we studied the matrix representation of images, where we presented two simple solutions for obtaining the attributes representation in each cell: point and area sampling.

 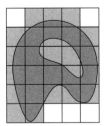 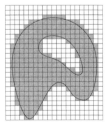

Figure 7.18. Matrix representation and topology of a planar region.

7.5 Rasterization

Region rasterization is of great importance in computer graphics and is fundamental to the process of visualizing graphics objects. The rasterization of a planar object, whose only attribute is color, provides a representation of that object by a digital image.

Given a grid on the plane and a planar graphics object $\mathcal{O} = (U, f)$ with geometric support $U \subset \mathbb{R}^2$, we call *rasterization*[3] the process of determining a matrix representation of object $\mathcal{O}$.

Given a grid $\Delta_{(\Delta x, \Delta y)}$ on the plane with $m \times n$ cells, an *enumeration* is a sequence $C_1, C_2, \ldots, C_n$ of cells of Δ. In other words, an enumeration consists of a (finite) subset of cells and the order of traversing those cells. The rasterization process consists of enumerating the grid cells, which forms a matrix representation of the graphics object. In this way, the union of the enumeration cells brings information about the topology of the graphics object and provides a good approximation of its geometry.

An important part of rasterization is determining whether a grid cell C_i should be enumerated in the process. A complete criterion to make this decision consists of verifying the intersection: C_i is a representation cell if and only if the intersection $C_i \cap U \neq \varnothing$; that is, the intersection between the cell and the graphics object support is nonempty (see Figure 7.19).

However, in general the intersection calculation is a computationally costly process. A simpler criterion consists of taking the centroid P_i of the cell C_i and then verifying if P_i is an element of U: C_i is a representation cell if and only if $P_i \in U$. However, this test is far too restrictive. In fact, it is enough to observe that a curve not passing through the centroid of any cell has empty rasterization.[4]

Certainly we should obtain an intermediate solution for the two criteria above. The details of which is a good choice is most directly related to implementation issues and will not be treated here. From now on, we will always assume the use of the most robust intersection criterion.

Rasterization methods use two basic strategies: incremental rasterization and rasterization by subdivision. For each of these strategies, the process either can be intrinsic to the geometry of the object or can use the geometry of the space in which the object is embedded (in our case the plane $\mathbb{R}^2$).

Figure 7.19. Choosing cells in the rasterization.

---

[3]What we are calling here matrix representation is a generalization of the concept of "raster graphics."
[4]Notice here, once again, our familiar point membership classification problem.

7.5.1 Intrinsic Incremental Rasterization

For intrinsic incremental rasterization, we have a grid defined a priori, and a method of visiting its cells. We visit grid points as we move along the points of the graphics object. As we visit the cells, we obtain the enumeration that defines the matrix representation. This method is appropriate for parametric curves $\gamma\colon [a,b] \subset \mathbb{R} \to \mathbb{R}^2$ because making the parameter t vary from a to b results in $\gamma(t)$ varying along the curve.

For implicit curves we can use the incremental rasterization method by using differential equations. Given a regular curve $\gamma\colon \mathbb{R}^2 \to \mathbb{R}$, defined implicitly by $\gamma = f^{-1}(0)$, the gradient vector

$$\operatorname{grad}(F) = (\frac{\partial f}{\partial x}, \frac{\partial f}{\partial y})$$

is perpendicular to γ at each point. Therefore, vector $T = (\frac{\partial f}{\partial y}, -\frac{\partial f}{\partial x})$ is tangent to the curve. In other words, we have the following system of equations:

$$\frac{dx}{dt} = \frac{\partial f}{\partial y};$$
$$\frac{dy}{dt} = -\frac{\partial f}{\partial x}.$$

The solution of this system provides a method for traversing approximately along curve γ, thus making incremental rasterization possible.

DDA-Bresenham algorithm. We will now discuss the intrinsic incremental rasterization problem for a straight line segment. Besides the importance of rasterizing a basic geometric object, this algorithm, in reality, provides an efficient method of traversing along a straight line in a discretized space. This aspect of the algorithm is used in several contexts.

Let us consider a uniform grid and a straight line segment defined by its extreme points (x_0, y_0), $x_1, y_1)$ (see Figure 7.20(a)). The equation of the straight line support of the segment is given by $y = y_1 + m(x - x_1)$, where $m = (y_1 - y_0)/(x_1 - x_0)$ measures the inclination (slope) of the straight line (the tangent of the angle between the straight line and the x-axis). This equation can be written in the form

$$y = mx + b, \quad \text{where} \quad b = y_1 - mx_1. \tag{7.1}$$

In the above notation, we use the letter y to represent the function defined by the straight line; that is, we write $y = y(x) = mx + b$.

Let us assume that $0 < m < 1$, as the other cases can be treated using symmetries and reflections on the plane $\mathbb{R}^2$. We will consider the pixel as being the center of the grid cell (marked with the symbol $+$ in the figure).

The simplest rasterization algorithm for a straight line segment consists of directly using the straight line equation in Equation (7.1):

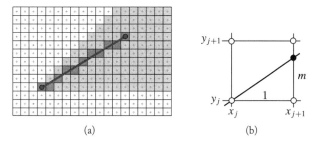

Figure 7.20. (a) Rasterization of a segment and (b) incremental calculation of y.

```
for x = x_0, x ≤ x_1 do
    y = mx + b;
    y = int(y + 0.5);
    Set_Pixel(x, y);
    x = x + 1;
end for
```

This algorithm produces bad results due to rounding errors. However, the main problem is the fact that it does not completely explore the incremental nature of the straight line equation and, for this reason, does unnecessary calculations. Note that this algorithm can be used for rasterizing any function $y = f(x)$.

In the case of the straight line we observe that, when moving from pixel $(x_j y_j)$ to pixel $(x_{j+1} y_{j+1})$, we have $x_{j+1} = x_j + 1$; therefore, from Equation (7.1) it follows

$$
\begin{aligned}
y(x_{j+1}) &= mx_{j+1} + b \\
&= mx_j + m + b \\
&= y_j + m,
\end{aligned}
\tag{7.2}
$$

as we illustrate in Figure 7.20(b). This fact leads us to the next incremental rasterization algorithm for a straight line:

```
for x = x_0, y = y_0, x ≤ x_1 do
    y = int(y + 0.5);
    Set_Pixel(x, y);
    x = x + 1;
    y = y + m;
end for
```

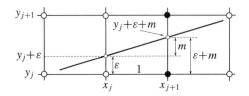

Figure 7.21. Incremental error calculation.

This algorithm is known as the *digital differential analyzer* (DDA).[5] DDA still uses floating point with roundings at each stage in the calculation of the value of y. In 1965, Jack Bresenham developed an incremental rasterization algorithm [Bresenham 65] for the straight line with the following characteristics: it calculates the value of y without the need for roundings and it is possible to implement the algorithm using only integer arithmetic. This second condition was essential at that time due to the nonexistence of floating point support in the processors.

The first important observation to understand Bresenham's algorithm is that when processing pixel (x_j, y_j), we increase x_j by $x_{j+1} = x_j + 1$ and, as $0 < m < 1$, we have only two possibilities of pixel choice: $(x_j + 1, y_j)$ or $(x_j + 1, y_j + 1)$. They are the black pixels in Figure 7.21.

Making the decision of which pixel to select is very simple. First take the middle point M of the segment $y_j y_{j+1}$. Then calculate the point y where the segment intersects the straight line $x = x_j + 1$. If $y < M$, select (x_{j+1}, y_j); otherwise, select (x_{j+1}, y_{j+1}).

To efficiently implement the above process of selecting pixels, we replace the calculation of M by an incremental process. We use the error made when rasterizing each pixel. When selecting pixel (x_j, y_j) we have an error ε_j given by the difference between the real value of y and the chosen value y_j: $\varepsilon_j = y - y_j$ (see Figure 7.21). Notice that ε varies between -0.5 and $+0.5$. Using ε we have the following selection process: select $(x_j + 1, y_j)$ if $y + \varepsilon_j + m < y + 0.5$. Otherwise, select $(x_j + 1, y_j + 1)$.

The important point is that the error can be calculated in an incremental form: if $(x_j + 1, y_j)$ is selected, then $\varepsilon_{j+1} = (y_j + \varepsilon_j + m) - y_j$. Otherwise the error is given by $\varepsilon_{j+1} = (y_j + \varepsilon_j + m) - (y_j + 1)$.

We then have Bresenham's line algorithm:

$y = \text{int}(y_0 + 0.5)$;
$\varepsilon = y - y_0$;
for $(x = x_0, x \leq x_1)$ **do**
    Set_Pixel(x, y);
    if $(\varepsilon + m < 0.5)$ **then**

---

[5]Differential Analyzer was the name of an analog computer built in MIT in the '40s, aimed at solving differential equations; the DDA algorithm produces a solution for the differential equation $y' = m$ in digital computers.

$$\varepsilon = \varepsilon + m$$
    else
        $y = y + 1;$
        $\varepsilon = \varepsilon + m - 1;$
    end if
  end for

Bresenham's line algorithm is also called that *middle point algorithm*. Because Bresenham's algorithm uses only integer arithmetic we should eliminate m in the above equation, but we left this task for the exercises at the end of this chapter.

7.5.2 Spatial Incremental Rasterization

For spatial incremental rasterization we traverse the grid cells enumerating the representation cells. A possible strategy consists of traversing each grid line starting from the origin. The rasterization methods using this strategy are known as *scanline rasterizations*.

In spatial incremental rasterization we might traverse a large number of grid cells not intersecting the object. Note that the rasterization of a straight line (DDA algorithm) in a grid with $n^2 = n \times n$ cells uses, approximately, only n cells. In general, for well-behaved curves, we traverse a number of cells, which is much smaller than the total number of existing grid cells.

7.5.3 Rasterization by Subdivision

Rasterization by subdivision can be intrinsic or extrinsic. In general, the subdivision method is done recursively for computational efficiency reasons.

Rasterization by intrinsic subdivision. In this method, we recursively subdivide the geometric support of the graphics object until each obtained subset is contained in a grid cell. The matrix representation of the object comprises the cells containing sets of the subdivision.

This method is quite useful since there are several graphics objects whose description naturally admits a process of intrinsic recursive subdivision. This is the case, for example, of a curve parametrically defined: we recursively subdivide the interval parameters by taking half of the length of the interval at each subdivision.

Rasterization by spatial subdivision. In this method, we consider a rectangle on the plane containing the graphics object and we subdivide it recursively in subrectangles. A simple method of subdivision consists of subdividing the rectangle in four subrectangles and continuing to subdivide each subrectangle analogously (see Figure 7.22).

The subdivision of each subrectangle stops when either the points of the graphics object do not exist within the subrectangle or the subrectangle has the size of a grid cell.

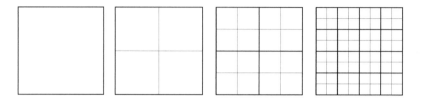

Figure 7.22. Spatial subdivision.

For the spatial subdivision method to be efficient, it should have a mode of deciding whether a given subrectangle intersects the geometric support of the graphics object.[6] In the case of an object implicitly defined by a function $F \colon \mathbb{R}^2 \to \mathbb{R}$, it is necessary to have an efficient method to verify if function F has roots inside an arbitrary subdivision subrectangle. While these methods exist, they will not be discussed in this book.

Figure 7.23 illustrates rasterization by spatial subdivision of the implicit curve given by the equation $y^2 - x^3 + x = 0$, defined in the region $[-2, 2] \times [-2, 2]$ on the plane. Note that, starting from the third stage, the subdivision process is halted for a large number of subrectangles.

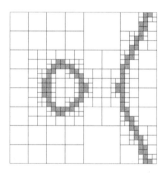

Figure 7.23. Rasterization by adaptive subdivision. (*Courtesy L. H. de Figueiredo.*)

7.5.4   Hybrid Rasterization

The rasterization method by spatial subdivision can be used for rasterizing regions, provided an efficient criterion is in place for deciding if a certain subrectangle of the subdivision intersects the geometric support of the graphics object. An interesting rasterization technique consists of combining intrinsic with spatial incremental rasterization methods.

Spatial incremental rasterization by lines successively rasterizes each grid line; the intersection of each of these lines with the boundary curves of the region is a union of intervals (see Figure 7.24), and therefore solving the point membership classification problem

---

[6]Notice here, once again, the point membership classification problem.

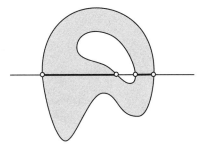

Figure 7.24. Rasterization of a line.

becomes very simple. We use intrinsic incremental rasterization to determine those intervals with a minimum computational effort.

The strategy consists of using a line-based rasterization of the region while avoiding the visit of unnecessary cells. The method performs an intrinsic incremental rasterization of the boundary curves and, in parallel, of the cells in each grid line. In greater detail, the method is as follows (see Figure 7.25):

1. We find the minimum ordered cell in the region and rasterize the line of this cell. That is, we determine all the cells of this line that are in the region.

2. We move to the next line, performing an intrinsic incremental rasterization of the boundary curves. This rasterization divides the line into a union of disjunct intervals.

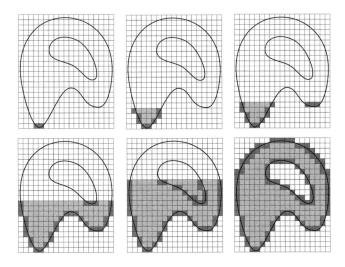

Figure 7.25. Region rasterization.

3. We perform an incremental rasterization of the line from the previous item to determine which intervals are contained in the region. Here we use the unidimensional point membership classification problem, which is simple.

4. The method continues until we obtain the rasterization of the entire region.

A particular case of great importance is the rasterization of polygonal regions. In this case we use intrinsic incremental rasterization of straight line segments, which can be done with great efficiency using the DDA-Bresenham algorithm.

7.6 Representation, Sampling, and Interpolation

Representation using a piecewise linear decomposition is commonly called *vector representation* and the object represented as such is often called a *vector object*. Similarly, it is common to refer to a graphics object represented by matrices as a *matrix object*.

When representing a parametric curve by a polygonal approximation, we obtain a sampling

$$p_i = \gamma(t_i), \ i = 1, \ldots, n$$

of n points in the curve γ. The polygonal representation is obtained by sampling the curve, structuring the samples (ordering), and then reconstructing by linear interpolation.

As we know, linear reconstruction provides only an approximation of the original curve. The above procedure is also used to obtain a polygonal representation of implicit curves.

We therefore unified the problem of representing implicit or parametric objects through the sampling and reconstruction processes. Moving from the mathematical to the representation universe is achieved by means of discretization. When the object is described by functions (as in the case of parametric and implicit descriptions), such discretization can be obtained using a point sampling method. The original object is reconstructed, starting from the representation samples, using an interpolation method. In this context, the interpolation methods are called *reconstruction* methods (see Figure 7.26).

The matrix representation also fits perfectly in the above paradigm. In fact, as we discussed in Chapter 6, there are several digital methods of reconstructing an image represented as a matrix. The discussion about reconstruction in that chapter applies to the present case as well. Figure 7.27 shows the function graph and the reconstruction function graph using Haar's interpolation kernel.

Creating a unified view of a problem, as we have done here for the problem of representation, is of great importance. It allows us to find new problem-solving methods. In

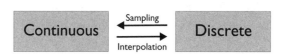

Figure 7.26. Discretization and reconstruction.

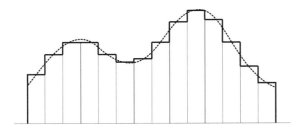

Figure 7.27. Matrix representation and reconstruction. The function graph is shown in stippled lines; the reconstruction function graph using Haar's interpolation kernel is shown in solid lines.

the case of object representation, we immediately see it is natural to obtain a representation where the interpolation methods use piecewise polynomial functions, with polynomials of high degree, instead of constant or linear ones.

7.7 Viewing Planar Graphic Objects

Graphics objects are displayed on *graphics output devices* such as monitors and printers. These devices have a 2D (planar) display apparatus. Planar graphics objects, and in particular curves and regions on the plane, are naturally mapped to the support surfaces of these display devices.

7.7.1 Graphics Output Devices

Graphics devices receive discrete graphics objects as input from a computational system and then reconstruct these object in their display support. These devices have their own representation space, so any graphics object to be displayed should be represented and its representation should be mapped to the representation space of the device. The device reconstructs the object, allowing its subsequent visualization. This process is described in the diagram below.

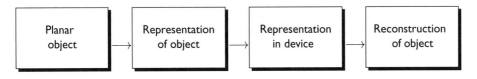

The representation space of the graphics output device has a 2D geometric support.[7] These devices are divided in two classes according to their representation space: vector and matrix devices.

---

[7]Some devices with 3D representation space are commercially available.

Vector devices. The representation space of vector devices has only points and vectors (oriented straight line segments). We associate a color attribute to each point and vector. These devices are appropriate for displaying planar graphics objects represented by a polygonal approximation (vector objects).

Matrix devices. The representation space of matrix devices has a uniform grid, and each grid cell has a color attribute. This way, these devices allow visualization of a digital image.

Each cell on the support surface of a matrix device is called a *pixel* (from "picture element"). The order $m \times n$ of the grid cell matrix is called the *device's spatial resolution*. These devices are appropriate for visualizing planar graphics objects represented in matrix form.

The device quality is determined by its spatial resolution, its resolution density (i.e., the number of pixels by length unit, usually measured in dots per inch (dpi)), and the distance between its cells ("dot pitch").

7.7.2 Viewing and Devices

Vector devices were predominant in computer graphics in the '70s. Today the vast majority of graphics devices are of the matrix type. There are a number of reasons for this shift.

To visualize a planar graphics object in a matrix device, we must find a matrix representation of the object with the same resolution as the device and then map the representation of the object in the representation space of the device, making each grid cell of the object representation correspond to a grid cell of the representation space of the device. The color attribute of each cell in the object representation is mapped to the color attribute of the corresponding cell in the representation space of the device.

From what we previously saw, to display a curve or a planar region in a matrix device, it is enough to perform a rasterization of the object in the device resolution and to apply a Euclidean plane transformation on the support surface of the device. This transformation requires a change of coordinate system, as we will explain in the next section.

7.7.3 Coordinates and Graphics Devices

Displaying graphics objects is an important application of coordinate system changing. Graphics output devices have a planar support surface in which several geometric object representations are manifested in the form of points and curved and planar regions. The support surface has a coordinate system, called the *device coordinate system*.

An example is the monitor. Its screen is manifested as a rectangle on the plane with an orthogonal coordinate system whose origin is located at some screen point (usually at the left, top, or bottom).

A graphics object has a coordinate system in which its points are defined. This coordinate system is called the *object coordinate system*. When we need to represent this object on a graphic device, we must perform a change from object to device coordinate system.

In general, rectangles defined in the object and device coordinate systems are called, respectively, *window* and *viewport*. The object is displayed by a change of coordinates,

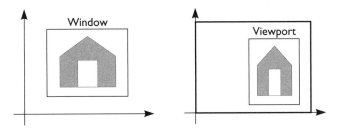

Figure 7.28. Devices and change of coordinates.

which transforms the rectangle defining the window into the rectangle defining the view-port (Figure 7.28).

You may be thinking that, according to the use of the windowing terminology of current computing systems (MS-Windows, Mac-OS, and UNIX's X-Windows), the correct name for what we above called "viewport" is really "window." For historical reasons, in computer graphics the term "window" is used for specifying a region in the object space.

A window is specified by two diagonally opposite vertices A and B, as shown in Figure 7.29. Suppose we have points $A = (x_0, y_0)$ and $B = (x_1, y_1)$. In an analogous way, the viewport is specified by vertices $C = (u_0, v_0)$ and $D = (u_1, v_1)$. The transformation we are looking for should transform the reference frame $(A, \{e_1, e_2\})$ (Figure 7.29(a)) into the reference frame $(C, \{e_1, e_2\})$ (Figure 7.29(d)) from the device representation space.

This transformation is obtained with three successive changes of coordinates:

1. Translate the window point A to the origin of the world coordinate system (Figure 7.29(b)). The matrix of this translation is given by

$$T(-x_0, -y_0) = \begin{pmatrix} 1 & 0 & -x_0 \\ 0 & 1 & -y_0 \\ 0 & 0 & 1 \end{pmatrix};$$

2. Perform a change of scale that transforms the new window rectangle into one congruent with the viewport rectangle (Figure 7.29(c)). The matrix of this scaling is

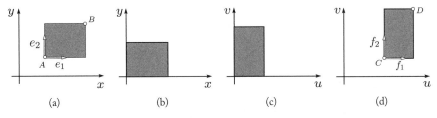

Figure 7.29. Window to viewport transformation.

given by

$$S\left(\frac{u_1 - u_0}{x_1 - x_0}, \frac{v_1 - v_0}{y_1 - y_0}\right) = \begin{pmatrix} \frac{u_1 - u_0}{x_1 - x_0} & 0 & 0 \\ 0 & \frac{v_1 - v_0}{y_1 - y_0} & 0 \\ 0 & 0 & 1 \end{pmatrix};$$

3. Translate the origin of the system (u, v) to the viewport point C (Figure 7.29(d)). This translation matrix is given by

$$T(u_0, v_0) = \begin{pmatrix} 1 & 0 & u_0 \\ 0 & 1 & v_0 \\ 0 & 0 & 1 \end{pmatrix}.$$

The final transformation, taking the window reference frame into the viewport reference frame, is given by the composition

$$L = T(x_0, y_0) \circ S\left(\frac{u_1 - u_0}{x_1 - x_0}, \frac{v_1 - v_0}{y_1 - y_0}\right) \circ T(-u_0, -v_0).$$

The change from the window coordinates (x, y) to the viewport coordinates (u, v) is given by the inverse transformation

$$L^{-1} = T(u_0, u_0) \circ S\left(\frac{u_1 - u_0}{x_1 - x_0}, \frac{v_1 - v_0}{y_1 - y_0}\right) \circ T(-x_0, -y_0).$$

or, by finding the matrix product,

$$M = \begin{pmatrix} \frac{u_1 - u_0}{x_1 - x_0} & 0 & \frac{u_1 - u_0}{x_1 - x_0}(-x_0) + u_0 \\ 0 & \frac{v_1 - v_0}{y_1 - y_0} & \frac{v_1 - v_0}{y_1 - y_0}(-y_0) + v_0 \\ 0 & 0 & 1 \end{pmatrix}.$$

and expressing the coordinates, $(u, v) = M(x, y)$:

$$u = \frac{u_1 - u_0}{x_1 - x_0}(x - x_0) + u_0;$$

$$v = \frac{v_1 - v_0}{y_1 - y_0}(y - y_0) + v_0.$$

This transformation is called a *screen transformation* and is a *2D viewing transformation*. Usually, to avoid numerical errors, this transformation is applied before rasterization.

7.8   2D Clipping

When we specify a window in the object space we are only interested in displaying objects on the plane that are located inside that window. Therefore, it is necessary to introduce an

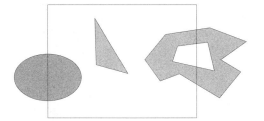 

Figure 7.30. Clipping and topological compatibility.

operation to eliminate the objects outside it. If the graphics object exists partially inside the window, as in Figure 7.30, the operation should then eliminate only the parts of the objects outside it. For this reason this operation is called *clipping*.

The clipping operation is justified in terms of computational efficiency: it avoids applying screen transformation to points of the graphics objects that will not be displayed.

To make clipping efficient, several aspects should be observed. An important factor is to use a proper coordinate system. For example, consider the clipping operation of a point P in relation to the rectangle in Figure 7.31. The rectangle region is delimited by the intersection of the 4 semiplanes defined by the straight lines of the sides. Therefore the clipping operation is solved with a maximum of 4 tests for classifying point P in relation to those semiplanes. This is a fact that does not depend on the rectangle position on the plane. However, if the rectangle is in a standard position as shown in Figure 7.31(b), the classification tests are reduced to a comparison of coordinates. Notice that we can move from the rectangle in (a) to the rectangle in (b) by a change of coordinates.

The operation of clipping an object is executed in three stages. First we calculate the intersection of the object with the window border. Computationally speaking, the calculation of the intersection is usually costly. One of the strategies to make clipping algorithms efficient is to solve the stage of point classification to avoid unnecessary intersection calculations.

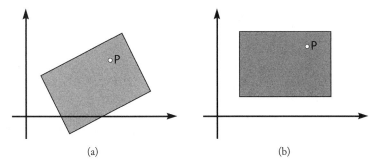

(a)   (b)

Figure 7.31. Clipping a point in relation to a rectangle.

The next stage is to classify the points of the object as interior or exterior. This is directly related with the familiar point membership classification problem. This stage divides the object $\mathcal{O}$ into a collection of subobjects $\mathcal{O}_1, \mathcal{O}_2, \ldots, \mathcal{O}_n$ such that

$$\mathcal{O} = \bigcup_j \mathcal{O}_j.$$

Each subobject $\mathcal{O}_j$ is contained in either the interior or exterior of the window (the border points are considered to be inside the window). It is key to maintain the correct topology of each subobject during the clipping calculation (Figure 7.30).

In practice, the most common approach for maintaining correct topology of the subobjects is to select clipping algorithms for the various objects through a hierarchy, taking into account the geometry and topology complexities of the object. We select algorithms for clipping in the following order: (1) points, (2) straight line segments, (3) polygonal curves, and (4) polygonal regions.

Point clipping is a point membership classification problem. As the window is a rectangle, with sides parallel to the coordinated axes, the classification of points is easily solved with a simple coordinate comparison.

Clipping methods for straight line segments have been widely researched. The clipping of polygonal curves is reduced, in an obvious way, to the clipping of straight line segments. The clipping of polygonal regions is more complex because, as we pointed out previously, we must maintain the correct topology of each clipped region.

The final stage in object clipping, once we have determined which points are inside and which are outside the window, is to eliminate those that are outside. Chapter 12 is dedicated to this stage.

7.9   Viewing Operations

Screen transformation, clipping, and rasterization are the essence of the process of displaying planar graphics objects in devices of the type matrix, as shown in Figure 7.32.

The order in which clipping, rasterization operations, and screen transformations are performed is of fundamental importance. This order is linked to both computational efficiency and numerical errors problems.

Deciding in what order to perform these operations requires us to choose whether to perform clipping in the continuous (mathematical universe) or in the discrete (representation universe) domain (i.e., should we perform rasterization before or after clipping?). Clipping after rasterization essentially reduces the clipping problem to the point membership classification problem (cell classification). On the other hand, rasterization can be computationally expensive. If the object geometry has a complex geometry description, then clipping in the continuous domain can be very computationally expensive due to the intersection calculation. In this case, it is more efficient to rasterize the objects before clipping. In practice, it is most efficient to perform clipping with rasterization. This strategy allows the use of clipping regions with more complex geometries.

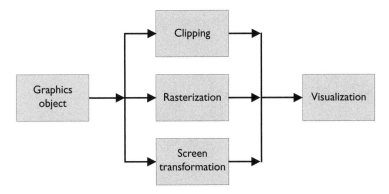

Figure 7.32. 2D viewing operations.

7.10   Comments and References

In this chapter we briefly discussed graphics output devices. We will not dedicate more space in the book to this topic. For a broader study on graphics devices, consult [Gomes and Velho 02].

The concept of graphics objects was introduced in [Gomes et al. 96], where it was used to unify the given coverage for the warping and morphing problems in computing graphics (including deformation of images, curves, surfaces, and volumetric data).

We have mentioned that a curve can be arbitrarily approximated by a polygonal line. The curves satisfying this property are called *rectifiable curves*. A complete coverage of curves in $\mathbb{R}^n$, with an exhaustive discussion on rectifiable curves, can to be seen in [Lima 83].

We also mentioned that the effect on regular value is an immediate consequence of the implicit function theorem covered in calculus courses. A complete description of this theorem can be seen in [Lima 83]. However, it is not an introductory text.

The technique of rasterization by spatial subdivision, as seen in Figure 7.22, is described in [de Figueiredo and Stolfi 96].

7.10.1   Additional Topics

This chapter could be called "2D Computer Graphics." In fact, it could be easily extended to become a complete course on 2D computer graphics by including such topics as clipping and rasterization algorithms, digital typography, Postscript, and painting systems, among others. Too often, 2D computer graphics receives little attention in curricula despite having great importance.

An important additional topic is the study of various representation and reconstruction methods of planar curves (e.g., splines, Bezier, Hermite, and subdivision curves). Another interesting and vast topic is the digital typography study, in which several methods and techniques described in this chapter are used.

Exercises

1. Why does the letter "X" not have the same topology as a straight line interval?

2. Discuss the unicity representation problem of a subset on the plane:

    (a) Is the polygonal representation of a curve unique?

    (b) Is the matrix representation of a subset unique?

    (c) Is the polygonal representation of a subset unique?

3. Describe a method for converting a matrix representation of a topological curve to a polygonal representation.

4. Write the pseudocode of the Bresenham algorithm using integer arithmetic. (Hint: remember that $m = \Delta y / \Delta x$. Multiply the test $\varepsilon + m < 0.5$ by $2\Delta x$, obtaining a test with integer arithmetic. Calculate the new error $\varepsilon' = \varepsilon \Delta x$ in an incremental way.)

5. Complete the pseudocode of the Bresenham algorithm for all cases beyond $0 < m < 1$.

6. Describe a method for converting a polygonal representation of a topological curve into a matrix representation.

7. Consider a parametric curve of class C^∞ $f \colon [a,b] \to \mathbb{R}^2$, $f(t) = (x(t), y(t)$. The vector $v(t) = f'(t) = (x'(t), y'(t))$ is called the *velocity vector* of f.

    (a) Define a parameterization change (i.e., change of coordinates) of f.

    (b) Show that a parameterization change alters the velocity vector of f.

8. Consider a parametric curve of class C^∞ $f \colon [a,b] \to \mathbb{R}^2$, $f(t) = (x(t), y(t)$. If $|f'(t)| = 1$, for every $t \in [a,b]$ we say the curve is *parameterized by arc length*.

    (a) Interpret this parameterization geometrically. Define a change of parameterization of f and show that the re-parameterization alters the velocity of the curve.

    (b) Show that if the velocity of f is never null, then a reparameterization of f by arc length exists.

9. Let $f \colon I \subset \mathbb{R} \to \mathbb{R}^2$ be a curve parameterized by arc length and $\mathbf{v}(t) = (x'(t), y'(t))$ its velocity vector.

    (a) Show that the vector $\mathbf{n} = (-y'(t), x'(t))$ is normal to the curve f.

    (b) Show that for each $t \in I$, there exists a constant $k(t)$ such that $\mathbf{t}'(t) = k(t)\mathbf{n}(t)$. (The function $k(t)$ is called the *curvature* of curve f.)

    (c) Calculate the curvature of circle $f(t) = (r\cos t, r \operatorname{sen} t)$.

(d) How would you define the curvature of a polygonal curve without performing reconstruction?

10. Consider a parametric curve of class C^∞ $f\colon [a, b] \to \mathbb{R}^2$. Describe a recursive algorithm for polygonizing f so that the polygonal segments have smaller length in regions of larger curvature.

11. Describe a hybrid rasterization algorithm for a triangle.

12. Describe a method to perform clipping of a straight line segment in a rectangular window whose sides are parallel to the coordinate axes on the plane. Your method should avoid unnecessary intersection calculations. How do you perform clipping of a straight line segment in relation to a rectangular window arbitrarily positioned on the plane?

13. Describe a clipping method for a straight line segment in relation to a window defined as an ellipse on the plane.

14. Consider the curves implicitly defined by $y + x^2 - 4 = 0$ and $x^2 + 4y^2 - 16 = 0$.

(a) Show that the curves are regular.

(b) The complement set of the union of those two curves comprises several regions on the plane. Determine each of those regions by giving its implicit description.

(c) Perform clipping of the straight line segment connecting the points $(-2, -4)$ and $(4, 0)$ in relation to each of those regions.

15. Present four different methods for obtaining a polygonization of the unit circle (you may include the methods discussed in the chapter).

16. Describe a representation method for the polygonal approximation of a regular implicit curve using the gradient of the implicit function. Does this method fit the sampling/reconstruction paradigm studied in this chapter? Discuss the advantages and disadvantages of this method in relation to the method described in this chapter.

17. Solve the questions below related to polygonization methods of planar implicit curves presented in this chapter:

(a) The method does not work in the case that the implicit curve passes through the vertices of a triangle from the decomposition. Why? How can this problem be solved?

(b) How can the described method be modified so that the vertices of the resulting polygonal curve are points of the implicit curve?

18. In this chapter we only defined parametric graphics objects of dimension 1. Define parameterized planar graphics objects of dimension 2 and give examples.

19. Describe a method for solving the point membership classification problem for a region on the plane defined by a triangle.

20. Describe a method of clipping a straight line segment in relation to a triangle on the plane.

21. Describe a method for solving the point membership classification problem for a region on the plane defined by a polygonal Jordan curve.

22. Describe an analytical method for clipping a circle in relation to a standard rectangle on the plane.

23. Write pseudocode for a polygon rasterization algorithm using the hybrid rasterization method.

24. Consider a nonuniform grid, fixed and associated to a quadtree structure. Describe an algorithm for rasterizing a straight line segment in this grid (quadtree structures will be described further in Chapter 9).

8 Spatial Graphics Objects

In Chapter 7 we studied planar graphics objects. In this chapter we return to the problem of describing and representing graphics objects with a focus on spatial graphics objects. The study of spatial graphics objects includes the mathematical modeling of objects from our physical world. The study of the geometry, topology, and representation of graphics objects is part of the area of geometric modeling. Geometric modeling was the first area in computer graphics that sought a robust formalization of its methods by introducing mathematical concepts.

8.1 Digital Geometry Processing

The four universes paradigm helps us understand the process of geometric modeling:

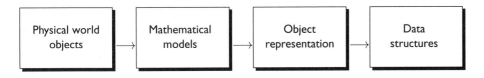

In Chapter 7 we saw that graphics objects are characterized by their support set and attribute function. More precisely, a graphics object $\mathcal{O}$ is a pair $\mathcal{O} = (U, f)$, where the function $f \colon U \subset \mathbb{R}^m \to \mathbb{R}^n$ defines the attributes. In this chapter, we are interested in spatial objects, in which the ambient space dimension is $m = 3$.

Some of the planar graphics objects ($m = 2$) we previously studied are listed in the table below.

Object Type	Object Dim
Planar curve	1
Planar region	2
2D solids	2
Image	2

We have the following spatial graphics objects ($m = 3$):

Object Type	Object Dim
Curves	1
Surfaces	2
Solids	3
3D image	3
Volumetric objects	3

In this chapter we will focus on surfaces and solids, exploring their description, representation, reconstruction methods, and computer coding.

If you are a bit baffled by the diversity of graphics objects presented in this table, there is no need to worry: we listed all these object types to pique your curiosity. You may be asking yourself, What is a volumetric object? How is it different from a solid object? What is a 3D image? While we will not provide a deep discussion about the answers to these questions in this book, you may want to study them further.

Recently, volumetric objects—objects that resemble images but also contain geometric and topological information—have come to play a significant role in geometry description. Representing such objects requires processing techniques that go beyond extending image processing techniques to the 3D space. The terms *digital geometry* and *digital geometry processing* describe the various methods and techniques required to solve these challenges. Our ability to create volumetric objects has increased thanks to advances in the technology used to capturing volumetric data, which have evolved considerably. Some of this technology includes magnetic resonance equipment for industrial and medical applications, computerized tomography, and lasers that measure the depth of a scene (laser scanners). There have also been advances in the methods of representing these objects, as well as in the conversion between different descriptions or representations.

8.2   Spatial Curves

A parametric curve in $\mathbb{R}^3$ is an application $g\colon I \subset \mathbb{R} \to \mathbb{R}^3$. Therefore, g is determined by its coordinates

$$g(t) = (x(t), y(t), z(t)), \quad t \in I.$$

These curves have a great variety of applications in computer graphics. For example, in modeling they are used as auxiliary elements for surface construction; in animation they are used as trajectories. The velocity vector of curve g is defined by the derivative

$$g'(t) = (x'(t), y'(t), z'(t)).$$

8.3 Surfaces

A *topological surface* is a subset S of the Euclidean space $\mathbb{R}^3$ that is locally homeomorphic to the Euclidean plane $\mathbb{R}^2$. More precisely, for each point $p \in S$ there exists a spherical neighborhood $B_\varepsilon^3(p) \subset \mathbb{R}^3$ with center in p such the set $B_\varepsilon^3(p) \cap S$ is homeomorphic to the unit disk

$$B_1^2(0) = \{(x, y) \in \mathbb{R}^2 \; ; \; x^2 + y^2 < 1\} \tag{8.1}$$

of the Euclidean plane (see Figure 8.1). Intuitively, this definition means that a surface is obtained by gluing deformed pieces of the plane. We obtain different classes of surfaces by enforcing different types of regularity on the homeomorphism defining S locally. A very common type of regularity consists of enforcing the homeomorphism to be a diffeomorphism, resulting, in this case, in a differentiable surface. As we will see later, differentiable surfaces are characterized by having a tangent plane at each point.

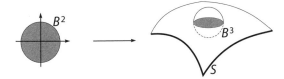

Figure 8.1. Definition of a surface.

Note that the above definition does not consider surfaces with borders. However, it can be modified to that end by considering, in the definition beyond the open disk of the plane in Equation (8.1), the unit semidisk

$$\tilde{B}_1^2(0) = \{(x, y) \in \mathbb{R}^2 ; x^2 + y^2 < 1 \quad \text{and} \quad y \geq 0\},$$

of the semiplane $\{(x, y) \in \mathbb{R}^2 \; ; \; y \geq 0\}$. In this case, at the border points the neighborhood $B_\varepsilon^3(p) \cap S$ is homeomorphic to $\tilde{B}_1^2(0)$, and at the interior points of the surface the neighborhood $B_\varepsilon^3(p) \cap S$ is homeomorphic to $B_1^2(0)$. This fact is illustrated in Figure 8.2(a).

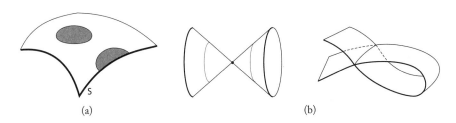

(a)   (b)

Figure 8.2. (a) Surface neighborhoods; (b) geometric objects that are not surfaces.

The purpose of taking such care to define a surface is to avoid having geometric objects such as the two shown in Figure 8.2(b), which have either vertices or intersections, defined as surfaces.

8.3.1   Description of a Surface

Surfaces are spatial graphics objects of dimension 2. There is some similarity between the study of surfaces of $\mathbb{R}^3$ and curves in $\mathbb{R}^2$, which comes from the fact that both of them have the same codimension 1 (codimension is the difference between the dimensions of the space and the object). As in the case of curves, there are essentially two methods of describing a surface: parametric and implicit descriptions.

Parametric surfaces. A parametric surface S is described by a transformation $f \colon U \subset \mathbb{R}^2 \to \mathbb{R}^3$, as shown in Figure 8.3. Geometrically, f defines a system of 2D curvilinear coordinates in the surface $S = f(U)$.

To avoid degenerate cases of parametric surfaces, we apply some conditions to function f. A natural condition is for f to be bijective in the interior of domain U, and its derivative has rank 2. Geometrically, this means the partial derivative vectors

$$\frac{\partial f}{\partial u} = \left(\frac{\partial f_1}{\partial u}, \frac{\partial f_2}{\partial u}, \frac{\partial f_3}{\partial u} \right),$$

and

$$\frac{\partial f}{\partial v} = \left(\frac{\partial f_1}{\partial v}, \frac{\partial f_2}{\partial v}, \frac{\partial f_3}{\partial v} \right),$$

are linearly independent. Notice these vectors form a vector basis of the plane tangent to the surface at the point.

Example 8.1 (Cylinder). A cylinder is the surface described as the set of points in space equidistant to a straight line, called the axis of the cylinder. The constant distance to the axis is the *radius of the cylinder*. Let us take the axis as being the z-axis in space $\mathbb{R}^3$, and let us assume the cylinder has radius R. If (x, y, z) is a point on the cylinder, then (x, y)

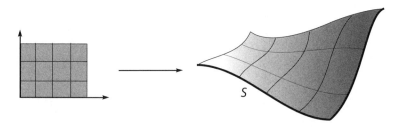

Figure 8.3. Parametric surface.

belongs to a circle of radius R with center in the origin of plane xy. The parameterization of this circle is given by

$$(x, y) = (R \cos u, R \sin u), \quad u \in \mathbb{R}.$$

We then have a parameterization of the cylinder $f \colon [0, 2\pi] \times \mathbb{R} \to \mathbb{R}^3$,

$$f(u, v) = (R \cos u, R \sin u, v).$$

This parameterization introduces *cylindrical coordinates* in the surface. (Actually, by varying the radius R we introduce the system of cylindrical coordinates in space $\mathbb{R}^3$). ❏

Implicit surfaces. An implicit surface $S \subset \mathbb{R}^3$ is defined by the set of the roots of a function $F \colon U \subset \mathbb{R}^3 \to \mathbb{R}$. That is,

$$S = \{(x, y, z) \in U \; ; \; F(x, y, z) = 0\}.$$

This set is indicated by the notation $F^{-1}(0)$, called the *inverse image* of set $\{0\} \subset \mathbb{R}$ by function f, and it defines a *level surface* of function F. This definition is illustrated in Figure 8.4, where the implicit function F is defined in a cube of $\mathbb{R}^3$. It is common to call the function $f \colon U \subset \mathbb{R}^3 \to \mathbb{R}$ the *scalar field* in U because, similar to a vector field, it associates the scalar $F(p)$ to each point $p \in U$.

As with parametric surfaces, we should apply conditions to function f to avoid having degenerate surfaces. A natural condition, analogous to the condition about the derivative in the parameterized case, consists of enforcing the gradient vector

$$\operatorname{grad}(f) = \left(\frac{\partial F}{\partial x}, \frac{\partial F}{\partial y}, \frac{\partial F}{\partial z} \right),$$

not to be canceled at the points in the surface $S = F^{-1}(0)$.

Example 8.2 (Cylinder). As in the example above, we use a cylinder, defined as the surface described by the set of points in space equidistant to its axis. Again we take the axis as

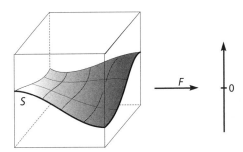

Figure 8.4. Level surface of function F.

being the z-axis of the coordinate system and assume the cylinder has radius $R > 0$. If (x, y, z) is a point of the cylinder, then

$$||(x, y, z) - (0, 0, z)|| = R,$$

from which follows

$$x^2 + y^2 - R^2 = 0.$$

Therefore, the cylinder is implicitly defined by the function $F \colon \mathbb{R}^3 \to \mathbb{R}$ given by

$$F(x, y, z) = x^2 + y^2 - R^2.$$

The cylinder is an algebraic surface of degree 2 and is part of the family of surfaces known as *quadrics*.   ❏

As we already observed for planar objects, we must remember that not every parametric surface defines a topological surface, and not all implicit equations define a topological surface.

8.3.2   Geometric Attributes of Surfaces

Consider a surface $S \subset \mathbb{R}^3$ and a point $p \in S$. A vector $v \in \mathbb{R}^3$ is tangent to S at point p if there exists a curve $\gamma \colon (-1, 1) \to S$ such that $\gamma(0) = p$ and $\gamma'(0) = v$ (see Figure 8.5(a)). In other words, v is the velocity vector of a curve contained in the surface. The set of all tangent vectors S at point p forms the *tangent plane* of surface S at point p, which we indicate by $T_p S$. Vector $n \in \mathbb{R}^3$ is normal to the surface S at point p if n is perpendicular to the tangent plane $T_p S$ of S at p (see Figure 8.5(b)).

In the case of a parametric surface

$$f(u, v) = (f_1(u, v), f_2(u, v), f_3(u, v)),$$

the normal vector S is calculated using the wedge product

$$\overrightarrow{n} = \frac{\partial f}{\partial u} \wedge \frac{\partial f}{\partial v},$$

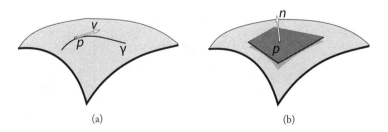

$$(a) \qquad\qquad\qquad\qquad (b)$$

Figure 8.5. (a) Tangent vector; (b) normal vector and tangent plane.

where

$$\frac{\partial f}{\partial u} = \left(\frac{\partial f_1}{\partial u}, \frac{\partial f_2}{\partial u}, \frac{\partial f_3}{\partial u}\right), \quad \text{and} \quad \frac{\partial f}{\partial v} = \left(\frac{\partial f_1}{\partial v}, \frac{\partial f_2}{\partial v}, \frac{\partial f_3}{\partial v}\right).$$

Note the previous condition, in which the derivative of the parameterization f has rank 2, guarantees the normal vector will not be canceled.

In the case that surface S is implicitly defined, $S = F^{-1}(0)$, by a function $F \colon \mathbb{R}^3 \to \mathbb{R}$, the normal vector is given by the gradient

$$\mathrm{grad}(F) = \left(\frac{\partial F}{\partial x}, \frac{\partial F}{\partial y}, \frac{\partial F}{\partial z}\right).$$

The normal vector is an essential attribute for calculating surface illumination (Chapter 14) and extracting features (e.g., silhouettes, creases, and other discontinuities) [Gooch and Gooch 01].

8.4   Volumetric Objects

A *volumetric object* is the 3D analogue of regions in the planar case. In other words, it is a graphics object having the same dimensions as the ambient space. These objects, also called *solids*, are used in many areas, especially in the manipulation and visualization of objects reconstructed from data captured through sampling devices, such as computerized tomography and magnetic resonance equipments. For this reason, these graphics objects are largely used in the areas of medical and seismic imaging. Solid objects are also extensively used for describing mechanical parts in the computer.

A solid is a limited subset $V \subset \mathbb{R}^3$ such that, for every point $p \in V$, there exists an open spherical neighborhood $B_\varepsilon^3(p)$ with center in p and radius $\varepsilon > 0$ in the space, so that $B_\varepsilon^3(p) \cap V$ is homeomorphic to the unit sphere

$$B_1^3(0) = \{(x, y, z) \in \mathbb{R}^3; x^2 + y^2 + z^2 < 1\},$$

in the space $\mathbb{R}^3$, or to the unit sphere

$$\tilde{B}_1^3(0) = \{(x, y, z) \in \mathbb{R}^3; x^2 + y^2 + z^2 < 1 \quad \text{and} \quad z \geq 0\}$$

in the semispace

$$\mathbb{R}_+^3 = \{(x, y, z) \in \mathbb{R}^3; z \geq 0\},$$

Points $p \in V$, whose open neighborhood $B_\varepsilon^3(p) \cap V$ is homeomorphic to $\tilde{B}_1^3(0)$, constitute the border points of V.

A volumetric object is a graphics object of dimension 3, embedded in the space $\mathbb{R}^3$. More generally, we can consider a graphics object of dimension n embedded in a Euclidean space $\mathbb{R}^n$, with $n \geq 2$. (In the case $n = 2$ we have a planar region (2D solid), which we studied in Chapter 7.)

In general, a volumetric object is described by its density function: a function of constant density characterizes the type of solids commonly used to describe mechanical parts. A function of variable density describes volumetric objects with variable opacities, such as those in medical imaging, where the density of tissue types varies (e.g., skin, bone, muscles).

8.4.1   Volumetric Object and Image

We should note that an image can be considered a 2D volumetric object embedded in the plane. The color attributes of each pixel can have different interpretations depending on the application. For example, in a grayscale image we can interpret the gray values at each pixel as being its density (provided by the density function).

While images may be considered 2D volumetric objects, there is greater value in distinguishing images from volumetric objects of dimension ≥ 3 (in the 3D case, the attribute function of a volumetric object determines the geometric and topological characteristics of the object), as this allows for an intersection between the study of volumetric objects and the area of geometric modeling (digital geometry).

8.4.2   Description of Volumetric Objects

We now discuss two methods for describing volumetric objects: by boundary and implicit functions.

Boundary description. In Chapter 7 we used Jordan's Theorem to characterize regions of the plane. This theorem extends to 3D space. The analogue of a closed and simple curve is a surface without boundary, closed and limited. These surfaces are called *compact surfaces* in topology. An example of a compact surface is the unit sphere of $\mathbb{R}^n$: $S^2 = \{x \in \mathbb{R}^n; |x| = 1\}$.

We may state Jordan's Theorem as follows: a compact surface M in $\mathbb{R}^3$ divides the space into two regions R_1 and R_2, one being limited and the other unlimited, of which M is the common boundary. The limited region S_1 defines a *solid* in space. According to this theorem, Figure 8.6 displays a solid limited by a surface.

The description of a solid in space by its boundary surface is called a *boundary description* of the solid. The boundary description of a solid consists of a description of the boundary surface followed by solution of the point membership classification problem. In other words, we should have a computational method to classify points of the space in relation to the two regions R_1 and R_2 from Jordan's Theorem. Depending on the method used to describe the boundary surface, this task is not easy. The boundary representation is not an efficient way of representing a volumetric object because we ought to first solve the point membership classification problem to decide if a certain point in space even belongs to the solid. What is more, this representation does not make possible a description of solids constituted of nonhomogeneous matter, which has variable density.

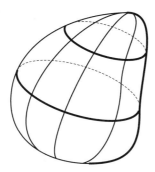

Figure 8.6. Solid limited by a surface.

Description by implicit functions. Let $F\colon \mathbb{R}^3 \to \mathbb{R}$ be a function dividing the points in space into three classes:

1. $\{(x, y, z) \in \mathbb{R}^3 \; ; \; F(x, y, z) > 0\}$,

2. $\{(x, y, z) \in \mathbb{R}^3 \; ; \; F(x, y, z) = 0\} = F^{-1}(0)$,

3. $\{(x, y, z) \in \mathbb{R}^3 \; ; \; F(x, y, z) < 0\}$.

Geometrically, the set $F^{-1}(0)$ defines an implicit surface M, and the other two sets define the interior and exterior of M. If surface M is limited, then function F describes a solid whose boundary is M. The solid is formed by the limited region of the space defined by $F(x, y, z) < 0$ or $F(x, y, z) > 0$, together with the surface M of the boundary.

Function F solves the point membership classification problem and also provides a computational method to calculate, respectively, the interior and external regions R_1 and R_2 given by Jordan's Theorem. However, $F^{-1}(0)$ can be a nonconvex set. In this case, the surface comprises convex connected components.

Function F solves the point membership classification problem for a solid implicitly defined. Furthermore, F has an interesting physical description: F can be interpreted as being the *density function* of the solid.

On the other hand, the density function of a solid $\mathcal{O}$ naturally defines an implicit description of $\mathcal{O}$. In reality, different parts of the object $\mathcal{O}$ can be characterized as implicit volumes $\{p \in \mathbb{R}^3 \; ; \; F(p) \leq c\}$: in other words, the set of points in space with a density function below a certain value.

When a volumetric object $\mathcal{O}$ has a constant density function, it can be described implicitly by

$$\mathcal{O} = \{(x, y, z) \in \mathbb{R}^3 \; ; \; F_\mathcal{O}(x, y, z) = 1\},$$

where $F_\mathcal{O}$ is the *characteristic function* of the geometric support. That is,

$$F_\mathcal{O}(p) = \begin{cases} 1, & \text{if } p \in \mathcal{O}; \\ 0, & \text{if } p \notin \mathcal{O}. \end{cases} \tag{8.2}$$

We have shown that volumetric objects and implicit functions are closely related. The implicit description of volumetric objects is a powerful tool for synthesizing volumetric objects of very complex geometric forms.

Level surfaces. Let us assume a volumetric object $\mathcal{O}$ is described by some implicit function F. For each $c \in \mathbb{R}$, the inverse image $F^{-1}(c)$ describes a level surface of the object. If F describes the object's density, the level set $F^{-1}(c)$ contains all the points in space with constant density c. Level surfaces are also called *isosurfaces* and are useful to the visualization, manipulation, and analysis of volumetric objects.

8.5   Triangulations and Polyhedral Surfaces

In Chapter 7 we saw that polygonal curves play an important role in the representation of planar curves. In this section we will describe the role of polyhedral surfaces in the representation of surfaces. Polyhedral surfaces are based on the concept of triangulation, as introduced in Chapter 7 for planar regions.

8.5.1   2D Triangulation

Three points $p_0, p_1, p_2 \in \mathbb{R}^3$ form a triangle in $\mathbb{R}^3$ if the vectors $p_1 - p_0$ and $p_2 - p_0$ are linearly independent. A *2D triangulation* in $\mathbb{R}^3$ is a collection $\mathcal{T} = \{T_i\}$ of triangles in $\mathbb{R}^3$ such that, given two triangles $T_i, T_j \in \mathcal{T}$, if $T_i \cap T_j \neq \varnothing$ then $T_i \cap T_j$ is a common vertex, or $T_i \cap T_j$ is a common edge.

Notice this definition is identical to that given for triangulation of a plane, in which we simply did not require the triangles to be contained in the plane $\mathbb{R}^2$. We will also define a 3D triangulation in $\mathbb{R}^3$ that will be useful further on.

8.5.2   3D Triangulation

A list of four points $\sigma = (p_0, p_1, p_2, p_3)$ with $p_i \in \mathbb{R}^3$ form a *tetrahedron* in $\mathbb{R}^3$ if they constitute an affine basis; in other words, the vectors $p_1 - p_0$, $p_2 - p_0$, and $p_3 - p_0$ are linearly independent. Geometrically, a tetrahedron is a pyramid with a triangular base, as shown in Figure 8.7.

A tetrahedron is therefore a volumetric triangle. The points p_0, p_1, p_2, and p_3 are called the vertices of the tetrahedron; the segments p_0p_1, p_1p_2, p_0p_2, p_0p_3, p_1p_3, and p_2p_3 are called the *edges* of the tetrahedron; and the triangles $\Delta p_0p_1p_2$, $\Delta p_0p_1p_3$, and $\Delta p_1p_2p_3$ are called the *faces* of the tetrahedron σ.

The tetrahedron σ has a coordinate system naturally defined by barycentric coordinates: a point p belongs to σ if and only if

$$p = t_0p_0 + t_1p_1 + t_2p_2 + t_3p_3, \quad \text{where} \quad 0 \leq t_i \leq 1.$$

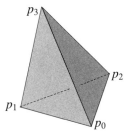

Figure 8.7. Tetrahedron in $\mathbb{R}^3$ space.

The list (t_0, t_1, t_2, t_3) is the barycentric coordinates of the point (see Chapter 2 for more details).

A *3D triangulation* or *volumetric triangulation* of the space is a finite set $\{\sigma_1, \ldots, \sigma_n\}$ of tetrahedra such that the intersection between two tetrahedra of the set is null or results in either a vertex, an edge, or a face.

8.5.3 Polyhedral Surfaces

We can use the concept of 2D triangulation to define the analogue of a polygonal curve: a surface. A *polyhedral surface* (see Figure 8.8) is a triangulation of the space that is a topological surface. As we have much freedom in positioning triangles in space, we must be careful to define polyhedral surfaces in a way that guarantees the topology of the triangular

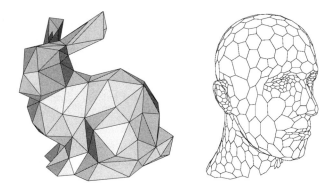

Figure 8.8. Polyhedral surface.

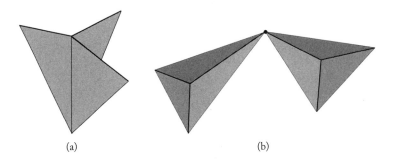

(a)   (b)

Figure 8.9. Triangulations that are not surfaces. (a) Three triangles with a common edge. Any neighborhood of the points of that edge is homeomorphic to a neighborhood of the Euclidean plane. (b) Two tetrahedra with a common vertex. No neighborhood of that vertex is homeomorphic to a neighborhood of the plane.

object has the topology of a surface. Figure 8.9 shows two cases we want to avoid of objects formed by triangles. As a surface should be locally homeomorphic to the plane, we want to avoid such triangulations. The case in Figure 8.9(a) can be avoided by enforcing the following condition that an edge is common to at most two triangles. When this condition is satisfied, the polyhedral is called a *pseudosurface*. Note, however, that degeneracies can still happen in a pseudosurface, as shown in Figure 8.9(b). A pseudosurface that is locally homeomorphic to the Euclidean plane is called a *polyhedral surface*. Local homeomorphism excludes situations like the ones illustrated in Figure 8.9(b).

Why surfaces? Why triangles? We have limited ourselves to studying surfaces because they are an object model extensively studied in mathematics with many results. That being said, there is no reason to ignore objects like those in Figure 8.9 that can be useful as geometric models. But in order to represent such objects we would need to use more general triangulated structures (called *simplicial complexes*) than surfaces, and these structures are outside the scope of this book.

We have also defined polyhedral surfaces solely with triangular faces, instead of with faces that are arbitrary polygons. This decision is motivated by simplicity. Triangular faces have the advantage that they are always planar polygonal curves, they have a linear coordinate system naturally associated to them (barycentric coordinates), and they extend naturally to $\mathbb{R}^n$ using the concept of *simplicial complexes*. The triangulations forming a topological surface are also relatively easy to represent and manipulate on a computer. The planarity of triangles facilitates calculations with polyhedral surfaces. On the other hand, the barycentric coordinate system can be used to define surface attributes in the vertices of each triangle and to perform interpolation on those coordinates. We should highlight that any polygon can be subdivided into triangles; therefore, we lose almost nothing by the decision to restrict ourselves to triangular faces.

To consider polyhedral surfaces with faces as arbitrary polygons we require, as in a triangulation, that two polygons to be either disjunct or intersect each other at a vertex or edge. We will use the term *polyhedral surface* for this general sense and *triangulated surface* when it is necessary to emphasize that face polygons are triangles.

8.5.4   Coding Polyhedral Surfaces

In this section we will treat the problem of coding the various geometric and topological structures of a polyhedral surface. This coding is directly related to the data structures associated to the triangulation of the surface.

Coding and graphs. There are several graphs naturally associated to a polyhedral surface. Among these graphs, the most common are the *vertex graph* and the *dual graph*.

Vertex graph. The vertices of this graph are made up of the collection of vertices of the triangles. Two vertices are connected if and only if they are edges of some triangle of the surface. In other words, this graph is constituted by the vertices and edges of the surface (see Figure 8.10(a)).

Dual graph. The vertices of this graph are formed by the surface triangles. Two vertices are connected if the corresponding triangles have one common edge. In Figure 8.10(b), we show the vertex graph in stippled lines and the dual graph in solid lines.

The problem of structuring a polyhedral surface is therefore a problem of coding these graphs. We will now describe three coding methods: explicit coding, coding through a vertex list, and coding through an edge list. We will use the pyramid in Figure 8.11, which has five vertices, five faces, and eight edges, to exemplify each method.

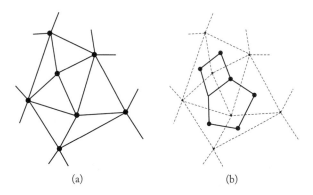

(a)   (b)

Figure 8.10. (a) Vertex graph; (b) dual graph.

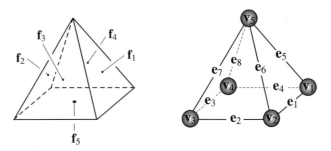

Figure 8.11. Quadrangular pyramid.

Explicit coding. This method explicitly codes each surface polygon by providing a list of its vertices and their coordinates. In the case of our example, the coding is given in the table below.

Explicit Coding
$\mathbf{f}_1 = ((x_1, y_1, z_1), (x_5, y_5, z_5), (x_2, y_2, z_2))$
$\mathbf{f}_2 = ((x_3, y_3, z_3), (x_2, y_2, z_2), (x_5, y_5, z_5))$
$\mathbf{f}_3 = ((x_3, y_3, z_3), (x_4, y_4, z_4), (x_5, y_5, z_5))$
$\mathbf{f}_4 = ((x_1, y_1, z_1), (x_4, y_4, z_4), (x_5, y_5, z_5))$
$\mathbf{f}_5 = ((x_1, y_1, z_1), (x_2, y_2, z_2), (x_3, y_3, z_3), (x_4, y_4, z_4))$

This coding has the advantage of being extremely simple; however, this simplicity is associated with some disadvantages. For example, we are not taking into account that vertices are generally shared by two or more triangles. This creates redundancy in our vertex coding, which can lead to problems. The redundancy occupies unnecessary storage space and in the visualization of the polygonal mesh, each shared edge is drawn twice, which is inefficient and inconvenient. Also, geometric operations with triangles can introduce numerical errors in the coordinates. As vertices are processed independently, those errors can result in having same vertices with different coordinates.

When manipulating a geometric model on the computer, we need topological and geometric information about the model. From this point of view, the surface representation is seen as a geometric database that receives several types of queries. As the surface elements are coded through graphs, the query problem becomes a graph search problem. For example, consider some common queries: find all edges incident to a vertex; find polygons sharing an edge or a vertex; find the edges delimiting a polygon; and visualize the surface.

Notice that these queries cannot be immediately answered when we have used explicit coding: instead we must compare coordinates, which is usually done with floating point numeric representation. This comparison can be slow for a surface with many triangles, and it is subject to errors, as seen already. Another complication proceeds from the fact that adjacency relations among vertices, edges, and polygons are not stored explicitly, which forces the execution of geometric calculations, an expensive process from the computational point of view and subject to errors.

Finding efficient codings is extremely important for geometric modeling. To avoid the problems found in explicit structuring, we need to avoid replication of vertices; and encode adjacency information so as to facilitate the queries about the surface geometry and topology. (Notice this is the familiar problem of finding an optimum balance between storage space and processing time.)

Coding with a vertex list. Coding with a vertex list is simpler and more efficient than explicit coding. In this method we create a vertex list and each surface polygon (face) is defined by reference to its vertices in this list. In the case of the pyramidal surface of Figure 8.11, the vertex and polygon (face) lists are given in the two tables below.

Vertex List
$\mathbf{v}_1 = (x_1, y_1, z_1)$
$\mathbf{v}_2 = (x_2, y_2, z_2)$
$\mathbf{v}_3 = (x_3, y_3, z_3)$
$\mathbf{v}_4 = (x_4, y_4, z_4)$
$\mathbf{v}_5 = (x_5, y_5, z_5)$

Face List
$\mathbf{f}_1 \rightarrow (\mathbf{v}_1, \mathbf{v}_5, \mathbf{v}_2)$
$\mathbf{f}_2 \rightarrow (\mathbf{v}_3, \mathbf{v}_2, \mathbf{v}_5)$
$\mathbf{f}_3 \rightarrow (\mathbf{v}_3, \mathbf{v}_4, \mathbf{v}_5)$
$\mathbf{f}_4 \rightarrow (\mathbf{v}_1, \mathbf{v}_4, \mathbf{v}_5)$
$\mathbf{f}_5 \rightarrow (\mathbf{v}_1, \mathbf{v}_2, \mathbf{v}_3, \mathbf{v}_4)$

Because each vertex is stored only once, this coding is more economical in terms of space. This coding also has the advantage that when one alters the coordinates of a particular vertex, all of its incident polygons are automatically altered. This facilitates the interactive manipulation of a surface. Despite these advantages, with this coding it is still difficult to identify polygons sharing an edge, and shared edges are still drawn twice.

Coding with edge list. The vertex list has greater flexibility and robustness than explicit coding. Coding with an edge list has even more advantages. In this coding, we define faces as references to the edges, and not directly to the vertices. With the edge list we can obtain dual graph information about the edge adjacencies.

Consequently, we now have three lists: the vertex list, containing the coordinates of all the polygon vertices which constitute the object, the edge list, in which each edge is defined by a reference to the vertices defining it, and the face list, in which each face is defined by a reference to the edges defining it. These lists, for the case of the pyramidal surface in Figure 8.11, are given in the tables below.

Vertex List
$\mathbf{v}_1 = (x_1, y_1, z_1)$
$\mathbf{v}_2 = (x_2, y_2, z_2)$
$\mathbf{v}_3 = (x_3, y_3, z_3)$
$\mathbf{v}_4 = (x_4, y_4, z_4)$
$\mathbf{v}_5 = (x_5, y_5, z_5)$

Edge List
Edges $\rightarrow$ Vertices
$\mathbf{e}_1 \rightarrow \mathbf{v}_1, \mathbf{v}_2$
$\mathbf{e}_2 \rightarrow \mathbf{v}_2, \mathbf{v}_3$
$\mathbf{e}_3 \rightarrow \mathbf{v}_3, \mathbf{v}_4$
$\mathbf{e}_4 \rightarrow \mathbf{v}_4, \mathbf{v}_1$
$\mathbf{e}_5 \rightarrow \mathbf{v}_1, \mathbf{v}_5$
$\mathbf{e}_6 \rightarrow \mathbf{v}_2, \mathbf{v}_5$
$\mathbf{e}_7 \rightarrow \mathbf{v}_3, \mathbf{v}_5$
$\mathbf{e}_8 \rightarrow \mathbf{v}_4, \mathbf{v}_5$

Face List
Face $\rightarrow$ Edges
$\mathbf{f}_1 \rightarrow \mathbf{e}_1, \mathbf{e}_5, \mathbf{e}_6$
$\mathbf{f}_2 \rightarrow \mathbf{e}_2, \mathbf{e}_6, \mathbf{e}_7$
$\mathbf{f}_3 \rightarrow \mathbf{e}_3, \mathbf{e}_7, \mathbf{e}_8$
$\mathbf{f}_4 \rightarrow \mathbf{e}_4, \mathbf{e}_8, \mathbf{e}_5$
$\mathbf{f}_5 \rightarrow \mathbf{e}_1, \mathbf{e}_2, \mathbf{e}_3, \mathbf{e}_4$

In this coding we have access to all of the edges, starting from the edge list, without having to traverse the border of the polygons. Finding the edges incident to a particular vertex still requires a geometric algorithm, but we can use more space and introduce, for each edge, the references for the two faces sharing it. In this way, it is now simple to determine the polygons that are incident to a particular edge.

Edge List
Edges → Vertices + Faces
$e_1 \rightarrow v_1, v_2, f_1, f_5$
$e_2 \rightarrow v_2, v_3, f_3, f_5$
$e_3 \rightarrow v_3, v_4, f_2, f_5$
$e_4 \rightarrow v_4, v_1, f_4, f_5$
$e_5 \rightarrow v_1, v_5, f_1, f_4$
$e_6 \rightarrow v_2, v_5, f_1, f_3$
$e_7 \rightarrow v_3, v_5, f_2, f_5$
$e_8 \rightarrow v_4, v_5, f_2, f_4$

The above codings still introduce many constraints related to the topology of the faces and to the geometry of the graphics object. This approach would lead us to more complete codings, such as the classic winged-edge, radial-edge, and half-edge data structures.

8.6 Representation of Parametric Surfaces

In this section, we explore representation techniques for surfaces defined parametrically.

8.6.1 Polyhedral Representation

Just as for curves, a natural method of representing a surface S consists of approximating S by a polyhedral surface $\tilde{S}$. From the sampling and reconstruction point of view, this representation method consists of point sampling surface S, then reconstructing it using piecewise linear interpolation, and finally structuring the samples in a way to obtain a triangulation (each sample is a vertex of a triangle).

As we already observed, proper sample structuring is key to obtaining the correct reconstruction. In the case of planar curves, the structuring is given by ordering the samples; in the case of surfaces, structuring is a more delicate problem, and we should use a more complex graph as we saw previously.

Given a parametric surface S, with parameterization given by $f : U \subset \mathbb{R}^2 \rightarrow \mathbb{R}^3$, the representation of f by a polyhedral surface whose faces are triangles is called a *polyhedral approximation of S*. We sometimes incorrectly call surface $\tilde{S}$ the triangulation of S. The triangulation of S is reduced to a triangulation of the parameterization domain U. In fact, if Δ_i is a triangle of U with vertices $\Delta_i = (p_{i1}, p_{i2}, p_{i3})$, then images $f(p_{i1})$, $f(p_{i2})$, and $f(p_{i3})$ of the vertices of Δ_i, for the parameterization f, are the vertices of a triangle that approximates surface S. The structuring of the triangulation of S is exactly the same as that used in the triangulation of U.

Polyhedral representations are quite common. The five regular polyhedrals (platonic polyhedrals) are classic examples of representations of a sphere. The representation of a surface by a polygonal approximation is intuitive enough; examples of such representation, outside the computer graphics arena, have existed for many years.

8.6.2 Representation by Parametric Subdivision

A topological surface S is said to be *piecewise parametric* if there exists a decomposition $S = \cup S_i$ of S in subsurfaces S_i, so each surface S_i is described by a parameterization $\varphi_i \colon U \to S_i$.

An example of a piecewise parameterization is given by a polyhedral surface S: S is a union of triangles Δ_i and each triangle $\Delta = P_0 P_1 P_2$ has a barycentric coordinate system naturally associated to it. This way, every point $P \in \Delta$ can be written in a unique way in the form

$$P = \lambda_0 P_0 + \lambda_1 P_1 + \lambda_2 P_2.$$

Barycentric coordinates naturally form a parameterization of each triangle of the surface. Therefore, a polygonal triangular representation defines a surface that is *piecewise linear*.

A piecewise parameterization is a representation method, called *representation by parametric subdivision* or *patch representation*. Surface S is first decomposed into subsurfaces S_i so that each subsurface holds a parametric description. Each subsurface S_i is then represented, together with a structuring of the decomposition of S, in the subsurfaces S_i. Finally, the reconstruction of S is performed using the structuring of the decomposition, together with a method of reconstructing the parameterization of each patch S_i.

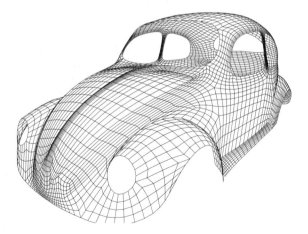

Figure 8.12. Representation by parametric subdivision.

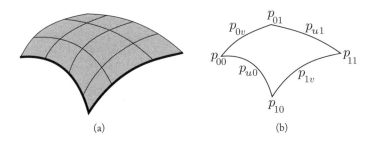

Figure 8.13. (a) Quadrangular mesh; (b) mesh patch.

Figure 8.12 shows a representation of a Volkswagen Beetle using a parametric subdivision. The parametric subdivision used for that illustration is the most common case: we define a square mesh of points in the surface, as shown in Figure 8.13(a). Each square element of the mesh represents subsurfaces of the representation, and each of these surfaces can be parametrically described. Figure 8.13(b) shows a subsurface of the mesh. Notice the mesh structure itself defines the structuring of a "curvilinear grid" of the subsurfaces.

Observe that we have eight elements in the subsurface of the mesh in Figure 8.13(b): four vertices p_{00}, p_{01} p_{11}, and p_{01} and four boundary curves p_{u0}, p_{u1}, p_{1v}, and p_{0v}. This means we have three representation methods of each patch S_i: representation by the vertices; representation by two boundary curves; and representation by four boundary curves. In the following sections, for each type of subsurface representation of the mesh, we describe the associated method of reconstruction.

Representation by vertices. In this representation, we use the sequence p_{00}, p_{01}, p_{11}, and p_{01} of four vertices to represent the subsurface. We then need a method of reconstructing the original patch (or its approximation) from the four vertices.

Consider the unit square $[0,1]^2 = [0,1] \times [0,1]$ and four points A, B, C, and D in the space $\mathbb{R}^3$. The values of a transformation in the vertices of the square are given by

$$T(0,0) = A, \quad T(1,0) = B, \quad T(1,1) = C, \quad T(0,1) = D. \tag{8.3}$$

We need an interpolation method that extends the transformation to the entire square.

To develop the interpolation method, we determine a parameterization $T \colon [0,1] \times [0,1] \to \mathbb{R}^3$ satisfying Equation (8.3). Of course, the solution to this problem is not unique. On the other hand, linear interpolation does not solve the problem if the points are not coplanar. A polynomial transformation of the second degree, providing a solution to the problem, can be obtained by *bilinear interpolation*.

The parameterization for bilinear interpolation is obtained by performing two linear interpolations. As illustrated in Figure 8.14, we first interpolate the sides AD and BC,

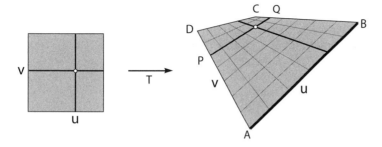

Figure 8.14. Bilinear interpolation.

obtaining points P and Q:

$$P = (1 - v)A + vD,$$
$$Q = (1 - v)B + vC.$$

Next, we interpolate the segment PQ using the parameter u:

$$T(u, v) = (1 - u)P + uQ.$$

The final transformation is given by

$$T(u, v) = (1 - u)[(1 - v)A + vD] + u[(1 - v)B + vC]$$
$$= (1 - u)(1 - v)A + (1 - u)vD + u(1 - v)B + uvC.$$

This parameterization can be written using matrix notation

$$T(u, v) = \begin{pmatrix} 1 - u & u \end{pmatrix} \begin{pmatrix} A & D \\ B & C \end{pmatrix} \begin{pmatrix} 1 - v \\ v \end{pmatrix}.$$

The reader can easily verify that a bilinear parameterization $f: [0, 1] \times [0, 1] \to \mathbb{R}^3$ is defined by $f(u, v) = (f_1(u, v), f_2(u, v), f_3(u, v))$, where each coordinate function $f_i: [0, 1] \times [0, 1] \to \mathbb{R}$ is a polynomial of degree 2 in the variables u and v. More precisely,

$$f_i(u, v) = a_i uv + b_i u + c_i v + d_i.$$

Bilinear interpolation has the following properties: if the points A, B, C, and D are coplanar, the resulting patch is a quadrilateral. Horizontal and vertical straight line segments on the plane (u, v) are transformed into straight line segments. Other segments on the plane (u, v) are transformed into second degree curves (hyperboles). A uniform subdivision on the sides of the unit square in the domain is taken into a uniform subdivision on the sides of the patch. Reconstruction using bilinear interpolation has the disadvantage of approximating the boundary curves of the patch by a single straight line segment.

Representation by two boundary curves. In this representation, we use the pair (p_{u0}, p_{u1}) or (p_{0v}, p_{1v}) of the boundary curves of the patch in the representation. The associated reconstruction requires an interpolation method of the two curves to create the original patch (or its approximation). A simple technique for doing this consists of performing a linear interpolation between the two curves. This technique is called *lofting* and is illustrated in Figure 8.15(a).

In reconstruction by lofting, two of the boundary curves from the original patch are approximated by a single straight line segment in the reconstruction. To solve this problem, we can use a representation constituted by these four boundary curves.

Representation by four boundary curves. In this method the patch is represented through the specification of the four boundary curves p_{u0}, p_{u1}, p_{0v}, and $p_{1,v}$ (see Figure 8.15(b)). We naturally associate a reconstruction method to this representation, starting from the four boundary curves. Next, we will study a classic solution for this problem, called *Coons parameterization*.

Consider the four vertices of the patch p_{00}, p_{10}, p_{01}, and p_{11} in $\mathbb{R}^3$, and the four boundary curves p_{0v}, p_{1v}, p_{u0}, p_{u1}. We therefore have (see Figure 8.15(b))

$$p_{u0}(0) = p_{0v}(0) = p_{00};$$
$$p_{u1}(0) = p_{0v}(1) = p_{01};$$
$$p_{u0}(1) = p_{1v}(0) = p_{10};$$
$$p_{u1}(1) = p_{1v}(1) = p_{11}.$$

The reconstruction challenge is to construct a parameterization $C \colon [0,1] \times [0,1] \to \mathbb{R}^3$ so the curves $p_{0v}(v)$, $p_{1v}(v)$, $p_{u0}(u)$, and $p_{u1}(u)$ are the boundary of the patch defined by c.

More precisely, the boundary curves of the unit square are mapped into the curves

$$C(0,v) = p_{0v}(v); \ C(1,v) = p_{1v}(v); \ C(u,0) = p_{u0}(u); \ C(u,1) = p_{u1}(u). \qquad (8.4)$$

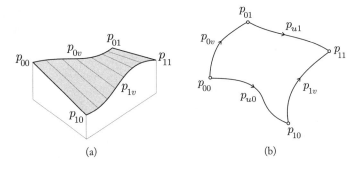

(a)   (b)

Figure 8.15. (a) Reconstruction with lofting; (b) boundary curves.

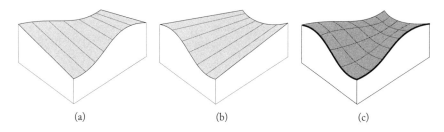

Figure 8.16. (a) Vertical and (b) horizontal lofting; (c) Coons patch.

The solution of this problem is certainly not unique. Besides, we should observe, if the curves $p_{0v}(v), p_{1v}(v), p_{u0}(u), p_{u1}(u)$ are straight line segments, the solution is then given by the bilinear parameterization studied previously. For this reason, it would be natural to seek for a solution reducing itself to a bilinear parameterization when the boundary curves are straight line segments.

Steven Anson Coons [Coons 74] introduced an extremely simple, elegant solution to solve this problem. It consists of combining several linear interpolations of the boundary curves, as described next.

Vertical lofting. We initially interpolate the curves p_{u0} and p_{u1} using linear interpolation:

$$(1-v)p_{u0}(u) + vp_{u1}(u).$$

In other words, we build a parameterization using lofting, as shown in Figure 8.16(a).

Horizontal lofting. This second stage consists of performing the linear interpolation of the other two boundary curves p_{0v} and p_{1v}:

$$(1-u)p_{0v}(v) + up_{1v}(v).$$

This interpolation is shown in Figure 8.16(b).

Adding the two loftings. In this stage we add the horizontal and vertical lofting operations from the two previous stages and obtain the following parameterization:

$$\widetilde{C}(u,v) = (1-v)p_{u0}(u) + vp_{u1}(u) + (1-u)p_{0v}(v) + up_{1v}(v). \qquad (8.5)$$

Note that the boundary

$$\widetilde{C}(0,v) = (1-v)p_{00} + vp_{01} + p_{0v}(v)$$

of this parameterization is formed by the boundary curve p_{0v} added to a linear interpolation $(1-v)p_{00} + vp_{01}$ of the vertices p_{00} and p_{01}. An analogous result is valid to the other boundary curves $\widetilde{C}(1,v)$, $\widetilde{C}(u,0)$, and $\widetilde{C}(u,1)$.

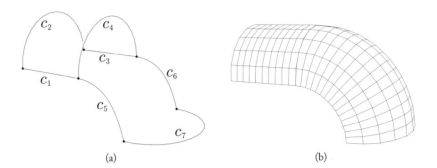

Figure 8.17. Surface defined by Coons patches. (*[Coelho 98] Courtesy of L.C.G. Coelho, PUC-Rio.*)

Bilinear subtraction. By subtracting the bilinear parameterization $B(u, v)$, defined by the vertices p_{00}, p_{01}, p_{10}, and p_{11} from the parameterization $\widetilde{C}(u, v)$, we obtain the following parameterization:

$$C(u, v) = \widetilde{C}(u, v) - B(u, v),$$

which satisfies all ours requirements in Equation (8.4). It is easy to see that the above transformation is reduced to a bilinear patch when the boundary curves are straight line segments. This parameterization is called *Coons patch* (see Figure 8.16(c)).

Note that if the boundary curves $p_{0v}(v), p_{1v}(v), p_{u0}(u), p_{u1}(u)$ are planar curves, the image of the Coons parameterization is also contained in the plane; therefore, $C(u, v)$ defines a transformation on the plane $C \colon [0, 1] \times [0, 1] \to \mathbb{R}^2$.

Figure 8.17(a) displays a set of seven curves $c_1, \ldots, c_7$. The surface shown in (b) is constructed using two Coons patches: the first patch is defined by the curves c_1, c_2, c_3, and c_4, while the second patch is formed by the curves c_4, c_5, c_6, and c_7.

8.6.3   Representation and Continuity

In the representation methods by parametric subdivision, in which the surface S is subdivided into subsurfaces S_i, $i = 1, \ldots, M$, the reconstruction of each parametric subsurface S_i is done separately. An important problem consists of controlling the degree of regularity when joining the several elements S_i to obtain surface S. This regularity depends on the reconstruction method used. Several types of regularity are required in different applications.

In the case of a piecewise linear representation (polygonal surface), we should require, at least, the continuity of the reconstructed surface to avoid cracks, as shown in Figure 8.18. A quite common case happens when the parameterized subsurface S_i is reconstructed using polynomials of degree n, $n \geq 1$. In this case, it is natural to require the reconstructed surface to be of class C^{n-1}.

The various classic families of splines (e.g., B-splines, Bezier) meet this continuity requirement in the joining process, when using a representation by parametric subdivision.

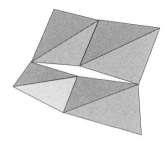

Figure 8.18. Joining with discontinuity.

8.7   Representation of Implicit Surfaces

We have two choices when working with implicit surfaces: to define the concepts and attributes in an implicit way or to seek for a nonimplicit representation of the surface. For example, the visibility and illumination calculations of an implicit surface can be done without the need of a nonimplicit representation; we can use, for instance, the ray tracing method, which we will study later in this book. On the other hand, the calculation of a nonimplicit representation, even an approximated one, gives us access to a larger number of tools for the study of the surface.

In general, the problem of obtaining a parametric representation associated to an implicit surface does not have a solution. The most common approximation method for a nonimplicit representation consists of obtaining a local parameterization by patches. A particular case is to obtain polygonal patches when we have a polygonization of the surface. We will study this case next.

8.7.1   Polygonization of Implicit Surfaces

If a solid $V \subset \mathbb{R}^3$ is implicitly described, $V = \{(x, y, z); F(x, y, z) \leq 0\}$, the implicit surface $S = F^{-1}(0)$ provides a boundary description of V. Our goal is to obtain a polyhedral representation for the boundary of the solid by performing a polygonization of surface S.

In this section we describe a method that naturally extends the polygonization technique of implicit planar curves previously studied in the chapter. There are three stages to this method. We first obtain a subdivision of the space. Then we solve the problem locally in each element of the subdivision. And finally we structure the local solutions to obtain the global one.

Space subdivision. Several subdivision strategies exist; here we will discuss subdivision by volumetric triangulation. To obtain a triangulation, we can define a uniform grid in space where the cells are cubes. We then triangulate each cube so the union of the triangulations of the grid cubes provides a triangulation of the space. There are several existing methods for cube triangulation that satisfy this condition.

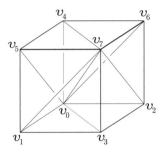

Figure 8.19. CFK triangulation of the cube.

One simple method for cube triangulation consists of the following stages:

1. Take the centroid O of the cube;

2. Construct six quadrangular pyramids whose vertex is O and whose basis is one of the faces of the cube.

3. Subdivide each of those pyramids into two tetrahedra, obtaining a cube triangulation by 12 tetrahedra.

We obtain a triangulation of the space by performing the subdivision of the pyramids into tetrahedra and similarly into adjacent cubes.

A second triangulation method can be obtained as follows (see Figure 8.19):

1. Fix a diagonal d in the cube and take its projection in each of the cube faces, therefore determining two triangles in each face.

2. Form a tetrahedron with each of those triangles, together with the vertex of the diagonal d, which does not belong to it. This results in a cube triangulation with six tetrahedra as shown in Figure 8.19.

This triangulation is called *Coxeter-Freudenthal-Kuhn triangulation* or *CFK triangulation*. The tetrahedra in this triangulation are given by

$$(v_0, v_1, v_3, v_7), \qquad (v_0, v_1, v_5, v_7),$$
$$(v_0, v_2, v_3, v_7), \qquad (v_0, v_2, v_6, v_7),$$
$$(v_0, v_4, v_5, v_7), \qquad (v_0, v_4, v_6, v_7).$$

A third method involves triangulation with only five tetrahedra, as shown in Figure 8.20. Initially, we construct four tetrahedra by taking, for each tetrahedron, one of the vertices v_7, v_1, v_3, and v_4, together with the three adjacent vertices. We then remove those tetrahedra from the cube and the resulting solid is another tetrahedra with vertices (v_0, v_2, v_5, v_6).

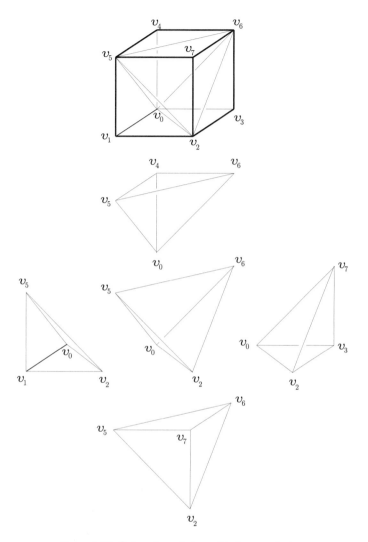

Figure 8.20. Cube triangulation with five tetrahedra.

Problem localization. An appropriate method for problem localization, when using triangulation, consists of performing a linearization in each tetrahedron: if the diameter of each tetrahedron of the triangulation is sufficiently small and the implicit function f is sufficiently regular, then f can be approximated by a linear function in each tetrahedron.

The local solution for the problem is reduced to solving a linear equation for each tetrahedron. A similar approach can be obtained using barycentric coordinates, as we explain next.

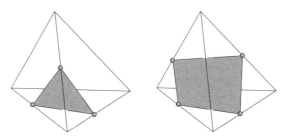

Figure 8.21. Inverse image of $\widetilde{F}$ in each tetrahedron.

Let p_0, p_1, p_2, and p_3 be the vertices of an arbitrary tetrahedron σ of the triangulation. Using barycentric coordinates, every point $p \in \sigma$ can be written in the form

$$p = \lambda_0 p_0 + \lambda_1 p_1 + \lambda_2 p_2 + \lambda_3 p_3,$$

where $\lambda_0 + \lambda_1 + \lambda_2 + \lambda_3 = 1$ and $\lambda_i \geq 0$. We define, in σ, the function $\widetilde{F} \colon \sigma \to \mathbb{R}$ by placing $\widetilde{F}(p_i) = F(p_i)$ at the vertices of the tetrahedron, and

$$\widetilde{F}(p) = \lambda_0 \widetilde{F}(p_0) + \lambda_1 \widetilde{F}(p_1) + \lambda_2 \widetilde{F}(p_2) + \lambda_3 \widetilde{F}(p_3),$$

at the other points of σ.

If the diameter of σ is small, and F has good regularity properties, then $\widetilde{F}$ approximates f in σ. In this case, the surface $S = F^{-1}(0)$ is approximated by the surface $\widetilde{F}^{-1}(0)$. The advantage of this approach is that the calculation of $\widetilde{F}^{-1}(0)$ is reduced to the solution of a linear systems in each tetrahedron σ of the triangulation.

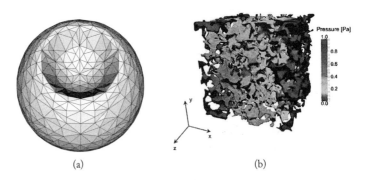

(a)   (b)

Figure 8.22. (a) Triangulated blob surface; (b) x-ray computed tomography (CT) scan of a rock pore-space geometry with computed fluid pressure distribution in the pores. (*Picture courtesy of Yan Zaretskiy, Heriot-Watt University, United Kingdom [Zaretskiy et al. 10]*. See Color Plate XIV.)

Geometrically, the solution of the equation $\widetilde{F}^{-1}(0)$ in each tetrahedron provides a triangle or a quadrilateral, as illustrated in Figure 8.21. As $\widetilde{F}$ is continuous, the solution of $\widetilde{F}^{-1}(0)$ consists of gluing those triangles and quadrilaterals, obtaining a polyhedral approximation. Of course, we should subdivide each quadrilateral into two triangles to obtain a triangulation.

The structuring of local linear solutions to obtain the global triangulation of the implicit surface naturally precedes the structuring of the triangulation. Figure 8.22 shows triangulations of two surfaces.

Final comments. Why not use the original decomposition of the space into cubes, instead of subdividing them into tetrahedra? We can give two reasons for this. The resulting geometry from intersecting a plane with a tetrahedron is simpler than from a cube. Also, linearization of the problem in the cube is more difficult due to the absence of barycentric coordinates.

Another strategy for solving the problem in each tetrahedron consists of first solving the equation $F(x, y, z) = 0$ in each edge of the tetrahedron (a unidimensional problem). We then use those solutions as vertices of the triangles approximating the surface in the tetrahedron; note we only have to correctly connect those vertices by studying the different cases.

We should highlight the pioneering computer graphics solution for the problem of implicit surfaces triangulation using a subdivision by cubes instead of a triangulation. For this reason, this algorithm is known as *marching cube*. The algorithm solves the implicit equation $F(x, y, z) = 0$ in each edge of the cube (unidimensional problem), aiming at discovering the points in the cube that are candidates to vertices in the polygons. Once those points are calculated, a connectivity table is used to determine the correct topology of the polyhedral surface in the cube. The marching cube method was introduced in [Lorensen and Cline 87] to obtain a boundary representation starting from a volumetric description with the goal of using it in the area of medical imaging. As is common in scientific discoveries, the polygonization technique using the CFK triangulation appeared independently in [Allgower and Schmidt 85].

Figure 8.23 shows the effect of increasing the grid resolution in the solution of the problem of implicit surfaces polygonization. With low resolution we obtain a model with

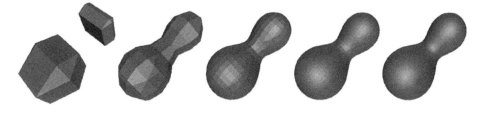

Figure 8.23. Increasing the grid resolution.

incorrect topology and with 70 faces; by increasing the resolution, we reach a good approximation of the geometry (the rightmost model has 27,000 faces). This example was calculated with the marching cube method.

Another solution to the problem of boundary representation of implicit surfaces using parametric approximations consists of approximating the function f in each element (cube or tetrahedron) by polynomial patches, resulting in representations with better regularity.

8.8   Representation of Volumetric Objects

A volumetric object is a 3D solid. A good analytical representation is to define it by an implicit function. For the sake of concept illustration, it is convenient to consider an image as being a 2D volumetric object, where the pixel color values represent the density of the solid. In this section, we will describe some methods of representing volumetric objects. The two main representation methods are: boundary representation and representation by decomposition.

8.8.1   Boundary Representation

The idea of boundary representation is based on Jordan's Theorem, which affirms that a solid is determined by its boundary surface. Note however that this fact applies only if the solid does not have attributes varying in its interior. For example, if the solid has a variable density function, we will not be able to represent it solely using its boundary. Examples include industrial volumetric objects such as mechanical parts.

Representations using the boundary to represent the solid are called *boundary representations*, or *Brep representations*. (We should point out that there is some terminology confusion in writings about geometric modeling, where it is common to use the expression "boundary representation" to indicate the polyhedral representation of a surface.)

To use a boundary representation, it is assumed we have a method to calculate the surface delimiting the boundary of the solid. This problem is known as *boundary evaluation*. Taking into account that implicit surfaces define volumetric objects, the previously studied polygonization methods for these surfaces are, in reality, boundary evaluation methods. This is how the image of the pore-space reservoir rock shown in Figure 8.22 was obtained: starting from a boundary representation of the volumetric data of the rock.

In the case of objects with variable density, calculating the boundary is important because it allows one to obtain level surfaces of the solid. Those surfaces correspond to subsets of the solid in which one or more attributes are constant (for instance, the surfaces where the density assumes a certain value). Those surfaces are very useful in the visualization of volumetric objects and are extensively used in medical imaging. (As we mentioned in the previous section, the marching cube algorithm was developed for this purpose.)

8.8.2   Representation by Decomposition

This representation consists of performing a subdivision of the space into a family of volumetric cells $v_0, v_1, \ldots, v_n$; the object is then represented by enumerating the cells intersecting it and by sampling the object attributes in that cell.

This is the most appropriate representation for volumetric objects with attribute functions varying in the entire volume (e.g., the density). This is because we can sample those attributes in each cell (exactly as we do in the case of an image).

We have two representation classes by decomposition: uniform and nonuniform representation. In the uniform representation, the space decomposition most used is the uniform grid, which gives rise to the *matrix* representation.

Matrix representation. This is the most common type of uniform representation, and it naturally extends to the matrix representation of 2D solids, as we saw in Chapter 7 (in particular, the matrix representation of images). We define the uniform partitions of the coordinate axes

$$\{k\Delta x; k = 0, \ldots, m\},$$
$$\{k\Delta y; k = 0, \ldots, n\}, \quad \text{and}$$
$$\{k\Delta z; k = 0, \ldots, p\},$$

and through them, we obtain a grid in space by finding the Cartesian product of these partitions.

Each grid cell (called *voxel*, from *volume element*) defines a parallelepiped in space. This cell holds information about the density of the solid, as well as other object attributes. The representation of the object is reduced to its representation in each of the grid cells. This can be achieved using a sampling process.

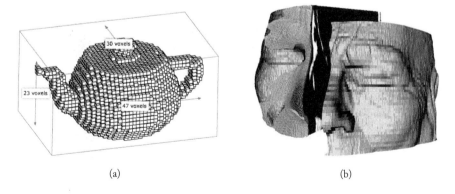

(a)   (b)

Figure 8.24. Matrix representation: (a) Utah teapot (after voxelizing its triangle mesh), (*imagery by Arjan Westerdiep www.drububu.com*); (b) volumetric human head acquired by CT scanning (integrated with laser scan data). (©*2004 Rdiger Marmulla.* See Color Plate XV.)

A volumetric graphics object, given by its matrix representation, is the 3D analogue of an image, where the voxels play the role of the pixels. For this reason, a volumetric object represented in the matrix form is also called *3D image*. The matrix representation of a volumetric object is also called *volumetric representation*. Figure 8.24(a) shows the matrix representation of a solid torus. In Figure 8.24(b), we show the matrix representation of a human head from a computed tomography (CT) scan. Raw voxels of the CT data set (in yellow) are clearly perceptible; rainbow colored surfaces from a laser scan are placed on top of the CT-skin surface.

This matrix representation is broadly used in volumetric objects and for this reason, it is very common to consider a volumetric object as being a solid given by its representation matrix. The conceptualization of a volumetric object given in this chapter is quite broad and flexible.

Nonuniform representation. The voxel representation for volumetric objects is broadly used for three main reasons. First, several image processing and analysis techniques can be extended for the case of volumetric objects. Second, the visualization of the matrix representation is easier due to its simple structuring. Third, this is the representation used by most equipment for capturing volumetric objects.

Robust representations using adaptive and nonuniform decomposition methods (e.g., adaptive representation using voxels of different dimensions) can be used with great advantages. Figure 8.25(a) shows the representation of an image using quadtrees. This representation extends to volumetric objects through the use of octrees. Figure 8.25(b) displays an adaptive representation of an image where both the dimensions of the decomposition elements (Voroni cells) and their geometry are variable.

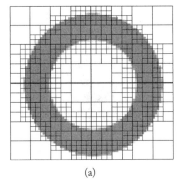

(a)   (b)

Figure 8.25. Nonuniform representations: (a) quadtree; (b) Voronoi cells (2,000 samples). (*Reprinted from [Darsa et al. 98], courtesy of L. Darsa, B. Costa, and A. Varshney, with permission from Elsevier.* See Color Plate XVI.)

For each type of application, one should seek for the most appropriate description and representation of a graphics object. In general, nonuniform representations allow for better data adaptation; however, they are more complex to manipulate.

8.8.3   Converting between Representations

Conversion methods between different representations play an important role in algorithm development. Those methods allow us to obtain the most convenient representation for each type of algorithm; however, this conversion is generally a very difficult problem in geometric modeling.

Two problems of great importance are the conversion from a boundary to a matrix representation and the problem of obtaining a boundary representation from a volumetric description (boundary evaluation).

Boundary to matrix representation. A correct boundary representation of a solid should provide a method of solving the point membership classification problem for the solid. From a mathematical point of view, this means we should provide a method to perform the calculations of the characteristic function of the solid defined by Equation (8.2). We can calculate a matrix representation of the solid by using the point membership classification algorithm together with a 3D version of a rasterization algorithm (see [Kaufman 94]).

The point membership classification problem can be solved by making an implicitization of the solid. In other words, we obtain an implicit function $F \colon \mathbb{R}^3 \to \mathbb{R}$ so the boundary surface is described by $S = F^{-1}(0)$ and the solid is described by the inequality $f \leq c$. Many choices exist for the function f, as we illustrate in Figure 8.26, in which the segment AB is defined by $f \geq 0.5$, $g \geq 0.5$, or $h \leq 0.5$.

The distance-oriented function has been used as a choice for the implicit function f: for each $p \in \mathbb{R}^3$, $f(p)$ is the Euclidean-oriented distance of p to the surface S, which describes the boundary of the solid. The distance function has also been used as an effective tool in the description of implicit volumetric objects (see [Bloomenthal and Wyvill 97]).

Volumetric to boundary description. The conversion between volumetric and boundary descriptions is related to the conversion problem between implicit and parametric descriptions. The conversion problem is very difficult and in most cases it does not have a solution.

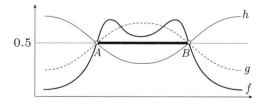

Figure 8.26. Distinct implicit descriptions of the "unidimensional solid" AB.

Even when the problem admits analytical solutions, they are computationally very expensive.

A practical alternative is to develop techniques for obtaining approximated conversions between implicit surfaces and parametric descriptions. The polygonization of implicit surfaces (Section 8.7.1) is a solution along these lines: an implicit surface is approximated by a polygonal one. This method provides an approximated conversion from a volumetric model to a boundary representation and can be applied to obtain boundary representations of volumetric objects defined in the matrix form.

8.9   Comments and References

Geometric modeling was one of the first computer graphics areas to receive an appropriate conceptualization for its various methods and techniques. This conceptualization was developed during the 1970s and was first published in [Requicha 80]. This pioneering work was driven by Aristides Requicha, a Portuguese researcher established in the United States. We recommend that the reader consult this work.

8.9.1   Additional Topics

A more complete study of coding polyhedral surfaces, mainly of the classic winged-edge [Baumgart 75], half-edge [Eastman and Weiss 82], and radial-edge [Weiler 82] data structures, is an important supplement to this chapter as well as a good starting point for a course that emphasizes geometric modeling.

In a more advanced course, and where time is available for more in-depth coverage, this chapter also provides a good place from which to introduce the construction methods for parametric surfaces using interpolation with splines, Bezier, etc. Another related and important topic is *subdivision surfaces*. A good reference is [Warren and Weimer 01]. The multiresolution representation of spatial objects (level of detail) is also very relevant.

The study of implicit objects can itself be organized into a specific course. This subject is well covered in the literature, including [Velho et al. 02] and [Bloomenthal and Wyvill 97]. In particular, the second reference provides a broad description of several polygonization methods of implicit surfaces. In this area, the study of algebraic surfaces deserves special attention. Semialgebraic sets is also an interesting and important subject.

Another direction is the representation of volumetric objects using wavelets, as well as representations by adaptive subdivision of the space (e.g., octrees). The problem of reconstruction of objects starting from scattered samples (scattered data interpolation) is useful to several applications and is directly related to the methods covered in this chapter.

Exercises

1.   Define a topological curve in $\mathbb{R}^3$.

2.   Discuss the problem of parametric description of volumetric objects.

Figure 8.27. Figure for Exercise 4.

3. Given the four vertices $A = (a_1, a_2)$, $B = (b_1, b_2)$, $C = (c_1, c_2)$, and $D = (d_1, d_2)$ of a quadrilateral Q on the plane, find the bilinear transformation $f : [0, 1] \times [0, 1] \to Q$. (Hint: use the fact that $f = (f_1, f_2)$ where each f_i is a polynomial of degree 2.)

4. Determine, with details, the tables of faces, edges, and vertices of the five platonic solids in Figure 8.27 (tetrahedron, cube, octahedron, icosahedron, and dodecahedron).

5. We call the *norm* of a triangulation the maximum value of the diameters of the circles circumscribed to the triangles of the triangulation.

    (a) Describe at least six methods for triangulating the sphere using triangulations that satisfy the following property: as the number of triangles grows, the triangulation norm decreases and approaches 0. (Hint: remember the platonic polyhedral.)

    (b) Describe a criterion to define what a "good triangulation" is and choose, among the sphere triangulations of item (a), the best one matching this criterion. (Hint: in a triangulation, avoid thin and elongated triangles.)

6. Define a trilinear transformation. Describe a solution for the problem of scattered volumetric data interpolation using trilinear interpolation.

7. It is possible to form nine types of relationships of different adjacencies by just considering vertices, edges, and faces. For instance, $F(A)$ refers to the set of edges around a face, $F(V)$ to the set of vertices around the face, etc. Which of these sets can be sorted?

8. How can we alter the data structure based on references (pointers) of an edge list so the query of which faces share a particular edge can be executed efficiently?

9. If $M = F^{-1}(1)$ is an implicit surface, what is the surface represented by the function $G^{-1}(1)$, where $G = 1/f$?

10. Define an implicit curve in $\mathbb{R}^3$.

11. Discuss the advantages and disadvantages of a boundary representation (which uses a parametric description of the solid) and a volumetric representation (which uses an implicit description of the solid). What applications better fit each of these representations? Why?

12. Show that a triangulation does not exist for a cube with four tetrahedra.

13. Extend the DDA-Bresenham algorithm (for rasterizing straight line segments) to rasterize a straight line segment in a volumetric matrix grid.

14. Discuss how to obtain the adjacency relationship, among vertices, edges, and faces, starting from the following data structures:

    (a) Face list.

    (b) Pointers to vertex list.

    (c) Pointers to edge list.

15. Construct a parameterization, $\phi : \mathbb{R}^2 \to \mathbb{R}^3$, for one triangle in space, given by the coordinates of the three vertices. (Hint: use barycentric coordinates).

16. A polygon p in space is a sequence $V_1, V_2, ..., V_k$ of different points located in the same plane such that $V_1 = V_k$. The representation schema LV associates p to the n-tuple $LV(P) = (V_1, V_2, ..., V_{k-1})$.

    (a) Specify a data structure to implement this representation.

    (b) Relate the operations associated to this data structure with the operations in the model.

    (c) Investigate the possible topological and geometric inconsistencies that can occur in this representation.

    (d) Describe procedures to identify the inconsistencies from the previous item.

17. Let M be a parametric surface $\varphi : U \to \mathbb{R}^3$ defined in a rectangular domain U. Consider the representation of M by point sampling in the vertices in a grid of the domain U.

    (a) Discuss this problem from the sampling and reconstruction point of view.

    (b) Discuss advantages and disadvantages of this representation in relation to the polyhedral representation we introduced in this chapter.

18. In the polygonization algorithm we should also sample the surface attributes in each polygon vertex. Describe, in detail, a method to calculate the surface normal in each vertex.

19. Extend the triangle concept for the space $\mathbb{R}^n$ and define a volumetric triangulation nD. (Hint: we already covered the case of $n = 2$ and $n = 3$.)

20. This exercise provides a more flexible and robust definition of Coons surface. Consider four boundary curves $c_1(u)$, $c_2(u)$, $d_1(v)$, $d_2(v)$ with $u, v \in [0, 1]$, as given in the definition of boundary curves for Coons surface. Yet, consider two pairs of functions $f_i(u)$ and $g_i(u)$, $i = 1, 2$, with these constraints: $f_1(0) = g_1(0) = 1$; $f_1(1) = g_1(1) = 0$; $f_1(u) + f_2(u) = 1$ and $g_1(v) + g_2(v) = 1$. Define the parametric surface

$$r(u, v) = \begin{pmatrix} f_1(u) & f_2(u) \end{pmatrix} \begin{pmatrix} d_1(v) \\ d_2(v) \end{pmatrix}$$

$$+ \begin{pmatrix} c_1(u) & c_2(u) \end{pmatrix} \begin{pmatrix} g_1(v) \\ g_2(v) \end{pmatrix}$$

$$- \begin{pmatrix} f_1(u) & f_2(u) \end{pmatrix} \begin{pmatrix} c_1(0) & c_2(0) \\ c_1(1) & c_2(1) \end{pmatrix}.$$

(a) Show this surface interpolates the four given curves:

$$r(u, 0) = c_1(u), \ r(u, 1) = c_2(u),$$
$$r(0, v) = d_1(v), \ r(1, v) = d_2(v).$$

(b) Show the functions f_1, f_2, g_1, g_2 can be used to obtain boundary conditions over the derivatives of the parametric surface.

21. Volumetric Coons surface: generalize the construction of Coons, as given in the text (or in the previous exercise), to obtain a parametric surface of dimension 3 $r \colon [0, 1] \times [0, 1] \times [0, 1] \to \mathbb{R}^3$, $r = r(s, u, v)$, whose boundary $r(s, u, 0)$, $r(s, u, 1)$, $r(s, 0, v)$, $r(v, 1, v)$, $r(0, u, v)$, and $r(1, u, v)$ are the six parametric surfaces previously given.

22. To gain a more in-depth understanding of the marching cube algorithm,

(a) Describe, in detail, the marching cube algorithm in the 2D case, that is, with implicit functions $F \colon U \subset \mathbb{R}^2 \to \mathbb{R}$.

(b) Describe the connectivity table of the marching cube algorithm in the 3D case. (Hint: look at the original work and correct some existing problems.)

9   Hierarchies

When we deal with complex models—either combining multiple graphics objects to form a single composition, such as an animation of a human or animal, or decomposing an object into simpler parts in order to solve problems in the spirit of "divide and conquer"—we rely on hierarchies to relate the various graphics objects to each other.

9.1   Objects with Hierarchy

In a broad sense, a *hierarchy* is a graph in which vertices are associated with graphics objects. We therefore have a set objects pairs $(\mathcal{O}_i, \mathcal{O}_j)$ as the edges. When two objects constitute an edge, we say they have a *linkage* relation in the hierarchy. Objects defined by a hierarchy can be classified as composed or articulated objects.

In *composed objects*, there is no relative motion between the subobjects $\mathcal{O}_i$ of the hierarchy; in other words, the linkages are rigid. Composed objects are very useful for describing scenes with many elements. In a house, for instance, we have a natural hierarchy, as illustrated in Figure 9.1. Hierarchical modeling can make scene viewing operations more flexible.

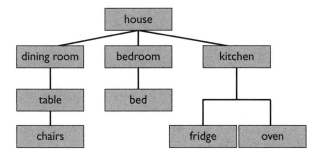

Figure 9.1. Hierarchy of a house.

247

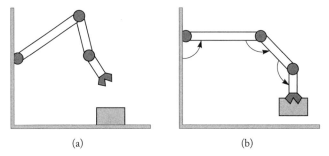

Figure 9.2. (a) Robotic arm and its end-effector, (b) the end-effector in contact with an object. (*[da Silva 98] Courtesy of F.W. da Silva.*)

In *articulated objects*, links are not rigid. These objects are made up of rigid parts connected by articulations, or joints, forming a linkage among the parts and thus allowing relative motion between them. The human body is an example of an articulated object in the physical world; robots are another important class of articulated objects. Figure 9.2 shows an articulated structure representing a robotic arm. Note that the "hand" of the robot is a rigid part that has an end extremity—that is, a place where a connection joint does not attach to any other rigid part of the structure. This extremity is called an *end-effector*. The end-effector is used for contacting between the structures and other objects, as shown in Figure 9.2(b). Of course, not every hierarchical structure has an *end-effector*—for example, a chain hung by the two ends.

The links connecting the various parts of articulated objects can be classified as either geometric or physical.

Geometric links. A geometric link is defined by a contact relation between linked objects. An articulation, for instance, can be used as a geometric link between two objects, creating an articulated structure with greater range of motion. A robot's arm is a practical example of the use of a geometric link between two objects.

Physical links. In contrast to geometric links, physical links are dynamic. Consider, for example, the links between the bodies in a planetary system, or the spring in a mass-spring system. The links between these objects are defined by attraction or repulsion force fields and proximity among objects is defined by a time-varying function. For example, the distance between bodies in a planetary system is modeled according to the gravitational force existing between them; a mass-spring system is simulated by establishing a function that models the attraction of a mass according to the deformation of a spring.

In animation, both geometric and physical links can be used as behavioral links between objects (actors). For example, the predator is always attracted to the prey. Such links are defined by behavior rules and are most often physical links.

Generally a hierarchy graph is *oriented*; that is, the edges are oriented straight line segments indicating the subordination of an object in relation to others. This subordination translates to geometric or physical properties, depending on the type of link. For instance, we can establish, if we apply a transformation T to an object, that the transformation extends to its subordinate objects in the hierarchy. An important particular case happens when the hierarchy is defined by a tree structure; in other words, the hierarchy graph does not have cycles (closed edge paths). We will always consider a tree as an oriented graph starting from the root and moving toward the leaves. (In robotics terminology, the leaves of the tree represent the end-effectors of the hierarchy.)

In this chapter, we will study the hierarchies of articulated and composed objects. Unless explicitly stated otherwise, a hierarchy will always have a tree structure. In the case of articulated objects, we will focus on the hierarchy of the human locomotion system due to its importance in animation.

9.2   Hierarchy of Articulated Objects

There are several types of articulations (or joints); in the human body there are two: revolution and spherical joints.

In *revolution joints*, objects are connected through an axis around which they can rotate (Figure 9.3(a)). A *spherical joint* (Figure 9.3(b)) is geometrically represented by a sphere connecting the rigid parts of the structure. Each rigid part can assume any position in the sphere; furthermore, each of the parts can rotate around its longitudinal axis.

For a revolution joint, the positioning of the articulation is determined by the rotation angle (θ in the figure). Therefore the motion space is a subset of the group of plane rotations $SO(2)$ which, as we know, is represented by the circle S^1. In this way, the revolution joint has one *degree of freedom* (DOF).

For a spherical joint, the rigid parts of the articulation can assume any orientation in space. Therefore, the motion space of this joint is a subspace of $SO(3)$, the group of rotations of $\mathbb{R}^3$. As the space $SO(3)$ has dimension 3, we say the spherical joint has three degrees of freedom. The parameterization can be performed using Euler angles or by using the exponential application, as we saw in Chapter 4.[1]

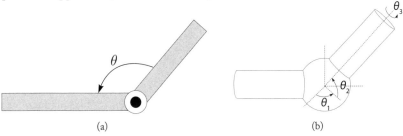

(a)   (b)

Figure 9.3. (a) Revolution and (b) spherical joints.

---

[1]In robotics, other types of parameterization are used, based in Lie groups.

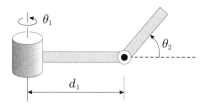

Figure 9.4. Articulated structure with two degrees of freedom.

In general, an articulated object with n joints has $d = 3 + \sum_{j+1}^{n} \mathrm{DOF}(j)$ degrees of freedom, where $\mathrm{DOF}(j)$ is the number of degrees of freedom of the j-th joint. The three additional degrees of freedom correspond to the positioning of the structures in the coordinate system in space. Therefore the motion of the structure is parameterized in a space of dimension d, called the *state space* or *configuration space* of the articulated object. Using Euler angles we see the state space can be parameterized by $\mathrm{DOF}(j)$, where each circle S^1 corresponds to a degree of freedom described through an Euler angle. For each state $(x, y, z, \theta_1, \ldots, \theta_d) \in M$, we have a configuration $\varphi(x, y, z, \theta_1, \ldots, \theta_d)$ of the articulated object in $\mathbb{R}^3$.

Parameterization by Euler angles is local and can bring problems. By representing the spherical joints with unit quaternions, we can obtain a global representation of the hierarchy. In this case, the motion representation space of that joint is the unit sphere $S^3 \subset \mathbb{R}^4$. Therefore the configuration space is parameterized by $\mathbb{R}^3 \times S^1 \times \cdots \times S^1 \times S^3 \times \cdots \times S^3$.

The number of degrees of freedom of an articulated structure is intimately associated with its motion capabilities in space. In other words, the larger the degree of freedom of the structure, the larger the number of possible configurations it can form in space. Figure 9.4 shows an example of a simple structure with two revolution joints allowing a rotation about either the cylindrical basis or the articulation axis of the two arms. We therefore have two degrees of freedom. It is easy to verify that the configuration space is the torus, $S^1 \times S^1$. In Chapter 3 we obtained the parameterization of the configuration space of some structures having the geometry of a robot arm.

9.2.1   Hierarchies, Transformations, and Motion

Hierarchies facilitate the implementation of object transformations, including hierarchy positioning and motion transformations thanks to the inheritance rules of hierarchy transformations. In the case of a house, we can establish the following inheritance rule: a transformation applied to the table is transmitted to the chairs.

We can describe the positioning of an articulated object by using an absolute reference frame in space; this reference frame is in relation to the position and the orientation of each rigid part of the object, both of which are explicitly provided. This positioning method in absolute coordinates works well if we are dealing with only a positional problem. However,

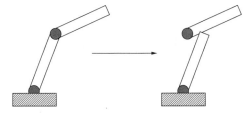

Figure 9.5. Rupture of the linkage structure. (*[da Silva 98] Courtesy of F.W. da Silva.*)

the need to specify the position of each articulation makes it difficult to guarantee the connectivity integrity of the set because small variations in the position of an articulation can destroy the links between the rigid parts of the structure (see Figure 9.5).

Organizing a structure by a hierarchical model solves the problem of maintaining the integrity of its structure. In such a model we define a system of local coordinates for each rigid part of the hierarchy. The positioning of each part is done in its coordinate system, and the global positioning of the structure is obtained by performing changes between the several coordinate systems. This way only the articulation of the top of the hierarchy needs to be positioned in space, while the rest of the structure is positioned using local transformations for coordinate changing. Therefore, as we apply a transformation T in one articulation, all of its subordinate articulations in the hierarchy will be affected by that transformation. We saw this type of procedure in Chapter 3.

To position an object defined by the hierarchy, a search is performed in the hierarchy graph passing through all the nodes, successively concatenating all the transformations found and applying those transformations to each subobject found in the hierarchy, according to the established inheritance rules. Unless stated otherwise, we will adopt a simple inheritance rule: every transformation applied to an object extends to its subordinated objects in the hierarchy. Figure 9.6 shows the hierarchy tree with 3 *end-effectors*, 5 rigid parts, and 3 articulations.

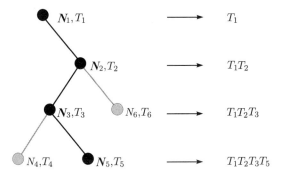

Figure 9.6. Hierarchy, tree, and transformations.

In the hierarchy shown in Figure 9.6, observe that each rigid part has a transformation T_i associated to it, which performs the change between its own coordinate system and the one of the part to which it is subordinated in the hierarchy. Each tree node N_i is therefore positioned in the global coordinate system, performing the product of transformations T_j corresponding to the ascending nodes until we reach the root of the tree. This transformation, called a *current transformation* of the node, should be applied in each rigid part of the hierarchy (tree node) in a way that positions itself in space.

While traversing the hierarchy, the current transformation is applied to each subobject found in order to obtain its correct positioning in the coordinate system in space. Consider the tree in Figure 9.6, where the nodes are enumerated from N_1 to N_6. We indicate the transformation of each node N_i by T_i. Consider node N_5. A path going from this node down to the root is given by (N_1, N_2, N_3, N_5) (bold letters in the illustration). The current transformation of node N_5 is $T_1 T_2 T_3 T_5$. The figure shows the current transformation of each intermediate node of this path.

Example 9.1. As an example, consider the hierarchy of the mechanical arm shown in Figure 9.7(a). Figure 9.7(b) shows the coordinate system of each rigid part.

We take as the root of the tree the rigid part of the hierarchy containing the cylindrical basis. Let us assume (1) the center of the cylinder in the origin of $\mathbb{R}^3$ and (2) originally, the reference frame basis of this part of the hierarchy coincides with the canonical basis. The change from the canonical basis to the reference frame of the arm of the cylindrical basis is given by one rotation by an angle θ_1 about the vector e_3:

$$T_1 = \begin{pmatrix} \cos\theta_1 & -\sin\theta_1 & 0 & 0 \\ \sin\theta_1 & \cos\theta_1 & 0 & 0 \\ 0 & 0 & 1 & 0 \\ 0 & 0 & 0 & 1 \end{pmatrix}.$$

The change from the reference frame of the cylindrical basis to the reference frame of the second rigid part is done by a translation of length d_1 along the new axis e_2, followed by a

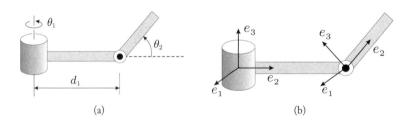

Figure 9.7. Articulated structure with two degrees of freedom.

rotation by an angle θ_2 about the vector e_1. The matrix is given by

$$
T_2 = \begin{pmatrix} 1 & 0 & 0 & 0 \\ 0 & 1 & 0 & d_1 \\ 0 & 0 & 1 & 0 \\ 0 & 0 & 0 & 1 \end{pmatrix} \begin{pmatrix} 1 & 0 & 0 & 0 \\ 0 & \cos\theta_2 & -\sin\theta_2 & 0 \\ 0 & \sin\theta_2 & \cos\theta_2 & 0 \\ 0 & 0 & 0 & 1 \end{pmatrix} = \begin{pmatrix} 1 & 0 & 0 & 0 \\ 0 & \cos\theta_2 & -\sin\theta_2 & d_1 \\ 0 & \sin\theta_2 & \cos\theta_2 & 0 \\ 0 & 0 & 0 & 1 \end{pmatrix}.
$$

The positioning of the arm is given by the product $T_1 T_2$. These calculations are given in detail in the study of this hierarchy covered in Chapter 3. ❑

9.2.2 Simplified Hierarchy of a Car

Consider an external simplification of the hierarchy of a car, as shown in Figure 9.8(a). This hierarchy is constituted by three parts: car body (B), front (FW) and back (BW) wheels. The hierarchy tree is shown in Figure 9.8(b). The articulation joints of the wheels with the car body are revolution joints, and their reference frames are given in (c) and (d).

The transformation from the canonical system in $\mathbb{R}^2$ to the car body system is given by a translation $T(0,1)$. The transformation from this system to the reference frame of the back wheel is given by a translation $T(2,0)$, followed by a rotation $R(\theta)$ of angle θ about the origin of the new system. Therefore the current transformation of the back wheel is given by $R(\theta)T(2,0)$. Similarly, the current transformation of the front wheel is $R(\theta)T(8,0)$. Notice the rotation angles of the front and back wheels are the same.

Figure 9.9(a) shows the hierarchy of the car with its transformations. To move the car we perform a horizontal translation in the coordinate system of its body. Notice, however, that the length to be translated depends on the turning of the wheels. In reality, if a wheel

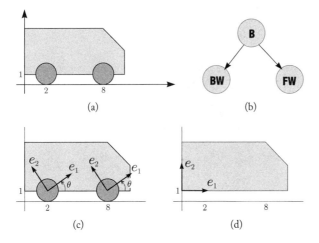

Figure 9.8. (a) Car, (b) its hierarchy tree, (c) the reference frame of the front and back wheels, and (d) the reference frame of the car body.

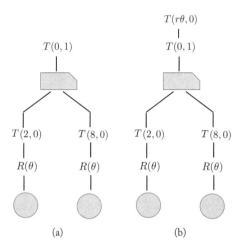

Figure 9.9. Car and hierarchy: (a) hierarchy of the car with transformations, (b) complete hierarchical structure.

(front or back) rotates by an angle θ, in radians, the car should then move by a length $r\theta$, where r is the radius of the wheel. Therefore, we can apply the translation $T(r\theta, 0)$ to the car body (root of the tree). Figure 9.9(b) shows the complete structure of the hierarchy.

This example presents an interesting situation. We are using a tree structure to represent the car hierarchy. In this case, the motion of each leaf of the hierarchy, represented by a tree node, should not influence higher levels of the hierarchy. However, observe in the case of the car that the vector used for translating the reference frame of the body (root of the tree) to move the car depends on the rotation angle of the wheel (leaf). Clearly the tree structure is not able to represent such a hierarchy.

If the motion of the car is modeled based on the laws of physics, the wheel with traction should transmit motion to the car body and, consequently, to the entire car. In this case, a correct representation of the car hierarchy would be given by the oriented graph shown in Figure 9.10.

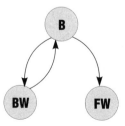

Figure 9.10. Hierarchy graph.

There are cases, as in the case of the car, where the hierarchical structure of an articulated object is better represented by an oriented graph, which can have cycles. This type of structure is very common when the motion of the articulated object is based on physical simulations.

9.3   Hierarchy of the Human Body

Modeling of the human body is a topic of great importance. It is useful in many applications, including virtual actor construction; character animation (3D cartoon modeling); choreography studies; motion coordination analysis (orthopedics); motion correction and improvement in sports; and virtual reality and video games.

Each of these applications demands different degrees of model complexity, but in all these applications there is a need for total control of body motion in order to obtain animations. With that purpose in mind, this section focuses on the modeling of the human body hierarchy. The skeleton of an adult has 208 bones (rigid parts of the hierarchy); these bones and their articulations, together with 501 muscles, constitute our locomotive system. Representing this hierarchy is a complex task.

While there are many details related to modeling the human body, in this chapter we will focus solely on the hierarchical aspects of human locomotion. For this, considering the hierarchy of a skeleton is sufficient. Modeling muscles and their motion is a physical simulation of the human locomotor system, and is outside the scope of this book. Also outside the scope of this book is the modeling of hair, skin, and muscle deformation effects, which are important factors in certain applications (e.g., correctly modeling deformation of facial muscles is essential to creating facial expressions such as sadness, happiness, and surprise).

9.3.1   Simplifying the Hierarchy

The articulations connecting bones have an extremely complicated geometry, with several contact points and countless degrees of freedom for rotation and even translation. This fact, together with the large number of bones, makes representation of the skeleton structure a very complex task. However, in many applications a very simplified representation of the skeletal structure is sufficient. The simplification allows a reduction of the articulated structure of the skeleton by reducing the number of joints as well as the number of degrees of freedom of several of those joints. While it is simplified, the basic structure we will examine here contains all the essential modeling elements needed for a study of more complex skeleton hierarchies.

Figure 9.11(a) illustrates a skeleton simplification for the arm and the hand. Notice that, in this simplified structure, the hand is incorporated into the lower part of the arm in a rigid way (without any articulation joint), thus reducing the whole set to one articulated object with two rigid parts (upper and lower parts of the arm) and one revolution joint. This revolution joint corresponds to the elbow, whose degrees of freedom we have reduced to one.

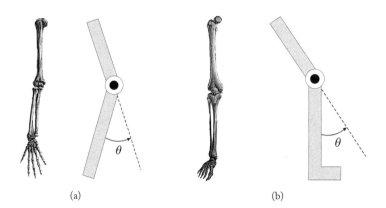

Figure 9.11. Simplification of the (a) arm and (b) the leg.

We can likewise simplify the leg by considering only its upper part (thigh) and by incorporating the foot to its lower part (see Figure 9.11(b)). We therefore obtain a hierarchy with two rigid elements—the upper and lower parts of the leg—and a joint (knee). We consider the knee a revolution joint (one degree of freedom).

The trunk (Figure 9.12), incorporating clavicle, ribs, spine, and hips, is another rigid part of our simplified structure. Besides having a connection joint with the neck (E), the trunk also maintains links with the arms (A and B) and the legs (C and D). Joint E is not an articulation of segments AE and BE of the clavicle, but only a neck articulation with the trunk. In other words, the simplified structure of the trunk, as shown in Figure 9.12, is completely rigid. We define only one degree of freedom for the trunk, allowing its front and back inclination, as shown in Figure 9.13(a).

The head and the neck are also considered rigid parts in our skeleton simplification, with a spherical joint connecting the neck to the trunk. The set head/neck can then tilt towards any direction. We can parameterize this orientation using Euler angles of roll

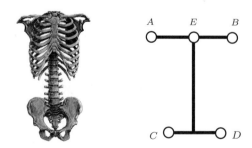

Figure 9.12. Simplified trunk and connection joints.

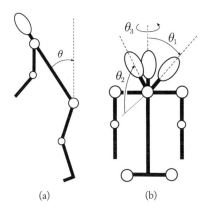

Figure 9.13. Degree of freedom of (a) the trunk and (b) the head.

(head inclined toward the sides), pitch (inclined forward), and yaw (head rotation about the vertical axis) (see Figure 9.13(b)).

In our simplified skeleton, each of the trunk's connection joints with the arms has two degrees of freedom. Similarly, the articulation joints between the legs and the trunk each have two degrees of freedom (see Figure 9.14).

Our simplified structure of the skeleton has ten rigid parts with nine joints. For future references, we will name this simplified hierarchy of the human body *Joe Stick*. The tree of the Joe Stick hierarchy is shown in Figure 9.15. The numbers beside each node indicate the number of degrees of freedom available for the motion of the corresponding part. We

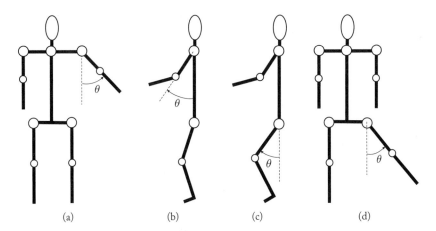

Figure 9.14. Degrees of freedom for the arm and legs: (a) and (d) lateral angular motion; (b) and (c) back and forth motion.

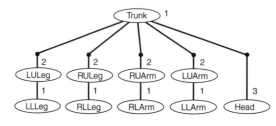

Trunk	= trunk (clavicle, spine, ribs and hips)	Head	= head (with the neck)
LULeg	= left upper leg	RULeg	= right upper leg
LLLeg	= left lower leg	RLLeg	= right lower leg
RUArm	= right upper arm	LUArm	= left upper arm
RLArm	= right lower arm	LLArm	= left lower arm

Figure 9.15. Hierarchy tree of Joe Stick.

have a total of 16 degrees of freedom associated with the positioning of the rigid parts. We need three additional degrees of freedom to position the hierarchy in space, which is obtained by positioning the trunk (the root of the hierarchy). We therefore have a total of 19 degrees of freedom.

Using the simplifications mentioned previously, we are able to create an articulated structure in the computer which simulates, with a reasonable degree of fidelity, the motions of the human body.

There are a number of simplified models of the human body. The goal of all these models is the same: to obtain a simple representation capable of being manipulated and visualized interactively on the computer that also reproduces, in a realistic way, the motion of human beings.

As we saw previously, the number of articulations (and, consequently, of degrees of freedom) of an articulated structure determines its range of representing motion in space. The minimum number of articulations that have been used in simplified models of the human body is between 15 and 20 articulations.

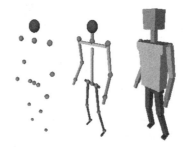

Figure 9.16. Possible models of Joe Stick. (*[da Silva 98] Courtesy of F.W. da Silva.*)

Modeling Joe Stick. Different geometric models can be incorporated into the Joe Stick hierarchy, depending on the desired visualization style. Figure 9.16 shows some of those possibilities. It is possible to use more complex objects in the visual representation of a virtual actor, making it more realistic, but at this stage we are not concerned with visual aspects of the model.

9.3.2 Current Transformations of Joe Stick

Our basic reference frame is the Cartesian coordinate system defined by the canonical basis $\{e_1, e_2, e_3\}$ of $\mathbb{R}^3$. We define the reference frame of each rigid part of Joe Stick so that we can determine the current transformations of those parts.

Figure 9.17(a) is an exploded view of Joe Stick showing the 10 reference frames associated with the rigid parts of the skeleton. Notice that in the arms we orient vector e_3 toward the hand; in the leg we orient it toward the feet. This choice, which is arbitrary, follows a convention used in robotics in which the vector e_3 at the end-effector points toward the extremity that has no articulation joint. In this way, when the end-effector gets closer to an object, it does so in the direction e_3 of the local coordinate system.

To determine the current transformations of each rigid part of Joe Stick, we need to know their dimensions. Those dimensions are given in Figure 9.18. For the motion of this hierarchy, we do not need the length of the lower part of the arm. This measure is necessary for activities in which the arm reaches some object in space (it is certainly necessary for creating the actual Joe Stick model). In the same way, the length of the lower part of the leg is necessary to appropriately positioning the structure on the plane. We will now determine the current transformations.

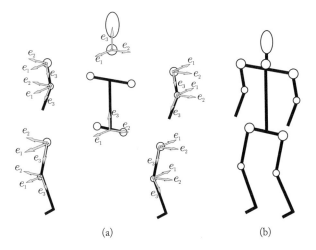

(a)   (b)

Figure 9.17. Reference frames of the Joe Stick hierarchy: (a) an exploded view; (b) complete model.

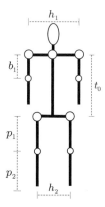

Figure 9.18. Dimensions of Joe Stick: h_1 and h_2 are the widths of the shoulders and hips, respectively; t_0 is the height of the trunk; b_1 is the length of the upper part of the arm; p_1 and p_2 are the lengths of the upper and lower parts of the leg, respectively.

Trunk. Assume the skeleton is initially positioned in space at the origin, over the plane xy, and oriented so the canonical basis of $\mathbb{R}^3$ coincides with the reference frame basis of the trunk. The origin of the reference frame for the trunk is obtained with the translation by the vector $(0, 0, p_1 + p_2)$, and Joe Stick is looking towards the positive direction of the e_1-axis.

The trunk has one degree of freedom for inclination, which is equivalent to a rotation about the e_2-axis. Therefore, the current transformation of the trunk is given by the change of reference frame from the canonical basis to the reference frame of the trunk, which can rotate about the vector e_2 to tilt the trunk forward or back. The sequence of transformations is given by

$$\mathrm{T}(0, 0, p_1 + p_2).$$
$$\mathrm{R}(e_2, \theta_1).$$

The current transformation of the trunk is given by the product (in this order)

$$T(0, 0, p_1 + p_2)R(e_2, \theta_1).$$

Head. The reference frame of the neck (which forms a rigid set with the head), is obtained from the reference frame of the trunk by performing a translation and later a change to orient the head. We have three degrees of freedom in this orientation. Using Euler angles, $\theta_2 = $ roll, $\theta_3 = $ yaw, and $\theta_4 = $ pitch, we have the sequence of transformations

$$\mathrm{T}(0, 0, t_0),$$
$$\mathrm{R}(e_1, \theta_2),$$
$$\mathrm{R}(e_2, \theta_3),$$
$$\mathrm{R}(e_3, \theta_4).$$

The current transformation is given by the product of these transformations, together with the transformations of the trunk (in this order):

$$T(0, 0, p_1 + p_2) R(e_2, \theta_1) \, T(0, 0, t_0) \, R(e_1, \theta_2) \, R(e_2, \theta_3) \, R(e_3, \theta_4).$$

To avoid being repetitive, we will continue by listing the transformations of each rigid part so the final transformation is obtained by doing a top-down product (execution order in a program):

$$
\begin{matrix}
T_1 \\
T_2 \\
\vdots \\
T_n
\end{matrix}
\qquad \Longleftrightarrow \qquad
T_1 T_2 \cdots T_n.
$$

Notice that the last transformation of the sequence is in reality the first to be applied to the object. To obtain the current transformation, we pre-multiply this product of the node transformations by the current transformation of the ascending node.

Left arm. The transformation of the upper part of the left arm is obtained by the transformation that changes from the reference frame of the trunk to the reference frame of the upper part of the arm:

$$
\begin{aligned}
&\mathrm{T}(0, h_1/2, t_0), \\
&\mathrm{R}(e_2, 180°), \\
&\mathrm{R}(e_2, \theta_5), \\
&\mathrm{R}(e_1, \theta_6).
\end{aligned}
$$

The transformation of the lower part of the arm is obtained from the reference frame of the upper part using the transformation

$$
\begin{aligned}
&\mathrm{T}(0, 0, b_1), \\
&\mathrm{R}(e_2, \theta_7).
\end{aligned}
$$

Right arm. For the upper part of the arm, we have

$$
\begin{aligned}
&\mathrm{T}(0, -h_1/2, t_0), \\
&\mathrm{R}(e_1, 180°), \\
&\mathrm{R}(e_1, \theta_8), \\
&\mathrm{R}(e_2, \theta_9).
\end{aligned}
$$

The lower part is given by

$$
\begin{aligned}
&\mathrm{T}(0, 0, b_1), \\
&\mathrm{R}(e_2, \theta_{10}).
\end{aligned}
$$

Right leg. The sequence of transformations of the upper part of the right leg is given by

$$T(0, -h_2/2, 0),$$
$$R(e_1, 180°),$$
$$R(e_1, \theta_{11}),$$
$$R(e_2, \theta_{12}).$$

The transformations of the lower part are obtained from the upper part, placing

$$T(0, 0, p_1),$$
$$R(e_2, \theta_{13}).$$

Left leg. The transformation of the upper part of the left leg is given by

$$T(0, h_2/2, 0),$$
$$R(e_2, 180°),$$
$$R(e_2, \theta_{14}),$$
$$R(e_1, \theta_{15}).$$

The sequence of transformations of the lower part is given by

$$T(0, 0, p_1),$$
$$R(e_2, \theta_{16}).$$

9.4   Current Transformation and Data Structure

Now that we've seen these examples of transformations, you may be asking what the most appropriate data structure is to implement the transformations in a hierarchy.

We first perform an in-depth tree traversal, visiting every node and applying the transformations with the goal of positioning the hierarchy in space. In Chapter 3 we saw that the use of a stack structure is an efficient way of implementing a sequence of applications in a hierarchy. This is because the transformations are applied in the inverse order in which they are specified. Consider the hierarchy represented in the tree of Figure 9.19.

We create a stack and start the search from the root of the tree with the empty stack and the current transformation $T_c = T_1$:

$$T_c = T_1, \quad \text{Stack} \rightarrow \boxed{\text{NULL}}$$

When passing to node N_2, we push the current transformation $T_c = T_1$ into the stack. The new current transformation is given by the product of the current transformation of the ascending node (N_1) and the transformation T_2, which performs the change of intrinsic coordinates from node N_1 to node N_2; that is,

$$T_c = T_1 T_2 \quad \text{Stack} \rightarrow \boxed{T_1 \mid \text{NULL}}$$

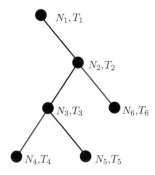

Figure 9.19. Tree of a hierarchy.

The next node in our search is N_3. In this node we repeat the operations applied to node N_2. We push the current transformation $T_c = T_1 T_2$ into the stack, and the new current transformation is given by the product of the current transformation of the ascending node $(T_1 T_2)$ and the transformation T_3, which makes the change of coordinates from node N_2 to node N_3. We then have

$$T_c = T_1 T_2 T_3 \qquad \text{Stack} \rightarrow \boxed{\; T_1 T_2 \;|\; T_1 \;|\; \text{NULL} \;}$$

Continuing to node N_4 we repeat the operations, resulting in

$$T_c = T_1 T_2 T_3 T_4 \qquad \text{Stack} \rightarrow \boxed{\; T_1 T_2 T_3 \;|\; T_1 T_2 \;|\; T_1 \;|\; \text{NULL} \;}$$

The node N_4 is a tree leaf, and our search has reached a maximum depth. We now return to the previous level to continue the search in other branches of the hierarchy; in this case, we return to level N_3. In this return, the current transformation becomes the transformation from the top of the stack, and we therefore have

$$T_c = T_1 T_2 T_3 \qquad \text{Stack} \rightarrow \boxed{\; T_1 T_2 \;|\; T_1 \;|\; \text{NULL} \;}$$

Notice this is automatically the correct position for node N_3, which is our current location in the tree. The final steps of our search are shown in the table below.

Node	T_c	Stack		
N_5	$T_1 T_2 T_3 T_5$	$T_1 T_2 T_3$	T_1	NULL
N_3	$T_1 T_2 T_3$	$T_1 T_2$	T_1	NULL
N_2	$T_1 T_2$		T_1	NULL
N_6	$T_1 T_2 T_6$	$T_1 T_2$	T_1	NULL

9.4.1   Implementation

There are several options for implementing the object transformations of the hierarchy using the stack structure. One option is to write a procedure performing an in-depth traversal of the tree, concatenate the transformations with the use of a stack, and apply the current transformations to each rigid part of the hierarchy.

This method is shown in the pseudocode below (the operation for pushing an element into a stack is called PUSH and the operation of removing an element at the top of the stack is called POP):

$\qquad$ Transform T_1

$\qquad$ Object N_1

$\qquad$ PUSH

$\qquad\qquad$ Transform T_2

$\qquad\qquad$ Object N_2

$\qquad\qquad$ PUSH

$\qquad\qquad\qquad$ Transform T_3

$\qquad\qquad\qquad$ Object N_3

$\qquad\qquad\qquad$ PUSH

$\qquad\qquad\qquad\qquad$ Transform T_4

$\qquad\qquad\qquad\qquad$ Object N_4

$\qquad\qquad\qquad$ POP

$\qquad\qquad\qquad$ PUSH

$\qquad\qquad\qquad\qquad$ Transform T_5

$\qquad\qquad\qquad\qquad$ Object N_5

$\qquad\qquad\qquad$ POP

$\qquad\qquad$ POP

$\qquad\qquad$ PUSH

$\qquad\qquad$ Transform T_6

$\qquad\qquad$ Object N_6

$\qquad\qquad$ POP

$\qquad$ POP

Another option for implementing the object transformations of the hierarchy using a stack structure is to create a procedural structure for each rigid part of the hierarchy. This procedure does all the stack operations for each element and calls its subordinate ones.

As an example of this method, the Joe Stick description using the stack structure can be implemented in the following way:

$$T(0, 0, p_1, +p_2),$$
$$R(e_1, \theta_1),$$
$$R(e_2, \theta_2),$$
$$R(e_3, \theta_3),$$
Trunk,
Head,
LeftLeg,
RightLeg,
LeftArm,
RightArm,

where each rigid element of the hierarchy (`Trunk`, `Head`, `LeftLeg`, `RightLeg`, `RightArm`, and `LeftArm`) is described using the stack structure. For example, using the names given in the tree of Figure 9.15, the subhierarchy of the left leg, `LeftLeg`, is described by

PUSH
$$T(x_0, y_0 z_0)$$
$$R(e_2, \theta_4)$$
$$R(e_1, \theta_5)$$
LULeg
PUSH
$$R(e_2, \theta_6)$$
LLLeg
POP
POP

We left the description of the remaining subhierarchies as an exercise.

9.5 Hierarchies of Composed Objects

So far we have focused on the hierarchy of articulated objects. In the remainder of this chapter, we will study hierarchies of composed objects. We are particularly interested on the hierarchies associated to subdivisions of space.

9.5.1 Hierarchy of Partitions

Let U be a subset of the space $\mathbb{R}^n$. A *polyhedral partition* of U is a family of polyhedra $V_1, V_2, \ldots, V_n$ such that the two following conditions are satisfied: the union of all polyhedra V_i is the set U and the intersection between two polyhedra is either empty or a common face.

Each polyhedron is called a partition cell. The partition is said to be *convex* if the polyhedra are convex. The polyhedra can be infinite for the case in which the set U is the entire space $\mathbb{R}^n$. A triangulation is an example of a convex polyhedral partition. Another classic example is the Voronoi diagram.

In mathematics, partitions can be considered in a much more general context; however, the topological concepts involved are much more elaborate and subtle (e.g., what is the border of a cell?). On the other hand, polyhedral partitions are more widely used in computer graphics.

A *hierarchy of partitions* of a subset $U \subset \mathbb{R}^n$ is a finite sequence of partitions $(\mathcal{P}_1, \mathcal{P}_2, \ldots, \mathcal{P}_m)$ such that

- ❑ Each element $\mathcal{P}_j$ of the sequence is a partition of U;

- ❑ The sequence is nested: $\mathcal{P}_j \supset \mathcal{P}_{j+1}$. This means every cell of the sequence $\mathcal{P}_{j+1}$ is contained in some cell of the predecessor sequence $\mathcal{P}_j$.

- ❑ If C_j is a cell of the partition $\mathcal{P}_j$, the collection D_{j+1}^k, $k = 1, \ldots, m$ of all of the cells of $\mathcal{P}_{j+1}$ is then contained in C_j, forming a partition of C_j.

From a computational point of view, a hierarchy should be easy to represent on the computer. For this, each cell should have a simple representation (which happens with polyhedral cells) and we should have a data structure supporting the hierarchy. Here, a natural structure is a tree where each cell in the sequence hierarchy is a node; the children of each node are the cells of the subsequent hierarchy that it contains.

Example 9.2 (Quadtree and octree). Consider a rectangular region on the plane (Figure 9.20). We can create a hierarchy by subdividing the rectangle into four subrectangles using the centroid. Figure 9.20 illustrates several subdivision levels of the hierarchy. Notice that the subdivision does not need to be uniform (i.e., the level of subdivisions in each subcell can vary). The data structure associated to this hierarchy is a quaternary tree (i.e., each nonleaf node has four children), and for this reason it is called a *quadtree*. This hierarchy extends to a 3D grid where each parallelepiped of the grid partition is subdivided into eight parallelepipeds. This hierarchy is called an *octree* because each nonterminal node

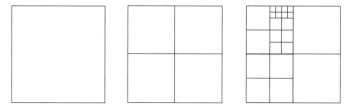

Figure 9.20. Quadtree.

in its associated data structure has eight children. These hierarchies are extensively used in to represent images or volumetric objects (3D images). ❏

Example 9.3 (Hierarchy of bounding volumes). One should not confuse the concepts of hierarchy of partitions with hierarchy of bounding volumes. By definition, in a *hierarchy of bounding volumes*, after we take any path in the hierarchy, from the root to a leaf, we obtain a sequence of bounding volumetric objects, i.e., $O_1 \supset O_2 \supset O_3 \supset \cdots \supset O_n$. However, these volumetric objects are not necessarily partition cells. Therefore every hierarchy of partitions is a hierarchy of bounding volumes, but the reverse is far from true. Hierarchies of bounding volumes are also extensively used in computer graphics. ❏

9.5.2 Properties and Applications

Consider a partition of the space, together with m objects $O_1, \ldots, O_m$. Also assume that each of those objects is contained in some partition cell (we can have more than one object per cell). In this case, we can state two fundamental properties of the partition:

1. Any object in a cell cannot intersect an object of another cell;

2. Given a viewing position, the objects contained in the same cell as the observer are visible in relation to the objects in any other cell.

These two properties (despite trivially resulting from the definition of a partition), constitute the basis of several partition-based applications in computer graphics. The first property allows us to attain great efficiency in operations aimed at determining relations among objects in a scene (e.g., clipping, Boolean operations); the second property plays a role in visibility (or orientation) operations of objects in space.

What does a hierarchy add to the two properties above? It allows us to explore three other properties in the solution of problems: *inheritance*, *level of detail*, and *sequencing*.

❏ **Inheritance.** A property is inherited when its validity for a hierarchy node results in validity for all of its subordinate nodes. For example, if a point P does not belong to a node cell, then it does not belong to the subnode cells.

❏ **Level of detail.** A path in the hierarchy tree, from the root to one of the terminal (leaf) nodes, is constituted by a sequence of nested sets. In this way, cells have smaller volumes along the several partition levels, meaning an increase in the level of detail. This fact can be used to represent objects in multi-resolution when using a representation by spatial decomposition.

❏ **Sequencing.** The solution to a problem at each level leads, cumulatively, to its solution globally.

Using those three properties, together with the two fundamental partition properties, we can use partition hierarchies to solve visibility, searching, and sorting problems, among others. As we saw in Chapter 6, the use of hierarchies is a solution for the color quantization problem (the median cut algorithm). We will have the opportunity to use partition hierarchies to solve other important problems including visibility, clipping in the virtual camera, and acceleration of the ray tracing method for calculating the illumination of a scene.

9.5.3   Construction of Hierarchies

Our next challenge is to determine efficient methods for constructing polyhedral partitions in space, as well as hierarchies of polyhedral partitions.

Notice, by the definition of a partition hierarchy itself, that each cell in the level $k+1$ is obtained by the partitioning (subdivision) of the cells in the level k. Therefore techniques based on subdivision methods are quite useful for constructing a hierarchy, as in the case with quadtrees and octrees. (This reinforces the fact that, in general, hierarchy applications for solving problems fit in the classic method of *divide and conquer*.)

9.6   Partitioning Trees (BSP-Trees)

Partitioning trees, also called BSP-trees (binary space partition trees), are the extension of binary search trees (broadly used in computer science as the solution to sorting and searching problems) to dimensions greater than one.

Partitioning trees are generated starting from a basic geometric property: a hyperplane h_1 divides the space $\mathbb{R}^n$ in two semispaces (e.g., a straight line divides the plane in two semiplanes and a plane in $\mathbb{R}^3$ divides the space in two semispaces). The semispace that the normal vector of the hyperplane points to will be indicated by h_1^+; the other will be indicated by h_1^-.

Given a region R, we use the above property to generate a partitioning operation of R resulting in two regions $R^+ = R \cap h_1^+$ and $R^- = R \cap h_1^-$. Indicating by $R^0 = R \cap h_1$, we have $R = R^+ \cup R^- \cup R^0$ (see Figure 9.21). The set R_0 is the boundary of each of the partitioning regions.

If, instead of a hyperplane, we have a list of hyperplanes $(h_1, h_2, \ldots, h_n)$, we can continue the process of binary partitioning in a recursive way: we use the hyperplane h_2 for partitioning the region R^+, the hyperplane h_3 for partitioning the region R^-, and so forth.

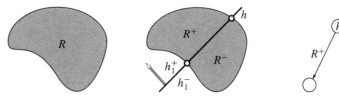

Figure 9.21. Binary partitioning.

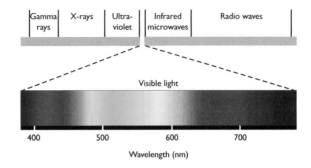

Gamma rays | X-rays | Ultra-violet | Infrared microwaves | Radio waves

Visible light

400   500   600   700

Wavelength (nm)

Plate I. Electromagnetic spectrum; top diagram has a logarithmic scale. (See Figure 5.1.)

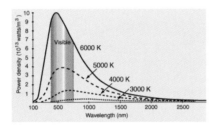

Plate II. The spectrum of sunlight is approximately that of a black body at around 5,800 K. (See page 111.)

Plate III. Experimental acquisition of color matching functions. (See Figure 5.11.)

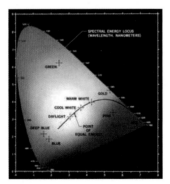

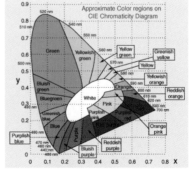

Plate IV. Chromaticity diagram in the CIE-XYZ system, showing in black the *Planck curve*, which represents the colors emitted by a black body at each temperature (left). (The colors shown are approximations; many colors in the diagram do not lie in the gamut of a printer or computer screen.) Common color names and the regions they signify (right). (See Figure 5.16.)

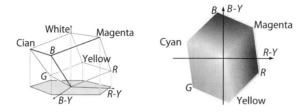

Plate V. Chrominance hexagon of the Y, R–Y, B–Y system. (See Figure 5.21.)

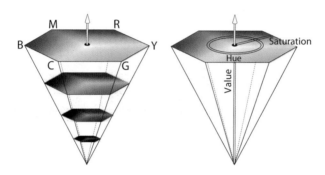

Plate VI. Color solid of the HSV system. (See Figure 5.23.)

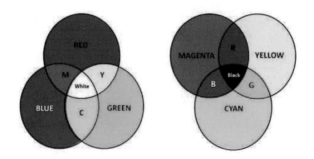

Plate VII. Additive and subtractive color systems (see Figure 5.24).

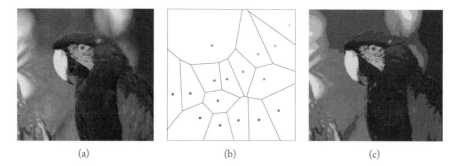

Plate VIII. 2D quantization. (*Original photo from Kodak Photo CD ⓒEastman Kodak Company.* See Figure 6.14.)

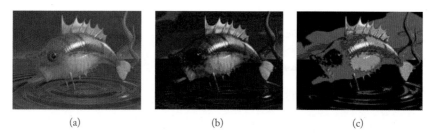

Plate IX. (a) Color image with 24 bits. (b) Uniform quantization of (a) with 8 bits. (c) Uniform quantization of (a) with 4 bits. (See Figure 6.19.)

Plate X. Populosity algorithm: quantization with (a) 8 and (b) 4 bits. (See Figure 6.20.)

(a)   (b)

Plate XI. Median cut algorithm: quantization with (a) 256 and (b) 16 colors. (See Figure 6.22.)

(a)   (b)

Plate XII. Quantization for (a) 256 and (b) 16 colors. (See Figure 6.24.)

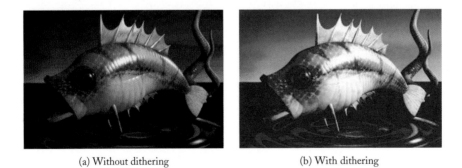

(a) Without dithering   (b) With dithering

Plate XIII. Quantizing from 24 to 8 bits. (See Figure 6.39.)

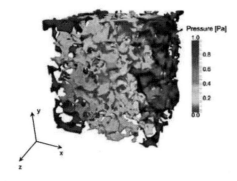

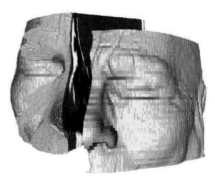

Plate XIV. X-ray computed tomography (CT) scan of a rock pore-space geometry with computed fluid pressure distribution in the pores. (*Picture courtesy of Yan Zaretskiy, Heriot-Watt University, United Kingdom [Zaretskiy et al. 10]. See page 236.*)

Plate XV. Matrix representation: volumetric human head acquired by CT scanning (integrated with laser scan data). (©*2004 Rdiger Marmulla. See page 239.*)

Plate XVI. Nonuniform representation: Voronoi cells (2,000 samples). (*Reprinted from [Darsa et al. 98], with permission from Elsevier. See page 240.*)

Plate XVII. Nautilus shell cutaway showing chambers. (*This Wikipedia and Wikimedia Commons image from user Chris 73 is freely available at http://commons.wikimedia.org/wiki/File:Nautilus CutawayLogarithmicSpiral.jpg under the creative commons cc-by-sa 3.0 license.1. See page 291.*)

(a) Waterfall　　　　　　　(b) Cloud　　　　　　　(c) Bromeliad

Plate XVIII. Procedural models. (a) Particle systems waterfall by Karl Sims, 1988. (*[Sims 90] ©1990 Association for Computing Machinery, Inc. Reprinted by permission.*) (b) Cloud model represented by a hypertexture with procedural details on the boundary and a homogeneous core. (*[Bouthors et al. 08] ©2008 Association for Computing Machinery, Inc. Reprinted by permission.*) (c) Model of a bromeliad resulting from the combination of L-systems parameterized using sketch-based modeling techniques. (*Reprinted from [Anastacio et al. 09], with permission from Elsevier. See Figure 10.7.*)

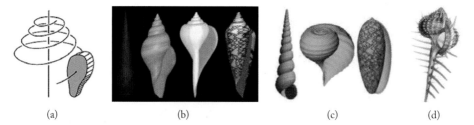

(a)　　　　　　　(b)　　　　　　　(c)　　　　　　　(d)

Plate XIX. (a) Logarithmic helix. Examples of eight seashells synthesized on the computer: (b) variation of the shell shape resulting from different generating curves. From left to right: turreted shell, two fusiform shells, and a conical shell. (*Reprinted from [Fowler et al. 92], courtesy of D. R. Fowler, P. Prusinkiewicz, and H. Meinhardt, ©1992 Association for Computing Machinery, Inc. Reprinted by permission.*) (c) From left to right: models of *Turrirella nivea*, *Papery rapa*, and *Oliva porphyria* shells. (*Reprinted from [Harary and Tal 11], courtesy of Harary and Tal, with permission from John Wiley and Sons.*) (d) Model of *Murex cabriti*. (*Reprinted from [Galbraith et al. 02], courtesy of C. Galbraith, P. Prusinkiewicz, and B. Wyvill, ©2002 Springer Science + Business Media. Reprinted by permission. See Figure 10.22.*)

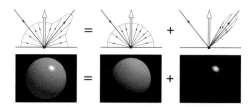

Plate XX. Reflection models of type "diffuse + specular-diffuse." (See Figure 14.4.)

Plate XXI. Motion blur generated using distributed ray tracing. (*[Cook et al. 84] ©1984 Association for Computing Machinery, Inc. Reprinted by permission.*) (See Figure 14.22.)

Plate XXII. Olaf rendered using cartoon shading. (Reprinted from *[Lake et al. 00] by permission of Intel Corporation. ©2000 Association for Computing Machinery, Inc. Reprinted by permission.*) (See Figure 14.23.)

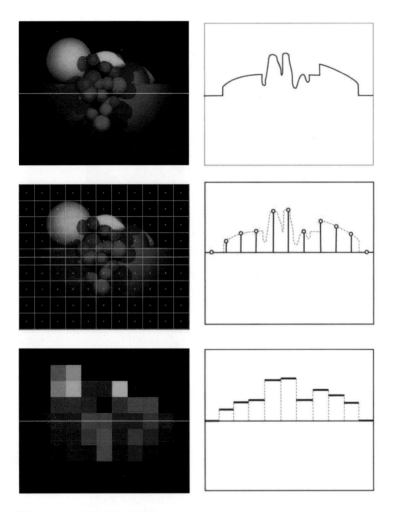

Plate XXIII. Sampling and reconstruction of a scene. Top left: the virtual screen with a projected scene, highlighting one scanline. Top right: the function graph of the associated shading on the scanline. Middle left: the pixels and the points at the center of the pixel where we calculate the shading function to obtain the color value of each pixel (point sampling). Middle right: the scanline samples. Bottom left: the reconstructed image. Bottom right: the scanline reconstruction. (*Left figures: ©Rosalee Wolfe. Used with permission.* See Figure 15.1.)

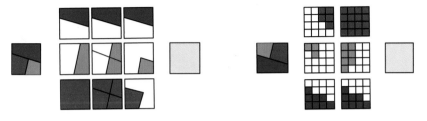

Plate XXIV. Analytical sampling of polygons. (See Figure 15.12.)

Plate XXV. A-buffer of one pixel with three polygon fragments. (See Figure 15.14.)

(a)

(b)

Plate XXVI. (a) Details created with texture mapping without geometry complexity. (*Image courtesy of Karin Eszterhas and 3DTotal.com, www.digitalgallery.dk, www.3dtotal.com.*) (b) Both images were obtained from the same scene containing 3,497 polygons; all the details of the image on the right were obtained using texture mapping. (*Image courtesy of Richard Tilbury and 3DTotal.com, www.richardtilburyart.com, www.3dtotal.com.* See Figure 16.14.)

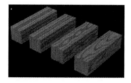

Plate XXVII. 3D texture mapping: 3D wood texture (*[Wolfe 97]* ©*Rosalee Wolfe. Used with permission.* See page 411.)

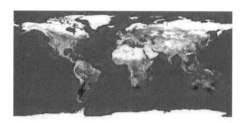 

Plate XXVIII. Texture mapping on a sphere. (See Figure 16.15.)

Plate XXIX. A stitched panoramic image and some of the photographs the image was stitched from. (*[Shenchang 95]* ©*1995 Association for Computing Machinery, Inc. Reprinted by permission.* See Figure 16.17)

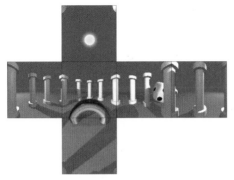
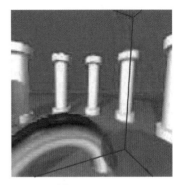

(a) Environment map.   (b) Visualization.

Plate XXX. Virtual panorama with cubic mapping [Gomes et al. 98]. (a) Unfolded cubical environment map. (b) Cube reprojection in a given viewing direction. (*Reprinted from [Darsa et al. 98], with permission from Elsevier.*) In (b) we show parts of the cube edges for reference purposes only. (See Figure 16.18.)

(a)   (b)

Plate XXXI. Examples of reflection mapping: (a) ray tracing approximation (*Courtesy of Castle Game Engine, http://castle-engine.sourceforge.net/*), (b) metal appearance (©*2011 Okino Computer Graphics, Inc.* See Figure 16.20).

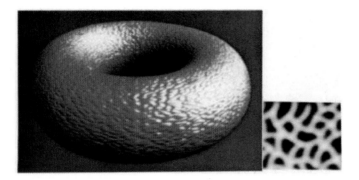

Plate XXXII. The result of bump mapping, obtained from the image at right, applied using mapping by parameterization. Hand drawn bump functions. (*[Blinn 78] ©1978 Association for Computing Machinery, Inc. Reprinted by permission.* See Figure 16.21.)

Plate XXXIII. Face of a coin generated with bump mapping. The texture was mapped using decal mapping with orthogonal projection. (See Figure 16.22.)

(a)   (b)

Plate XXXIV. A deformed Utah teapot using the same texture for (a) bump and (b) displacement mappings. (*From [Wolfe 97], ©Rosalee Wolfe. Used with permission.* See Figure 16.23)

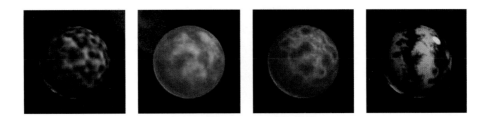

Plate XXXV. Spheres with 3D textures defined with the Perlin noise function. (©*2001 Ken Perlin.*) (a) Applying noise itself to modulate surface color. (b) Using a texture that consists of a fractal sum of noise calls: $\sum 1/f(\text{noise})$. (c) Using a fractal sum of the absolute value of noise: $\sum 1/f(|\text{noise}|)$. (d) Using the turbulence texture from (c) to do a phase shift in a stripe pattern, created with a sine function of the x coordinate of the surface location: $\sin(x + \sum 1/f(|\text{noise}|))$. (See Color Plate 16.39.)

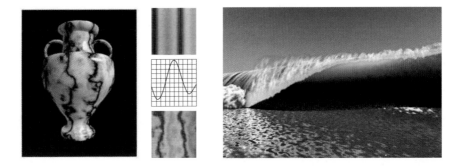

Plate XXXVI. Procedural textures. (a) 3D marble vase (*[Perlin 85]* ©*1985 Association for Computing Machinery, Inc. Reprinted by permission.*) (b) Marble texture obtained by using a sinusoid. (c) Water textures applied to a breaking wave model. (*Figure courtesy of Manuel Gamito and Ken Musgrave.* See Figure 16.40.)

Plate XXXVII. A procedural texture simulating flames. (©*2001 Ken Perlin.* See Figure 16.41.)

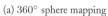
(a) 360° sphere mapping   (b) Reflection mapping

Plate XXXVIII. Sphere mapping in the teapot. (*Image appears in online Panda3D Manual, Panda3D open source 3D game engine, http://panda3d.org.* See Figure 16.42.)

Plate XXXIX. Image and alpha channel. (See Figure 17.1.)

Plate XL. Composition with the alpha channel. (See Figure 17.2.)

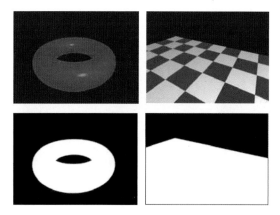

Plate XLI. Images used to illustrate the composition operations. (See page 460.)

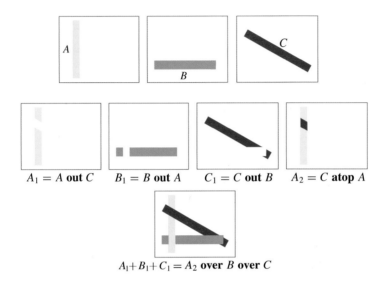

$A_1 = A$ **out** C $\qquad$ $B_1 = B$ **out** A $\qquad$ $C_1 = C$ **out** B $\qquad$ $A_2 = C$ **atop** A

$A_1 + B_1 + C_1 = A_2$ **over** B **over** C

Plate XLII. Resolving cycles with composition operations. (See Figure 17.21.)

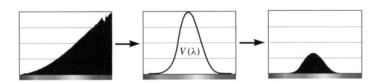

Plate XLIII. Filtering by the function of luminous efficiency. (See Figure 18.13.)

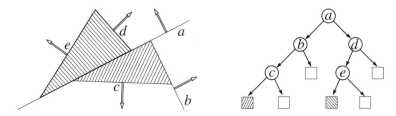

Figure 9.22. Partitioning tree.

When the subdivision process is stopped, the nonpartitioned regions are called *cells*. The boundary of those cells is formed by parts of the partitioning hyperplanes. Each cell is a convex polyhedron, obtained by successive intersections of semispaces. Clearly, as we divide the region R, we also divide its complementary region $\mathbb{R}^n - R$. We illustrate the process in Figure 9.22, where the region R is formed by two triangles intersecting each other along the boundary. We use the straight lines defined by the sides of those triangles as partitioning straight lines, forming the list of separating straight lines (a, b, c, d, e). The region R of the plane, defined by the two triangles, can be obtained by the selection of the partitioning straight lines and the union of the resulting cells, together with their boundaries.

From a structural point of view, the above process of binary partitioning creates a binary tree (partitioning tree). Each tree node represents a partitioning hyperplane, and the leaves represent the cells of the region R or of the complementary region $\mathbb{R}^n - R$. Figure 9.22 also shows the partitioning tree.

In the above example we see that the partitioning tree can be used to represent the region R by the two triangles. In fact, the point membership classification problem, which characterizes the region, can be solved solely using the tree.

Figure 9.23 illustrates another construction of a partitioning tree. We have four planar polygonal regions O_1, O_2, O_3, O_4, and we use three partitioning straight lines (P_1, P_2, P_3). We begin constructing the partitioning tree with the straight line P_1, dividing the plane in

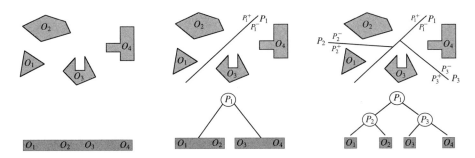

Figure 9.23. Object separation by partitioning.

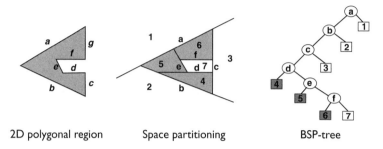

| 2D polygonal region | Space partitioning | BSP-tree |

Figure 9.24. Partitioning tree and geometry.

two semiplanes: the semiplane P_1^+ containing the regions O_1 and O_2, and the semiplane P_1^-, containing the regions O_3 and O_4. Next, the semispace P_1^+ is partitioned by the straight line P_2, creating two convex cells: the cells P_2^+ and P_2^- containing the regions O_1 and O_2, respectively. Similarly, the plane P_3 subdivides the semispace P_1^- into two cells, each containing the polygonal regions O_3 and O_4. At the end of the process, we obtain four cells, each containing one of the polygonal regions in its interior.

We have now seen two distinct uses of a partitioning tree: in the first example, it was used to describe a region, in the second example to separate four regions on the plane.

A natural question is how to choose the family of partitioning planes. In general, this choice depends on the application. As seen in the first example, the choice of the partitioning straight line is done so as to coincide with the sides of the triangles. This is the appropriate choice when we want to represent a polyhedral region by a partitioning tree. We show one more example of this application in Figure 9.24.

Despite being application-dependent, we can make a general observation regarding the choice of the partitioning planes: the size of the cells is reduced after each partitioning by a new plane; therefore, we should choose planes in such a way that, when traversing a BSP-tree path from the root to the leaves, we are appending more details from the scene or object. In this case we can see that partitioning trees can be used as representations, where the level of details of the geometry is variable.

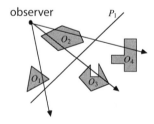

Figure 9.25. Fundamental properties of a partitioning tree.

Partitioning trees simultaneously present a search structure and a geometric representation. In the case of partitioning trees, we can reformulate the properties given at the beginning of Section 9.5.2 as follows (see Figure 9.25):

1. Any object on a side of the plane is not able to intersect an object on the other side;

2. Given a viewing position, the objects of one same side of the observer's plane are visible in relation to the objects on the other side.

The pseudocode below describes the construction algorithm of a BSP-tree. The function `make_bsp` receives as input a list of polygons (plist) and uses the supporting planes of each polygon to create the partitioning tree. The function works recursively, creating, at each call, the two subnodes of the tree associated with each processed polygon by calling the function `combine`.

```
Make_bsp(plist)
if plist == NULL then
    return NULL;
end if
root = select(plist);
for all p ∈ plist do
    if p on the '+' side of root then
        add(p, frontlist);
    else if p on the '-' side of the root then
        add(p, backlist);
    else
        split_poly(p, root, fp, bp);
        add(fp, frontlist);
        add(bp, backlist);
    end if
end for
return combine(make_bsp(frontlist), make_bsp(backlist));
```

9.7   Classification and Search using BSP-Trees

Partitioning trees can be used to solve classification problems. In fact, as seen previously, in the case of polyhedral regions we can use a partitioning tree to solve the point membership classification problem. In this way, the partitioning tree completely characterizes the geometry of the region. In fact, this example shows that partitioning trees can be used to perform Boolean operations among objects.

Just as binary trees are used in traditional computing for classification and search problems, BSP-trees can play a similar role in Euclidean space. When we need to partition the space and efficiently locate points in the partition cells, the partitioning cells provide an

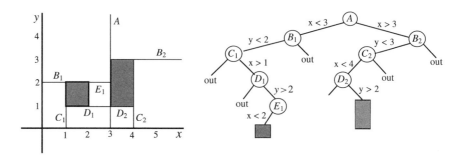

Figure 9.26. Search tree.

appropriate structuring to solve the problem. Figure 9.26 illustrates this fact. The particular case of a partitioning tree with splitting planes parallel to the coordinate planes is called a *kd-tree* (*k*-dimensional trees).

The median cut algorithm for color quantization, described in Chapter 6, uses a *kd*-tree structure to subdivide the color space into quantization cells. An efficient implementation of this algorithm can be made using the same *kd*-tree structure to locate the cell to which a certain color c belongs, thus obtaining the quantization of c.

A BSP-tree is constructed using recursive cell subdivision, starting with a cell at the root representing the entire world space. This way we see this structure representing a hierarchy of partitions whose data structure is a binary tree. Figure 9.27 illustrates the way this structure can be used to solve the problem of classifying a point a in relation to several objects: we construct a BSP-tree so that each cell contains only one object. Using the tree structure and performing simple tests, we efficiently find which cell the point is in. At this stage, we have solved the point membership classification problem for the object contained in this cell (of course, if the object is a polyhedral, this last step is not necessary).

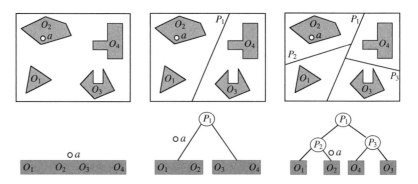

Figure 9.27. Point locating problem.

In Chapter 13, we will see how partitioning trees can be used to solve the problem of object visibility in a scene.

9.8   Comments and References

The master's thesis of Fernando Wagner [da Silva 98] discusses the problem of the human body hierarchy related to motion capture. In addition to some of his figures used in this chapter, discussions with Fernando were very useful. Partitioning trees were introduced in [Fuchs et al. 80].

9.8.1   Additional Topics

We recommend that the reader consult references on robotics to gain a more in-depth knowledge on hierarchies of articulated objects. A basic reference is [Craig 89]; another excellent reference is [Murray et al. 94]. A particularly interesting topic in this area is the study of complex articulations, such as prismatic joints.

Expanding this chapter to explore animation, especially character animation, could lead to the creation of an entire course. Several books cover different aspects of modeling and animation of the human body. Two broad references are [Badler et al. 91] and [Badler et al. 93].

The fundamentals of hierarchies of composed objects that we presented in this chapter is only the tip of the iceberg in the area of spatial subdivision and data structures, which has many applications in computer graphics and GIS (geographic information systems). A basic reference for partition hierarchies and spatial subdivisions, with the associated data structures is [Samet 90].

Partitioning trees have many applications in computer graphics and correlated areas. They are certainly a topic deserving more attention. We refer the reader to [Samet 05] for more details.

Exercises

1.   Extend the hierarchy of Section 9.2.2 for the space $\mathbb{R}^3$ and design a structure that allows making turns with the car.

2.   Describe the hierarchy of Joe Stick using the stack notation (PUSH and POP).

3.   Consider the articulated structure with two degrees of freedom shown in Figure 9.28, with $L_1 = 2, L_2 = 1$.

  (a) If $\theta_1 = 30°$ and $\theta_2 = 45°$, calculate the position $P = (x_1, y_1)$ of the end-effector.

  (b) Determine the region on the plane constituted by the points that are reachable by the end-effector of the arm.

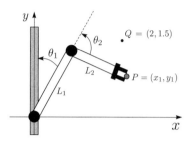

Figure 9.28. Articulated structure with two degrees of freedom (Exercise 3).

    (c) Show that the point of coordinates $Q = (2, 1.5)$ is reachable by the arm. Calculate a possible configuration for θ_1 and θ_2 so that the end-effector reaches that point. Is this the only configuration?

4.   Define prismatic joints and discuss their degrees of freedom and configuration space. (Hint: consult a book on robotics.)

5.   Consider a planetary system consisting of one sun, one planet, and one satellite (assume circular orbits with uniform motion). For this planetary system, describe the structure of its hierarchy, including its transformations.

6.   Place three degrees of freedom in the trunk of the hierarchy of Joe Stick (split the hips and the upper part of the trunk).

7.   Add three degrees of freedom to the joint connecting the legs with the trunk in the hierarchy of Joe Stick. Do not forget to take into account the fact that in the motion of the legs, each foot should always point ahead.

8.   A *balltree* is a hierarchy of bounding volumes, associated to a binary tree with the following property: each intermediate node is associated to a ball of the $\mathbb{R}^n$, which is the ball of smaller radius containing all the balls of the children of this node. Figure 9.29 displays a set of points (small balls), the associated balltree, and the tree structure. (This exercise is drawn from [Omohundro 89].)

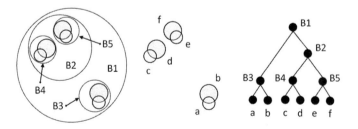

Figure 9.29. A balltree structure (Exercise 8).

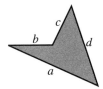

Figure 9.30. A 2D polygonal region (Exercise 12).

    (a) Describe a procedure to build a balltree associated to a set of points.

    (b) Describe how the balltree structure can be used to solve the nearest neighbor problem (i.e., given a discrete dataset U and a point $p \in U$, find the closest point $q \in U$ to p).

9.   Consider a plane P in $\mathbb{R}^3$ of equation $ax + by + cz + d = 0$, and a point $O \in \mathbb{R}^3$.

    (a) Describe the required procedure to determine the position of the point in relation to the plane.

    (b) If the plane P is specified by three points in space, what does change in its description?

10.   Is a partitioning tree invariant to affine transformations? To projective transformations?

11.   Prove that if two objects are convex, there is a plane in space splitting these objects; that is, the objects are contained in distinct semispaces.

12.   Consider the polygon shown in Figure 9.30.

    (a) Construct, on paper, the partitioning tree associated to the polygon, using as the root the supporting straight line of side a.

    (b) Repeat the same exercise using the supporting straight line of side b as the root.

13.   Color quantization algorithms generally make use of the search problem (i.e., given a partition of a region U and a point $p \in U$, determine the partition cell to which the point belongs).

    (a) Study the median cut algorithm for color quantization (see Chapter 6).

    (b) Prove that the partitioning structure of the color gamut for the image used by the algorithm is a BSP-tree. (The particular type of BSP-tree used in the algorithm is known as a kd-tree.)

    (c) Write a pseudocode for the median cut algorithm using this hierarchy.

10 Geometric Modeling

In the previous two chapters we studied planar and spatial graphics objects. In this chapter we will study the specification, construction, and representation of geometric models using planar or spatial objects. This area is part of geometric modeling. We will examine the basic concepts of representation and reconstruction and will explore modeling systems and some modeling techniques.

10.1 Modeling and Representation

Modeling deals with creating and manipulating the geometry and topology of graphics objects on the computer. In this area, graphics objects are generically called *models*. According to the four universes paradigm, we should look for methods to represent the topology and geometry of the model so that we can prepare for implementation.

Models are characterized by the dimension of the graphics object they represent: points represent dimension zero; curves represent dimension one; surfaces represent dimension two; and solids represent dimension three.

The representation of a model is the discretization of its geometry and attributes. One representation is a correspondence (relation) R of a set $\mathcal{O}$ of graphics objects in some representation space $\mathcal{R}$ (see Figure 10.1). Remember that a correspondence, or relation,

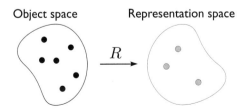

Figure 10.1. Representation of graphics objects.

277

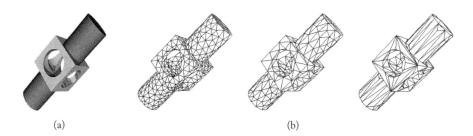

(a)   (b)

Figure 10.2. Approximated geometries with the same topology.

is a subset of $\mathcal{O} \times \mathcal{R}$. The set of graphics objects varies depending on the application; however in general this set is materialized in some function space. The representation space intrinsically depends on the space of graphics objects to be represented.

The reconstruction of a model, starting from its representation, is an operation of fundamental importance. This reconstruction is related to the inverse representation R^{-1}, and it associates a graphics object O_λ to each representation r_λ in the representation space. The semantics of the model (represented by the graphics object) is actually encapsulated in the reconstruction. The reconstruction is *exact* when it recovers the geometry and the topology of the model starting from its representation, that is, when the representation R is invertible.

Several variations exist for the reconstruction problem. An important case is the approximated reconstruction that constructs the correct topology of the model but with only an approximation of its geometry. In Figure 10.2(b) shows three approximations of the geometry of the mechanical part shown in (a), by using a polygonal B-rep. Notice the topology is correct in all three approximations.

There are reconstructions which, besides approximating the geometry, construct an incorrect topology of the original model. This is acceptable in some applications. For example, if an object is observed from a very distant position, its exact geometry and topology can be irrelevant to the visualization problem.

10.1.1   Wireframe Representation

In the early days of computer graphics (1960s through the mid 1970s) it was quite common to represent models using the *wireframe* method. In this representation method, we take pairs of points p_i, p_j, $i \neq j$, of the model, and the representation is given by the union of the segments $\overline{p_i p_j}$ connecting those pairs. In other words, the representation is given by a collection of straight line segments of the model. A common case of this representation consisted of obtaining a polyhedral approximation of the surface but using only the edges and vertices to obtain the representation. Figure 10.3 displays a wireframe representation of an aircraft.

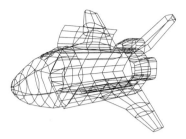

Figure 10.3. Wireframe representation.

In general, the correct choice of the pairs of points allows one to obtain a wireframe representation whose visualization provides a reasonable fidelity to the represented object. But the wireframe representation does not have an appropriate structuring to represent either the topology or the geometry of the model. The wireframe representation in Figure 10.4(a), for instance, can correspond to any one of the objects shown in Figure 10.4(b), (c), or (d).

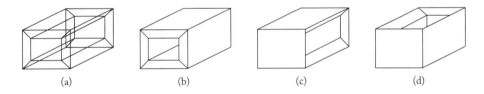

Figure 10.4. Different models with the same wireframe representation.

The wireframe method also allows the representation of objects whose geometry cannot be reconstructed in the 3D Euclidean space. This is the case of the object shown in Figure 10.5.

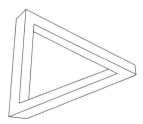

Figure 10.5. Impossible geometry represented in wireframe.

10.1.2   Representation Unicity and Ambiguity

A representation R has the property of *unicity* when each model has only one representation, that is, if $R(O_1) \neq R(O_2)$, then $O_1 \neq O_2$. Usually the unicity condition is difficult to obtain. The wireframe representation introduced in Example 10.1.1 does not have unicity. Even good representations for graphics objects, such as matrix and polygonal boundary representations, which we studied in Chapter 8, do not present unicity. Figure 10.2(b), for instance, displays three different model representations.

Ambiguity is another important concept related to representation: a representation R is *ambiguous* if we can reconstruct different models from a representation $R(O)$ of a graphics object. In other words, R is nonambiguous if, whenever $R(O_1) = R(O_2)$, we have $O_1 = O_2$ (injectivity). The example in Figure 10.4 shows that the wireframe representation is extremely ambiguous, because any of the models, (b), (c), or (d), can be reconstructed starting from the representation in (a). If a representation is ambiguous, we cannot associate a semantics to the represented model; in other words, we are able to determine neither the geometry nor the topology of the model starting from its representation.

10.1.3   Representation and Data Structure

People often do not clearly distinguish between the representation of a model and the data structure used to encode that representation. For example, representation by adaptive spatial subdivision shown in Figure 10.6 is called "modeling with representation by quadtree" because the quadtree is an appropriate data structure for implementing that particular representation method by subdivision.

However, according to the four universes paradigm, the data structure is part of the implementation universe and is not intrinsically associated with the representation: a single representation can use different data structures in its implementation.

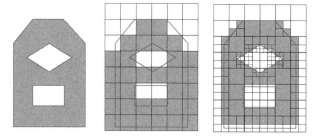

Figure 10.6. Representation by quadtree.

10.1.4   Representation Methods

Representation methods are directly related to the various methods of representing graphics objects that we have previously studied and can be divided into three categories: representation by decomposition (which includes both intrinsic and spatial decompositions); representation by construction; and hybrid representations.

When we use a representation by spatial decomposition, we have a vast range of possibilities for data structures, *spatial data structures* (see Chapter 9), associated to each representation. An important example of representation by decomposition is polyhedral surfaces.

Representation by construction makes use of the fact that complex models can be constructed starting from the combination of simpler models by means of operations. An example of this fact is the geometric models of molecules that can be built by combining cylinders and spheres, properly positioned in space.

A historical perspective. Up until the mid 1970s, the only representation used was the wireframe (see Section 10.1.1). As technology advanced and it became possible to use computational models in design and manufacturing (CAD/CAM), it became necessary to formalize the theory of geometric modeling and, in particular, the representation of models. The pioneer in this endeavor was Aristides Requicha [Requicha 80].

CAD/CAM required the representation of solid objects. One of the representation methods consisted of representing a solid by a polygonization of its boundary. This method was called *boundary representation*, or B-rep. Clearly, this was a representation by intrinsic decomposition. Aristides Requicha introduced another method for representing solids using a constructive representation, which he called *constructive solid geometry*, or *CSG representation*.

Procedural representation. While *geometric modeling*, or *solid modeling*, is very useful for modeling manufactured and other objects with well-defined geometric forms, it is not necessarily suitable for modeling all kinds of objects. In the physical universe there are many objects such as clouds, rain, smoke, fire, and water, that have highly complex geometry. For example, they may have significantly irregular boundaries (fractal) or time-varying ill-defined boundaries.

These objects are described and represented by *procedural modeling* or *algorithmic modeling*. The name comes from the fact the model representation is given, in this case, by an algorithm described in some virtual machine (e.g., Turing machine):

$$\text{Object} = \text{Algorithm(input, parameters)}.$$

Usually this algorithm is recursive and the input is an object with a simple geometry. The parameters allow one to control certain characteristics of the object. The semantics of the object is obtained by executing the algorithm. In Figure 10.7(a), we show the image of a procedurally modeled waterfall using a technique called a *particle system* [Reeves 83]. In Figure 10.7(b), we show a cloud modeled with the use of hypertextures [Perlin and Hoffert 89] and the Perlin noise function (this function will be studied in detail in Chapter 16.) In Figure 10.7(c), we show a Bromeliad modeled by combining L-systems [Prusinkiewicz and Lindenmayer 96] parameterized using sketch-based modeling techniques [Olsen et al. 09].

(a) Waterfall   (b) Cloud   (c) Bromeliad

Figure 10.7. Procedural models. (a) Particle systems waterfall by Karl Sims, 1988. (*[Sims 90]* ©*1990 Association for Computing Machinery, Inc. Reprinted by permission.*) (b) Cloud model represented by a hypertexture with procedural details on the boundary and a homogeneous core. (*[Bouthors et al. 08]* ©*2008 Association for Computing Machinery, Inc. Reprinted by permission.*) (c) Model of a bromeliad resulting from the combination of L-systems parameterized using sketch-based modeling techniques. (*Reprinted from [Anastacio et al. 09], with permission from Elsevier.* See Color Plate XVIII.)

10.2   CSG Representation

CSG representation uses three basic ingredients: geometric primitives, space transformations, and Boolean operations.

10.2.1   Geometric Primitives

Geometric primitives are the basic building blocks of models. In Figure 10.8 we show examples of geometric primitives on the plane (the unit square (a) and disk of radius 1 (b)) and in the space (the unit cube (c) and a spherical solid (d)). Geometric primitives are simple to describe and represent on the computer. For instance, the sphere is completely described by its center (x, y, z) and radius r, and can therefore be represented by the vector $(x, y, z, r) \in \mathbb{R}^4$.

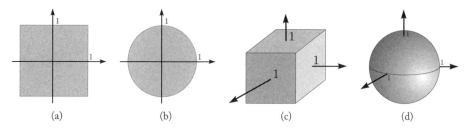

(a)   (b)   (c)   (d)

Figure 10.8. Geometric primitives on the plane (a) and (b), and in the space (c) and (d).

10.2.2   Space Transformations

Transformations are used in CSG representation to either position the primitives in space or modify their geometry. Rigid motions in space (rotation and translation) are used for positioning the primitives. It is useful to consider an intrinsic coordinate system for each primitive. The geometric positioning transformations perform the change between the coordinate system of the primitive and the global coordinate system of the space where the model should be constructed (Figure 10.8 shows the coordinate system of each primitive).

When modifying geometry, transformations assume the role of allowing the construction of several geometric forms starting from a single primitive. In particular, the scale transformation is widely used to modify the geometry of a primitive:

$$(x, y, z) \mapsto (\lambda_1 x, \lambda_2 y, \lambda_3 z),$$

which allows a change on the dimensions of the primitives. Using this transformation, the primitive given by the unit square can be transformed to obtain rectangles of varied dimensions. Starting with a square and using linear transformations, we can obtain any parallelogram on the plane. Using projective transformations, we can obtain any quadrilateral on the plane. These possibilities are shown in Figure 10.9. The use of transformations to modify primitives reduces the number of primitives needed.

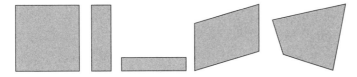

Figure 10.9. Changes to a square by projective transformations.

10.2.3   Boolean Operations

Finally, after properly positioning the primitive in the space, the CSG system uses Boolean operations to combine the various primitives and create the final model. The Boolean operations are the *union* ∪, *intersection* ∩, and *difference* −, of sets. These operations are illustrated in Figure 10.10.

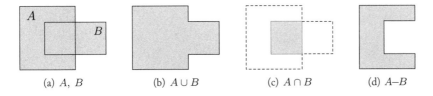

(a) A, B   (b) A ∪ B   (c) A ∩ B   (d) A–B

Figure 10.10. Boolean operations.

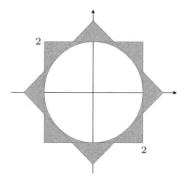

Figure 10.11. 2D solid.

Example 10.1. We will describe the construction of a 2D solid using CSG, as shown in Figure 10.11. The primitives used are the unit disk and square shown in Figure 10.8 (a) and (b), respectively.

1. We do a scaling $(x, y) \mapsto (2x, 2y)$ of the primitive defined by the unit disk, obtaining a disk D.

2. We do a scaling $(x, y) \mapsto (2x, 2y)$ of the primitive defined by the square, obtaining a square Q_1.

3. We do a scaling $(x, y) \mapsto (2x, 2y)$ of the primitive defined by the square, followed by a rotation of $45°$ about the origin. We obtain a square Q_2.

The final model is obtained by the operations $Q_1 \cup Q_2 - D$. ❏

We can use a simple notation, based on lists, to describe the operations for constructing the model:

$$(\mathrm{S}(2, 2), Q) \cup (\mathrm{R}(45°), \mathrm{S}(2, 2), Q) - (\mathrm{S}(2, 2), D).$$

Each graphics object is represented by a list containing the object transformations. The list should be evaluated starting from the right. That is,

$$(T_n, T_{n-1}, \ldots, T_1, \mathcal{O}) = T_n(T_{n-1}(\ldots (T_1(\mathcal{O})))).$$

10.2.4   Regularized Boolean Operations

In the CSG system, we use solid primitives to build solids. The underlying idea is Boolean operations with solids result in solids. However, Figure 10.12 demonstrates that this is not always true: here the intersection of two 2D solids does not result in a solid due to the unidimensional segment that appears in the resulting set.

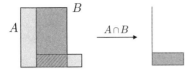

Figure 10.12. Intersections of solids might not be a solid.

We will use topological regularization as a way to discard those undesired elements. The topological regularization of a subset $A \subset \mathbb{R}^n$ is the set obtained by the successive application of two operations: the operation for obtaining the interior $i(A)$ of the set A (i.e., the points inside the boundary), followed by the operation of taking the closure $k(i(A))$ of the resulting set. The regularization is illustrated in Figure 10.13. Intuitively, the operation of taking the interior eliminates all the boundary points, including the points corresponding to the undesired geometric elements. The operation of taking the closure recovers the "good" boundary points so as to constitute a solid.

Using the operation of topological regularization, we define the regularized Boolean operations $\cup^*$, $\cap^*$, and $-^*$, placing $A \cup^* B = ki(A \cup B)$, $A \cap^* B = ki(A \cap B)$, and $A -^* B = ki(A - B)$. In other words, we apply the usual Boolean operation followed by a topological regularization.

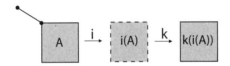

Figure 10.13. Regularization operation.

10.2.5 CSG Hierarchy

Construction of a CSG model has a naturally associated hierarchy of composed objects. This hierarchy is represented by a binary tree structure, called a *CSG tree*. The final model

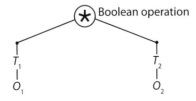

Figure 10.14. Tree representation of a Boolean operation.

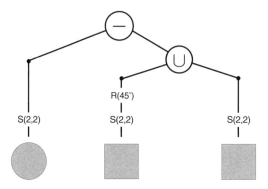

Figure 10.15. CSG tree of Example 10.1.

is at the root of the tree and the primitives at the leaves. The tree has two types of inter-mediate nodes: the *operations* node and the *geometric models* node.

Each operation node is associated to one of the Boolean operations (union, intersec-tion, or difference). This node has two descending nodes, each corresponding to objects that should be combined by the operation of the ascending node. Each node of a geomet-ric object has an associated transformation in space (it can be the identity transformation). This fact is illustrated in the diagram of Figure 10.14, showing a Boolean operation $*$, which should be applied to the objects $T_1(O_1)$ and $T_2(O_2)$, after applying the transforma-tions T_1 and T_2 to O_1 and O_2, respectively.

When the operation node is the difference between sets, we stipulate that the object of the left node be subtracted from the object of the right node. The tree of the CSG model created in Example 10.1 is shown in Figure 10.15. We leave the verification to the reader.

Besides being very important, the CSG representation is used as a paradigm to create representation systems by construction. In fact, we can replace the regularized Boolean operations with other operations among sets. The complexity of the CSG tree is directly related to the number and geometry of the primitives in the system. If we increase the number of primitives, we reduce the complexity of the tree, and vice versa. An extreme case is an object defined by an implicit function, which can be considered a CSG model with only one primitive.

10.3   Conversion between Representations

From our study on planar and spatial graphics objects, we know that an ideal representa-tion scheme does not exist. For example, the CSG method makes creating solid models easier and makes possible the use of Boolean operations between two different models, but it cannot represent solids with variable density and does not provide easy access to boundary information. On the other hand, the B-rep method does not work for solids with variable density and furthermore does not favor the execution of Boolean operations

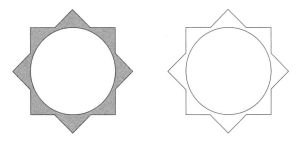

Figure 10.16. Converting from CSG to B-rep.

between models, but it does provide good control over the geometry of the model through an explicit description of its boundary. (Combining the CSG and B-rep method results in an interesting hybrid representation.)

Clearly we need methods that allow conversion between different representations. For example, converting from a CSG to a boundary representation consists of determining the boundary surface of a solid represented by a CSG tree, as illustrated in Figure 10.16.

This is a particular case of a problem we discussed in Chapter 8: *boundary evaluation*. In this particular case, the boundary of the CSG model is constituted by parts of the primitive boundaries. The solution to the problem first consists of partitioning the boundary of the model primitives and then classifying each element of the resulting partition as being from either the border or the interior. This procedure involves the clipping (intersection) operation and passes through our well-known point membership classification problem.

The inverse—converting from B-rep to CSG—is far more difficult. In the case of a model with a polyhedral B-rep representation, an interesting approach for converting this model into a solid one, with a CSG tree, is to use partitioning trees, as we discussed in Chapter 9.

In general, the problems of converting between different representations of graphics objects are very difficult to treat. However, some important particular cases can be solved. In Chapters 7 and 8 we studied the problem of polygonizing implicitly defined curves and surfaces, which is in reality a conversion (from an implicitly defined solid to a polyhedral B-rep model). This is a particular case of a CSG conversion to a polygonal B-rep, where the implicit solid is the primitive itself in the CSG system.

Another important case of conversion between representations happens when we want to obtain polyhedral representations with different resolutions, that is, different geometric approximations (see Figure 10.17). This requires converting from a higher-resolution representation (more polygons) to a lower-resolution representation (with less polygons). Converting from a higher-resolution to a lower-resolution representation is called *simplification*; converting from a lower-resolution model into a higher-resolution one is called *refinement*. In other words, simplification methods obtain representations that less closely approximates the geometry of the model while refinement methods obtain a representation that better approximates the model's geometry.

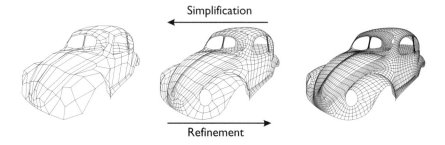

Figure 10.17. Simplification and refinement of a polyhedral representation.

10.3.1   Representation by Partitioning Trees

In Chapter 9 we studied partitioning trees (BSP-trees). We saw that partitioning trees represent a solid object by the use of a binary tree, in a similar way to a CSG representation. However, the BSP-tree representation is not a CSG representation: at each node of the partitioning tree we have a splitting plane, while the nodes of a CSG tree contain Boolean operations and transformations in space. At the leaves of the CSG tree we have the primitives of the representation, and at the leaves of the partitioning tree we have the object cells.

Notice the object cells at the leaves of a partitioning tree are disjoint, and their union (including the boundaries) forms the object. In this way, considering each cell as a primitive, we have a natural conversion from a partitioning tree to a CSG representation. Details of this conversion can be found in [Thibault and Naylor 87], which also discusses conversion from a polyhedral B-rep to a partitioning tree representation. We have already seen some examples of this conversion in Chapter 9 (refer to Figure 9.24, showing a partitioning tree associated to a polygonal B-rep region).

The conversion from a representation by partitioning trees to a polyhedral B-rep representation can be seen in [Comba and Naylor 96]. Of course, this can be done through a conversion to a CSG representation; however, [Comba and Naylor 96] provides a straightforward, more efficient method for directly exploring the partitioning tree structure in the calculation of vertices, edges, and faces of the B-rep representation.

10.4   Generative Modeling

A modeling technique is a combination of a user's specification method and an associated reconstruction technique. The basic elements in the model specification are points, vectors (oriented segments), and curves. We can construct and represent surfaces and solids from those elements.

In Chapter 8 we saw several techniques for describing surfaces from points or curves, including bilinear interpolation, lofting, and Coons patch. In this section we will study a simple and powerful modeling technique called *generative modeling*. The basic idea of this technique consists of moving a planar graphics object along a curve in $\mathbb{R}^3$ to create new graphics objects. Generally this motion, or displacement, describes a graphics object that has one more dimension than the planar object being displaced. Therefore, if the planar graphics object is a curve, we obtain a surface; if it is a planar region, we obtain a 3D solid. We will now examine three examples of generative models: revolution, extrusion, and tubular.

10.4.1 Models of Revolution

Starting from a curve, this technique allows one to build a surface with coaxial symmetry. Consider a straight line r in space and a curve γ contained in the plane passing through r (see Figure 10.18(a)). The surface is obtained by rotating γ about the straight line r.

The curve γ, the straight line r, and the obtained surface are called, respectively, the *profile curve*, *rotation axis*, and *surface of revolution*, (or *surface of rotation*). If, instead of the profile curve, we consider a region on the plane that passes through the axis (i.e., a 2D solid), we then obtain a solid of revolution (see Figure 10.18(b)). Several solids, such as spheres, cylinders, and cones, can be generated with this technique.

The surface of revolution is an example of a generative model: it is generated by the displacement of the profile curve along a circle. For this reason, this technique is also known as a *rotational sweep*.

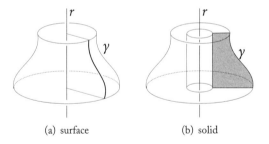

(a) surface   (b) solid

Figure 10.18. Surface and solid of revolution.

10.4.2 Models of Extrusion

The technique of *extrusion surfaces* or *extrusion solids* is a particular case of generative modeling in which the path along which we displace the planar object is a straight line segment. For this reason, this technique is also known as a *translational sweep*. This technique is illustrated in Figure 10.19, where a planar object is displaced along the direction of vector v.

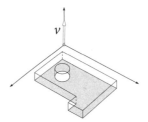

Figure 10.19. Modeling by extrusion.

10.4.3 Tubular Models

This is the most general case of generative modeling and includes the techniques of surface of revolution and extrusion described above. Despite its simplicity, it is a powerful modeling technique. Consider two curves $\gamma\colon [0,1] \to \mathbb{R}^3$ and $\alpha\colon [0,1] \to \mathbb{R}^3$, such that the path of the curve α is contained in the plane Π, normal to the curve γ at the point $\gamma(0)$ (see Figure 10.20(a)).

A *tubular surface* is obtained by displacing the plane Π (together with the curve α) along the curve γ. In other words, for each point $\gamma(t)$ of the curve γ, we consider the plane normal to the curve at that point and we take a "copy" of the curve α in this plane (see Figure 10.20(b)). The result is a surface that contains the curve γ as its "axis." This surface is called the *tubular surface* of γ. The curve γ is called a *guiding curve*, and the curve α is called a *section*. If, instead of the curve α, we consider a planar solid on the plane Π, we obtain a tubular solid.

As we previously saw, the extrusion technique is a particular case where γ is a straight line. The technique of surface of revolution is the particular case in which γ is a circle.

We can augment the technique of tubular surfaces by applying intrinsic transformations on plane Π as we displace it along the guiding curve (e.g., scaling, rotations, etc.). This fact allows us to obtain a great array of models by varying three parameters: the guiding curve, the section, and the transformations on the plane Π of the section.

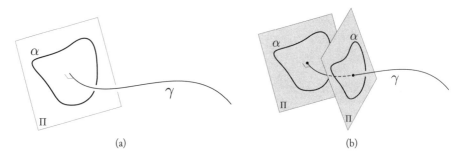

Figure 10.20. Tubular surfaces: (a) guiding curve, (b) sections.

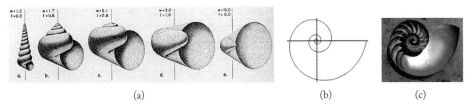

Figure 10.21. Geometry of seashells. (a) Hypothetical snail forms drawn from cross sections made by the computer method (**w** corresponds to the rate of enlargement of the generating curve and **t** to the rate of translation.) (*From [Raup 62]. Reprinted with permission from AAAS.*) (b) The logarithmic spiral, depicting the foundation of a seashell shape. (c) Nautilus shell cutaway showing chambers. (*This Wikipedia and Wikimedia Commons image from user Chris 73 is freely available at http://commons.wikimedia.org/wiki/File:NautilusCutawayLogarithmicSpiral.jpg under the creative commons cc-by-sa 3.0 license.1.* See Color Plate XVII.)

Modeling seashells. The modeling of seashells is an interesting example illustrating the technique of tubular surfaces. As shown in Figure 10.21(a), shells have a spiral form. It is widely known in biology that the orthogonal projection of this spiral, in the direction of the spiral axis, produces a logarithmic spiral (see Figure 10.21(b)). This fact suggests we can generate a conic as a generative model of tubular surface, where the guiding curve is a logarithmic helix (see Figure 10.22(a)).

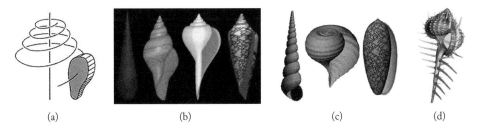

Figure 10.22. (a) Logarithmic helix. Examples of eight seashells synthesized on the computer: (b) variation of the shell shape resulting from different generating curves. From left to right: turreted shell, two fusiform shells, and a conical shell. (*Reprinted from [Fowler et al. 92], courtesy of D. R. Fowler, P. Prusinkiewicz, and H. Meinhardt, ⓒ1992 Association for Computing Machinery, Inc. Reprinted by permission.*) (c) From left to right: models of *Turrirella nivea*, *Papery rapa*, and *Oliva porphyria* shells. (*Reprinted from [Harary and Tal 11], courtesy of Harary and Tal, with permission from John Wiley and Sons.*) (d) Model of *Murex cabriti*. (*Reprinted from [Galbraith et al. 02], courtesy of C. Galbraith, P. Prusinkiewicz, and B. Wyvill, ⓒ2002 Springer Science + Business Media. Reprinted by permission.* See Color Plate XIX.)

The geometry of the section varies according to the type of shell we want to model. Also notice that as we displace the section along the guiding curve, the section should be scaled down to reduce its size. The parametric equation of the logarithmic helix is given by

$$\theta = t;$$
$$r = r_0 a^t;$$
$$z = z_0 b^t.$$

The constants r_0, z_0, a, and b allow us to alter some characteristics of the helix. Figure 10.22 shows images of shells produced with this method.

10.4.4   Generative Modeling and Group Action

The concept of generative modeling can be generalized if we consider the concept of a group action in a Euclidean space. In the case of a surface of revolution, the essence of the technique is rotation about an axis. In other words, we have the group $SO(2)$ of rotations in $\mathbb{R}^3$ about the r-axis. Supposing r is the z-axis, the elements of this group are the matrices in the form

$$R(\theta) = \begin{pmatrix} \cos\theta & -\sin\theta & 0 \\ \sin\theta & \cos\theta & 0 \\ 0 & 0 & 1 \end{pmatrix}.$$

By varying θ, we obtain a representation of the group $SO(2)$ by matrices. Notice we can define a transformation $\varphi\colon SO(2) \times \mathbb{R}^3 \to \mathbb{R}^3$, placing

$$\varphi(R(\theta), P) = R(\theta)P, \quad \theta \in \mathbb{R}.$$

This transformation is called an *action* of the group $SO(2)$ in $\mathbb{R}^3$. As we have a natural operation $h\colon \mathbb{R} \to SO(2)$, $h(\theta) = R(\theta)$, satisfying $h(\theta_1 + \theta_2) = R(\theta_1 + \theta_2) = R(\theta_1)R(\theta_2)$,

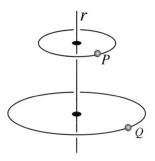

Figure 10.23. Orbits of P and Q.

the above example gives us an action $\phi\colon \mathbb{R} \times \mathbb{R}^3 \to \mathbb{R}^3$, defined by

$$\phi(t, P) = h(t)P = R(t)P. \tag{10.1}$$

Notice that by fixing $P \in \mathbb{R}^3$ and varying t in Equation (10.1), we obtain a circle (1) that is contained in a plane perpendicular to the rotation axis r, (2) that contains P, and (3) whose center is in the r-axis (see Figure 10.23). This circle is called the *orbit* of the point P by the group action. From this group action point of view, a surface of revolution is formed by the orbits of the points from the profile curve.

Viewing a surface of revolution as a set of orbits, by the action of one group, has the advantage of allowing the generalization of the class of generative models. In fact, given an action $\varphi\colon \mathbb{R} \times \mathbb{R}^3 \to \mathbb{R}^3$, we define the orbit of a point P by the curve $\gamma(t) = \varphi(t)P$. Generally speaking, the orbit of a point is a curve, the orbit of a curve is a surface, and the orbit of a surface is a solid. Therefore an action provides a good modeling technique when it allows us to move from a low dimensional object to higher dimensional objects.

In the case of the extrusion surface, we take a direction $v = (v_x, v_y, v_z)$ in space, and we consider the group of translations in that direction. This group is represented in homogeneous coordinates by the group of matrices

$$T(t) = \begin{pmatrix} 1 & 0 & 0 & tv_x \\ 0 & 1 & 0 & tv_y \\ 0 & 0 & 1 & tv_z \\ 0 & 0 & 0 & 1 \end{pmatrix}, \quad t \in \mathbb{R}.$$

Of course, the orbit of point P, by the action of T, is the straight line of equation

$$r(t) = P + tv, \quad t \in \mathbb{R}.$$

As the orbits are not limited, we take orbit segments $t_0 \le t \le t_1$ to construct models.

10.5 Modeling Systems

Modeling systems are programs, or sets of programs, whose objective is to provide the user with several methods for creating and manipulating, on the computer, the geometry and/or the topology of graphics objects. In a simplified way, a modeling system can be characterized by three basic modules: the creation of, operation with, and visualization of models.

Model visualization will be covered later in the book (see Chapters 11–16). In this section, we will focus on creating models and how this relates to operations with them, as illustrated in the diagram below:

10.5.1   Creation and Representation of Models

Model creation is performed by the user, which means the creation process is directly related to the user interface. However, our emphasis will be not on the interface but on the techniques of creating models by the user. *Modeling techniques* refer to the various methods of model creation.

The user creates models using several data input devices. Those devices range from the keyboard or mouse up to more specialized graphics input devices such as tablets or laser ranger scanners. The user specifies a graphics object by using a finite number of parameters that results in a representation of the desired object. Starting from that representation the object is reconstructed, acquiring a semantics and making possible the application of several operations. A more complete diagram of the modeling process is shown in Figure 10.24. Notice that operations are executed on either the representation or the reconstructed model.

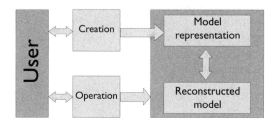

Figure 10.24. The modeling process.

10.5.2   Representation and Modeling Systems

Usually, modeling systems have only one representation for the models created by the user. The use of just one representation certainly hinders some model operations. Systems that use more than one representation type are of two classes: hybrid systems and multirepresentation systems.

Hybrid systems. In hybrid systems, several model representations exist simultaneously, and both model creation and manipulation can be done in any of the representations. This

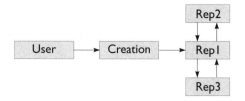

Figure 10.25. Hybrid system.

requires that the internal consistency of the model be maintained so that operations done in one representation are reflected in the others. Consequently hybrid systems must have robust algorithms, allowing for conversion between the several representation methods used in the system. This is a difficult task and can be achieved by only considerably limiting the universe of the objects that can be modeled by the system. Figure 10.25 illustrates the architecture of those systems.

An interesting example is a system using a polygonal B-rep and implementing simplification and refinement operations to convert models of different resolutions. In those systems, known as *multiresolution systems*, we can seamlessly work simultaneously with several approximations of the model's geometry.

Multirepresentation systems. In these systems we have a main representation of the model and several secondary representations used to solve specific problems on the model. The system converts between the main representation and each of the secondary ones. Modifications cannot be made in the secondary model representation. The diagram in Figure 10.26 illustrates the architecture of these systems. They are simpler than hybrid systems, as it is not necessary to maintain consistency between the representations.

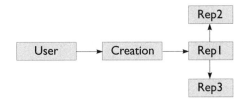

Figure 10.26. Multirepresentation system.

10.6   Operations with Models

A graphics object is determined by its geometry and topology. The topology of a model determines its form, independent of metric properties. The geometry of the model is responsible for determining the various metric properties associated with the model (e.g., length, area, volume, curvature).

Operations with models can be classified as *topological* or *geometric*. Topological operations work on the topology of the models; geometric operations make changes to the points of the model, transforming them in space and creating new models. Transformations in space are an example of geometric operations. They can reposition objects in space, as in the case of rigid motions, or they can deform objects by altering their geometry entirely. Algebraic operations among models, such as the Boolean operations, are also a type of geometric operation.

10.7 Comments and References

The work of [Requicha 80] undeniably had great influence on the conceptual evolution of geometric modeling. This work introduced the concept of model representation, presented the first conceptualization of representation methods, and introduced solid modeling methods using CSG. Furthermore, this work was used as the basis for the four universes paradigm used in [Gomes and Velho 95] and in this book.

10.7.1 Additional Topics

Computational geometry techniques and spatial data structures are important tools for those interested in geometric modeling. A very good reference in this area is [de Berg et al. 97]. Additional interesting topics include CSG representation, boundary calculation, and the visualization of CSG models. There are many writings on this subject; one good reference is [Hoffmann 89].

Interactive modeling techniques (using interpolation surfaces, splines, Bezier, etc.) are very important. These methods constitute an efficient modeling technique, making possible the construction of curves and surfaces starting from the interpolation of control points, which allows great interactivity with the user. A good reference on this subject is [Farin 93].

Another important technique is subdivision surfaces [Warren and Weimer 01] [Andersson and Stewart 10], which allows the representation of a surface with different levels of detail. The Volkswagen presented in this chapter (see Figure 10.17), was obtained with this technique.

An important topic, briefly mentioned in this chapter, is procedural modeling. There are several procedural techniques, including physical modeling. A broad coverage of this area can be found in [Ebert et al. 02]. A conceptual model of the area, using the abstraction paradigm of the four universes, is presented in [Gomes et al. 93].

Exercises

1. Consider the universe of the convex polygons of n sides. Explain the geometric representation methods used to describe the elements of that universe using

    (a) a constructive scheme.

    (b) a decomposition scheme.

    (c) an approximation scheme.

    Discuss the conversion between these systems.

2. Describe in detail the implementation of a "digital LEGO" system using CSG.

3. In this chapter we mentioned a hybrid representation involving the polyhedral B-rep and CSG representations.

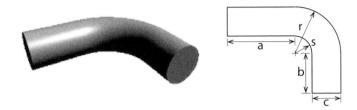

Figure 10.27. A mechanical part (Exercises 5 and 6).

    (a) Describe how such a representation can be achieved.

    (b) This representation holds the advantages of both the CSG and B-rep representations. Enumerate and discuss some of those advantages.

4.   Let $F_i : \mathbb{R}^n \to \mathbb{R}$ be functions determining an implicit object as the inverse image of the zero point (assuming zero is a regular point of F_i). Prove the following equalities:

    (a) $F_1 \cap F_2 = \max\{F_1, F_2\}$;

    (b) $F_1 \cup F_2 = \min\{F_1, F_2\}$;

    (c) $F_1 - F_2 = \max\{F_1, -F_2\}$.

Describe a representation system with implicit primitives based on the results of the previous items.

5.   Describe in detail the modeling of the mechanical part shown in Figure 10.27 using CSG (assume the CSG system has the following primitives: sphere, cylinder, cone, torus, and cube).

6.   Describe in detail the modeling of the mechanical part shown in Figure 10.27 using generative modeling.

7.   A representation scheme for polygonal curves using straight line segments associates to a polygonal curve L the set

$$\{(x_0, y_0, x_1, y_1), (x_1, y_1, x_2, y_2), ..., (x_{n-1}, y_{n-1}, x_n, y_n)\},$$

where $(x_i, y_i, x_{i+1}, y_{i+1})$ represent the initial and final coordinates of each straight line segment composing the polygonal curve.

    (a) Discuss this representation in relation to the properties of ambiguity and unicity.

    (b) Define a proximity notion among polygonal curves and discuss the representation in relation to this concept.

8.   Write the parametric equations of a surface of revolution given by the rotation of the curve $\varphi(t) = (y(t), z(t))$ about the z-axis. Calculate the normal vector to this surface at an arbitrary point.

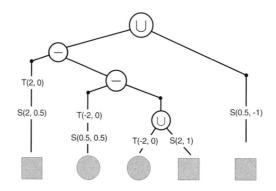

Figure 10.28. CSG tree (Exercise 10).

9. Consider the CSG expression

$$(T(1,1), \mathrm{D}) - \{(S(2,2), \mathrm{Q}) - (\mathrm{D}) \cup (T(2,2), \mathrm{D}) \cup (T(0,2), \mathrm{D}) \cup (T(2,0), \mathrm{D})\},$$

where $\mathrm{D} = \{(x,y) \in \mathbb{R}^2; \ x^2 + y^2 \leq 1\}$, and $\mathrm{Q} = \{(x,y) \in \mathbb{R}^2; \ |x| \leq 1 \ \mathrm{e} \ |y| \leq 1\}$, where $T(u,v)$ indicates translation by the vector (u,v), and $S(a,b)$ indicates the scale transformation $S(x,y) = (ax, by)$. Draw the associated CSG tree and make a sketch of the region on the plane defined by the expression.

10. Consider a 2D CSG representation system using the primitives of unit square and disk from the previous exercise. Draw the geometric object that is represented by the CSG tree of Figure 10.28, where T and S are the transformations defined in the previous exercise.

11. The Minkowski sum of two sets $A, B \subset \mathbb{R}^n$ is the set $A \oplus B$, defined by

$$A \oplus B = \{a + b \, ; \, a \in A, \ \mathrm{e} \ b \in B\}.$$

Interpret geometrically the set $A_p = A \oplus \{p\}$, and show that

$$A \oplus B = B \oplus A = \bigcup_{b \in B} A_b = \bigcup_{a \in A} B_a.$$

Then conclude that

$$A_p \oplus B_q = (A \oplus B)_{p+q}.$$

What is the geometric meaning of this result?

12. Show that the Minkowski sum is associative and commutative and has neutral elements. Show, however, that the equation $X \oplus A = B$ does not always present a solution.

13. Show that

$$\partial(A \oplus B) \subset \partial A \oplus \partial B,$$

where ∂ represents the boundary of the set in $\mathbb{R}^n$. Give an example showing that the equality is not true in general.

14. Show, if A and B are convex sets, that the Minkowski sum $A \oplus B$ is also a convex set.

15. The *convex closure* of a set A, $\text{Conv}(A)$, is the intersection of all the convex sets containing A.

    (a) Show that $\text{Conv}(A)$ is the smallest convex set containing A;
    (b) Show that $\text{Conv}(A \oplus B) = \text{Conv}(A) \oplus \text{Conv}(B)$.

16. Let $P = (v_1^P, v_2^P, \ldots, v_m^P)$ and $Q = (v_1^Q, v_2^Q, \ldots, v_n^Q)$ be two convex polygons with vertices v_i^P and v_j^Q, respectively. Show that

    $$P \oplus Q = \text{Conv}(v_i^P + v_j^Q \; ; \; i = 1, \ldots, m \text{ and } j = 1, \ldots, n)$$

17. Use the previous exercise to describe an algorithm for obtaining the Minkowski sum of two polygons P and Q. What is the computational complexity of that algorithm?

18. Use the Minkowski sum to describe a method that allows us to obtain a continuous deformation between two sets A and B in space.

19. Show how we can use the Minkowski sum to obtain an approximation of the boundary of a set. (Hint: use the sum with spheres, and the complementation operation of a set.)

20. Show how to define the Minkowski sum in objects given by their matrix representation. Use this result to describe an algorithm that determines region boundaries in a binary image.

21. Describe a geometric representation system using Minkowski operations.

22. Consider a fixed vector $v = (v_x, v_y, v_z) \in \mathbb{R}^3$ (which can be the z-axis), and let G be the group of matrices where the elements are given in homogeneous coordinates by

    $$\begin{pmatrix} \cos\theta_1 & -\text{sen}\,\theta_1 & 0 & tv_x \\ \text{sen}\,\theta_1 & \cos\theta_1 & 0 & tv_y \\ 0 & 0 & 1 & tv_z \\ 0 & 0 & 0 & 1 \end{pmatrix}.$$

    (a) Interpret geometrically the action of this group.
    (b) Describe the parametric equation of the orbit of a point by the action of the group.
    (c) What type of models can be created using the action of this group?

23. Describe a coordinate system adapted to correctly position the plane Π along the guiding curve γ in the technique of modeling for tubular surfaces. (Hint: use the Frenet trihedron from differential geometry).

24. Consider the surface of revolution S obtained by rotating the planar curve $f(t) = (x(t), y(t), 0)$ about the x-axis.

    (a) Obtain the parametric equations of S.
    (b) Obtain the expression of the normal vector to S.

25. Describe a method to obtain a polygonization for each of the generative models studied in this chapter.

26. Describe a CSG system to implement the toy LEGO on the computer.

11 Virtual Camera

Projection is the basic operation of both real and virtual cameras. Therefore, from the point of view of the four universes paradigm, projective geometry plays an important role in the search for a virtual camera model:

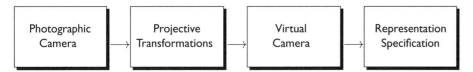

From the projective geometry point of view, the most generic projection possible is given by a projective transformation $T \colon \mathbb{RP}^3 \to \mathbb{RP}^2$ which, in homogeneous coordinates, can be represented using matrices:

$$\begin{pmatrix} y_1 \\ y_2 \\ y_3 \end{pmatrix} = \begin{pmatrix} a_{11} & a_{12} & a_{13} & a_{14} \\ a_{21} & a_{22} & a_{23} & a_{24} \\ a_{31} & a_{32} & a_{33} & a_{34} \end{pmatrix} \begin{pmatrix} x_1 \\ x_2 \\ x_3 \\ x_4 \end{pmatrix}.$$

We therefore have 11 degrees of freedom to define a virtual camera model using this transformation. In this model, several types of cameras are possible, including perspective (using the conical projection), affine, weak-perspective, and orthographic, among others. In this chapter we will study a camera model based on the conical projection, which is adapted for image synthesis.

11.1 A Basic Model

A camera model is very much related to its application in computer graphics. We have two areas where the use of good camera models are important: analysis and synthesis. In terms of analysis, we must determine the intrinsic and extrinsic parameters of the camera,

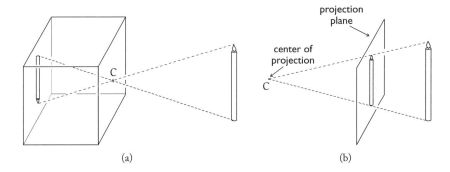

Figure 11.1. (a) Pinhole camera; (b) camera model.

starting from information present on the image; in terms of synthesis, those parameters are defined a priori, and aim to generate an image, that is, to take a virtual photograph.

In this chapter our goal is to describe a robust and flexible camera model to be used in the problem of image synthesis. First we should look for a mathematical model of a photographic camera; then we should determine a representation for this model, which will be obtained by the use of projective transformations (see Chapter 2). Our model will be called a *virtual camera*.

In creating our virtual camera model, we will not take into account the optic components of the photographic camera. Instead, we will base our virtual camera on the *pinhole camera* model, shown in Figure 11.1(a): light passes through the hole O in one of the sides of the box and projects the object's image on the opposite side of the box.

In our model, to avoid having the projected image inverted, we displace the projection plane by positioning it between the *center of projection C* and the object to be visualized. This model is illustrated in Figure 11.1(b).

From the point of view of geometry, our camera model is therefore reduced to a conical (or perspective) projection. In the next sections we will describe how to specify this projection to allow an efficient and flexible implementation of the viewing transformations.

11.2 Viewing Coordinate Systems

The viewing transformation aims at mapping surfaces of objects in a scene to a display device. The entire process involves the use of seven different coordinate systems (or reference spaces), associated to the object, world, virtual camera, image, normalized, visibility, and device spaces. Objects must be transformed to these coordinate systems, making the execution of the inherent tasks of each stage in the process more natural and convenient. The main purpose of the viewing transformation is to provide the most efficient execution of the stages of the viewing process.

❏ **Object space.** This is the intrinsic space associated to each object in the scene. This space has a coordinate system associated to the geometry of the object.

❏ **World space.** This is the global coordinate system in which the scene objects are positioned and oriented, some in relation to others, including the virtual camera. The dimensions are usually given in a particular application-dependent scale standard.

❏ **Virtual camera space.** This is the space defined by a coordinate system associated to the conical projection. This system is used to define the parameters of the virtual camera (position, orientation, focal distance, etc.).

❏ **Image space.** This space is defined by a coordinate system in the projection plane, where the virtual screen is located.

❏ **Normalized space.** This space is introduced so the clipping operations are efficiently applied to objects outside the camera's field of view.

❏ **Visibility space.** This space has a coordinate system that facilitates the visibility operation. This operation consists of determining, for any two objects in the scene, which one is visible from the point of view of the center of projection of the camera.

❏ **Device space.** Also called *screen space*, this is the space associated to the display surface of the graphics output device.

The coordinate systems of the object and the world space are directly related to scene modeling; device space was already discussed in relation to planar graphics objects (Chapter 7). We will therefore now discuss the virtual camera, image, normalized, and visibility spaces.

11.2.1   Virtual Camera Space

The camera space allows us to specify the position and orientation of the camera in relation to the world space. In this system, we define the intrinsic parameters of the virtual camera (virtual screen, focal distance, etc.). The position of the virtual camera is given by the *center of projection C*, corresponding to the optic center. We need to define a reference frame to specify the orientation. Initially, we define the optical axis (longitudinal direction) of the camera by specifying a vector $\mathbf{N}$ called the *optical axis vector*. This vector, together with the center of projection C, defines a ray r called the *optical axis* of the camera. The optical axis r and its vector $\mathbf{N}$ are shown in Figure 11.2(a).

The *projection plane* Π of the virtual camera is the plane perpendicular to the optical axis, located at a distance d from the center of projection (see Figure 11.2(a)). Therefore, the normal vector to the projection plane is the vector $\mathbf{N}$ of the optical axis. The distance d is called *focal distance*.

We now choose a vector $\mathbf{V}$, noncollinear with the vector $\mathbf{N}$ of the optical axis, as shown in Figure 11.2(a). This vector is called the *inclination vector*. The vectors $\mathbf{N}$ and $\mathbf{V}$

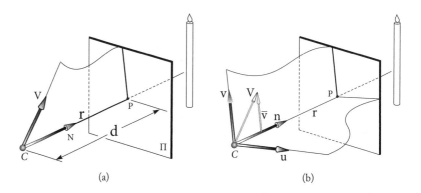

Figure 11.2. (a) View vector, up vector, and focal distance; (b) coordinate system of the virtual camera.

determine the *vertical-longitudinal plane* of the camera. (Intuitively, rotation of this plane about the optical axis determines the inclination of the camera in relation to the scene space.)

We now use the optical axis and inclination vectors $\mathbf{N}$ and $\mathbf{V}$, respectively, to construct the reference frame of the virtual camera. For this we normalize the optical axis vector $\mathbf{N}$, obtaining the vector $\mathbf{n} = \mathbf{N}/\|\mathbf{N}\|$. (This normalized vector will also be called the vector of the optical axis of the camera reference frame.) Next, starting from the inclination vector $\mathbf{V}$, we obtain a unit vector $\mathbf{v}$, normal to vector $\mathbf{n}$, using the usual process of vector orthonormalization (Gram-Schmidt): we project $\mathbf{V}$ on $\mathbf{n}$, subtract this projection from the vector $\mathbf{V}$, and then normalize the resulting vector. That is,

$$\mathbf{v} = \frac{\mathbf{V} - \langle V, \mathbf{n}\rangle \mathbf{n}}{|\mathbf{V} - \langle V, \mathbf{n}\rangle \mathbf{n}|}.$$

This operation is illustrated in Figure 11.2(b).

The vector $\mathbf{v}$ is called the *up vector* of the reference frame. The last vector $\mathbf{u}$ of the camera reference frame is obtained by taking the wedge (cross) product between the optical axis vector and the up vector:

$$\mathbf{u} = \mathbf{n} \wedge \mathbf{v}.$$

The reference frame $(C, \{\mathbf{u}, \mathbf{v}, \mathbf{n}\})$ defines the coordinates in the virtual camera space. Notice the orientation of the system is opposite that of the canonical reference frame $(O, \{\mathbf{e}_1, \mathbf{e}_2, \mathbf{e}_3\})$ of $\mathbb{R}^3$. The coordinates of a point Q in this system will be indicated by (Q_u, Q_v, Q_n).

11.2.2 Image Space

Let P be the point where the optical axis r pierces the projection plane Π (in computer vision, P is called the *main point*). The orthonormal reference frame $(P, \{\mathbf{u}, \mathbf{v}\})$, formed by

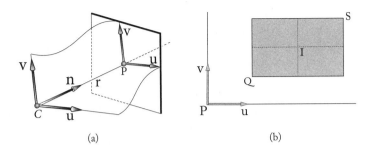

Figure 11.3. (a) Image space; (b) virtual screen.

the first two vectors of the camera reference frame, determines an orthonormal coordinate system of Π (see Figure 11.3(a)). The projection plane Π in conjunction with this coordinate system constitutes the *image space*. If a point Q has coordinates (Q_u, Q_v, Q_n) in the virtual camera coordinate system, then its orthogonal projection has coordinates (Q_u, Q_v) in the image space.

In the image space we define a rectangular window constituting the *virtual screen*. This window is specified by the left-bottom and top-right vertex coordinates, given by $Q = (u_{\min}, v_{\min})$, and $S = (u_{\max}, v_{\max})$, respectively (see Figure 11.3(b)). The width and height of the virtual screen are given by, respectively, $2s_u = u_{max} - u_{min}$ and $2s_v = v_{max} - v_{min}$. Therefore, the coordinates of the center I of the rectangle, defining the virtual screen, are

$$I_u = u_{min} + s_u;$$
$$I_v = v_{min} + s_v. \tag{11.1}$$

Viewing volume. The virtual screen, together with the center of projection C, determines a pyramid of rectangular basis in space, called the *viewing pyramid*, as shown in Figure 11.4(a).

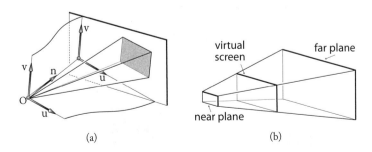

Figure 11.4. (a) Virtual screen in the image space; (b) viewing volume.

The ray, with origin at the center of projection and pointing toward the center of the virtual screen, is called the *viewing axis*. Notice the viewing pyramid might not be straight because the viewing axis might not be collinear with the optical axis of the camera (that is, the viewing axis is not necessarily perpendicular to the projection plane).

Of course, the conical projection defined by the virtual camera is not defined at the point of projection O. Besides, points too close to point O can result in numerical problems during their projection (division by a number close to zero). Similarly, numerical problems can also happen for points very far from the camera. For this reason, it is convenient to limit the viewing pyramid by two planes parallel to the projection plane, called the *near plane* and the *far plane*. The distance from the near and far planes to the projection center is denoted by n and f, respectively. The truncated pyramid, obtained by slicing off the viewing pyramid with these two planes, is called the *viewing volume* (see Figure 11.4(b)).

11.2.3 Normalized Space

We previously saw how to define the camera space, and consequently the viewing volume. The objects outside this volume will not be present in the final image. Similarly to the 2D case (Chapter 7), those elements should be clipped as part of the viewing operations.

For efficient clipping, we should use an appropriate space where the viewing volume has normalized coordinates. In the normalization operation, the viewing volume is transformed into the volume shown in Figure 11.5, defined by

$$-z \leq x \leq z, \quad -z \leq y \leq z, \quad z_{min} \leq z \leq 1.$$

Notice the camera's optical axis is transformed into the z-axis, and the far and near planes are respectively transformed into planes $z = 1$ and $z = z_{min}$. Next we will calculate z_{min} as a function of n and f. This viewing volume is called the *canonical volume* or *normalized volume*.

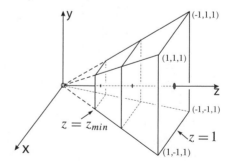

Figure 11.5. Normalized viewing volume.

11.2.4 Visibility Space

Given two points P and Q in the world (scene), we will need to determine which is closer to the projection point of the camera. To do so, we must first verify if P and Q are on the same projection straight line and then compare, along the projection straight line, the distances of P and Q to the center of projection.

Without an appropriate coordinate system, the solving this problem can be computationally expensive. Therefore we define a visibility space: the sorting space is defined so the projection straight lines are parallel among themselves and orthogonal to the projection plane, as shown in Figure 11.6(b).

Notice that in the visibility space two points $P = (p_x, p_y, p_z)$ and $Q = (q_x, q_y, q_z)$ are on the same projection straight line if and only if $p_x = q_x$ and $p_y = q_y$. Besides, P is closer to the projection plane if and only if $p_z < q_z$. Therefore, visibility is reduced to a problem of comparing coordinates. We can do this because any conical projection can be decomposed as a projective transformation followed by an orthogonal projection.

Geometrically, the change of coordinates to the visibility space is obtained using a projective transformation that maps the center of projection to a point on the infinite of the optical axis (which coincides with the viewing axis in normalized coordinates). Of course, that transformation is not unique. We obtain unicity by imposing two conditions: the near plane $z = z_{min}$ is transformed in the plane $z = 0$ and the far plane $z = 1$ is left fixed by the transformation (see Figure 11.12). The image is obtained from the visibility space by doing an orthogonal projection in the image space.

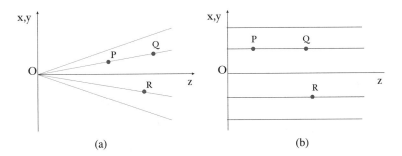

Figure 11.6. Visibility space.

11.3 Virtual Camera Parameters

Having defined a virtual camera and the associated reference frames composing the process of visualizing a scene, we now need to calculate the viewing transformations that perform the change of coordinates between those systems. Before, we will pause and review the parameters defining the virtual camera.

We define six extrinsic parameters (three scalars for C and three scalars $\mathbf{u}, \mathbf{v}, \mathbf{n}$):

- $C = (C_x, C_y, C_z)$—center of projection (position);
- $\{\mathbf{u}, \mathbf{v}, \mathbf{n}\}$—reference frame (orientation),

and have seven intrinsic parameters (d, n, f, and two scalars each for Q and S):

- d—focal distance.
- $Q = (u_{min}, u_{max})$, $S = (u_{max}, u_{max})$—virtual screen.
- n—distance of the near plane.
- f—distance of the far plane.

We still have some auxiliary parameters, obtained as a function of the above parameters. The most important of them are the dimensions of the virtual screen and the coordinates of the center of the virtual screen, given, respectively by

$$2s_u = u_{max} - u_{min}; \qquad I_u = u_{min} + s_u;$$
$$2s_v = v_{max} - v_{min}. \qquad I_v = v_{min} + s_v.$$

The *aspect ratio* of the virtual screen is the quotient s_v/s_u. The aspect ratio of the virtual screen is directly related to the aspect ratio of the device where the image will be displayed.

We still have the following three camera axes: *optical*, *vertical*, and *lateral* axes, respectively defined by reference frame vectors $\mathbf{n}$, $\mathbf{v}$, and $\mathbf{u}$.

11.3.1   Focal Distance and Field of View

Notice that the dimensions of the virtual screen, together with the focal distance, determine the field of view of the camera. Because the dimensions of the virtual screen are fixed, the smaller the focal distance, the larger the field of view and vice versa. We can normalize the virtual camera, taking into account the relation between the field of view and the focal distance of a real lens (specialized publications provide tables of such relations).

It is common to consider a more restricted camera model in which the viewing axis coincides with the optical axis. That is, the center of the virtual screen coincides with the main point, which is the origin of the image space: $I_u = I_v = 0$. In this case, the focal distance d is determined by the *view angle*, which is the preferred parameter in some camera models. This is the angle α defined by the center of projection and by the width of the virtual screen (some systems measure the angle in relation to the height of the virtual screen). The relation between the view angle and the focal distance (see Figure 11.7(a)) is given by

$$\alpha = 2 \arctan\left(\frac{PA}{CP}\right) = 2 \arctan\left(\frac{s_u}{d}\right) \quad \Longleftrightarrow \quad d = \frac{s_u}{\tan(\alpha/2)}.$$

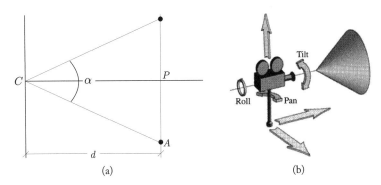

Figure 11.7. (a) Focal distance and view angle; (b) orientation, position, and focus.

11.3.2 Position and Orientation

Notice, in our camera model, that the essential parameters are the ones determining position, orientation, and focal distance. Position is the center of projection $C = (C_x, C_y, C_z)$. Orientation is determined by the reference frame $\{\mathbf{u}, \mathbf{v}, \mathbf{n}\}$. From our study of rotations, we know the orientation is determined by Euler angles. In the present context, they are called *roll*, *pan*, and *tilt* (see Figure 11.7(b)), which are, respectively, the rotation angles about the optical, vertical, and lateral axes of the camera. Therefore the essential specification of the virtual camera has seven degrees of freedom (three from the position, three from the orientation, and one from the focal distance). Those parameters are illustrated in Figure 11.7(b).

An interesting problem consists of determining the appropriate space to parameterize the camera. This problem is directly related to the study of the space of rotations $SO(3)$ (see Chapter 4). A classic parameterization is given by the Euler angles, as we saw above. Some systems also use quaternions to define the orientation of the virtual camera.

11.4 Viewing Operations

The viewing operations consist of a succession of changes between coordinate systems in the spaces previously introduced, together with the execution of some operations with the objects in the scene. In this section, we will explicitly determine the matrix of each change of coordinates.

11.4.1 Changing from World to Camera Space

The objective of this change is to replace the world reference frame by the camera reference frame, thus placing the camera in a standard position in space, with the center of projection at the origin and the optical axis pointing along the z-axis. Notice the objects are specified in the world system, and for this reason we need to change from world to camera

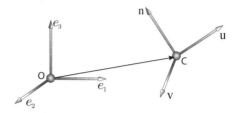

Figure 11.8. Change from world to camera space.

coordinates, meaning we need to change from the camera reference frame $(C, \{\mathbf{u}, \mathbf{v}, \mathbf{n}\})$ to the reference frame of canonical coordinates in the Euclidean space $(O, \{\mathbf{e}_1, \mathbf{e}_2, \mathbf{e}_3\})$ (Figure 11.8).

Let us indicate by V the transformation performing this change of reference frame. The inverse of V, V^{-1}, is the transformation obtained after performing a translation T by the vector $\overrightarrow{OC} = (C_x, C_y, C_z)$, which maps the origin to the center of projection, followed by a rotation R that transforms the canonical basis $\{\mathbf{e}_1, \mathbf{e}_2, \mathbf{e}_3\}$ into the camera reference frame $\{\mathbf{u}, \mathbf{v}, \mathbf{n}\}$. That is, $V^{-1} = TR$.

To calculate the transformation matrix, let us assume that the Cartesian coordinates of the vectors $\mathbf{u}, \mathbf{v}, \mathbf{n}$ and $\overrightarrow{OC}$ are given by $\mathbf{u} = (u_x, u_y, u_z)$, $\mathbf{v} = (v_x, v_y, v_z)$, $\mathbf{n} = (n_x, n_y, n_z)$, and $\overrightarrow{OC} = (C_x, C_y, C_z)$. We then have

$$
V^{-1} = TR = \begin{pmatrix} 1 & 0 & 0 & C_x \\ 0 & 1 & 0 & C_y \\ 0 & 0 & 1 & C_z \\ 0 & 0 & 0 & 1 \end{pmatrix} \begin{pmatrix} u_x & v_x & n_x & 0 \\ u_y & v_y & n_y & 0 \\ u_z & v_z & n_z & 0 \\ 0 & 0 & 0 & 1 \end{pmatrix} = \begin{pmatrix} u_x & v_x & n_x & C_x \\ u_y & v_y & n_y & C_y \\ u_z & v_z & n_z & C_z \\ 0 & 0 & 0 & 1 \end{pmatrix}.
$$

$$(11.2)$$

Note that the change of basis transformation V is an isometry. Therefore, distances are preserved when we move from world to camera space.

It is easy to verify that the inverse matrix is given by

$$
V = (TR)^{-1} = \begin{pmatrix} u_x & u_y & u_z & -\langle \overrightarrow{OC}, u \rangle \\ v_x & v_y & v_z & -\langle \overrightarrow{OC}, v \rangle \\ n_x & n_y & n_z & -\langle \overrightarrow{OC}, n \rangle \\ 0 & 0 & 0 & 1 \end{pmatrix}.
$$

$$(11.3)$$

We therefore obtain the first of the viewing transformations:

$$\text{World} \xrightarrow{V} \text{Camera}.$$

The camera coordinate system has a different orientation from the one given by the basis $\{\mathbf{e}_1, \mathbf{e}_2, \mathbf{e}_3\}$ of the world Cartesian coordinate system. When we change to the camera system, the x-axis is oriented to the right of the observer, the y-axis is oriented upwards, and the z-axis points toward the observed point in the world. It is natural to work with this system, which has a negative orientation. The distance of an object to the center of projection is given by the z coordinate, and the larger the value of z, the farther away the object is.

11.4.2 Viewing Volume Normalization

After transforming from world coordinates to camera space, the viewing volume is in a position in which the center of projection C is at the origin, and the optical axis is the z-axis. Furthermore, we have the projection plane, as well as the near and far planes, all parallel to the xy plane (Figure 11.9(a)). Notice the viewing pyramid is not straight because the viewing axis does not necessarily coincide with the optical axis of the camera.

The normalization operation of the viewing volume consists of applying a deformation in the camera space to transform the oblique viewing volume into the normalized one.

$$-z \leq x \leq z, \quad -z \leq y \leq z, \quad z_{min} \leq z \leq 1. \tag{11.4}$$

This volume, also called the *canonical volume*, is shown in Figure 11.5. After this deformation, the center of projection continues to be at the origin; the viewing axis points towards the direction of the z-axis (and both coincide with the optical axis of the camera); and the far and near planes are respectively transformed into the planes $z = 1$ and $z = z_{min}$.

This normalization deformation is applied in two stages. Initially we apply a shearing transformation C along the xy plane in order to align the viewing axis with the z-axis (optical axis of the camera). This transformation is illustrated in Figure 11.9(a). Next, we apply a scaling transformation S so the new viewing volume is mapped into the normalized volume defined in the three inequalities in (11.4). This scaling is illustrated in Figure 11.9(b).

The final deformation N will be given by the product $N = SC$. Let us now determine C and S, obtaining the matrices in the canonical basis of the camera space.

Shearing calculation. Let (a_u, a_v, f) be the point where the viewing axis pierces the far plane. We then have

$$\begin{aligned} C(1,0,0) &= (1,0,0); \\ C(0,1,0) &= (0,1,0); \\ C(a_u, a_v, f) &= (0,0,f). \end{aligned} \tag{11.5}$$

To have the matrix we should calculate $C(0,0,1)$. We have

$$(0,0,1) = \frac{1}{f}(a_u, a_v, f) - \frac{a_u}{f}(1,0,0) - \frac{a_v}{f}(0,1,0).$$

Applying C to the two members and using the given values in Equation (11.5), we obtain

$$\begin{aligned} C(0,0,1) &= \frac{1}{f}C(a_u, a_v, f) - \frac{a_u}{f}C(1,0,0) - \frac{a_v}{f}C(0,1,0) \\ &= (0,0,1) - (\frac{a_u}{f}, 0, 0) - (0, \frac{a_v}{f}, 0) = (-\frac{a_u}{f}, -\frac{a_v}{f}, 1). \end{aligned}$$

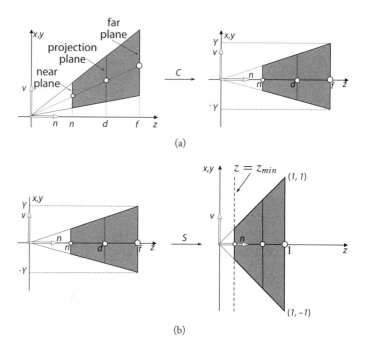

(a)

(b)

Figure 11.9. Transforming the viewing volume by (a) shearing and (b) scaling.

We need to determine a_u and a_v as a function of the parameters of the virtual camera. We know that (a_u, a_v) are the coordinates of the point where the viewing axis pierces the far plane (see Figure 11.10). If (I_u, I_v) are the coordinates of the center of the virtual screen, it follows from Figure 11.10 that

$$\frac{a_v}{f} = \frac{I_v}{d} \quad \Rightarrow \quad a_v = \frac{I_v f}{d},$$

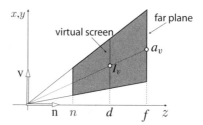

Figure 11.10. Calculation of a_v.

and, similarly,

$$a_u = \frac{I_u f}{d}.$$

Therefore

$$C(0,0,1) = (-\frac{a_u}{f}, -\frac{a_v}{f}, 1) = (-\frac{I_u}{d}, -\frac{I_v}{d}, 1).$$

Finally, the matrix of C in the canonical basis has the vectors $C(1,0,0)$, $C(0,1,0)$, and $C(0,0,1)$ as columns:

$$C = \begin{pmatrix} 1 & 0 & -\dfrac{I_u}{d} & 0 \\ 0 & 1 & -\dfrac{I_v}{d} & 0 \\ 0 & 0 & 1 & 0 \\ 0 & 0 & 0 & 1 \end{pmatrix}. \tag{11.6}$$

Scaling calculation. Let us assume, after the shearing, that the rectangle of the far plane has dimensions $[-X, X] \times [-Y, Y]$ (see Figure 11.9(a)). We can verify that

$$S(0,0,f) = (0,0,1) \Rightarrow fS(0,0,1) = (0,0,1) \Rightarrow S(0,0,1) = (0,0,\frac{1}{f});$$

$$S(0,Y,0) = (0,1,0) \Rightarrow YS(0,1,0) = (0,1,0) \Rightarrow S(0,1,0) = (0,\frac{1}{Y},0);$$

$$S(X,0,0) = (1,0,0) \Rightarrow XS(1,0,0) = (1,0,0) \Rightarrow S(1,0,0) = (\frac{1}{X},0,0).$$

We need to calculate the values of X and Y as a function of the parameters of the virtual camera. The window in the far plane extends from $-X$ to X in the u-axis, and from $-Y$ to Y in the v-axis. Therefore, in Figure 11.11, we have $P'A' = X$, $P'B' = Y$, $PA = s_u$, $PB = s_v$, $OP = d$ and $OP' = f$. Using the similarity of triangles ΔOPA

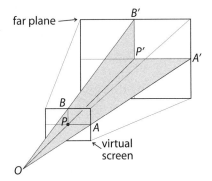

Figure 11.11. Straight viewing volume.

and $\Delta OP'A'$, we obtain

$$\frac{P'A'}{PA} = \frac{OP'}{OP} \quad \Longrightarrow \quad X = PA' = \frac{s_u f}{d}. \tag{11.7}$$

Analogously, using the similarity of triangles ΔCPB and $\Delta CP'B'$, we obtain

$$Y = PB' = \frac{s_v f}{d}. \tag{11.8}$$

We therefore have

$$S(1,0,0) = (\frac{1}{X},0,0) = (\frac{d}{s_u f},0,0);$$

$$S(0,1,0) = (0,\frac{1}{Y},0) = (0,\frac{d}{s_v f},0);$$

$$S(0,0,1) = (0,0,\frac{1}{f}).$$

It follows that the deformation matrix by scaling S is given by

$$S = \begin{pmatrix} \dfrac{d}{s_u f} & 0 & 0 & 0 \\ 0 & \dfrac{d}{s_v f} & 0 & 0 \\ 0 & 0 & \dfrac{1}{f} & 0 \\ 0 & 0 & 0 & 1 \end{pmatrix}. \tag{11.9}$$

Calculating the normalization deformation. From the matrices of shearing C and scaling S we obtain the desired deformation N:

$$N = SC = \begin{pmatrix} \dfrac{d}{s_u f} & 0 & 0 & 0 \\ 0 & \dfrac{d}{s_v f} & 0 & 0 \\ 0 & 0 & \frac{1}{f} & 0 \\ 0 & 0 & 0 & 1 \end{pmatrix} \begin{pmatrix} 1 & 0 & -\dfrac{I_u}{d} & 0 \\ 0 & 1 & -\dfrac{I_v}{d} & 0 \\ 0 & 0 & 1 & 0 \\ 0 & 0 & 0 & 1 \end{pmatrix} = \begin{pmatrix} \dfrac{d}{s_u f} & 0 & -\dfrac{I_u}{s_u f} & 0 \\ 0 & \dfrac{d}{s_v f} & -\dfrac{I_v}{s_v f} & 0 \\ 0 & 0 & \frac{1}{f} & 0 \\ 0 & 0 & 0 & 1 \end{pmatrix}.$$

We therefore obtained the second of the viewing transformations:

$$\text{World} \xrightarrow{V} \text{Camera} \xrightarrow{N} \text{Normalization}$$

The transformation N has two characteristics: the shearing transformation does not preserve angles and the scaling transformation is not an isometry. Therefore, model operations requiring angle or length measurements in the world space should be performed before applying the normalization deformation. Notice that shearing does not deform the image since its constraint of having each plane parallel to the xy plane is an isometry (exercise). However, scaling makes a distortion on the virtual screen, provided it is not a square. This distortion will be corrected in the coordinate transformation for the graphics device (the last transformation in the viewing pipeline), given that the aspect ratio of the device should be the same as the virtual screen.

In normalized coordinates, the near plane is given by $z = z_{min}$. Now we can calculate the value of z_{min}. In fact, it is enough to see that transformation N maps the plane z_{min} to the plane $z = n$. Given that transformation N performs a scaling, along the z-axis, by the factor f (see the matrix of N), it follows that

$$f z_{min} = n; \qquad \text{therefore} \qquad z_{min} = \frac{n}{f}.$$

11.4.3 Changing from Normalized to Visibility Space

We are looking for a projective transformation T that converts the normalized viewing volume

$$-z \le x \le z, \quad -z \le y \le z, \quad \frac{f}{n} \le z \le 1,$$

into the parallelepiped

$$-1 \le x \le 1, \quad -1 \le y \le 1, \quad 0 \le z \le 1,$$

as shown in Figure 11.12, and called *viewing parallelepiped.*

The projective transformation T maps the origin $(0, 0, 0, 1)$ to the point on the infinite $(0, 0, 1, 0)$, so the straight lines passing through the origin and not contained in the xy plane and are transformed into straight lines parallel to the z-axis (see Figure 11.13).

A projective transformation, satisfying the conditions above, is not unique. Unicity is obtained with these requirements: T converts perpendicular planes to the z-axis into the same perpendicular planes to the z-axis; T leaves the far plane $z = 1$ fixed; and T

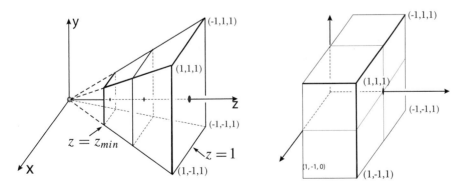

Figure 11.12. Viewing volume in the visibility space.

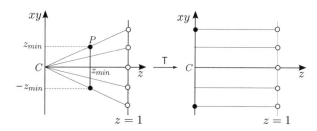

Figure 11.13. Points on the plane $z = 1$ are not altered by T.

transforms the square $[-z_{min}, z_{min}] \times [-z_{min}, z_{min}]$ of the near plane $z = z_{min}$ into the square $[-1, 1] \times [-1, 1]$ of the plane $z = 0$.

The second condition above can be mathematically translated by

$$T[x, y, 1, 1] = [wx, wy, w, w], \quad w \neq 0. \tag{11.10}$$

From the last condition above, it follows that T maps the near plane $z = z_{min}$ onto plane $z = 0$. Notice, from the constraint of T, that each perpendicular plane to the z-axis is a scaling. Therefore it also follows from the third condition that the constraint of T to the plane $z = z_{min}$ is a scaling by the factor $1/z_{min}$. In mathematical terms, we can translate this fact into the following equation:

$$T[x, y, z_{\min}, 1] = \frac{1}{z_{\min}}[w'x, w'y, 0, w']. \quad w' \neq 0. \tag{11.11}$$

To determine T, let us consider its matrix

$$T = \begin{pmatrix} a_{11} & a_{12} & a_{13} & a_{14} \\ a_{21} & a_{22} & a_{23} & a_{24} \\ a_{31} & a_{32} & a_{33} & a_{34} \\ a_{41} & a_{42} & a_{43} & a_{44} \end{pmatrix}. \tag{11.12}$$

We will determine the coefficients a_{ij} of T. Using Equation (11.10) and the matrix of T in Equation (11.12), we obtain

$$a_{11}x + a_{12}y + a_{13} + a_{14} = wx \tag{11.13}$$
$$a_{21}x + a_{22}y + a_{23} + a_{24} = wy \tag{11.14}$$
$$a_{31}x + a_{32}y + a_{33} + a_{34} = w \tag{11.15}$$
$$a_{41}x + a_{42}y + a_{43} + a_{44} = w. \tag{11.16}$$

These equations are valid for any values of x and y. Taking $x = y = 0$ in Equation (11.13), we obtain

$$a_{13} + a_{14} = 0.$$

Taking $x = 1$ and $y = 0$ in the same equation, we obtain

$$a_{11} + a_{13} + a_{14} = w;$$

therefore $a_{11} = w$. Likewise, taking from Equation (11.13) $x = 0$ and $y = 1$, we obtain $a_{12} = 0$.

Following up, as we did above with Equations (11.14), (11.15), and (11.16), we obtain

$$a_{21} = a_{31} = a_{32} = a_{41} = a_{42} = 0$$

$$a_{22} = w, \quad a_{24} = -a_{23} \quad a_{23} + a_{34} = a_{43} + a_{44}.$$

Therefore, the matrix of T in Equation (11.12) can be written in the form

$$T = \begin{pmatrix} w & 0 & a_{13} & -a_{13} \\ 0 & w & a_{23} & -a_{23} \\ 0 & 0 & a_{33} & w - a_{33} \\ 0 & 0 & a_{43} & w - a_{43} \end{pmatrix}. \tag{11.17}$$

Using Equation (11.11) and the matrix of T above, we obtain the equations:

$$wx + a_{13}(z_{\min} - 1) = \frac{w'x}{z_{\min}} \tag{11.18}$$

$$wy + a_{23}(z_{\min} - 1) = \frac{w'y}{z_{\min}} \tag{11.19}$$

$$a_{33}(z_{\min} - 1) + w - a_{33} = w' \tag{11.20}$$

$$a_{43}(z_{\min} - 1) + w - a_{43} = w'. \tag{11.21}$$

Again, these equations are valid for arbitrary values x and y. Taking $x = 0$ in Equation (11.18), we obtain $a_{13} = 0$. Substituting this value of a_{13} back in Equation (11.18) and taking $x = 1$, we obtain $w' = wz_{\min}$.

Following up, similarly with Equations (11.19), (11.20), and (11.21), we obtain

$$a_{23} = 0, \quad a_{33} = \frac{w}{1 - z_{\min}}, \quad \text{and} \quad a_{43} = w.$$

Substituting these values in the matrix of T given in Equation (11.17), we obtain

$$T = \begin{pmatrix} w & 0 & 0 & 0 \\ 0 & w & 0 & 0 \\ 0 & 0 & \frac{w}{1-z_{\min}} & \frac{-z_{\min}w}{1-z_{\min}} \\ 0 & 0 & w & 0 \end{pmatrix}.$$

The transformation T is the same for any value of w ($w \neq 0$). Taking $w = 1$, we finally obtain the desired matrix:

$$T = \begin{pmatrix} 1 & 0 & 0 & 0 \\ 0 & 1 & 0 & 0 \\ 0 & 0 & \frac{1}{1-z_{\min}} & \frac{-z_{\min}}{1-z_{\min}} \\ 0 & 0 & 1 & 0 \end{pmatrix}.$$

Substituting the value $z_{\min} = n/f$ in the above matrix, we obtain

$$T = \begin{pmatrix} 1 & 0 & 0 & 0 \\ 0 & 1 & 0 & 0 \\ 0 & 0 & \frac{f}{f-n} & \frac{-n}{f-n} \\ 0 & 0 & 1 & 0 \end{pmatrix}.$$

We therefore obtained the third viewing transformation:

$$\text{World} \xrightarrow{V} \text{Camera} \xrightarrow{N} \text{Normalization} \xrightarrow{T} \text{Visibility}$$

The new visibility viewing volume is called the *viewing parallelepiped*.

We use affine transformations to map points to the normalized viewing volume; in this way, we would not need to use homogeneous coordinates. The transformation for the visibility space is projective and we should necessarily use homogeneous coordinates. We can perform clipping in the visibility space, however, we should use an algorithm for clipping in homogeneous coordinates, since the clipping should be done before the projection onto the virtual screen.

11.4.4   Changing to Device Space

Starting from the visibility space, we move to a volumetric viewport in the device space. Initially, we move to a volumetric viewport in the Euclidian space, passing from homogeneous coordinates to coordinates in the $\mathbb{R}^3$ (division by the last coordinate). This volumetric viewport should be mapped into the parallelepiped

$$X_{min} \leq x \leq X_{max}, \; Y_{min} \leq y \leq Y_{max}, \; Z_{min} \leq z \leq Z_{max},$$

which defines the volumetric viewport in the device coordinate space. (Notice some devices store the depth of the scene z in a depth buffer. This z value is useful for the visibility calculation and for composting operations, as we will see later.)

Mapping in the device viewport is done in three stages. First we apply a transformation K, which translates the bottom-left vertex $(-1, -1, 0, 1)$ of the viewing parallelepiped to the origin and performs a scaling by the scale factor $1/2$ in the x and y coordinates. The transformation K maps the viewing parallelepiped to the cube $[0, 1] \times [0, 1] \times [0, 1]$. We then apply a scaling L, which operates in two steps. Initially, L transforms this parallelepiped into the volumetric viewport of the device. (That is, it performs a scaling by the factors $X_{max} - X_{min}$, $Y_{max} - Y_{min}$ and $Z_{max} - Z_{min}$ in the x, y and z axes, respectively.) Next, L translates the parallelepiped by the vector $(X_{min}, Y min, Z_{min})$. Finally, we round the coordinates by the coordinates of the closest virtual pixel.

The transformation matrix K is given by

$$K = \begin{pmatrix} 0.5 & 0 & 0 & 1 \\ 0 & 0.5 & 0 & 1 \\ 0 & 0 & 1 & 0 \\ 0 & 0 & 0 & 1 \end{pmatrix}.$$

The transformation matrix L is given by:

$$L = \begin{pmatrix} X_{max} - X_{min} & 0 & 0 & X_{min} \\ 0 & Y_{max} - Y_{min} & 0 & Y_{min} \\ 0 & 0 & Z_{max} - Z_{min} & Z_{min} \\ 0 & 0 & 0 & 1 \end{pmatrix}.$$

The rounding is obtained by adding 0.5 to the coordinates and truncating. This can be obtained by the transformation M given by

$$M = \begin{pmatrix} 1 & 0 & 0 & 0.5 \\ 0 & 1 & 0 & 0.5 \\ 0 & 0 & 1 & 0.5 \\ 0 & 0 & 0 & 1 \end{pmatrix}.$$

The matrix D, which changes from the coordinates of the normalized parallelepiped to the coordinates of the volumetric viewport, is given by the inverse $D = (KLM)^{-1}$. Using the notation $\Delta_i = (i_{max} - i_{min})$ and $\nabla_i = (i_{max} + i_{min})$, we have

$$D^{-1} = (KLM)^{-1} = \begin{pmatrix} \Delta_X/2 & 0 & 0 & \frac{\nabla_X}{4} + 1 \\ 0 & \Delta_Y/2 & 0 & \frac{\nabla_X}{4} + 1 \\ 0 & 0 & \Delta_z & \frac{\Delta_z}{2} + Z_{min} \\ 0 & 0 & 0 & 1 \end{pmatrix}.$$

Taking the inverse of this matrix, we obtain

$$D = \begin{pmatrix} 2/\Delta_X & 0 & 0 & -(\nabla_X + 4)/(2\Delta_X) \\ 0 & 2/\Delta_y & 0 & -(\nabla_Y + 4)/(2\Delta_Y) \\ 0 & 0 & 1/\Delta_z & -(\Delta_Z + 2Z_{min})/\Delta_z \\ 0 & 0 & 0 & 1 \end{pmatrix}.$$

We therefore obtain the fourth, and last, viewing transformation:

$$\text{World} \xrightarrow{V} \text{Camera} \xrightarrow{N} \text{Normalization} \xrightarrow{T} \text{Visibility} \xrightarrow{D} \text{Device}$$

The matrix of the complete mapping, from the world to the device space, is then given by the product $DTNV$. After applying this matrix to a point, it is enough to truncate the resulting coordinates to obtain the integer coordinates of the point in the device. Of course, the inverse mapping from device to world space is given by the inverse matrix $(DTNV)^{-1} = V^{-1}N^{-1}T^{-1}D^{-1}$.

11.5 Other Camera Models

There are other virtual camera models that use different coordinate systems, or different projections, to specify the intrinsic and extrinsic parameters of the virtual camera. The model presented in this chapter is quite generic and flexible so as to assist various applications in image synthesis.

11.6   Camera Specification

Camera specification is directly related to the user interface. In the case of interactive specification, we should establish some parameter constraints when using the mouse (with only two degrees of freedom). The dimensions of the virtual screen are usually specified a priori, based on the characteristics of the display device. The same happens with the near and far planes (some systems establish an automatic specification of these planes). Actually, we end up with seven parameters to be specified: three for position, three for orientation, and the focal distance parameters. The methods of camera specification are separated in two categories: direct and inverse specification. In direct specification, the seven camera parameters are explicitly declared as part of the specification. In inverse specification, the user indirectly determines the seven camera parameters.

11.6.1   Direct Specification

The focus is specified by a number, so we will deal with the specification of the position and orientation of the camera. We will describe three methods of direct specification: the "roll+look-at," OpenGL, and Euler angles methods.

Roll + Look-at specification. A simple and effective specification consists of determining the camera parameters based on the following questions: where am I? where am I looking? what is the camera inclination?

We use these questions to set the parameters of the virtual camera. The location of the observer who will photograph the scene corresponds to projection point O. If the observer is looking at a point of interest O' in the scene (*look-at* point), then vector $\overrightarrow{OO'}$ determines the viewing vector $\mathbf{N}$ and, consequently, the normalized vector $\mathbf{n}$ of the camera reference frame.

Starting from the camera inclination angle θ provided by the user (*roll* angle), and the vector $\mathbf{n}$, we obtain the orthonormal camera reference frame $\{O, \mathbf{u}, \mathbf{v}, \mathbf{n}\}$.

We will assume the observer's horizontal plane is the xy plane (see Figure 11.14). We take a unit vector $\overline{\mathbf{u}}$ parallel to the horizontal plane and perpendicular to the viewing vector $\mathbf{N} = \overrightarrow{OO'}$. Notice that if the viewing vector $\mathbf{N}$ is not perpendicular to the horizontal plane, there are only two options for choosing the vector $\overline{\mathbf{u}}$. We therefore choose this vector so that the basis $\{\overline{\mathbf{u}}, \mathbf{n}, \overline{\mathbf{u}} \times \mathbf{n}\}$ has the same orientation of the canonical basis $\{\mathbf{e}_1, \mathbf{e}_3, \mathbf{e}_2\}$ of the $\mathbb{R}^3$. (Notice, in this specification, that if the vector $\mathbf{n}$ is perpendicular to the horizontal plane, we have a singularity: infinite possibilities for choosing $\overline{\mathbf{u}}$.)

After choosing $\overline{\mathbf{u}}$, the vector $\mathbf{u}$ of the camera reference frame is obtained by performing a rotation of angle θ of $\overline{\mathbf{u}}$ about the optical axis (determined by the vector $\mathbf{N} = \overrightarrow{OO'}$). Vector $\mathbf{v}$ of the reference frame is obtained by taking the wedge (cross) product

$$\mathbf{v} = \mathbf{u} \wedge \mathbf{n}.$$

As we can observe above, the camera reference frame $\{O, \mathbf{u}, \mathbf{v}, \mathbf{n}\}$ is not defined in the case that the viewing vector $\mathbf{N}$ is perpendicular to the horizontal plane xy. This is a

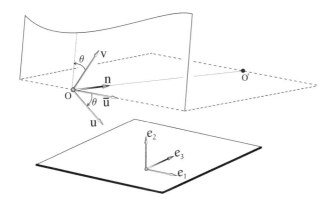

Figure 11.14. Construction of the camera reference frame.

disadvantage in specifying the camera reference frame through the inclination angle instead of the inclination vector **V**. Despite this disadvantage, the *roll + Look-at* specification is convenient when we need to walk with the camera in a virtual environment, capturing details of that environment.

OpenGL camera. In the OpenGL graphics system, the camera specification is obtained by providing the center of projection, the observation point (*look-at* point) and the inclination vector **V** (called *up vector*). Therefore, it is the same specification we used in our virtual camera model. However, in OpenGL, the camera reference frame $\{\mathbf{u}, \mathbf{v}, \mathbf{n}\}$ is positive, that is, the optical axis vector in OpenGL points in the direction opposite to the one defined in our model (as we increase the scene depth, the coordinate z decreases).

The specification is made by calling the routine below:

```
gluLookAt(camera_pos_x, camera_pos_y, camera_pos_z,
          look_at_x, look_at_y, look_at_z, up_x, up_y, up_z);
```

Likewise, in the OpenGL camera model, the viewing axis coincides with the optical axis, and the focal distance is specified by the viewing angle (called *field of view*), which is measured in relation to the height of the virtual screen (see Figure 11.15). Therefore, there

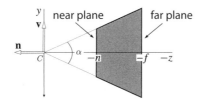

Figure 11.15. Parameters of the OpenGL camera.

is no need for the normalization deformation, since the viewing volume is normalized a priori.

Another difference between the OpenGL model and ours is that the visibility space obtained from the normalized viewing volume is given by $-1 \leq x, y, z \leq 1$. We leave it to the reader to show that, in this case, the matrix transforming the normalized volume in the visibility space is given by

$$
N = \begin{pmatrix} a & 0 & 0 & 0 \\ 0 & b & 0 & 0 \\ 0 & 0 & c & d \\ 0 & 0 & -1 & 0 \end{pmatrix},
\tag{11.22}
$$

where

$$
a = b = 1/\tan\left(\frac{\alpha}{2}\right), \quad c = \frac{f+n}{f-n} \quad \text{and} \quad d = \frac{2fn}{f-n}.
$$

(α is the field of view, f and n are the distances of the far and near planes, respectively, as shown in Figure 11.15.)

Specification by Euler angles. Another method of camera specification consists of considering the camera as being a glider, in which the user has control of its position and orientation (the focus and the other intrinsic camera parameters are fixed). The orientation is given by the Euler angles of *roll*, *pan*, and *tilt* (see Chapter 4). This specification is widely used in video games and in some VRML (virtual reality modeling language) browsers. Notice the user can navigate with only two of the angles (usually the *pan* and *roll*), and these angles then can be interactively specified with the two degrees of freedom from the mouse.

11.6.2 Inverse Specification

To obtain certain camera framings, you have to set the specifications. But direct specification of the camera parameters to obtain a certain framing is generally difficult. With inverse specification, the seven degrees of freedom from the camera are indirectly specified by the user. The idea of is to determine the camera positions using the "paradigm of the camera man": the user (in the director's role) specifies one framing in the image, and the system adjusts the camera parameters to get the desired framing.

Inverse specification facilitates camera specification by the user in exchange for the solution of a quite difficult mathematical problem to solve. To give an idea of that complexity, consider how difficult the placement of the problem illustrated in Figure 11.16 would be: a point P in space is fixed, and with the observer at point O, point P is projected in point A of the screen. We should obtain the camera parameters so point P is projected onto point B, which is located at the center of the screen.

Everything happens as if the user requested one framing with the point A at the center of the screen. Figure 11.16 displays one new position O' and one new camera orientation, which solves the problem (observe that the position of the camera is not unique).

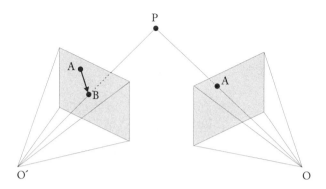

Figure 11.16. Inverse specification for positioning a point.

Mathematically speaking, we have a transformation $T_P \colon \mathbb{R}^7 \to \mathbb{R}^2$, where $y = T_P(x)$ is the projection of the camera parameterization space on the Euclidean plane. The solution to our problem is given by the inverse $T_P^{-1}(B)$. However, the transformation T_P is not linear and, in general, does not have an inverse. The nonlinearity of the problem makes its solution even more complex. Therefore, the solution should be sought within the spirit of the optimization theory; in other words, we should seek the "best solution."

11.7   Comments and References

As we observed in the beginning of the chapter, there are other virtual camera models besides the one introduced in this chapter. The definition of the parameters of the virtual camera described in this chapter is based on the implemented model from LucasFilms and is described in [Smith 84].

11.7.1   Additional Topics

The camera model introduced in this chapter is only projective; that is, it is not a realistic camera model that takes into account the optical components of the camera, the physical properties of the film (or of the sensors, in the case of a digital camera), and the problem of exposure with radiance. In particular, we do not have several intrinsic parameters such as focus, depth of field, etc. An interesting work presenting a camera model along these lines is [Kolb et al. 95].

In several applications it is important to synchronize a virtual camera according to the parameters of a real camera (an example is the virtual viewpoint replay for a soccer match, where a virtual camera is aligned with the real camera to produce drawings of the field and track player positions). This is called *calibration* and is an extremely relevant topic that complements this chapter.

The techniques of camera specification we covered in this chapter, in particular inverse specification, constitute another interesting topic for more advance studies. A discussion about using this technique in virtual reality can be seen in [Albuquerque 99].

Exercises

1. Show that if (1) the center of projection is at the origin, (2) the projection plane is the plane $Z = f$, and (3) the optical axis is the axis Z, the perspective projection is then given by the projective transformation $T \colon \mathbb{RP}^3 \to \mathbb{RP}^2$, $T(X, Y, Z, 1) = (x, y, 1)$, where

$$\begin{pmatrix} x \\ y \\ 1 \end{pmatrix} = \begin{pmatrix} f & 0 & 0 & 0 \\ 0 & f & 0 & 0 \\ 0 & 0 & 1 & 0 \end{pmatrix} \begin{pmatrix} X \\ Y \\ Z \\ 1 \end{pmatrix}.$$

2. Obtain, by following the virtual camera model of this chapter, the viewing transformations of a camera with parallel projection.

3. State and explain at least two problems we can face if we do not define the near plane in the virtual camera model. (Hint: one of the problems was mentioned in the text.)

4. Define a shearing transformation along a plane Π of the space $\mathbb{R}^3$. Show that shearing leaves the parallel planes to Π invariant and the constraint to each of those planes is an isometry.

5. In some virtual camera models, the far plane is placed in the infinite, $f = \infty$ (that is, we do not have a far plane).

   (a) What problems can this fact generate?

   (b) Do the camera transformations of this chapter work by making $f = \infty$? Why or why not?

   (c) Suggest a camera transformations model with the far plane in the infinite.

6. Algebraically calculate the matrix $DTNV$ given by the product of the matrices D, T, N and V, defined in this chapter, as well as its inverse $(DTNV)^{-1}$. These matrices do the transformation from the virtual camera space to the image space, and vice versa. In which situation can we use this matrix directly in the viewing process?

7. Obtain the expression of the matrix in Equation (11.22), which transforms from the normalized space to the visibility space in the OpenGL camera model.

8. Compare the camera transformation defined in this chapter with the one presented in the book by Foley and Van Dam [Foley et al. 96]. What is the difference? State an advantage of the parameters definition for the virtual camera introduced in this chapter.

9. Discuss viewing transformation processing for parametric and implicit models.

10. Show the projective transformation that changes from normalized coordinates to the visibility space, maps a point $P = (x, y, z)$ with $z < 0$, into a point with $z > 1$. In other words, points located behind the camera are transformed into points in front of the camera but out of the viewing volume.

11. True or false?

    (a) Given two viewing pyramids corresponding to different parameters of the virtual camera, there exists a projective transformation which maps a pyramid into the other.

    (b) Given two different images I_1 and I_2, there exists a projective transformation T such that $T(I_1) = T(I_2)$;

12. In the camera coordinate system we introduced on this chapter, the virtual screen cannot be centered. What is the visual effect in the image when we use a noncentered screen? What type of real-world camera is this model trying to simulate?

13. Show that if

$$
A = \begin{pmatrix} a & 0 & 0 & t_1 \\ 0 & b & 0 & t_2 \\ 0 & 0 & c & t_3 \\ 0 & 0 & 0 & 1 \end{pmatrix} \quad \text{then} \quad A^{-1} = \begin{pmatrix} 1/a & 0 & 0 & -t_1/a \\ 0 & 1/b & 0 & -t_2/b \\ 0 & 0 & 1/c & -t_3/c \\ 0 & 0 & 0 & 1 \end{pmatrix}.
$$

12 | Clipping

Clipping is a very important operation in computer graphics with many applications. In this chapter we will focus on the use of clipping during the viewing process. Clipping is the first operation in the viewing operations and performed in parallel with camera transformation.

12.1 Classification, Partitioning, and Clipping

Jordan's Theorem, in its broader form, affirms that a closed and connected surface M of dimension $n - 1$ in $\mathbb{R}^n$ partitions the space in two complementary regions, of which M is the common border. More precisely, $\mathbb{R}^n - M = A \cup B$, where A and B are disjunct, and the border of A and B is the surface M. A closed and connected curve partitions the plane in two regions (Figure 12.1(a)); a surface partitions the space $\mathbb{R}^3$ in two 3D regions; a straight line partitions the plane in two half-planes; and a plane partitions the space in two half-spaces.

Given the conditions of the previous paragraph, with $R^n - M = A \cup B$, we can cite two important problems: classification and clipping.

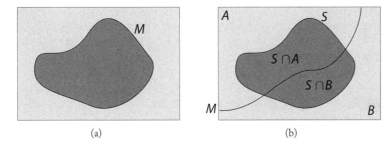

(a) (b)

Figure 12.1. Space partitioning.

We can state the classification problem this way: let M be a closed and connected surface of dimension $n - 1$ in $\mathbb{R}^n$ and $\mathbb{R}^n - M = A \cup B$. If $S \subset \mathbb{R}^n$ is a graphics object, determine if $S \subset A$ or $S \subset B$. We have already seen a particular case of a classification operation: the point membership classification problem.

We can state the clipping problem this way: let M be a closed and connected surface of dimension $n - 1$ in $\mathbb{R}^n$ and $R^n - M = A \cup B$. If $S \subset \mathbb{R}^n$ is a graphics object and $S \cap M \neq \varnothing$, determine the objects $S \cap (A \cup M)$ or $S \cap (B \cup M)$.

In each of these problems, surface M is called either a *partition* or *clipping surface*. When $n = 3$, M is a closed and connected surface in $\mathbb{R}^3$ and the clipping is known as 3D or three-dimensional clipping. When $n = 2$, M is a closed and connected curve on the plane $\mathbb{R}^2$, and the clipping is 2D. If M is closed but not connected, a similar result is valid; however, the complement $\mathbb{R}^n - M$ is constituted by more than two connected regions. The classification and clipping problems can be easily extended for this particular case.

In general, the computational complexity of the classification problem is much smaller than the clipping one because it does not involve intersection calculations. For this reason, classification generally precedes clipping to make this last operation more efficient: initially, we try to classify, and on the impossibility, we perform clipping. In other words, the strategy used in several algorithms, whenever clipping is required, consists of trying to replace intersection operations with classification ones. The clipping problem is solved in three stages: intersection, classification, and structuring.

Intersection. The intersection calculation $S \cap M$, between the object S and the clipping surface M, aims at determining the points of the object that are in the boundary. In general, this stage involves solving equations using numerical methods, which is costly from the computational point of view (see Figure 12.2(a)).

Classification. After (or simultaneously with) clipping, we classify in which of the two complementary regions of the partition surface the clipped object is located (classification problem). This operation involves performing tests and, in general, has lower complexity than the intersection operation.

Structuring. The intersection operation changes not only the geometry of the object, but its topology as well. This way, the structuring stage consists of building up the correct topology of each part of the object on each of the resulting clipping regions (see Figure 12.2(c)). This operation can be done in parallel with clipping or in a subsequent stage (post-processing).

A geometric illustration of these three stages is given in Figure 12.2. (In some clipping applications, one of these stages might not be present.) Notice that in the structuring it is necessary to introduce nonexistent edges in the original polygonal region.

The complexity of the clipping operation is present in one of the three stages described above, depending on the geometric and topological complexity of the object to be clipped, as well as on the clipping surface.

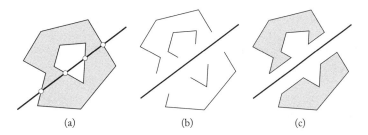

Figure 12.2. Stages of the clipping operation. (a) A polygonal object and a straight line on the plane, which is the clipping surface. (b) The partitioning points of the polygonal region, after the intersection calculation. (c) The structuring of the points in a way to form two polygonal regions, resulting from the clipping operation.

12.2 Clipping Applications

Classification and clipping problems have application in several areas in computer graphics. For instance, in animation, to detect collision of objects; in image synthesis, to accelerate ray tracing algorithms; and in visibility, to determine the elements inside the viewing volume of the virtual camera.

12.2.1 Clipping, Boolean Operations, and Interface

Boolean operations between planar and spatial graphics objects are of great importance in geometric modeling. We can highlight, for instance, their use in the CSG operations as we saw in Chapter 2. Among those operations we can highlight *union*, $A \cup B$, *intersection*, $A \cap B$, *difference*, $A - B$ or $B - A$, and the symmetrical difference, $(A - B) \cup (B - A)$, as illustrated in Figure 12.3. These operations are implemented as clipping ones. For

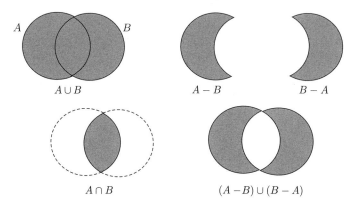

Figure 12.3. Clipping and Boolean operations.

example, the difference $A - B$ between two graphics objects is obtained by clipping A in relation to B, obtaining $A \cap B$, followed by clipping B in relation to the set $A \cap B$.

Clipping is also used as a standard implementation method for selecting geometric elements in an interactive graphical interface (*pick* operations). For this, a small rectangle surrounding the cursor, called a *pick* window, is used to determine which primitives are in the cursor neighborhood. Most of the primitives are trivially rejected and, among the accepted, the one with highest priority is selected.

In a windows-based interface system, clipping operations are used to determine the visible and invisible parts of overlapping windows. In postscript, clipping operation between planar graphics objects play a fundamental role.

12.2.2   Clipping and Viewing

One of the important applications of the clipping operation is in viewing graphics objects. In Chapter 11 we saw the virtual camera defining the viewing volume, which corresponds geometrically to a truncated pyramid, as shown in Figure 12.4.

Usually some world (scene) objects are in the interior of the viewing volume, others are just partially contained in the viewing volume, and others are entirely outside the viewing volume. In viewing the world, we can either project all the world objects by the virtual camera transformation, rasterize the projected object, and then decide which pixels are within the region on the virtual screen, or we can first clip the world objects in relation to the viewing volume, and then perform the projection.

In the first case, clipping is performed after rasterization in the discrete pixel space (2D); in the second case, clipping is performed in the 3D space, before the projection. Which is more efficient? Note that if many objects exist in the exterior of the viewing volume, the first method involves unnecessarily rasterizing many pixels (a computationally costly operation). On the other hand, in this same situation, the second option involves a 3D clipping of the objects in relation to the viewing volume, which is an operation that can also be computationally very costly. However, there are two advantages of the second method: we can have view errors (e.g., points behind the camera that are mapped to the screen) in the projection when using a 2D clipping; 3D clipping can be more efficient by using acceleration techniques.

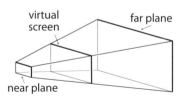

Figure 12.4. Viewing volume of the virtual camera.

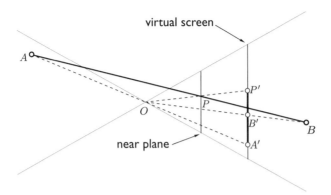

Figure 12.5. Projection of segment AB.

Viewing error in the projection followed by 2D clipping. Consider Figure 12.5, which shows, in the virtual screen, the center of projection O of the straight line segment AB. Points A and B are projected into points A' and B', respectively. Therefore, the projection of segment AB is the segment $A'B'$. This result is obviously wrong because there is not any object in the field of view delimited by the segment $A'B'$. By performing the clipping of segment AB in the near plane, we obtain the segment BP. The point P is projected into the point P'; thus, the projection of segment BP is the segment $B'P'$, and this is the correct result. This example demonstrates that clipping should be performed to avoid having points behind the center of projection O projected on the virtual screen.

To obtain the correct result, it would certainly be enough to perform clipping in relation to the viewing pyramid. However, we use the near plane so that very close points to the center of projection O are not projected. As the projection at point O is not defined, during the projection transformation, the projection of points very close to O results in a division by numbers close of 0, and this can produce numerical errors in the viewing process.

The previous example shows that we cannot simply project the objects in the virtual screen and then perform a 2D clipping. However, we could adopt a hybrid strategy: perform a 3D clipping in relation to the near plane and then project the resulting elements in the virtual screen to perform a 2D clipping. But we previously saw this is not the best strategy unless we have efficient 3D clipping methods.

12.3 Clipping Acceleration

Clipping acceleration techniques replace clipping by classification so as to efficiently discard the largest possible number of objects that could potentially be clipped, that is, to efficiently classify the objects (or parts of objects) that are completely outside or completely inside the viewing pyramid. Next we will describe three acceleration techniques: bounding objects, hierarchy of bounding objects, and spatial subdivision.

12.3.1 Bounding Objects

Given a graphics object O (which can have a very large number of polygons or can be several grouped objects), we look for a volumetric object V (i.e., a surface bounding a finite volume in space), satisfying two conditions:

(a) $O \subset V$;

(b) The geometric support of V is simple enough to allow its efficient classification in relation to the viewing volume.

This way, instead of testing each element of object O in relation to the viewing pyramid, we initially test the object V (from (b); this test is fast and efficient): if V is outside (or inside) the viewing pyramid, then from condition (a), the object O will also be outside (or inside), as shown in Figure 12.6.

Common bounding objects include spheres, parallelepipeds, and convex polyhedra.

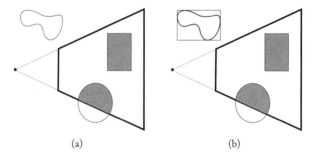

(a)   (b)

Figure 12.6. Bounding box and clipping.

12.3.2 Hierarchy of Bounding Objects

The method of bounding objects is unable to appropriately handle cases where only a part of the object is contained within the viewing volume. To treat this case we can use a hierarchy of bounding objects. Each node of the hierarchy has a bounding volume containing part of the object. We start at the root of the hierarchy, testing if the bounding object associated to this node is in the exterior (or interior) of the viewing volume; if this is true, the entire object contained in the hierarchy will be in the exterior (or interior). Otherwise we continue testing each node of the hierarchy; if we find a subnode outside (or inside) of the viewing volume, the entire part of the scene object contained in the bounding object of that subnode will also be outside (or inside) the viewing pyramid. This method is very efficient; the key to its success is the construction of good hierarchies.

12.3.3   Hierarchical Spatial Subdivision

When world objects are static we can perform a hierarchical spatial subdivision, which increases the clipping efficiency; in this way, we can determine the objects of the hierarchy that are inside or outside of the viewing volume (see Chapter 10).

There is a pre-processing cost to perform the hierarchy calculations; however, it becomes fixed once the world objects are static. We can perform a uniform spatial subdivision with volume elements (voxels) or an adaptive voxel subdivision (using octrees, for instance). We can also use partitioning trees (BSP-trees), which generate quite efficient hierarchies.

A standard example, using spatial subdivision methods, happens when we perform a camera motion inside a building (walkthrough). This method is illustrated in Figure 12.7, where we show a hierarchy based on partitioning trees. As we saw previously, partitioning trees can be useful not just for the clipping problem, but also for the visibility calculation itself. We will see more on this topic further on.

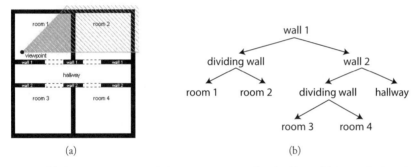

(a)   (b)

Figure 12.7. (a) Camera position and clipping region inside a building; (b) hierarchy based on partitioning trees.

12.4   Clipping Methodology

Clipping involves the stages of intersection, classification, and structuring. However, if the object does not intersect the clipping surface, we only need to perform classification and, possibly, structuring. A basic requirement for an efficient clipping algorithm is to avoid unnecessary intersection calculations.

In classification, we should explore several geometric and topological homogeneity properties of both the clipping surface and the graphics object to be clipped. As an example of homogeneity, we can cite (a) the linearity of the elements (straight line, planes, etc.); (b) the convexity of the clipping regions and the objects to be clipped; and (c) the continuity of the objects, among other examples.

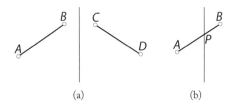

Figure 12.8. Clipping segments on the plane.

Example 12.1 (Clipping of a segment against a straight line). Consider the clipping of a straight line segment AB on the plane $\mathbb{R}^2$ in relation to a vertical straight line $x = x_0$. The clipping regions are the two semiplanes $\{(x, y); x \leq x_0\}$ and $\{(x, y); x \geq x_0\}$, thus being convex regions. Therefore, a segment is contained within the region if and only if their ends belong to the region. We therefore see that a simple test can be done before intersecting the segment with the clipping straight line. In Figure 12.8(a), if we are interested on the segments of the region $x \leq x_0$, we trivially accept AB and reject CD. The segment of Figure 12.8(b) has each of their ends located in different semiplanes. We therefore calculate the intersection P, accept AP, and reject PB. ❏

Depending on the application, clipping can operate in different types of geometric data. The most common types are points, straight lines, straight line segments, planes, polygons, and polyhedra.

The clipping calculation can be exact or approximate. Most of the time, it is not necessary to have a precise intersection calculation. Recursive clipping methods usually operate within a certain tolerance. The example below illustrates clipping with an approximate calculation.

Example 12.2 (Recursive intersection of a straight line segment). Once we have an efficient algorithm for point classification in relation to the clipping surface, we can obtain a recursive algorithm to calculate the intersection between a straight line segment with that surface. Consider the intersection of a straight line segment AB in relation to a hyperplane in space (see Figure 12.9(a)).

We will assume the segment intersects the clipping hyperplane at point P. We need to determine P in a way that divides AB into two segments: AP and PB. A recursive

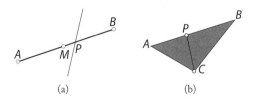

Figure 12.9. (a) Clipping of segment AB; (b) clipping and attributes.

algorithm, which gives an approximate solution of point P, can be obtained by calculating the midpoint $M = (A+B)/2$ of the segment AB, obtaining the segments AM and MB. One of these segments is contained in one of the clipping semispaces (AM in the figure). We take the segment MB and repeat the operation of determining the midpoint. The algorithm continues until $ax_m + by_m + c < \varepsilon$, where $ax + by + c = 0$ is the equation of the clipping straight line and (x_m, y_m) are the coordinates of midpoint M.   ❏

12.4.1   Clipping and Attributes

A graphics object $O = (U, f)$ is defined by the geometry of its geometric support U and by the attribute function f. When performing the clipping of object O, we obtain a new object $O' = (U', g)$, where $U' \subset U$ and $g = f|U'$ are the constraints of function f to the new geometric support U'. In practice, the set U is described by a representation and the attribute function is also defined in this representation. When performing the clipping operation, a new representation is obtained for the support U'; therefore, we should recalculate the attribute function in this new representation.

Consider, for instance, a triangle ABC with one color attribute function f (see Figure 12.9(b)). It is very common to represent f by its value at the vertices A, B, and C of the triangle, and to do its reconstruction at the interior points using linear interpolation with barycentric coordinates. If the triangle is clipped in two triangles ACP and BCP, we calculate the color function f at the new vertex P.

12.5   2D Clipping

In terms of what object is being clipped, clipping a point is simplest. In terms of increasing geometric complexity, we can consider straight line segments, followed by polygonal regions (in other words, regions whose border is constituted by nonintersecting polygonal lines). In this section, we will study some important clipping algorithms, giving emphasis to clipping during the viewing operation.

12.5.1   Clipping a Point

The clipping of a point $P = (x_p, y_p)$ involves only the classification stage. This stage consists of determining the region on the plane to which the point belongs. This is the point membership classification problem discussed in Chapter 7.

If the clipping curve is implicitly described by the equation $f(x, y) = 0$, this classification is simply obtained by evaluating the sign of the expression $f(x_p, y_p)$, obtained by replacing the point coordinates in the implicit equation of the curve.

In the case of classifying a point $P = (x_p, y_p)$ in relation to a straight line with equation $ax + by + c = 0$, we look at the sign of the expression $ax_p + by_p + c$. Note that this test involves two products. If the straight line has equation $x = x_0$ (vertical straight line) or $y = y_0$ (horizontal straight line), the classification is reduced to comparing the point coordinate x_p (or y_p) with x_0 (or y_0) and therefore does not involve arithmetic calculations. To achieve robust and efficient clippings, we should explore the particularities on the description of the elements in the clipping operation.

The clipping operation is far more difficult when the curve is parametrically defined, or more usually, when it is described piecewise. An important case is the clipping of a point P in relation to a polygonal region on the plane. The algorithm below describes the general classification of a point in relation to a semiplane.

Example 12.3 (Classification of a point in relation to a semiplane). Consider semiplane H, defined by a straight line r on the plane, as shown in Figure 12.10(a). Let n be the normal vector appearing outside of semiplane H. Given a point P of the plane, a classification of P in relation to H can be obtained by taking an arbitrary point P_0 of the straight line r and calculating the inner product $\langle n, P - P_0 \rangle$. We leave it to the reader to verify that if this inner product is negative, the point is inside H; if the product is positive, P is outside semiplane H; and if the product is 0, point P is exactly over the straight line r, which is the boundary of semiplane H.   ❏

Once we have the clipping algorithm in relation to a semiplane, we can use partitioning trees to obtain the classification of a point in relation to an arbitrary planar polygonal region. Clipping algorithms for some particular polygonal regions can be more efficiently obtained by exploring the geometry of the region. This is the case of clipping against the rectangle on the virtual screen.

Clipping a point against the virtual screen. The virtual screen is defined by a rectangle whose sides are parallel to the coordinate axes (see Figure 12.10(b)). We saw that the pick operation (to select points in an graphics interface) takes place in this same clipping region. Therefore, the clipping operation in relation to this polygonal region is important in the viewing process.

As the sides are parallel to the coordinate axes, clipping is achieved simply by comparing the coordinates of the point with the coordinates of the horizontal and vertical straight lines defining the sides of the rectangle. If the diagonal of the rectangle is given by the

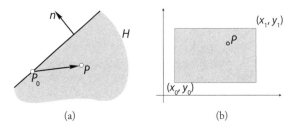

(a)   (b)

Figure 12.10. (a) Classification of a point in relation to a semiplane; (b) clipping a point against a rectangular region.

points (x_0, y_0) and (x_1, y_1), the four straight line have equation $x = x_0$, $x = x_1$, $y = y_0$, and $y = y_1$.

12.5.2   Clipping a Straight Line Segment

The clipping operation on a segment, due to the simplicity of its topology, focuses more on the intersection and classification stages than on structuring. The subdivision of a segment AB at a point P results in two segments: AP and BP.

In this section, we are interested in the particular case of clipping a segment against one of the semiplanes defined by straight line r. As the semiplane is convex, the classification of a segment AB is reduced to classifying its extreme points A and B. Therefore, the clipping of a segment is reduced to the intersection calculation (when it exists). This point calculation can be done by solving a system of two linear equations. A robust solution of this system is computationally costly, but a naive solution can result in clipping errors.

Recursive clipping. In Example 12.2, we saw an approximate recursive intersection calculation. This method can be used to develop a recursive clipping algorithm for a straight line segment according to Figure 12.11. This algorithm is simple; however, there are analytical methods for calculating the point in a robust and efficient way, mostly in specific, common situations. We will describe two of those methods below.

Horizontal and vertical clipping. Clipping in relation to a horizontal or vertical straight line is important due to several existing clipping applications in relation to the rectangle

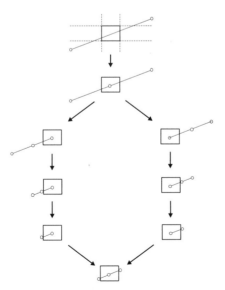

Figure 12.11. Recursive clipping of a straight line segment.

on the virtual screen. Consider a segment AB with $A = (x_a, y_a)$ and $B = (x_b, y_b)$. The parametric equation of this segment is given by $P(t) = A + t(B - A)$, that is,

$$x = x_a + t(x_b - x_a),$$
$$y = y_a + t(y_b - y_a),$$

with $0 \leq t \leq 1$. If the horizontal straight line has equation $y = y_1$, from the second equation above we obtain

$$t_0 = \frac{y_1 - y_a}{y_b - y_a}.$$

Therefore the coordinates (x_p, y_p) of the intersection point P are given by

$$x_p = x_a + t_0(x_b - x_a),$$
$$y_p = y_1.$$

Clipping with a vertical straight line is calculated in an similar way. Notice that the particular cases in which the segment AB is either horizontal or vertical can be trivially treated with this method.

Parametric clipping. Consider one of the semiplanes H defined by a straight line r on the plane, as shown in Figure 12.12. Let n be the normal vector appearing outside the semiplane H. Given a segment AB, the parametric equation of the straight line support is given by $P(t) = A + t(B - A)$. As we saw in Example 12.3, if t_0 is the solution of equation $\langle n, P(t) - P_0 \rangle = 0$ and $0 \leq t_0 \leq 1$, then segment AB intersects the straight line r at point $P(t_0)$.

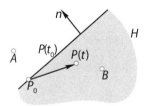

Figure 12.12. Clipping a straight line segment against a semiplane.

12.6   Clipping a Segment against the Virtual Screen

The clipping of a segment in relation to the rectangle on the virtual screen could be treated within a more general context of clipping a segment in relation to a convex polygonal region on the plane. However, due to the importance of this particular clipping scenario, more efficient algorithms exploring the geometric particularities of the rectangle on the virtual screen have been developed. In this section, we will discuss some of these algorithms.

12.6.1 Cohen-Sutherland Algorithm

Let us assume the sides of the rectangle are defined by four straight line parallel to the coordinate axes: $y = y_1$, $y = y_0$, $x = x_1$, and $x = x_0$ (see Figure 12.13(a)). These straight lines divide the plane into eight semiplanes. $y \geq y_1$, $y \leq y_1$, $y \geq y_0$, $y \leq y_0$, $x \geq x_1$, $x \leq x_1$, $x \geq x_0$, $x \leq x_0$.

To efficiently classify the ends of the segments, we associate to each point P on the plane a list $\mathrm{Boole}(p) = (b_1, b_2, b_3, b_4)$ of four Boolean variables that classify point P in relation to the eight semiplanes, in the following order:

1. The variable b_1 corresponds to the straight line $y = y_1$. If $b_1 = 1$, point P is on the semiplane $y \geq y_1$. If $b_1 = 0$, point P is on the semiplane $y \leq y_1$.

2. The variable b_2 corresponds to the straight line $y = y_0$. If $b_2 = 1$, point P is on the semiplane $y \leq y_0$. If $b_2 = 0$, point P is on the semiplane $y \geq y_0$.

3. The variable b_3 corresponds to the straight line $x = x_1$. If $b_3 = 1$, point P is on the semiplane $x \geq x_1$. If $b_3 = 0$, point P is on the semiplane $x \leq x_1$.

4. The variable b_4 corresponds to the straight line $x = x_0$. If $b_4 = 1$, point P is on the semiplane $x \leq x_0$. If $b_3 = 0$, point P is on the semiplane $x \geq x_0$.

The eight semiplanes divide the plane into nine regions, and in each of those regions the Boolean list of points is constant. Figure 12.13(b) shows the nine regions with the constant value of the Boolean list of each of them. Note that the points on the virtual screen are characterized by having the *Boole* vector $(0, 0, 0, 0)$ as null.

Given a segment AB, if $\mathrm{Boole}(A) = \mathrm{Boole}(B) = (0, 0, 0, 0)$ then segment AB is contained in the clipping rectangle and should be trivially accepted. It is also easy to verify, by inspecting Figure 12.13(b), that if $\mathrm{Boole}(A) \wedge \mathrm{Boole}(B) \neq (0, 0, 0, 0)$, then the segment extremities are then in one of the four semiplanes $y \geq y_1$, $y \leq y_0$, $x \geq x_1$, or $x \leq x_0$, and it should be trivially rejected.[1]

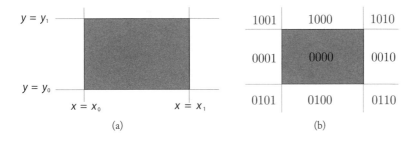

Figure 12.13. Stages of the Cohen-Sutherland clipping algorithm.

---

[1] We define Boolean operations between two lists by performing the operation between the corresponding elements in the lists. In particular, the set operation *"and"* ($\wedge$) is given by $(b_1, b_2, b_3, b_4) \wedge (b'_1, b'_2, b'_3, b'_4) = (b_1 \wedge b'_1, b_2 \wedge b'_2, b_3 \wedge b'_3, b_4 \wedge b'_4)$.

If the segment is neither trivially accepted nor rejected using the two tests above, then we proceed with the intersection calculation of the segment against the sides of the rectangle to perform the clipping. In this case, the segment can be clipped into several subsegments that are trivially rejected, where one of them can be accepted. It is necessary to impose an ordering to the clipping of a segment into subsegments. We will apply the order we previously used to define the Boolean list of points: clipping against the straight lines $y = y_1$, $y = y_0$, $x = x_1$, and $x = x_0$, in that order. The clipping algorithm is described in four steps below:

1. If the segment is trivially either accepted or rejected, the clipping problem is resolved and the algorithm stops.

2. Using the lists $\text{Boole}(A)$ and $\text{Boole}(B)$, we determine the extreme point of the segment that is outside the clipping rectangle (there will be at least one). Without loss of generality, we can assume this extreme is the point A, as illustrated in Figure 12.14.

3. Using $\text{Boole}(A)$, we determine the straight lines segment AB intersects. In the case of Figure 12.14, we have $\text{Boole}(A) = (1, 0, 0, 1)$ and we conclude the segment intersects the straight lines $y = y_1$ and $x = x_0$.

4. We calculate the intersection point P of the segment with each of the straight lines according to the ordering previously given (top, bottom, right, and left). In the case of Figure 12.14, we initially perform the intersection calculations with the straight line $y = y_1$, obtaining the point P. The intersection point calculation is done using the method described in Section 12.5.2.

5. We return to the first stage of the algorithm, with the clipped segments AP and PB. Segment AP is trivially rejected. In the case of Figure 12.14, segment PB is neither trivially accepted nor rejected, and the algorithm continues with its clipping: point B is external to the clipping rectangle and, as $\text{Boole}(B) = (0, 0, 1, 0)$, the segment intersects the straight line $x = x_1$. By calculating point R, we obtain two segments: PR and RB. Segment BR is trivially rejected and segment RQ is trivially accepted.

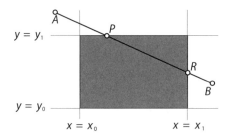

Figure 12.14. Stages of the Cohen-Sutherland clipping algorithm.

As we have only nine regions on the plane, the algorithm stops after a finite number of iterations. Notice, however, that usually several intersection operations are performed to clip out a segment when it is neither trivially accepted nor rejected.

For this reason, the Cohen-Sutherland algorithm is particularly efficient in cases where most of the segments are either trivially accepted or rejected. This happens when the clipping rectangle is either very large (it contains almost all of the segments), or very small (almost all of the segments are outside the rectangle). The case of a small rectangle happens in the selection operation (*pick*) in a graphical interface. The case of a large rectangle happens in the viewing process when most of the world objects are within the field of view of the camera.

12.6.2   Cyrus-Beck Algorithm

As we previously saw, the Cohen-Sutherland algorithm performs unnecessary intersection calculations when the segment is neither trivially accepted nor rejected. The Cyrus-Beck algorithm allows us to clip out those segments with a smaller number of operations. Consequently, this algorithm is more efficient than the Cohen-Sutherland algorithm for cases in which many segments are not trivially clipped out.

Given a segment AB, its straight line support intersects the four lateral straight lines of the rectangle in four points (some of those points can coincide). Using the parametric equation of the straight line

$$P(t) = A + t(B - A), \tag{12.1}$$

and the parametric clipping method seen in Section 12.5.2, we calculate the four values of the parameter t, in which the straight line intersects the four lateral straight lines of the clipping rectangle. Taking a point P_0 over the straight line (this can be one of the rectangle vertices), the value of t is given by the solution of the equation

$$\langle n, P(t) - P_0 \rangle = 0.$$

Placing the value of $P(t)$, given in Equation (12.1), in the above equation, we obtain

$$t = -\frac{\langle n, A - P_0 \rangle}{\langle n, P_1 - P_0 \rangle}. \tag{12.2}$$

After calculating the value of t for the four straight lines, we discard the values of t outside the interval $[0, 1]$, as those points do not belong to the segment AB.

To describe the algorithm, we will call the semiplanes $y \leq y_1$, $y \geq y_0$, $x \leq x_1$, and $x \geq x_0$, *positive semiplanes*. The intersection of these four semiplanes defines the clipping rectangle.

When the straight line $P(t)$ enters any of these semiplanes at an intersection point $P(t_0)$, it is identified with the $+$ sign; if it exits at the intersect point $P(t_0)$, we identify it with the $-$ sign. Figure 12.15(a) shows several segments with the classification of the input and output points of the semiplanes.

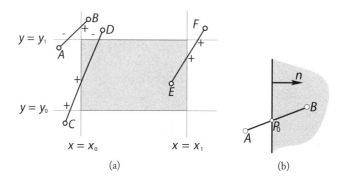

(a)   (b)

Figure 12.15. (a) Classification of the intersection points; (b) calculation of the sign of a point.

Algebraically, the sign of a point is obtained using the sign of the inner product $\langle n, B - A \rangle$ (see Figure 12.15(b)): if $\langle n, B - A \rangle > 0$, the point is positive $(+)$, otherwise it is negative $(-)$. Note that this inner product has already been calculated when we performed the calculation of the intersection parameter t in Equation (12.2).

With this algorithm, we first determine the four values of parameter t where the straight line $P(t)$ intersects the border of the four positive semiplanes. Then we determine the $+$ and $-$ signs of the intersection points. We next sort the intersection points by the value of parameter t and we discard the points with $t < 0$ or $t > 1$. Finally, the clipped out segment is PQ, where P is the positive point $(+)$ (with the largest value of parameter t), and Q is the negative point $-$ (with the smallest value of parameter t).

12.7 Polygon Clipping

Clipping is not merely for polygon vertices and edges. In Figure 12.16(a), all vertices and edges are outside the clipping rectangle; however, the polygon is not in the external region to the rectangle: the clipping is the rectangle itself.

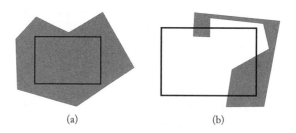

(a)   (b)

Figure 12.16. Clipping a polygon against a rectangle.

In the case that the object to be clipped is a point, the only operation to perform is classification. If the object is a straight line, we may also calculate intersections as a means of classifying each resulting segment. In neither of these cases do we have to deal with structuring. But in the case of polygonal regions, clipping involves: intersection, classification, and structuring.

In general, to clip regions we perform a boundary clipping and then determine the boundary polygonal curve of each resulting region from the clipping operation. Because the boundary clipping is reduced to recording a sequence of straight line segments, it does not present any difficulty—the real challenge is in structuring the data to obtain the boundary of each region resulting from the clipping. The simplest case of polygon clipping is clipping a triangle, which we will study next.

12.7.1   Clipping Triangles

A triangle is a convex set; therefore, if its three vertices are all located on a particular semiplane, the entire triangle is contained on it, and it will be trivially clipped out (see Figure 12.17(a)).

If there are triangle vertices located in different semiplanes, the triangle should be subdivided for clipping. We have two cases to consider, as shown in Figure 12.17(b) and (c). In the case of (b), two edges of the triangle intersect the clipping straight line. An obvious choice would be to first determine the intersection points P and Q of edges AC and BC with the clipping straight line, and then subdivide the triangle into quadrilateral $BPQC$ and triangle APQ. This method has the disadvantage of generating quadrilaterals in the clipping process.

In case (c), the triangle has a vertex, C, over the clipping straight line. In this case, we determine the intersection point P between the clipping straight line with the opposite edge to vertex C, and we subdivide the triangle into two triangles, APC and BPC, each contained in one of the semiplanes. This triangle clipping with a vertex over the clipping straight line is important for the recursive triangle clipping algorithm we will describe next.

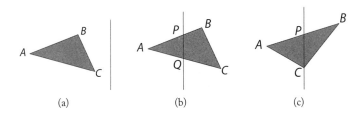

Figure 12.17. Clipping a triangle. Three different positions of a triangle in relation to the clipping straight line.

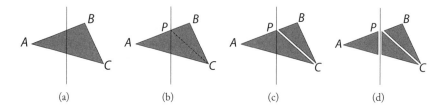

Figure 12.18. Recursive clipping of a triangle.

Recursive clipping of triangles. Consider triangle ABC. Let us assume vertices A and B are contained in opposite semiplanes, as shown in Figure 12.18(a). We determine the intersection point P on side AB with the clipping straight line (Figure 12.18(b)). Next we connect point P to the vertex opposite edge AB, thereby subdividing triangle ABC into two triangles: APC and BPC (Figure 12.18(c)).

If vertex C belongs to the clipping straight line, the triangle is clipped out in two triangles APC and BPC (Figure 12.17(c)). Otherwise, vertex C belongs to one of the semiplanes. Let us assume C is in the same semiplane as vertex B, as shown in Figure 12.18(b). In this case, the triangle BPC is contained in the same semiplane and is trivially clipped out. We then apply the same algorithm to the other triangle, APC. Given it has the vertex P on the clipping straight line, it will be clipped out in two triangles, each contained in a particular semiplane (see Figure 12.18(d)). This completes the clipping.

The recursive triangle clipping algorithm can be used to clip out a triangle in relation to the rectangle on the virtual screen. For this, we perform successive clippings in relation to each of the sides, classifying the resulting subtriangles. This clipping is illustrated in Figure 12.19.

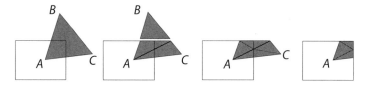

Figure 12.19. Recursive clipping against the virtual screen.

12.7.2   Sutherland-Hodgman Algorithm

The Sutherland-Hodgman polygon clipping algorithm [Sutherland and Hodgman 74] was a very important contribution in the development of computer graphics. Until then, polygon clipping algorithms were structured in a way similar to the clipping algorithms for straight line segments: each edge was analyzed in relation to the clipping region as a whole. This approach resulted in a loss of connectivity of the clipped polygon. The solution to this topological problem was to invert the order of the operations and consider the intersection of each clipping plane in relation to the polygon as a whole. The limitation of this algorithm is that it works only if the clipping region is convex and the disconnected parts of the resulting polygon are linked by straight line segments. The algorithm therefore requires a post-processing to obtain the final polygon.

There are other more general algorithms that perform the clipping of any polygon in relation to another arbitrary polygon. The oldest of them is the Weiler-Atherton algorithm [Weiler and Atherton 77], which was developed to solve the visibility problem of polyhedral surfaces. Another algorithm for arbitrary polygon clipping is [Vatti 92]. The Greiner and Hormann algorithm [Greiner and Hormann 98] also performs clipping of arbitrary polygons. It is much simpler than the two previous algorithms (e.g., it uses a doubly-linked vertex list as the polygon data structure), and furthermore allows self-intersections within the polygonal regions.

12.8   3D Clipping

As we previously saw, the most correct procedure for viewing 3D objects is to perform the object exclusion and clipping operations in the 3D space, where the clipping surface is the viewing volume.

Several of the algorithms previously seen extend naturally to 3D clippings. Among them we can mention the Cohen-Sutherland algorithm for clipping straight line segments. In the 3D case, the clipping surface is a parallelepiped or a truncated pyramid. Each face is extended, creating 27 regions. The Boolean code now has six bits. At the most, six intersections are calculated, one for each side of the viewing volume.

The Lyan-Barsky algorithm is based on the algorithm of point classification in relation to a semiplane, as described in Example 12.3. This algorithm immediately extends to a point classification in relation to a semispace in the Euclidean $\mathbb{R}^3$ space. In this way, we also have a natural extension of the Lyan-Barsky algorithm to obtain the 3D clipping of a straight line segment in relation to a parallelepiped whose sides are parallel to the coordinate planes. The recursive clipping algorithm of straight line segments also extends to 3D clipping. The classification of a point in relation to an arbitrary polyhedral region can be obtained using partitioning trees.

Both the Weiler-Atherton and Sutherland-Hodgman algorithms can also be extended to 3D clipping. We have a natural extension of the recursive clipping algorithm of triangles, which allows us to perform the clipping of a triangle against a plane. In fact, it is enough to observe this algorithm depends solely on the topology of the triangle and on the clipping of straight line segments against a plane.

Observe that even if the graphics object is described by a polygonal B-rep representation (where the polygons are not triangles), we can subdivide each polygon of the representation, obtaining a triangulation of the object. This new object (which has the same geometry) can be clipped out using a triangle clipping algorithm.

An important observation about the recursive clipping of triangles is that we should take special care on the vertex classification in relation to the plane. Unlike sequential algorithms, such as the Sutherland-Hodgman, the recursive algorithm results in vertices that are not in a generic positioning. This happens because the subdivision of an edge produces vertices located exactly on the clipping plane. Therefore, during the algorithm recursion, the classification of vertices should take into account this possibility. Notice this type of care should also be taken in generic clipping algorithms to avoid inconsistencies between the geometry and topology of the model.

12.9   Clipping and Viewing

We have two options when performing clipping in relation to the viewing volume of the virtual camera: to perform the clipping in relation to the viewing volume in the normalized space (see Figure 12.20(a)) or to transform the viewing volume to the visibility space and then perform the clipping (see Figure 12.20(b)). In the first case the clipping surface is a truncated pyramid with rectangular basis; in the second case this surface is a parallelepiped whose sides are parallel to the coordinate planes. Both the clipping and exclusion operations are simpler and more efficient in the visibility space. However, we should take into account the additional cost of transforming all world objects to the visibility space. We have three options: we can perform clipping and exclusion operations in the camera space; we can perform exclusion in the camera space and clipping in the visibility space; or we can perform both operations in the visibility space.

There are very efficient exclusion tests in the camera space when we use bounding objects defined by parallelepipeds. For instance, refer to [Greene 94] or [Green and Hatch 95], [Assarsson and Möller 00]. Note that changing the coordinate systems from camera to visibility space is a projective transformation (see Chapter 11). In this way, to perform clipping in the visibility space, we should work with the clipping in homogeneous coordinates, since clipping should be performed before the projection (division by the last coordinate to move from homogeneous to Euclidean coordinates). We will briefly describe this clipping in the next section.

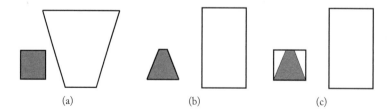

(a)   (b)   (c)

Figure 12.20. Clipping (gray shaded polygon) and the virtual camera. Truncated pyramid and parallelepiped correspond to visibility and normalized spaces, respectively.

12.9.1   Clipping in Homogeneous Coordinates

In Euclidean space, the parametric equation of a straight line segment passing through points P and Q in the projective space is given by

$$r(t) = P + t(Q - P), \quad t \in [0, 1].$$

Remember that to obtain the corresponding segment in Euclidean space $\mathbb{R}^3$, we project, in projective space, the segment on the plane $w = 1$, as shown in Figure 12.21(a).

An unusual case happens if one of the ends of the segment has the coordinate w with a negative sign. In this case, the projected segment passes through the point on the infinite and we obtain a solution with incorrect topology, as shown in Figure 12.21(b).

By observing the matrix that performs the change from the normalized to the visibility space, we see this case happening in the viewing problem with all of the points that are placed behind the camera position. We have two ways to avoid this problem: we can perform clipping in relation to the near plane before moving to normalized coordinates, or we can change the point coordinates (x, y, z, w) by $(-x, -y, -z, -w)$ before the clipping.

Taking into account what we discussed above, clipping in homogeneous coordinates is no different from 3D clipping—we just have one extra coordinate. The canonical viewing volume in the visibility space is given by

$$-1 \le x \le 1, \quad -1 \le y \le 1, \quad 0 \le z \le 1.$$

Replacing x, y, and z by x/w, y/w, and z/w, respectively, we have

$$-1 \le \frac{x}{w} \le 1, \quad -1 \le \frac{y}{w} \le 1, \quad 0 \le \frac{z}{w} \le 1. \tag{12.3}$$

If w is positive, we multiply the inequalities in (12.3) by w, obtaining the inequalities

$$-w \le x \le w, \quad -w \le y \le w, \quad 0 \le z \le w.$$

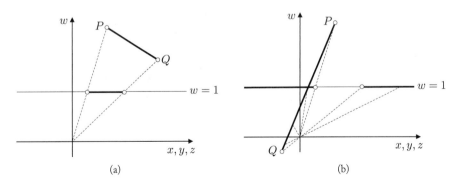

(a)   (b)

Figure 12.21. Clipping and the virtual camera. (a) Projecting the segment on the plane; (b) projected segment passes through the point on the infinite.

If w is negative, the multiplication of the inequalities in (12.3) by w inverts the direction of the inequalities, resulting in

$$-w \geq x \geq w, \quad -w \geq y \geq w, \quad 0 \leq z \geq w.$$

We therefore have six regions.

12.10   Comments and References

The techniques of spatial subdivision and bounding volumes to accelerate the clipping algorithms are also used in other areas of computer graphics. In this book, we will revisit both techniques when we discuss ray tracing methods in Chapter 14. These techniques are also used in the area of visibility to calculate the visible surfaces of a scene (world), as we will see in Chapter 13.

The Cohen-Sutherland clipping algorithm was never published. It was broadly described after being presented in [Foley et al. 96]. The Cyrus-Beck algorithm appeared in [Cyrus and Beck 78]. In 1984, Liang and Barsky [Liang and Barsky 84] published a more efficient version of the Cyrus-Beck algorithm, valid only for clipping in relation to the rectangle on the virtual screen. The algorithm avoids a priori calculation of the four values of parameter t, only calculating the values when necessary. In [Foley et al. 96], the reader can find pseudocodes for the Cohen-Sutherland algorithm and Liang-Barsky's version Cyrus-Beck algorithm.

Another interesting and efficient clipping algorithm for straight line segments was published in [Nicholl et al. 87]. However, it does not extend to 3D clippings and moreover is only used for clipping in relation to the rectangle on the virtual screen.

12.10.1   Additional Topics

In this chapter, due to space limitations, we did not discuss the details of many clipping algorithms. The study of those algorithms is appropriate to a more advanced course. Several libraries, with open source code, are available for implementing the clipping algorithms described or cited in this chapter.

Numerical precision is the Achilles heel of clipping operations, primarily in the case of Boolean operations for geometric modeling, as numerical errors can lead to an incorrect structuring. There are several works on this subject. The techniques of bounding volumes or hierarchies of bounding volumes are also used to solve the problem of collision detection and interference among objects.

Exercises

1.   Extend the algorithm of Example 12.3 to obtain a classification algorithm of a point $p \in \mathbb{R}^3$ in relation the semispace of $\mathbb{R}^3$. Use this result to extend the Lyan-Barsky algorithm to $\mathbb{R}^3$.

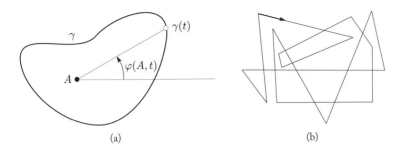

Figure 12.22. (a) Closed and (b) sketched curves (Exercise 3).

2. Show that two convex polyhedra do not intersect if and only if there exists a straight line in space such that the projection of these polyhedra on that straight line consists of two disjunct intervals (*separation axis test*). Is the result valid for nonconvex polyhedra?

3. Consider a closed curve $\gamma\colon [0, 1] \to \mathbb{R}^2$ on the plane (self-intersections allowed), and let A be a point inside and not over the curve (see Figure 12.22(a)). For each point $\gamma(t) \in \gamma$, the segment $A\gamma(t)$ forms an angle $\varphi(A, t)$ with the x-axis. The function $\varphi(A, t)$ is called the *angle function* of the curve γ associated to point A. The *rotation number* of curve γ in relation to point A is defined by

$$n_\gamma(A) = \frac{\varphi(A, 1) - \varphi(A, 0)}{2\pi}.$$

Intuitively, $n_\gamma(A)$ measures the "number of turns" that a point, moving along the curve, makes around point A. Point A is inside curve γ if and only if $n_\gamma(A)$ is odd (therefore A is an exterior point if and only if the rotation number is even).

(a) Determine the rotation number of each region determined by the curve sketched in Figure 12.22(b).

(b) Using the rotation number, describe an algorithm to determine whether a point is on the interior or exterior of a closed polygonal curve on the plane.

4. Write the pseudocode of the recursive clipping algorithm for a straight line segment, as illustrated in Figure 12.11.

5. Give an example of a polygon P and clipping region R in which all the vertices of P are inside R, but P is not inside R.

6. Describe an algorithm to perform the clipping of a segment in relation to a convex polygonal region on the plane.

7. Show that the Cyrus-Beck algorithm extends for clipping segments in relation to convex polygonal regions.

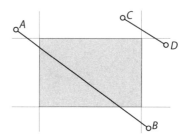

Figure 12.23. Clipping two straight line segments (Exercise 11).

8. In this book we have used barycentric coordinates to solve some important problems. Here is another opportunity to use those coordinates:

    (a) Using barycentric coordinates, describe a clipping algorithm for a point in relation to a triangular region.

    (b) Use the method of item (a) to extend the Cohen-Sutherland algorithm for clipping a segment in relation to a triangle on the plane.

9. Determine the clipping of segment AB with $A = (2, 3)$ and $B = (4, -6)$ in relation to the planar region defined by the inequality $y^2 - x^2 \geq 0$.

10. What is the maximum number of intersections that should be calculated to clip out a segment in the 2D Cohen-Sutherland algorithm?

11. Consider the straight line segments AB and CD in Figure 12.23.

    (a) Sketch all the stages of the Cohen-Sutherland algorithm to determine the clipping of the straight line segments.

    (b) Repeat item (a) for the Cyrus-Beck algorithm.

12. Given a triangle ABC in $\mathbb{R}^3$ and one ray r with origin $O = (0, 0, 0)$ and direction $v = (r_1, r_2, r_3)$, describe an algorithm to determine if r pierces the triangle ABC. (Hint: use projections on the coordinate planes, or barycentric coordinates.)

13. Consider a triangle ABC and an attribute function f represented by their values $f(A)$, $f(B)$, and $f(C)$ at the vertices of the polygon.

    (a) Describe how f can be reconstructed using barycentric coordinates.

    (b) Another reconstruction method is to use three successive interpolations (see Figure 12.24): given a point P inside the triangle, we consider the horizontal straight line segment passing through P and intersecting the sides AB and BC at the points Q and R. We calculate $f(Q)$, $f(R)$, $f(P)$, respectively, by a linear interpolation between $f(A)$ and $f(B)$, $f(C)$ and $f(B)$, and between $f(Q)$ and $f(R)$.

        Show that the interpolated value obtained is the same as that in item (a). Discuss the advantages and disadvantages of the two calculation strategies.

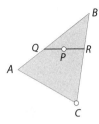

Figure 12.24. Attribute function reconstruction by three successive interpolations (Exercise 13).

14. A polygon $P_1 P_2 \ldots P_n$ on the plane is positive if, while traversing their edges in the vertex order, we find the interior of the polygon always to our left.

   (a) Consider the polygon contained in plane xy of space $\mathbb{R}^3$, and let $e_3 = (0,0,1)$. Show this polygon is positive if and only if $\langle P_i P_{i+1} \wedge P_{i+1,i+2}, e_3 \rangle > 0$.

   (b) Show the area of a triangle ABC on the plane with $A = (x_a, y_a)$, $B = (x_b, y_b)$, and $C = (x_c, y_c)$ is given by the absolute value of the expression

   $$\frac{1}{2} \operatorname{Det} \begin{pmatrix} x_a & y_a & 1 \\ x_b & y_b & 1 \\ x_c & y_c & 1 \end{pmatrix}.$$

   In addition, show that the sign of the determinant provides the sign of the triangle orientation.

   (c) Show that two segments AB and CD on the plane intersect each other given the following necessary and adequate condition: $\operatorname{sign}(CAB) \neq \operatorname{sign}(DAB)$ and $\operatorname{sign}(ACD) \neq \operatorname{sign}(BCD)$.

   (d) Without drawing, verify whether segments AB and CD intersect each other, where $A = (-3, 2)$, $B = (4, -1)$, $C = (-1, -5)$, and $D = (3, 7)$.

   (e) Let ABC be a triangle with positive orientation and Q a point on the plane. Show that point Q is inside the triangle if and only if $\operatorname{sign}(QAB)$, $\operatorname{sign}(QBC)$, and $\operatorname{sign}(QCA)$ are all greater than or equal to 0.

15. Describe an extension of the recursive triangle clipping algorithm to clip out a triangle in relation to an arbitrary polygon on the plane.

16. Show the recursive triangle clipping algorithm extends to clip out convex polygons. What is the difficulty in extending this algorithm to clip out arbitrary polygons?

17. Several 2D clipping algorithms extend to 3D clipping.

   (a) Describe the extension of Cohen-Sutherland's clipping algorithm to obtain the clipping of a segment in relation to the visibility space of the virtual camera.

   (b) Repeat item (a) for the Cyrus-Beck algorithm.

18. We can use a clipping algorithm for the triangulation of a simple polygonal region. One such method is the *ear clipping*. Given a polygonal region with vertices $(v_0, v_1, \ldots, v_{n-1})$, an "ear" is a sequence of three vertices $(v_{t_1}, v_{t_2}, v_{t_3})$ forming a (nondegenerate) triangle; in addition, no polygon vertex belongs to the triangle.

(a) Show that the triangulation of a polygon with n vertices has at least $n - 2$ triangles;

(b) Describe a triangulation algorithm that consists of performing (1) a search for polygon ears and then (2) a clipping of those ears;

(c) What is the computational complexity of this triangulation algorithm?

13 | Visibility

The surfaces of a 3D scene (world) are projected on the image plane by the viewing process. As this projection is not bijective, it becomes necessary to solve the conflicts which happen when several surfaces are mapped into the same pixel. For this, we use the visibility concept. In this chapter, we will study the problem of calculating the visible surfaces of a 3D scene.

13.1 Visibility Foundations

A scene can be constituted by graphics objects of dimensions 1, 2, or 3, corresponding to curves, surfaces, or volumetric elements (we are excluding fractal objects). In this chapter we will focus on surface visibility, with a few references to curve visibility.

The visibility problem essentially consists of determining the elements of the scene geometry that are (1) inside the field of view of the camera (viewing pyramid) and (2) closer to the camera. In general, those are the only visible elements in the scene (provided there are no objects with transparent material). This problem can be seen as an ordering problem. Notice we are interested in partial ordering, that is, up to the first opaque surface along a viewing ray.

The visible surface calculation is related to all the other operations in the viewing process because the visibility algorithms need to structure the viewing operations in order to reach a particular solution. Within the scope of this structuring, the following relations are particularly relevant: viewing transformations should be performed so as to map the objects to a coordinate system that allows for efficient ordering. The rasterization (see Chapter 14) can be combined in several ways with the visibility algorithm used. Once the visible surfaces are determined, the lighting calculation should be executed to produce the color of the image elements.

One of the starting points for developing visibility algorithms is analyzing properties of the scene, including the *coherence* of the scene, meaning the degree of similarity between objects in the scene; the *geometry* of the scene, including the homogeneity (relative size) of objects in the scene and how they have been assembled; and the *order* of the scene, meaning the distribution of groups of objects in the scene. This allows us to determine interference between objects and the depth complexity of the scene.

13.1.1 Scene Properties and Coherence

To achieve computational efficiency, visibility algorithms explore several types of coherence relations associated with the intrinsic aspects of the scene objects. There are three types of coherence among scene elements: geometric, topological, and positional (or hierarchical).

Our choice of which type of coherence to use depends on the complexity of the scene (i.e., the number of objects in the scene). In a scene with low complexity (i.e., few objects), a particular surface occupies several pixels, which lends itself to geometric or topological coherence. As scenes increase in complexity, the situation is reversed: instead of having an object occupy several pixels, we have several objects contained in a single pixel. In this case, the use of geometric or topological coherence is less efficient, and hierarchical coherence is more effective.

13.1.2 Representation and Coordinate Systems

Another important point to consider is the geometry of objects. Visibility algorithms can be either general or specific, accepting homogeneous or heterogeneous geometric descriptions of the objects in the scene.

The types of geometric descriptions most used in visibility algorithms include

- ❏ Polygonal B-reps (e.g., polygonal meshes);

- ❏ Polynomial parametric surfaces (e.g., spline, Bezier patches);

- ❏ Algebraic implicit surfaces (e.g., quadrics, superquadrics);

- ❏ Implicit CSG models (e.g., constructive solid geometry);

- ❏ Procedural models (e.g., fractals).

Another issue related to geometry has to do with the internal representation adopted by the visualization system. Generalization affects both the complexity and performance of the algorithm. To simplify this problem, some visualization systems assemble several specific procedures in charge of individual parts of the operations, producing a common representation that can be combined in a subsequent integration stage. An extreme case of this strategy consists of converting, from the beginning, the geometry of all the objects to a common type (e.g., to convert to a polygonal approximation). Another extreme consists of combining, at the last stage, the images from groups of objects (e.g., using image compositing).

Visibility algorithms can be classified as operating on either the world (scene) or image space. World space algorithms work directly with the representation of the objects, thus calculating the exact solution. We therefore have a sorted list of subobjects projected on the plane of the virtual screen. In this case, visibility is the first operation performed. This type of algorithm generally uses a parametric description.

Algorithms working on image precision aim at the correct solution to the problem for a certain resolution level (which is not necessarily the same from the image). In this case, the algorithm tries to solve the problem for each pixel, analyzing their relative depths along the viewing ray; therefore, visibility is usually postponed until the very last stage. This type of algorithm generally uses an implicit description.

13.1.3 Visibility and Ordering

To calculate visibility, we must also consider the ordering of objects: given a center of projection O, we must order (sort) the scene objects so any ray, with origin at point O, intersects the scene objects in an order always compatible with the order of the objects.

The center of projection can represent any number of things, such as the optical center of a camera (the observer's position) or a point light source. Consequently, determining order of visibility is important in a number of areas of computer graphics, including visibility of a scene, which we are studying in this chapter.

Visibility algorithms can be classified according to the ordering method used to determine the visible surfaces from the camera's point of view.

The ordering structure of visibility algorithms is intimately related to the rasterization operation, which determines the region in the image corresponding to the objects in the scene. Objects occupying disjunct areas of the image are independent in terms of visibility. Rasterization can also be seen as an ordering process from which we perform a spatial enumeration of the pixels occupied by each object. Essentially, rasterization results in an ordering along the following directions: x and y (horizontal and vertical), and z (depth), in the camera coordinate system.

The computational structures of the visibility algorithms use the following ordering sequences (the parentheses indicate the combined ordering operation):

$(YXZ) \rightarrow$ visibility integrated with rasterization;

$(XY)Z \rightarrow$ visibility after rasterization;

$Z(XY) \rightarrow$ visibility before rasterization.

Algorithms of types (YXZ) and $(XY)Z$ calculate visibility locally, while processing either the scene objects or each individual pixel. In contrast, $Z(XY)$ algorithms calculate visibility globally by processing the ordering along Z based on the geometry of the objects. In the next three sections we describe the most important examples of these three types of algorithms.

13.2   (YXZ) Algorithms: Visibility with Rasterization

Algorithms of type (YXZ) calculate the visibility in image precision, reducing the sorting problem to a neighborhood of the image elements (pixels). They calculate visibility while processing the scene objects.

13.2.1   Z-buffer

The z-buffer algorithm stores, for each pixel, the distance up to the closest surface to that point on the image, which is the actual visible surface. This algorithm basically corresponds to a bucket sort. The volumetric grid given by the Cartesian product of the virtual screen by the interval $[0, z_{max}]$, which contains the depth of the scene, is called the *frame buffer*.

Pseudocode of the z-buffer algorithm is shown below:

```
Initialize frame buffer (RGBZ)
for all surfaces S of the scene do
    for all pixels on the projection of the surface do
        if Z of surface smaller than Z of buffer then
            shade the pixel with attributes of S;
        end if
    end for
end for
```

The z-buffer algorithm is an integral part of the firmware of graphics boards having 3D acceleration. The algorithm is very simple, but it has some deficiencies: the same pixel can be shaded several times, and storing the z coordinates occupies memory space. A current problem is the lack of appropriate resolution of the z coordinate.

13.3   (XY)Z Algorithms: Visibility after Rasterization

Algorithms of type $(XY)Z$ calculate the visibility in image precision, reducing the sorting problem to a neighborhood of the image elements (pixels). They calculate the visibility for each pixel.

13.3.1   Ray Tracing

The ray tracing method solves the ray shooting problem: given a finite set of objects $O_1, O_2, \ldots, O_n$ of the Euclidean space, and a ray $r = (0, \mathbf{v})$, determine whether an intersection point exists between r and the objects that are closer to the origin 0 of the ray.

The ray r is called the *viewing ray* and is determined by the center of projection and by the center of each pixel. In ray tracing, for each pixel we first compute the viewing ray leaving the center of projection center and passing through the center of the pixel; we then

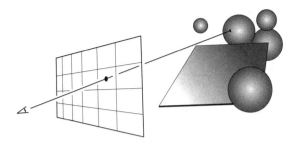

Figure 13.1. Viewing ray of a pixel.

calculate the intersection between the viewing ray and all the 3D scene objects, selecting the intersection point closest to the origin of the camera (see Figure 13.1).

The ray tracing method simplifies the visibility calculation by reducing it to a unidimensional sorting problem along of the viewing ray. However, the intersection calculation between the ray and every object in the scene is computationally very expensive.

The pseudocode of the algorithm is given below:

for all pixels of the image **do**
   calculate ray passing through the pixel
   for all surfaces of the scene **do**
      calculate intersection of the ray with the surface
      determine the closest intersection
      shade the pixel using the corresponding surface
   end for
end for

Ray tracing also plays an important role in the lighting calculation, as we will see later.

Intersection with primitive objects. In the ray tracing method we calculate, for each pixel, the intersection between the corresponding ray and every object in the scene. In general, those objects are represented by graphics primitives (e.g., polygons, spheres, cubes). The visible surface corresponds to the intersection with the smallest positive parameter t along the ray. This process corresponds to ordering by selection. Intersection calculation is generally computationally expensive. It is therefore essential to use several types of coherence to avoid, as much as possible, unnecessary intersection calculations (notice this is similar to the idea of clipping algorithms, where we try to discard the largest possible number of objects before performing the actual clipping). We will discuss this problem in more detail in Chapter 14, where we will use ray tracing to calculate the shading function at each pixel.

Ray tracing and CSG models. The ray tracing method is ideal for visualizing objects defined by Boolean operators (e.g., a CSG modeling system) because the calculation of

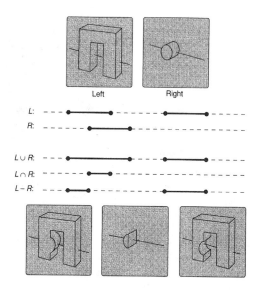

Figure 13.2. Boolean operations and ray tracing combining ray classifications. (*Reprinted from [Roth 82] with permission from Elsevier.*)

relations between primitives is reduced to unidimensional Boolean operations, i.e., operations between intervals on a straight line.

The intersection is given by a list of points where the ray crosses the surface delimiting the solid. Starting from this list, we can determine intervals along the ray corresponding to interior and exterior points to the solid. Those intervals are combined by the CSG operation, resulting in a new list. This process is illustrated in Figure 13.2 and corresponds to a merge-sort procedure. For more details, please refer to the original work on the subject [Roth 82], or [Glassner 89].

13.3.2 Warnock Algorithm

The *Warnock algorithm*, also known as the *recursive subdivision algorithm*, performs a subdivision of the virtual screen using a quadtree structure (see Chapter 9). The visibility problem is solved as described below (Figure 13.3).

We start by setting the region of the virtual screen as the root of the quadtree. The geometric configuration on the region of a quadtree cell is said to be *simple* if one of the following cases happens:

1. There exists a visible polygon covering the entire region of the cell;

2. The region of the cell does not have projected polygons;

3. There exists only one polygon contained within or intersecting the region.

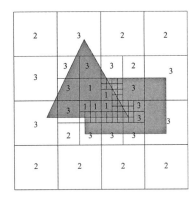

Figure 13.3. Recursive subdivision in Warnock algorithm.

In Figure 13.3, we illustrate the classification by presenting three levels of quadtree subdivision. The numbers in the cells indicate the *simplicity* of the cells, according to the three conditions discussed above.

When a cell configuration is simple, we can easily solve for visibility in the following way: in case (1) every cell pixel has the color of the polygon; in case (2) we assign the background color to every pixel on the cell; and in case (3) we clip the polygon against the cell (if necessary) and perform rasterization.

If a cell configuration is not simple, we subdivide the cell, repeating the process for each subcell in the quadtree. We continue the quadtree subdivision until we obtain cells matching the dimensions of the pixel (which is determined by the desired image resolution and the dimensions of the virtual screen).

If we reach a level of maximum subdivision (i.e., quadtree cells with pixel dimensions) and the cell configuration is still not simple, then we select the closest polygon to the camera in the cell, assigning its color to the cell color.

The advantage of this algorithm is that it does not process polygons unnecessarily (as in the case of the z-buffer algorithm); furthermore, its structure allows the anti-aliasing calculation in a natural way: if a cell reaches the dimensions of the pixel and its configuration is not simple, we continue the subdivision process, therefore performing a supersampling. (Refer to Chapter 15 for more details on anti-aliasing).

The pseudocode below describes the recursive subdivision algorithm.

```
recursive_subdivision(plist, quadrant)
for P in plist do
   if P in r then
      classify P;
   else
      remove P from plist;
   end if
```

```
end for
if configuration == SIMPLE then
   render r;
else
   recursive_subdivision(plist, quadrant 1);
   recursive_subdivision(plist, quadrant 2);
   recursive_subdivision(plist, quadrant 3);
   recursive_subdivision(plist, quadrant 4);
end if
```

13.4 Z(XY) Algorithms: Visibility before Rasterization

Algorithms of type $Z(XY)$ precisely calculate visibility by processing the ordering along Z, globally, using the geometry of the objects directly.

13.4.1 Painter's Algorithm

The z-sort, also known as the *painter's algorithm*, has two stages: (1) the scene components are sorted in relation to the virtual camera, and (2) they are rasterized from the farthest scene component to the closest one. There are two possible implementations for the painter's algorithm: the methods of approximated and complete z-sort.

Approximated z-sort. In the method of approximated z-sort, polygons are sorted based on a distance value from the polygon to the projection point. This value can be estimated starting from the centroid, or even from one of the polygon vertices. Because this method uses only one distance value for each polygon, it does not guarantee that the order of the polygons will always be correct, from the point of view of the visible surfaces calculation. However, the method is simple to implement, works well for most cases, and is also the first stage of the complete z-sort method.

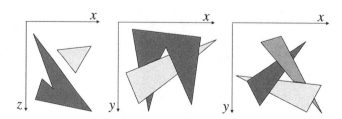

Figure 13.4. A cycle of polygons.

Complete z-sort. Complete z-sort begins with a sorted polygon list produced by the approximated z-sort. After that, the algorithm traverses this list, verifying whether the polygon order is correct from the point of view of visible surfaces calculation. If two polygons P and Q are not in the correct order, their positions are swapped in the list.

The criterion for determining whether the visibility order is correct is that polygon Q (to be painted after P) cannot occlude P. This criterion is determined by a sequence of tests of growing complexity, involving, for instance, tests with the bounding boxes of P and Q, with splitting a polygon by the support plane of the other, and with intersecting the projections of P and Q on the screen. The pseudocode of the complete z-sort algorithm is shown below.

```
Sort l by the centroid in Z (approximated z-sort);
while l ≠ ∅ do
    select P and Q;
    if P does not occlude Q then
        continue;
    else if Q marked then
        resolve cycle;
    else if Q does not occlude P then
        swap P with Q;
        mark Q;
    end if
end while
Paint l in the order;
```

There exist some situations where it is not possible to establish a solid visibility order, as shown in Figure 13.4. In those cases, it is necessary to subdivide one or more polygons to solve the problem.

13.4.2 Visibility by Spatial Subdivision

Space partition algorithms classify scene objects independently of the virtual camera parameters, creating a data structure which, once the camera position is specified, can be traversed so as to indicate the correct visibility order of the objects.

The structure most used for this goal is the partitioning tree we studied in Chapter 9. The use of partitioning tree structures for visibility has two steps: pre-processing, in which we construct the tree structure, and visibility, in which we traverse the tree structure based on the camera's position.

Construction of the partitioning tree was studied in detail in Chapter 9, where we provided the pseudocode of one particular algorithm. In this section, we will concentrate on the use of partitioning trees to calculate visibility.

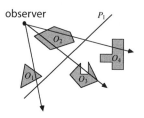

Figure 13.5. Space partitioning and visibility.

BSP-trees, ordering, and visibility. When solving the visibility ordering problem, we establish priority for the scene objects so that any ray shot from the center of projection, and intersecting two objects, always intersects first the object with greater priority (i.e., the visible object).

The use of partitioning trees to solve the visibility problem is based on the following property: if a plane splits a group of objects, it creates a visibility ordering—the objects of the semispace where the center of projection is have priority over the objects of the other semispace (see Figure 13.5).

We do not completely solve the visibility problem using a single plane because we cannot establish priority for objects contained in the same semispace as the center of projection. So we add more planes to completely solve the visibility ordering problem. The brute force approach for solving this problem, i.e., determining a plane to split the objects two by two, requires a very large number of planes. A better strategy is to recursively subdivide the space by planes. The objects are initially split by a plane into two groups; each of those two groups is then further split independently by two planes into two subgroups and so forth. The subdivision continues until each object is contained in a subset of the subdivision.

This process generates a tree partitioning. Now, how can this partitioning tree be used to solve the visibility ordering problem? Given a center of projection, we first determine on which side of the root plane it is positioned. Once this semispace is located, we know that the objects of its associated subtree have priority over the objects of the other subtree. We now need to order the objects that are in the same semispace as the center of projection. Notice that this procedure is like that of z-sort, using the root plane of the subtree containing the center of projection. In this way, we recursively continue this process, obtaining the desired visibility order by traversing the nodes of the subtree in which the center of projection is located.

Figures 13.6 (a) and (b) illustrate the procedure described in the previous paragraph. In (a), we place the center of projection in the positive semispace of the root plane; in (b), the center of projection (the observer's position) is located in the negative semispace. In both cases we indicate below the tree leaves a number indicating the position of the leaf object in the ordering sequence. We also indicate the cell where the center of projection is located.

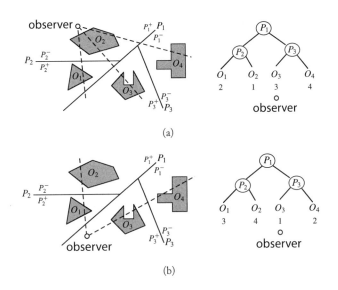

(a)

(b)

Figure 13.6. Visibility and BSP-trees.

This process has some restrictions. It cannot guarantee the existence of a splitting plane between two objects in the scene—we can only guarantee the existence of that plane for convex objects. It also does not consider the ordering of the actual faces of objects to solve self-occlusion problems, which is necessary if objects are not convex.

The solution to the first restriction is simple: if the splitting plane intersects one of the objects (i.e., the object intersects the two semispaces), then we clip the object against the plane, thus creating two objects, each contained in a semispace.

The second restriction can be solved by considering a partitioning tree where the polygons (constituting the faces of the objects) determine splitting planes of the partitioning tree. However, in this case we will have faces on the splitting planes, and the ordering relation will have three types of classification, *near* $\rightarrow$ *on* $\rightarrow$ *far*, instead of just *near* $\rightarrow$ *far* (Figure 13.7). In other words, polygons in the same semispace as the center of projection have higher priority than polygons on the splitting plane, which in turn have greater priority than polygons in the complementing semispace.

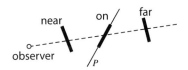

Figure 13.7. Ordering and BSP-trees.

The pseudocode below describes the visibility calculations using BSP-tree. The function receives two parameters: the observer's position, c, and a pointer to the BSP-tree structure of the scene, t.

```
bsp_traverse(c, t)
if t == NULL then
    return;
end if
if c in front of t.root then
    bsp_traverse(c, t->back);
    render(t->root);
    bsp_traverse(c, t->front);
else
    bsp_traverse(c, (t->front);
    if backfaces then
        render(c, t->root);
    end if
    bsp_traverse(c, t->back);
end if
```

13.4.3 Visibility by Recursive Clipping

The recursive clipping algorithm determines a set of visible areas of the objects, which are disjunct on the image plane. The pseudocode below describes the algorithm.

```
recursive_clipping(list)
if list == empty then
    return;
end if
sort approximately in z;
select front polygon P;
split list in inside and outside, clipping in relation to P;
while inside != empty do
    select Q;
    if P in front of Q then
        remove Q;
    else
        swap P and Q;
    end if
end while
render P;
recursive_clipping(outside);
```

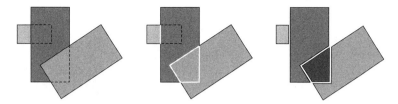

Figure 13.8. Recursive clipping.

Figure 13.8 illustrates the algorithm for a scene with three polygons. The recursive clipping results in four visible polygons. To implement this algorithm, we would obviously need any clipping algorithm between two polygons, such as the Weiler-Atherton or the algorithm described in [Vatti 92].

With some modifications, this method can be applied for shadows (notice the problem of calculating shadows can be seen as a visibility study from the point of view of the light sources in the scene); furthermore, it is also used in the beam tracing algorithm [Heckbert and Hanrahan 84].

13.5   Comments and References

The first systematic analysis of algorithms for visible surfaces calculation was made by [Sutherland et al. 74]. This work proposed a characterization of the visibility topic according to the ordering criterion, as we described in Section 13.1.3.

There are several proposed method for calculating visibility in scenes with heterogeneous geometry: [Crow 82] suggests that the visibility calculation for groups of objects be later combined in a post-processing image compositing; [Cook et al. 87] propose a preprocessing stage to convert every surface in the scene into micropolygons, thus allowing for efficient visibility calculation.

The z-sort algorithm is described in [Newell et al. 72]. The space partition algorithm was initially introduced by [Shumacker et al. 69] for flight simulation systems and later adapted for more general applications by [Fuchs et al. 83].

Recursive clipping was developed by Weiler and Atherton for the calculation of visible surfaces [Weiler and Atherton 77] and shadows [Atherton et al. 78]. A variation of this method was used in the beam tracing algorithm by [Heckbert and Hanrahan 84] (a description of this algorithm is given in Chapter 14).

The algorithm of recursive image subdivision was introduced by [Warnock 69]. The scanline visibility algorithm, a type of (YXZ) algorithm, was independently developed by [Watkins 70] and [Bouknight and Kelly 70].

13.5.1   Additional Topics

Using the ray tracing method to calculate visibility was first done by [Group 68]. A complete description of this method can be found in [Roth 82]. The ray shooting problem used in this method is actually a particular case of a problem known in computational geometry as the *range search problem*: let S be a set of n geometric objects in $\mathbb{R}^n$ and $U \subset \mathbb{R}^n$ one subset. Determine the elements of $S \cap U$.

An important problem is how to determine the visibility of large geometric databases (e.g., walkthrough). Real-time visibility is fundamental to some applications, mainly in the areas of simulation and games development. Within this scope, an important topic consists of the development of *occlusion culling* methods: to eliminate, a priori, scene objects which, despite being contained in the viewing volume, will not be visible (e.g., their view can be occluded by other objects). *Back face removal* is a simple and classic example of this technique.

Exercises

1.   Using the z-buffer, describe an algorithm to perform Boolean operations between solids in the screen space.

2.   The method of *back face removal* is frequently used to reduce scene complexity when elements are represented using a polygonal B-rep (polygon meshes). Describe this method. What type of coherence does it use?

3.   Describe a simple visibility algorithm that works for scenes constituted solely by convex surfaces. What types of coherence did you use in the algorithm?

4.   Define a silhouette curve for an object in the scene. Determine a visibility algorithm for silhouette curves. Cite examples of applications using visualization algorithms for silhouette curves.

5.   From a geometric point of view, the shadow of an object in a scene can be determined by solving the visibility problem starting from the light source. Based on this fact, define the regions of umbra and penumbra.

14 Illumination

An image is complete when color information is assigned to objects in the image. In the case of 3D scenes, this correspondence can be established by calculating the illumination (lighting). This is a natural step because it simulates our visual perception of the physical world. Our challenge in creating natural-looking lighting is to determine the illumination of the scene: given a point $p \in S \subset \mathbb{R}^3$, a light source ℓ, and an observer at a point O, determine the light L originating at point P and its interaction with the given point at the surface, which is noticed by the observer.

Calculating the lighting of a scene is closely linked to determining the color of a scene. In fact, the whole process of illumination calculation traditionally revolves around colorimetry (see Chapter 5) rather than radiometry (see Chapter 18). This is apparent in the way we construct graphics models: the color of an object is specified by three wavelengths, $R(\lambda = 620\text{nm})$, $G(\lambda = 550\text{nm})$, and $B(\lambda = 440\text{nm})$; and in most systems, light sources are specified by their RGB values without any reference to radiant quantities. Using colorimetry instead of radiometry does not bring any issues, provided the generated images do not have any bond to physical reality (e.g., the images will not be used for virtual studies of the illumination of an environment).

14.1 Foundations

The illumination problem can be better understood using the four universes paradigm. In the physical universe, illumination is related to the transport of luminous energy and its interaction with objects in the scene. In the mathematical universe, we describe this phenomenon through illumination models. In the representation universe, we solve equations to calculate the various model representations. In the implementation universe, we establish the computational methods that allow us to implement the shading function:

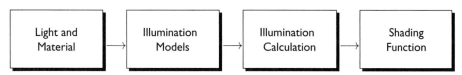

This provides us with a well-established road map for this chapter: (1) study the physical information about propagation of light in an environment; (2) determine a mathematical model for this phenomenon; (3) set up an equation for the illumination problem; and (4) calculate its result. In other words, the solution to the illumination problem consists of obtaining an illumination equation

$$K(L, O, S, p, \ell) = 0$$

and solve it to determine the light L.

The observer's position O and the surface point P define a direction $\mathbf{u} = (p - O)/|p - O|$ in space; therefore, the solution to the equation determines the function $f(O, \mathbf{u})$, which calculates the perceived color for an observer located at point O looking toward direction $\mathbf{u}$. This function is called the *shading function*. Notice, it is the function we use to perform calculations to shade (i.e., paint, colorize) each pixel of the virtual screen, generating the image of the scene. (In this case, point O is the position of the camera and, for each pixel, vector $\mathbf{u}$ is defined by the ray connecting O to the center of the pixel.)

14.2   The Nature of Light

Illumination is the study of light propagation through ambient space. A 3D scene is composed of objects, which emit, reflect, or transmit light.

Light has a dual nature: it behaves like a beam of particles and like a wave. The *particle* model of light postulates that the flow of energy along a ray is quantized by particles in motion called *photons* (which can to be thought as small wave packets with frequency, speed, and wavelength). On the other hand, the *wave* model of light describes luminous energy through the combination of two fields: electric and magnetic. Some illumination phenomena are described by the particle model; others by the wave model. The wave model is governed by *Maxwell's equations* of electromagnetism.

14.2.1   Light Propagation Models

To obtain an illumination equation, we need to establish a physical-mathematical model of the phenomenon of luminous energy propagation. We can create three possible models corresponding to three different areas: geometric optics, physical optics, and radioactive transport.

Geometric optics. The area of *geometric optics* assumes that light spreads along a straight line with constant speed. It models the interaction of this propagation in the boundary surfaces between two media, assuming these surfaces are optically perfect. This model appropriately describes the macroscopic effects of light transport such as linear propagation, reflection, and refraction; however, it does not take into account the concepts of radiometry that allow us to measure all quantities involved.

Physical optics. The area of *physical optics* extends geometric optics by incorporating Maxwell's equations in the modeling. Besides the classic phenomena, this method can model other phenomena, such as scattering, interference, and diffraction, that happen when the geometric scale of the scene is smaller than the wavelength of the light being transported. While physical optics is more versatile than geometric optics, it cannot quantify the transport of light in environments where the geometric scale is greater than the wavelength of the light.

Radioactive transport. The area of *radioactive transport*, a particular case of transport theory, interprets the illumination problem as transport of radiant energy. It essentially combines the theory of geometric optics with thermodynamics (early radiometry methods in computer graphics used heat transfer methods from thermal engineering [Goral et al. 84]).

Of the three models of light propagation, the radioactive transfer model is the most appropriate for synthesizing images in computer graphics. To use it, we need to understand a number of radiometry concepts. These concepts are explained in Chapter 18, which supplements this chapter.

Using the transport theory of radiant energy, we can obtain the general illumination equation. However, in this chapter we will cover simpler illumination models with simplified illumination equations that represent rough approximations of physical models. There are three advantages to starting with these models: they are simple and relatively easy to understand; they are sophisticated enough to give readers a good understanding of the illumination problem; and in practice they are good enough for the majority of computer graphics applications. Chapters 18 and 19 present a more complete study of the subject.

14.2.2   Propagation Laws

If we assume that the propagation medium does not participate in the illumination, the relevant illumination phenomena are related to the interactions between light and surfaces that determine the boundary between two media. Such phenomena depend on both the geometry of the surfaces and the material of the objects. These factors determine the path and the radiant energy propagated on each interaction between the two media.

A basic principle governing light propagation is energy conservation: energy arriving at a separation surface between two media is either reflected, transmitted, or absorbed. The sum of the reflected, transmitted, and absorbed energies is equal to the incident energy.

In this section, we will study the interaction of light with some ideal surfaces and determine the propagation laws in those surfaces. Three types of material are considered perfect in terms of their interaction with light: the perfect specular reflector, the perfect transmitter, and the perfect Lambertian diffuser.

First, the notation (see Figure 14.1): if S is a surface and $p \in S$, $\mathbf{n}$ is the normal vector to S in p. Given an observer at a point O, the unit vector $\omega_r = (O-p)/|O-p|$ is called the *eye vector*. If a luminous ray is incident upon S at point P, the angle between the ray and the normal vector $\mathbf{n}$ is called the *angle of incidence* θ_i. The *vector of incidence*, represented by

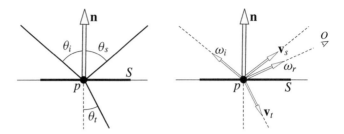

Figure 14.1. Angles of incidence, reflection, and transmission.

ω_i, is the unit vector with opposite direction to the incident ray. Notice $\cos \theta_i = \langle \omega_i, \mathbf{n} \rangle$.[1] The plane formed by the vector of incidence ω_i and the normal vector $\mathbf{n}$ to the surface is called the *plane of incidence*. The other angles and vectors shown in Figure 14.1 will be defined further on.

Perfect specular reflector. A surface S is a *perfect specular reflector* when it satisfies the *law of specular reflection*: for each incident ray, there exists a single reflection ray that is contained in the plane of incidence; furthermore, its angles of incidence and reflection are equal, $\theta_i = \theta_s$. The unit vector in the direction of the specularly reflected ray will be indicated by $\mathbf{v}_s$. Figure 14.1 illustrates the geometry of specular reflection. Notice $\cos \theta_s = \langle \mathbf{n}, \mathbf{v}_s \rangle$.

The constants k_i and k_t in the Snell-Descartes law are called *indices of refraction* of the media delimited by the surface. The fact that the angles of incidence and reflection are equal in the case of perfect specular reflection is described by the Snell-Descartes law about two equal media. It is common to call this equality the *Snell-Descartes law of reflection*.

According to the Snell-Descartes law, both the reflected and the transmitted rays are contained in the plane of incidence. This means that if we rotate the surface about its normal vector, either the reflected or transmitted ray remains constant. This property characterizes a class of surfaces called *isotropic*. Examples of anisotropic materials include brushed stainless steel and hair.

To model these objects we here assume another simplified surface property: the reflected ray emanates from the surface at the same point the incident ray struck the surface, meaning there is no subsurface scattering.

Perfect transmitter. A surface S is a *perfect transmitter* if the transmission of light satisfies the *Snell-Descartes law*: for each incident ray there exists a single transmission ray that is contained in the plane of incidence; furthermore, the angles of incidence and transmission satisfy the equation $k_i \sin \theta_i = k_t \sin \theta_t$. The unit vector in the direction of the transmitted ray will be indicated by $\mathbf{v}_t$. Figure 14.1 illustrates the geometry of transmission.

---

[1]Instead of the function $\cos \theta_i = \langle \omega_i, \mathbf{n} \rangle$, it is more correct to use the function $\max\{0, \langle \omega_i, \mathbf{n} \rangle\}$ to avoid illumination from light sources that are behind the surface (backface illumination).

Lambertian diffuser. A surface S is a *perfect diffuser* when the reflected energy of an incident ray, with constant irradiance, has constant radiance (or luminance) in every direction (i.e., in any solid angle). Thus a change in the observer's position does not alter the perceived radiance of the surface. Chalk is an example that closely approximates a perfect diffuser.

Perfect diffusers are also called *Lambertian* because they satisfy *Lambert's law*: if the incident ray has radiant intensity I_i, the reflected radiant intensity is given by $L_r = L_i \cos \theta_i = L_i \langle \omega_i, \mathbf{n} \rangle$. From Lambert's law we know that as the incident ray distances itself from the normal, the luminous intensity reflected by the surface (which is constant), decreases proportionally to the cosine of the angle with the normal. The concept of a Lambertian surface can be extended to surfaces that transmitting light (translucent), as well as to surfaces that emit their own light (light sources).

14.2.3   Material Classification

In terms of radiant energy, surfaces are characterized by the bidirectional reflectance distribution function (BRDF). However, a simple material classification is useful for formulating our illumination model. In a simplified view, materials can be of three types: dielectric, metallic, or composite.

Dielectric materials (e.g., glass) are translucent and are electrical insulators. In the boundary between air and a translucent dielectric material, most of the light is transmitted (a small amount is absorbed; for instance, transformed into heat).

Metallic materials (e.g., copper, gold, and aluminum) are opaque and are electrical conductors. In the boundary between air and a metallic material, most of the light is reflected (a small amount is absorbed).

Composite materials (e.g., plastics and paints) are formed by opaque pigments in suspension in a transparent substratum. In the boundary between air and a composite material, a combination of effects takes place: light is reflected, transmitted, and absorbed in different proportions.

14.2.4   Geometric Classification

The geometry of the surface can be optically smooth or rough (see Figure 14.2). A *smooth surface* is locally modeled by its tangent plane. In the case of smooth surfaces, light propagates toward either the reflection or the transmission directions. An ideal smooth surface would be a perfect specular reflector.

A *rough surface* does not have a well-defined tangent plane at any point (and therefore does not have a well-defined normal vector). An ideal rough surface would be a perfect diffuser. Several mathematical models allow us to obtain the BRDF of a *rough surface*. One well-known example is the Torrance-Sparrow model, which uses probabilistic distributed microfacets. It was introduced to the computer graphics community in [Cook and Torrance 81] and [Blinn 77].

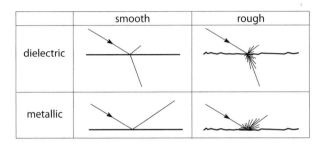

	smooth	rough
dielectric		
metallic		

Figure 14.2. Materials and patterns of reflection and transmission.

When a ray strikes a surface that is neither very rough nor very smooth, the light rays are scattered in a preferential direction: the direction of specular reflection. This type of reflection is called *specular-diffuse*. Specular-diffuse reflections are very common in our daily life. Consider a surface with a certain degree of polishing (i.e., a certain degree of smoothness). Recall that, depending on your position in relation to the surface, you can see a region with a very strong light reflection. This reflection falls off toward the surface's extremities. The region of strong reflected light intensity is called the *highlight*.

Reflectance. The *reflectance* function, indicated by ρ, relates incident and reflected energies at a point x. Given a direction of incidence ω_i, ρ measures the fraction of radiant flux density reflected by the surface in every direction, in relation to the incident radiant flux density on the surface toward direction ω_i.

$$\rho = \frac{\text{density of reflected radiant flux}}{\text{density of incident radiant flux}} = \frac{dB_r}{dE_i}.$$

The BRDF function, described in Chapter 18, is another function that establishes relations between the radiance emitted by a surface with the incident irradiance, thus characterizing materials in terms of their interaction with radiant energy. Both functions play a similar role: they compare reflected radiant energy (radiosity) with incident radiant energy (irradiance).

Values of the radiance function are within the interval $[0, 1]$ (given the energy conservation principle, a surface is not able to reflect more energy than the incident energy unless it emits energy). In contrast, the BRDF assumes values within the interval $[0, +\infty)$. What is more, because the radiance function involves the quotient of the same radiometric quantities, it can measure reflectance in relation to other quantities such as radiant intensity or flux density.

The reflectance function is simpler and more intuitive than the BRDF. But replacing the BRDF with the reflectance function results in a great loss: the BRDF function captures all the luminous phenomena of the interaction between the energy and the surface (e.g., scattering, surface anisotropy); the reflectance coefficient is only a ratio between the incoming and outgoing energies, without taking into account the interaction with the microgeometry of the surface.

We should also mention that there are several types of reflectance functions, depending on the integration region for measuring reflected energy. The most common type is *directional hemispherical reflectance function*, where we calculate the total radiosity in the superior hemisphere. The *bidirectional hemispherical reflectance function* performs calculations for each incident and reflected direction, exactly like the BRDF. Context should always determine the type of reflectance that is used.

The total radiosity of the radiant energy reflected by the surface is given by B:

$$dB_r = \int_\Omega L_r(x, \theta_r, \varphi_r) \cos \theta_r d\omega_r.$$

Therefore,

$$
\begin{aligned}
\rho(x, \theta_r, \varphi_r) = \frac{dB_r}{dE_i} &= \frac{\int_\Omega L_r(x, \theta_r, \varphi_r) \cos \theta_r d\omega_r}{dE_i} \\
&= \int_\Omega \frac{L_r(x, \theta_r, \varphi_r)}{dE_i} \cos \theta_r d\omega_r \quad\quad (14.1) \\
&= \int_\Omega f_r(x, \theta_i, \varphi_i, \theta_r, \varphi_r) \cos \theta_r d\omega_r.
\end{aligned}
$$

(The term dE_i is constant because it represents the irradiance on the surface along the (preset) ω_i direction.) It is common for people to use the term *reflectance function*, or simply *reflectance*, for the BRDF: please pay attention to the context to avoid confusion.

Reflectance and perfect diffusers. On a Lambertian surface (perfect diffuser), the BRDF f_r is a constant f_d. Using Equation (14.1), we have the reflectance ρ_d of the surface given by

$$\rho_d = \int_\Omega f_d \cos \theta_r d\omega_r = f_d \int_0^\pi \int_0^{2\pi} \cos \theta_r d\omega_r = \pi f_d.$$

Reflectance and specular reflection. In practice, specular reflection is not modeled; instead, it is well approximated by a mirror surface of good quality. If ω_r is the reflected ray, the reflectance function is given by the function δ:

$$\delta(\omega, \omega_i) = \begin{cases} 1 & \text{if } \omega = \omega_i, \\ 0 & \text{otherwise.} \end{cases}$$

Reflectance and specular-diffuse reflection. In the case of the specular-diffuse function, reflectance is characterized by a scattering of reflected energy about the specular reflection vector. Figure 14.3 shows the reflectance graph for a certain direction of incidence. (There exists a solid angle out of which the reflected energy is null.)

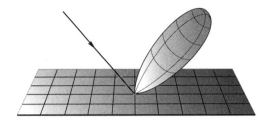

Figure 14.3. Specular-diffuse reflection model.

14.2.5   Light Sources

Surfaces in a scene that emit light are called *light sources*. Particularly prominent light sources are commonly called *primary light sources*; objects that indirectly emit light by reflection are called *secondary light sources*.

Primary light sources. In computer graphics, we can classify the most common types of light sources into four categories: directional lights, point lights, spotlights, and area lights.

Directional lights, such as the sun, are located at an infinite (i.e., very distant) location. A directional light spreads itself toward a particular direction and their energy is not attenuated with distance. This type of light is specified by the direction of propagation and by the emitted radiant energy (radiosity).

A *point light*, such as a candle or an incandescent lamp, is a localized light source at a point in the scene that irradiates in all directions in an isotropic way. Its energy is attenuated with distance. This type of light is specified by its location point and by the emitted value of radiant energy.

A *spotlight*, such as a table lamp or theater reflector, is a point light with a cone of illumination: emitted radiant energy is null outside this cone. This source lights up a delimited region of the scene. It is specified by its position in the scene, its emitted energy, and its cone of luminous propagation (this can be parameterized in a way to have a focus of variable solid angle).

The source of an *area light* is a nonnull, finite area. These light sources produce regions of illumination of total shadow (umbra) and of partial shadows (penumbra) in the scene. An example is a fluorescent tube.

Secondary light sources. In a real scene, most objects work as secondary light sources. However, modeling this sort of light is very difficult. In computer graphics we approximate this effecting by introducing *ambient light*. This light has a constant irradiation value and is nondirectional: its light, when reflected by the objects in the scene, has the same irradiance value, independent of objects' position and orientation or the observer's position. We will discuss this, and other simplifications, in the Section 14.3.

14.3   A Simple Illumination Model

We will now establish some simplifying hypotheses that will allow us to obtain a very simple illumination model. Although this model is simple, it will allow us to obtain virtual images of excellent quality and will give us a basis for understanding the more complex illumination models typically used in computer graphics.

In a real scene, the light striking surfaces comes from both primary and secondary sources. Indirect illumination is very difficult to compute, so we will instead introduce ambient light, as we described in the previous section. We will not consider transmitted energy (i.e., we will not have translucent objects), and we will assume that all radiant energy, striking surfaces from a direct light source, is partially reflected as both diffuse and specular-diffuse reflections. Given these premises, we can write

$$\text{Reflected intensity} = \text{Ambient} + \text{Diffuse} + \text{Specular-Diffuse}. \qquad (14.2)$$

The top row of Figure 14.4 shows a vertical section of a surface's reflectance graph. According to the above reflection equation, this reflectance graph is the sum of the reflectance of a Lambertian diffuser with a material having a specular-diffuse reflection (we exclude ambient light reflection, which is constant and nondirectional). The section of the graph shown in the figure is obtained by the plane defined by the incident vector and the normal, which contains the specular reflection vector. Because we are assuming the surface is isotropic, this graph is independent of the direction of incidence. The bottom row of Figure 14.4 shows the generation of a synthetic image using this model. An illustration of the reflectance surface is shown in Figure 14.5 for two different directions of incidence.

To translate Equation (14.2) into mathematical terms, consider a single (direct) light source ℓ. I_i is the incident light intensity coming from the light source ℓ. I_r is the total intensity of reflected light. I_a indicates the intensity of the ambient light. I_d is the intensity of the diffuse reflection, and I_s is the intensity of the specular-diffuse reflection.

The three terms to the right in Equation (14.2) can be mathematically expressed by

$$\begin{aligned}
\text{Ambient reflection} &= k_a I_a; \\
\text{Diffuse reflection} &= k_d I_d; \qquad (14.3) \\
\text{Diffuse-Specular reflection} &= k_s I_s.
\end{aligned}$$

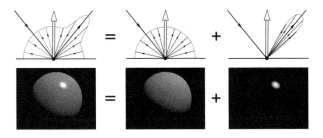

Figure 14.4. Reflection models of type "diffuse + specular-diffuse." (See Color Plate XX.)

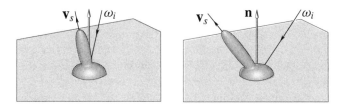

Figure 14.5. Reflectance of the model "diffuse + specular-diffuse."

The constants k_a, k_b, and k_s characterize the material of the surface. The constant k_a is the *coefficient of ambient reflection*. It varies within the interval $[0, 1]$ and indicates the percentage of ambient light being reflected. The constant k_d is the *coefficient of diffuse reflection*. It varies within the interval $[0, 1]$ and it indicates the percentage of incident energy diffusely reflected. The constant k_s is the *coefficient of specular reflection*. It varies within the interval $[0, 1]$ and indicates the percentage of incident energy being reflected in a specular-diffuse way.

By placing the values given in Equation (14.3) in (14.2), we obtain

$$I_r = k_a I_a + k_d I_d + k_s I_s. \tag{14.4}$$

The term I_a is constant and is specified by the user. We now need to calculate the terms I_d and I_s. Assuming the surface is a perfect diffuser, the diffuse reflection is calculated through Lambert's law:

$$I_d = I_i \cos \theta_i = I_i \langle \omega_i, \mathbf{n} \rangle. \tag{14.5}$$

We now determine the term I_s of the specular-diffuse reflection. Unlike the diffuse component, our solution will be empirical. Consider a phenomenon present in daily life. If you look at a polished metallic surface, there will be a highlight that is bright in the center and fades towards the edges. In computer graphics terms, the highlight region of a specular-diffuse reflection has a strong intensity at a central point that radially declines. This happens because there is a scattering of the reflected energy around the direction of specular reflection, given by the Snell-Descartes law.

Even without turning to physics, we can easily find a mathematical model for this phenomenon. Consider an observer's positioned along the direction of the reflected light, and let α be the angle between the direction of the observer ω_r and the direction of specular reflection $\mathbf{v}_s$ (see Figure 14.6). The maximum reflected intensity noticed by the observer happens for $\alpha = 0$. As α increases, the intensity decreases. This fact can be modeled by a differentiable function $\gamma : [\pi/2, \pi/2] \to \mathbb{R}$ that satisfies the following conditions: $\gamma(0) = 1$; $\gamma(\pi/2) = 0$; and γ is symmetrical, i.e., $\gamma(-\alpha) = \gamma(\alpha)$. Given the function $\gamma(\alpha)$, the component of specular-diffuse reflection can be calculated by placing $I_s = I_i \gamma(\alpha)$.

The function

$$\gamma(\alpha) = \cos^e \alpha = \langle \mathbf{v}_s, \omega_r \rangle^e, \qquad e \in [0, \infty)$$

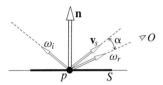

Figure 14.6. Observation angle.

satisfies the above conditions. Figure 14.7 shows the graph of function γ for some values of the exponent e.

We can therefore write

$$I_s = I_i \gamma(\alpha) = I_i \langle \mathbf{v}_s, \omega_r \rangle^e.$$

Inputting in Equation (14.4) the value of I_s given by the above equation, and the value of I_d given by Equation (14.5), we obtain:

$$\begin{aligned} I_r &= k_a I_a + k_d I_i \cos \theta_i + k_s I_i \cos^e(\alpha) \\ &= k_a I_a + k_d I_i \langle \omega_i, \mathbf{n} \rangle + k_s I_i \langle \mathbf{v}_s, \omega_r \rangle^e. \end{aligned} \tag{14.6}$$

The exponent e controls the decay rate of the function defining the highlight. When $e \to +\infty$, the function approaches the δ function, which means we will have a perfect specular reflection (the highlight is reduced to a point). On the other hand, when $e \to 0$, the highlight area increases, meaning the surface is rougher (it increases the scattering of the reflection). From a physical point of view, the exponent e controls the roughness of the surface. It is called the *roughness exponent* or *Phong exponent*.

The reflected energy

$$I_r = I_r(p, k_a, k_d, k_s, I_i, I_a, \mathbf{n}, \omega_i, \omega_r),$$

is a function of the parameters that define the geometry of the surface, the type of material, and the light sources. If instead of a single light source we have n sources, $\ell_1, \ell_2, \dots, \ell_n$ and

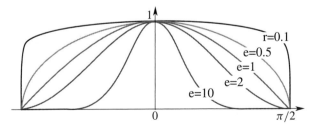

Figure 14.7. Coefficient of specularity.

we ignore occlusions, we extend the above equation to take into account the contribution of each light source. We then obtain

$$I_r = k_a I_a + I_i \left(+k_d \sum_{j=1}^{n} \langle \omega_i^j, \mathbf{n} \rangle + k_s \sum_{j=1}^{n} \langle \mathbf{v}_s^j, \omega_r \rangle^e \right). \tag{14.7}$$

This is the solution we sought for the illumination problem described in the beginning of this chapter. It is called the *Phong illumination equation* or *local illumination equation*. The illumination model on which it is based is called the *Phong model*, in honor of the computer graphics researcher and pioneer B.T. Phong.

14.3.1 Phong BRDF Function

In this section we will discard the ambient energy component in the Phong equation and focus only on reflection by direct illumination: diffuse and specular-diffuse reflections.

Up to this point, we have not considered any of the radiometric quantities of the incident and reflected energies in the Phong equation. In reality, this does not matter. Given that all of the elements of the equation (except I_i and I_r) are dimensionless, we can use any radiometric quantity to measure the incident energy. The reflected energy will be expressed in the same quantity.

We can, for instance, write the equation in radiance terms:

$$L_r = L_i \left(k_d \langle \omega_i, \mathbf{n} \rangle + k_s \langle \mathbf{v}_s, \omega_r \rangle^e \right).$$

This way, the term

$$\rho(\omega_i, \omega_r) = k_d \langle \omega_i, \mathbf{n} \rangle + k_s \langle \mathbf{v}_s, \omega_r \rangle^e, \tag{14.8}$$

is the reflectance of the Phong model. The above equation can be rewritten as

$$L_r = \rho(\omega_i, \omega_r) L_i. \tag{14.9}$$

To calculate the BRDF of the Phong illumination model, we need to write the incident radiance L_i in terms of the irradiance E_i. As is shown in Chapter 18,

$$L_i = \frac{E_i}{\cos \theta_i d\omega_i} = \frac{E_i}{\cos \langle \mathbf{n}, \omega_i \rangle d\omega_i}.$$

If we input this expression into Equation (14.9), we obtain

$$L_r = \frac{\rho(\omega_i, \omega_r)}{\cos \langle \mathbf{n}, \omega_i \rangle d\omega_i} E_i.$$

Hence we conclude that the Phong BRDF, $f_r(\omega_i, \omega_r)$, is given by

$$f_r(\omega_i, \omega_r) = \frac{L_r}{E_i} = \frac{\rho(\omega_i, \omega_r)}{\cos \langle \mathbf{n}, \omega_i \rangle d\omega_i} = \frac{\rho(\omega_i, \omega_r)}{\cos \theta_i \, d\omega_i}.$$

Or, using Equation (14.8),

$$f_r(\omega_i, \omega_r) = \frac{k_d \langle \omega_i, \mathbf{n} \rangle + k_s \langle \mathbf{v}_s, \omega_r \rangle^e}{\cos \langle \mathbf{n}, \omega_i \rangle d\omega_i} = \frac{k_d \langle \omega_i, \mathbf{n} \rangle + k_s \langle \mathbf{v}_s, \omega_r \rangle^e}{\cos \theta_i \, d\omega_i}.$$

14.3.2 Parameter Specification

Writing a spectral model of Phong equation is straightforward: incident energy of a certain wavelength results in reflected energy of the same wavelength.[2] In the spectral Phong equation, the coefficients of ambient and diffuse reflections depend on the wavelength $k_a = k_a(\lambda)$, $k_d = k_d(\lambda)$. If $C = C(\lambda)$ is the spectral distribution of the surface color, and $L_i = L_i(\lambda)$ is the spectral distribution of the incident light, we have

$$L_r(\lambda) = L_a(\lambda) k_a(\lambda) C(\lambda) + L_i(\lambda) \left(k_d(\lambda) C(\lambda) \langle \omega_i, \mathbf{n} \rangle + k_s \langle \mathbf{v}_s, \omega_r \rangle^e \right).$$

Notice that the coefficient of specular reflection k_s does not depend on the wavelength. What is more, the color of the object $C(\lambda)$ does not have any influence in the component of specular-diffuse reflection. This means the highlight color is the same as the incident light $L_i(\lambda)$ (which is correct for most materials presenting a high degree of specularity). Of course we can change this equation so the color of the object influences the color of the highlight, and we can even assign two colors to the object: one for the diffuse reflection and another for the specular reflection. Since the model is empirical, we can make several changes to the equation to obtain specific illumination effects.

14.3.3 Wrapping Up

Now that we have the Phong model and its illumination equation, what do we do next? We need to calculate the illumination of the scene (i.e., shade the pixels); improve the Phong model (i.e., improve the BRDF of Phong); and look for an illumination model that better captures the phenomenon of energy propagation (i.e., a model conceptually based in physics).

The remainder of this chapter will describe how we illuminate the scene. How to improve the Phong model is outside the scope of this book, but much has been written on the subject. The pioneer works in this area were from Cook-Torrance [Cook and Torrance 81] and Jim Blinn [Blinn 77], who based their work on the Torrance-Sparrow model, which was used in physics. This model calculates the BRDF by modeling the microgeometry of the surface using microfacets.

This book will address the challenge of developing an illumination model that better captures the phenomenon of energy propagation (see Chapter 19).

---

[2]Here we have another simplification, given that the wavelength from the reflected radiation can be different.

14.4   Illumination Calculation

Now that we have an illumination equation, our next step is to obtain a solution to determine the shading function. The Phong equation explicitly provides the light reflected by each surface. In other words, that equation is already solved. But to calculate the reflected energy at a point we must know the normal vector to the surface at that point. How do we find the normal vector? In the case of polyhedral B-rep objects, the normal vector field is represented by the normal vector at each of the polygon vertices. Therefore, we need a reconstruction method to determine the illumination function at an arbitrary point in a polygon (see Figure 14.8).

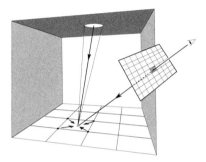

Figure 14.8. Reconstruction of the illumination function.

We have two options: we can perform the calculation in the world (scene) coordinate system or we can perform the calculation in the device coordinate system. In other words, we can perform the calculation before or after applying the camera transformations. If we calculate the shading function after the camera transformation, we should take into account that perspective transformation does not preserve some geometric properties (e.g., the normal vectors are no longer normal after the transformation).

We have three methods for calculating the shading function: constant shading, Gouraud shading, and Phong shading.

14.4.1   Constant Shading

Constant shading, also known as *Bouknight shading* or *flat shading*, comes from the pioneering work of Bouknight [Bouknight 70]. He considered polygonal B-rep surfaces and used only the diffuse component of the reflected energy (Lambert's law), with directional light sources. For calculating the shading function in a polygon, Bouknight considered the normal vector to the polygon; therefore, the diffuse component of the reflected illumination is constant throughout the entire polygon (see Figure 14.9(a)). In other words, the method calculates a constant illumination function on each face of the B-rep representation. It consists of (1) calculating the illumination function of each face, (2) projecting the face in the camera transformations and (3) using the obtained value to shade each projected face in the rasterization process.

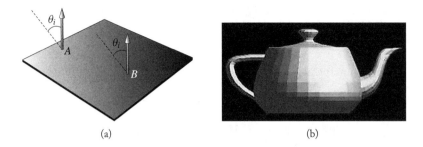

(a)   (b)

Figure 14.9. (a) Constant diffuse reflection; (b) image shaded with the Bouknight method (flat shading).

Bouknight shading is not useful for generating realistic images because it emphasizes the faceted surface aspect of polyhedral representations, as shown in Figure 14.9(b).

14.4.2   Gouraud Shading

The first nonconstant reconstruction method of the shading function was implemented by the computer scientist Henri Gouraud [Gouraud 71] and is known as *Gouraud shading*. He used B-rep polygonal surfaces, directional light sources, and only the diffuse component of reflected energy (Lambertian diffuser). The method assumes each polygon of the representation has a normal vector at each vertex. It calculates the illumination function at each vertex using the Phong equation without the specular-diffuse component. Then it reconstructs the illumination function at the interior points of the polygon using the interpolation method below.

The Gouraud reconstruction is performed in screen space during rasterization and uses a scanline rasterization. Consider a point P, belonging to a scanline MN, as illustrated in Figure 14.10(a). The value of the shading function f in P, $f(P)$, is calculated as follows:

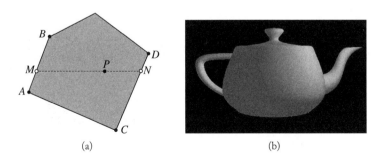

(a)   (b)

Figure 14.10. (a) Gouraud reconstruction; (b) image shaded with the Gouraud method.

❏ Given a scanline MN, determine the side of the polygon containing the point M (side AB in the illustration), and the side containing the point N (side CD in the illustration);

❏ We linearly interpolate $f(A)$ and $f(B)$ to obtain the value of f at the point M;

❏ We linearly interpolate $f(C)$ and $f(D)$ to obtain the value of f at the point N;

❏ Finally, we linearly interpolate $f(M)$ and $f(N)$ to obtain the sought value $f(P)$ at the point P on the scanline.

As we can see, the reconstruction calculates the illumination function at the vertices and interpolates those values. Now, observe the following: if the surface has a high degree of specularity and an eventual highlight is located in the interior of a polygon, such a highlight will not be reconstructed. For this reason, the Gouraud method is known as an illumination method for diffuse reflection. Figure 14.10(b) illustrates the teapot shaded with the Gouraud method.

The problem with the Gouraud reconstruction method is that it does not have unicity; that is, depending on the orientation of the polygon on the screen, the interpolated value at point P can change. However, we leave for an exercise the following fact: if the polygon is a triangle, the Gouraud reconstruction method coincides with the interpolation method using barycentric coordinates—therefore, we have unicity.

14.4.3 Phong Shading

After developing the illumination equation, Phong noticed the need for a more precise reconstruction method that takes into account specular-diffuse reflection. The method he developed is known as *Phong shading* [Phong 75].

As in Gouraud, the Phong method assumes the surface is represented using polyhedral B-rep, and it uses the same interpolation methodology for reconstructing the illumination in each polygon in screen space (by scanline rasterization). However, the Phong

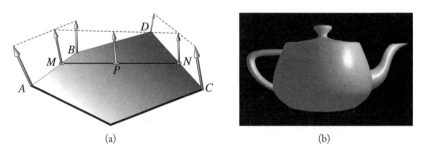

(a)   (b)

Figure 14.11. (a) Phong reconstruction; (b) image shaded with the Phong method.

method does not calculate the illumination function at the vertices followed by a reconstruction inside the polygon. Instead, it reconstructs the normal vector field in the interior
of the polygon and then performs the shading function calculation at each point using the
reconstructed normal vector. This reconstruction is illustrated in Figure 14.11(a). Figure 14.11(b) shows the teapot shaded using the Phong method.

14.5   Ray Tracing

Up to this point, we have not worried about the transmitted component of radiant energy;
the methods for calculating the shading function that we have seen thus far do not treat
the transmitted component.

The Phong model was developed for reflected energy, but it can be extended to include
transmitted energy, if we incorporate an additional term in the Phong equation. The transmitted radiant energy is modeled similarly to how we modeled specular-diffuse reflection.
In other words, we have the direction of ideal transmission $\mathbf{v}_t$, given by the Snell-Descartes
law, with a scattering of the transmitted ray around this direction (specular-diffuse transmission). By introducing a *transmission coefficient*, k_t, the transmitted energy along the
direction ω_t is given by

$$L_t = L_i \, k_t \langle \mathbf{v}_t, \omega_t \rangle^q.$$

As in the case of the specular-diffuse reflection, the exponent q controls the scattering degree of the transmitted rays. This model was introduced in the literature by Turner
Whitted in [Whitted 80], who also introduced the ray tracing method to calculate illumination, taking into account the transmitted component. We previously saw the use of
the ray tracing method to determine the visibility of a surface: here we will see how this
method can be extended to calculate illumination.

Ray tracing is the real incarnation of the methods in geometric optics for solving the
illumination problem. In a real scene, light sources emit luminous rays that are either reflected or transmitted by objects in the scene. Some of those rays reach the observer's eye,
making the visualization of the scene possible from the observer's point of view. Computationally, this process is very difficult to model:[3] thousands of the traced rays would
be wasted as they do not reach the observer's eye, but we would know which rays were
unneeded only after performing the calculations. This fact is illustrated in Figure 14.12,
where some rays reach the camera and others do not.

A simple method for tracing only the rays that arrive at the observer, is to follow the
inverse path: trace the rays *from* the observer *into* the scene and follow their path with
the Snell-Descartes laws of reflection and refraction. This method traces only the rays of
interest.

---

[3]Because our processors do not yet work at the speed of light. Maybe with quantum computing...

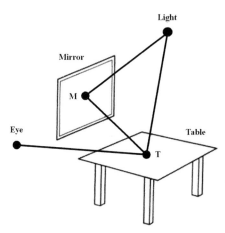

Figure 14.12. Rays traced in a scene starting from the observer. The ray tracer knows that if point T can "see" the light, it will contribute to its illumination. However, the ray tracer is completely unaware of light arriving by the indirect light route of the ray from point M to point T. This ray does not reach the observer's eye.

14.5.1   Illumination and Ray Tracing

The ray shooting problem is this: given a finite set of objects $O_1, O_2, \ldots, O_n$ of the Euclidean space, a unit vector $\mathbf{v}$, and a ray $r = (0, \mathbf{v})$, determine, if it exists, an intersection point, between r and the scene objects, that is closest to the origin 0 of the ray.

In Chapter 13 we saw how a solution to this problem can be used to determine the visibility of objects in the scene. In this case, for each pixel, we solve the ray shooting problem for the viewing ray of the pixel, i.e., the ray whose origin is the observer's eye (optical center of the virtual camera) pointing toward the direction of the pixel center. This ray is called the *viewing ray*, or *primary ray*.

The use of ray tracing for the illumination calculation is more complex. When determining the visible point of the primary ray, we use the local illumination equation to calculate the radiance value of the pixel. Assuming the surface is made of a specular material, we have a reflected ray. If it is also made of a translucent material (e.g., glass), we have a transmitted ray. We therefore solve the ray shooting problem for each ray, calculating the illumination at the intersection points and continuing successively. The final radiance of the pixel is the sum of all of the accumulated radiances.

To illustrate this procedure, consider the scene in Figure 14.13, which shows four objects: O_1 and O_3 (transparent), O_2 and O_4 (opaque). We assume all of them are made of materials with high degrees of specularity. The primary ray, when finding object O_1, is reflected in ray r_1 and transmitted in ray t_1; ray r_1 then reaches object O_3, producing reflected ray r_2 and transmitted ray t_2; ray t_1 reaches surface O_2, which is opaque, producing reflected ray r_3.

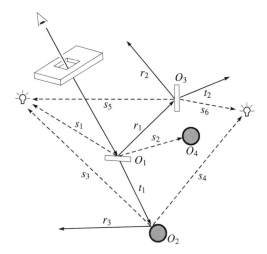

Figure 14.13. Path of a ray in a scene.

Besides determining the visibility and illumination calculations, the ray tracing method allows us to generate shadows in a very simple way. We trace secondary rays, called *shadow rays*, from each intersection point P toward the light sources. We solve the ray shooting problem for each shadow ray to determine the visibility of the light source in relation to the point: if the distance between point P and the light source is larger than the distance between P and the intersection point, the light source is obstructed by the surface of intersection. If the light source is obstructed by some object in the scene, P is in a shadow region and affected only by ambient light.

The shadow ray should take into account translucent surfaces.[4] In Figure 14.13, the shadow rays are shown in dashed lines (note that shadow ray s_2 is obstructed by object O_4). In the above explanation, we assume we have point light sources; therefore the shadow does not have a penumbra region (only umbra). For non–point light sources, the calculation is more complicated because we must sample the light source to calculate the umbra and penumbra regions.

From a data structure point of view, this procedure generates a tree. Each intersection point is a tree node with two children corresponding to the transmitted and reflected rays. Figure 14.14 (left) shows the corresponding tree of the scene in Figure 14.13; the illustration on the right shows the same tree with the shadow rays. Therefore, from the programming point of view the ray tracing is implemented as a search algorithm in this tree where, at each node, a routine is called to solve both the ray shooting and the illumination calculation problems.

---

[4]To be strictly accurate, we should also have taken into account light sources visible from the reflections of the shadow rays; however this fact is generally neglected in implementations.

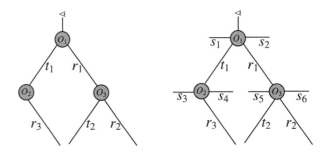

Figure 14.14. Ray tracing tree of the scene in Figure 14.13.

The pixel color is calculated by adding the radiance calculated at each intersection point (tree node). At each of these points we have

$$I = I_{\text{local}} + I_{\text{reflected}} + I_{\text{transmitted}}, \tag{14.10}$$

where I_{local} is the radiance directly calculated at the intersection point using the Phong model. The terms $I_{\text{reflected}}$ and $I_{\text{transmitted}}$ represent the radiance arriving at the intersection point due to the light reflection of another object, or to the transmission in the case that the object is translucent.

In short, the ray tracing method uses the *local* illumination equation; however, it performs a *global* calculation of the illumination, integrating the equation along the paths traversed by the primary ray. For this reason, it is common to consider ray tracing a global illumination method. However, a *complete* global illumination method would take into account an illumination equation that really represents the exchange of energy among the elements in the scene (mainly the radiosity originating from the light reflection of the several objects); in the local equation, this is roughly approximated by ambient light.

From the above exposition, we can see that the surfaces should have a material with a high degree of specularity for the ray tracing method to work satisfactorily.

14.5.2 Pseudocode

We divide the pseudocode of the ray tracing algorithm into a main program, Main, and two functions: Trace and Shade. The Main program simply reads each pixel (x, y), calls the function Trace, which returns the value of the pixel color, and writes this value in the image file. The pseudocode of the main program is given below:

Main{
   for each pixel (x, y) **do**
      $ray = (O, (x, y))$;
      radiance(x, y) = Trace(ray);
      write(radiance(x, y));

   end for

}

   The function Trace has a ray as an argument. It solves the ray shooting problem for the ray passed as a parameter and calls the function Shade to perform the radiance calculation at the intersection point. The pseudocode of the function Trace is given below:

Trace(ray){
   Calculate the intersection P between the ray and the closest object;
   Determine the type of material of the intersected object;
   Return = Shade(ray, P, material);

}

   The function Shade receives the ray, the intersection point P, and the material of the surface. It determines the radiance at P, originating from every light source, and then recursively calculates the contributions of the reflected and transmitted rays to the radiance at point P.

Shade(ray, P, material){
   rad = reflected ambient Radiance;
   for each light source L **do**
      if L visible **then**
         Calculate radiance, rad, at P;
         rad = rad + Radiance at P;
      end if
   end for
   if material == specular **then**
      ray = reflected Ray;
      rad = rad + Trace(ray);
   end if
   if material == translucent **then**
      ray = refracted Ray;
      rad = rad + Trace(ray);
   end if
   Return(rad);

}

   Note that we did not establish a stop criterion for the recursion. A simple criterion consists of establishing a maximum number of reflections of a ray (that is, the maximum number of calls to the Trace routine by the Shade routine). We will return to this subject further on.

14.6   Ray Tracing Acceleration

Looking at the pseudocode of the previous section, it is really impressive that such a simple algorithm gives such good results. In fact, the ray tracing algorithm performs both the visibility and the illumination calculations and even produces shadows.

However, there is no free lunch: the algorithm is extremely expensive from a computational point of view. The solution to the ray shooting problem, if executed in brute force mode (solving for both primary and secondary rays of each pixel), can make the algorithm unfeasible. For this reason, there is a great deal of research devoted to finding more efficient methods to solve the ray shooting problem and more effective solutions for the illumination calculation. (Notice we are performing point sampling for the illumination and shadow calculations).

Methods aimed at increasing the efficiency of solving the ray shooting problem are called *acceleration methods*. These methods try to either optimize the intersection calculation or trace a smaller number of rays. In the following sections we will study some techniques in each category.

14.6.1   Optimizing the Intersection Calculation

Existing optimization techniques use hierarchical structures associated with the scene objects. The underlying idea is that the ray shooting problem is, in fact, a searching problem, and its computational complexity is of type $O(n)$, where n is the number of objects in the scene. Using hierarchical structures, the goal is to reduce that complexity to $\log(n)$. In this section we will describe the following commonly used techniques: bounding objects (individual and in hierarchies), spatial subdivision (by voxels and octrees), and other subdivision methods.

Bounding objects. This technique aims at reducing the processing time of the intersection routine between a ray and an object. Similar to what we did to optimize clipping algorithms when the geometry of an object is very complex, we can use bounding objects to efficiently decide if a ray is not intersecting a certain object. Given a scene object O, let us consider another object O_1 that satisfies two conditions:

1. $O \subset O_1$;

2. O_1 allows for a more efficient intersection test than the original object O.

For the first condition, object O_1 is called the *bounding object* and O the *bounded object*. (The bounded object can be disconnected; that is, we can have several scene objects associated to a single bounding object.) Once the bounding object is defined, we first test whether the ray intersects that object and only later (if the result is positive), do we test for the intersection between the ray and the bounded object.

For the second condition, the geometry of the bounding objects should allow for an efficient intersection test. Examples of bounding objects include spheres and parallelepipeds (especially if the faces of the parallelepiped are parallel to the coordinate planes). If a ray

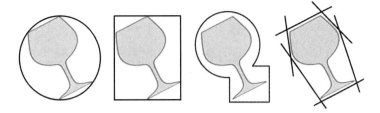

Figure 14.15. Bounding objects.

does not intersect the bounding object, it certainly does not intersect the bounded object. However, if a ray intersects the bounding object, nothing can be said regarding its intersection with the bounded object. In this way, to minimize the occurrence of false-positive tests (i.e., cases in which the ray intersects the bounding object but does not intersect the bounded object), the ideal is to have the geometry of those bounding objects defined by the geometry of the bounded object. For example, if the bounded object is long and thin, a sphere would not be a good bounding object because it would give many false-positive results; in this case, a parallelepiped or other convex polyhedron would be a better solution. Figure 14.15 illustrates the geometry of bounding objects with some examples.

Hierarchies of bounding objects. Rather than using a single bounding object to reduce the number of intersection between rays and objects, a more efficient method consists of using a hierarchy of wrapping volumes of bounding objects (see Chapter 9). In this case, we include a scene object (which can be the whole scene) into a finite family of bounding objects composing a hierarchical structure of wrapping volumes.

The intersection test is processed in the following way: initially we test the intersection between the ray and the root of the hierarchy; if there is no intersection, the ray does not intersect the bounding object, otherwise, we recursively repeat the test in the nodes of the hierarchy. Figure 14.16 illustrates the process using a 2D hierarchy where the objects in the hierarchy are rectangles. (Notice that use of the hierarchy avoids the intersection between *ray 1* with nine objects (0,1,2,3,4,5,6,9, and 0), also avoiding the intersection of its reflected

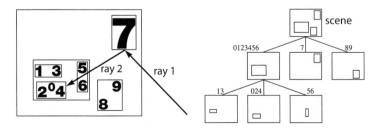

Figure 14.16. Hierarchy of bounding objects.

ray, *ray 2*, with four objects (1,3,8, and 9.)A list of properties that should be satisfied for a hierarchy of bounding objects is given in [Kay and Kajiya 86]:

1. The subtrees of the hierarchy should contain objects that are closer to the bounded object; the deeper the subtree is, the larger this proximity should be;

2. The volume of the object in each node should be the smallest possible one;

3. The sum of the volumes of all bounding objects of the hierarchy should be the smallest possible one;

4. Greater attention should be given to the nodes of the hierarchy near the root; eliminating those nodes will avoid intersections with more complex objects than would eliminating nodes with larger depth;

5. Time spent building the hierarchy should be significantly less than time gained from using the hierarchy.

We left as an as exercise the justification of each of the above statements.

Spatial subdivision by voxels. This technique partitions space into a uniform grid of volume elements (voxels), i.e., parallelepipeds whose faces are parallel to the coordinate planes. For each voxel V, we build a list of objects intersecting this V. For the method to be efficient, we should achieve a balance between having few objects in the list of each voxel and not having voxels with very small volumes (i.e., the ideal is to have few voxels with few scene objects in each voxel).

In this method, the intersection calculation between the ray and the objects is only accomplished for the voxels that are intersected by the ray. The ray traced along the voxels (voxel traversal) continues until the first intersection point is found.

As the spatial subdivision is uniform, there are extremely efficient methods of tracing the ray along the voxels grid (this process is a volumetric rasterization of a ray). In fact, we can use the extension of DDA-Bresenham (straight line rasterization on the plane) for a 3D grid. This extension is called the 3D DDA-Bresenham algorithm. Figure 14.17 shows the same scene as Figure 14.16, now processed with one spatial subdivision by voxels. In

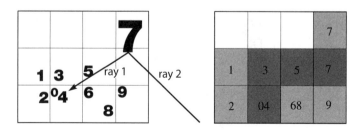

Figure 14.17. Spatial subdivision with 12 voxels.

the illustration to the right, we show the objects in each voxel and the voxels that are "rasterized" along the path of the ray. Notice that ray 1 calculates only one intersection (with object 7) in the first-traversed voxel; ray 2 calculates an intersection with object 5 in the second-traversed voxel, with object 3 in the third voxel, and with objects 0 and 4 in the fourth voxel (the intersection with object 2 is not accomplished because the algorithm stops before any calculation is performed).

Spatial subdivision by octree. Spatial subdivision by voxels is uniform and nonhierarchical, so it is difficult to reach a balance between not having tiny voxels and at the same time avoiding voxels intersecting many scene objects. To achieve a good balance of these conditions, we adapt a hierarchy of partitions in voxels of variable sizes: wherever there are few objects, the voxels will have larger volume, and in regions with many objects, the voxels will have smaller volume.

The octree data structure supports this type of spatial subdivision. Each voxel is recursively divided into eight subvoxels, forming a tree in each node (voxel) having eight subnodes (subvoxels), as illustrated in Figure 14.18(a).

Figure 14.18(b) shows the scene of Figure 14.17, now using an octree (actually a quadtree, as the illustration is 2D). Notice ray 1 is only intersecting object 7, and ray 2 is only intersecting objects 5 and 4, which represents a substantial gain in relation to the uniform voxel grid method. In this example, something interesting is happening: object 4 intersects three voxels of the octree, and those three voxels are part of the rasterization of ray2. How can we avoid that the intersection between this ray and object 4 be computed for each of the voxels (i.e., three times)? We left this question as an exercise.

We can see that the method of spatial subdivision by voxels satisfies the conditions we previously looked for: on average, the number of objects per voxel is constant (not considering the empty voxels). What is more, by controlling the number of subdivisions, we can regulate the size of the voxels and the number of objects per voxel. However, this method presents an additional problem in relation to the uniform subdivision method: we need an efficient algorithm to perform the rasterization of the ray in the octree space, where the size of the voxels is variable (as we saw, in the case of uniform subdivision, we can use the 3D DDA-Bresenham algorithm).

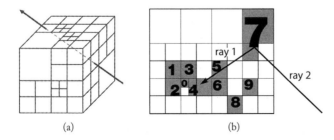

(a)   (b)

Figure 14.18. Recursive subdivision of the space with the octree structure.

Other spatial subdivision methods. Whenever we have a hierarchical structure of space partitions, we can use it to develop a ray tracing acceleration method. Examples of other published structures include subdivision using kd-tree (k-dimensional trees) and also hierarchies defined by partitioning trees (BSP-tree). Note that the construction method of a BSP-tree to accelerate the ray tracing algorithm is *not* the same method presented in Chapter 10. First, the scene objects cannot be polyhedral; second, the goal here, in using the structure, is not to solve the visibility problem (which is already accomplished by shooting rays), but instead, to accelerate the process. Another important point is that we should traverse the BSP-tree in an order inverse to the one used in the visibility BSP-tree (i.e., front to back).

It is difficult to say which of these methods is superior to the others. The performance of each subdivision method strongly depends on the scene to be visualized. When choosing an acceleration method based on spatial subdivision, the following characteristics will influence its efficiency:

❏ The pre-processing cost to create the structure;

❏ The cost of calculating the ray propagation along the cells of the structure;

❏ The dimensions of the decomposition cells (we should avoid very small cells);

❏ The intersections of the cells (we should have cells that intersect a large number of objects);

❏ In the case of dynamic scenes, the cost of updating the spatial subdivision structure.

14.6.2   Tracing Fewer Rays

In addition to optimizing intersection calculations, the other acceleration method for increasing the efficiency of ray tracing is to trace a smaller number of rays. There are several simple approaches to decreasing the number of secondary rays shot at each pixel. In this section, we will discuss three methods: adaptive depth, z-buffer visibility, and single path.

Adaptive depth. The adaptive depth method consists of performing an adaptive control of the ray tracing tree depth. We usually establish a maximum tree depth as a criterion to stop the recursion of the algorithm (i.e., we limit the number of reflections of the primary rays). However, it is more efficient to perform an adaptive control, which avoids unnecessary processing performed to reach the maximum pre-established level.

A simple method consists of establishing a threshold L so that when the illumination contribution of a particular node for the pixel is smaller than L, we stop processing the ray reflections. For example, if $K_r = 0.5$ for a first intersection hit, the contribution would be 0.5 of the calculated illumination; in a second interaction by reflection, it would be $0.5^2 = 0.25$; in a third interaction it would be $0.25 \times 0.5 = 0.125$; in a fourth it would be 0.0625; and so on.

We can do the same for the transmitted ray. The falloff here is even faster because there is a decay of the luminance when the ray passes through a translucent surface. Notice we can still use, for both reflections and transmissions, the distance traveled by the ray as an attenuation factor.

Z-buffer visibility. This method consists of replacing the initial visibility calculation using the primary ray with the z-buffer algorithm. First, each scene object is encoded with a single color. Then we execute the z-buffer algorithm. The final buffer contains the code (color) of the visible object (actually, we can place in this buffer a pointer to the visible object). Once this pre-processing is done, we no longer need to trace the primary rays, but only the secondary ones.

Of course, instead of the z-buffer, we can use any other visibility algorithm that is more efficient than the visibility method by ray tracing.

Single path. A great deal of the computational effort of the ray tracing algorithm is expended in solving the ray shooting problem for the secondary rays, but many of those rays contribute little to the final radiance of the pixel.

We can avoid this problem by following only a single path of secondary rays in the ray tracing tree. The choice of this single path in each tree node should be made randomly.

14.7   Sampling and Ray Tracing

So far we have presented the ray tracing method entirely in terms of the ray shooting problem, which uses point sampling at each intersection point. (There are several inherent problems in this type of sampling, which we will address in Chapter 15.)

Instead of point sampling we could use supersampling, which is much simpler but computationally very costly. For example, if we subdivide each pixel in $n \times m$ subpixels, then we should solve, for each pixel, the ray shooting problem for $n \times m$ primary rays and their secondary rays: a substantial increase in complexity.

We can lessen this problem by using an adaptive supersampling approach. The idea is to have supersampling occur only in regions of high frequencies, thus reducing the number

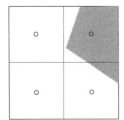 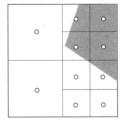 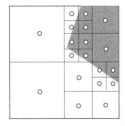

Figure 14.19. Supersampling by adaptive subdivision.

of additional rays at each pixel. One such approach, introduced in [Whitted 80], is to subdivide each pixel into four subpixels, calculate the value of the illumination at each of the subpixels, and then compare whether those values are too discrepant to continue the quadruple subdivision process in the pixels representing nonnull values. This method is illustrated in Figure 14.19.

14.7.1   Tracing Beams of Rays

Several methods have been developed to minimize sampling problems in the ray tracing method and to improve the quality of the image. These methods trace beams of rays instead of a single ray. Well known methods include *beam tracing*, *cone tracing*, and *distributed ray tracing*.

Beam tracing and cone tracing. Beam tracing and cone tracing replace the collection of individual rays with *volumetric rays*: the beam tracing approximates the rays by a truncated pyramid; cone tracing approximates them with conical rays (Figure 14.20).

In addition to minimizing aliasing problems and improving the quality of the generated image as a whole, beam tracing and cone tracing also accelerate the execution of the ray tracing algorithm. (In fact, these techniques are often classified as ray tracing acceleration techniques.) In general terms, this is because the surface approximating the beam of rays constitutes an intrinsic bounding object to the ray itself, so using these methods allows us to process several objects each time we shoot a beam. Of course, these methods can also be combined with the acceleration methods previously studied.

Some of these methods of tracing beams are difficult to implement and impose constraints to the geometry of the scene objects (e.g., in the original beam tracing algorithm the objects should be polygons). A very interesting description of a beam tracing algorithm can be seen in [Ghazanfarpour and Hasenfratz 98], where the Warnock algorithm is used in the virtual screen space to create a hierarchy of beams, thereby increasing the efficiency of the method. What is more, the scene can be constituted by any convex polyhedral object.

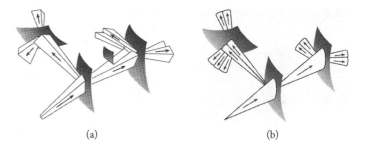

(a)   (b)

Figure 14.20. (a) Beam tracing uses a truncated pyramid; (b) cone tracing uses conical rays.

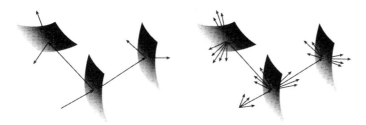

Figure 14.21. Distributed ray tracing.

Distributed ray tracing. The distributed ray tracing method combines the supersampling method with the idea of tracing beams of rays, without replacing each beam by a volume. The rays of the beam are shot using stochastic sampling: we perform first a subpixel supersampling and then a random perturbation of each subpixel, and finally we shoot a ray at each perturbed subpixel. This process is illustrated in Figure 14.21. We will discuss the process of stochastic sampling in more detail in Section 15.3.4.

To use this method, we must decide which probability distribution to use for the perturbation in the subpixels. A uniform distribution was used in the original implementation of the algorithm. Certainly, a Gaussian distribution can be used, but studies indicate that a Poisson Disk distribution is more suitable.

The distributed ray tracing algorithm is very efficient for aliasing elimination in the sampling process, including temporal aliasing (whose anti-aliasing is known as motion blur). Figure 14.22 illustrates the effect of motion blur in an image of a billiard ball in motion.

Figure 14.22. Motion blur generated using distributed ray tracing. (*[Cook et al. 84] © 1984 Association for Computing Machinery, Inc. Reprinted by permission.* See Color Plate XXI.)

14.8   Comments and References

The simplest local illumination model was introduced by Henri Gouraud [Gouraud 71]. Subsequently, Bui Tui Phong extended the Gouraud model by including an empirically calculated specular component [Phong 75].   More sophisticated models were proposed by Jim Blinn [Blinn 77] and by Rob Cook and Ken Torrance [Cook and Torrance 81]. The pioneering work using ray shooting to solve the illumination problem was presented in [Whitted 80]. The beam tracing algorithm was introduced in [Heckbert and Hanrahan 84]. The cone tracing algorithm was published in [Amanatides 84]. The distributed ray tracing algorithm was published in [Cook et al. 84]. A vast amount of information on the techniques and methods discussed in this chapter can be found in the webpage compiled by Eric Haines [Haines 03].

14.8.1   Additional Topics

Some of the topics in this chapter that deserve a more in-depth study include the classic illumination models we mentioned in the text (e.g., Cook-Torrance), ray tracing acceleration methods, and simulation of illumination effects (e.g., dispersal, diffraction, etc.). We left completely out of the chapter the illumination problem of volumetric objects, a topic of great importance in scientific visualization. We also did not cover the case of illumination with participating media. An area that has recently gained importance is estimating the ambient illumination based on real pictures (image-based illumination estimation). This area has great importance in virtual reality and in virtual scenes combining synthetic and real world objects (augmented reality).

Exercises

1.   Describe in detail the illumination equation and the local illumination model for the transmitted energy, in a similar way to what we did for the reflected energy. What is the BRTF of this model?

2.   If ω_i is the unit vector along the direction of incidence, and $\mathbf{n}$ the unit normal vector at a point P on the surface, show that the specular reflection unit vector $\mathbf{v}_s$ is given by

$$\mathbf{v}_s = \omega_i - 2\langle \omega_i, \mathbf{n} \rangle \mathbf{n},$$

3.   Blinn introduced an illumination model similar to Phong's, by replacing the specular reflection vector $\mathbf{v}_s$ by the 'half' vector $H = (\omega_i + V)/2$, between the vectors of incidence ω_i and the direction of reflection ω_r. Discuss the advantages and disadvantages of his model in relation to Phong's.

4.   Does the Phong equation change if the scene has point light sources? What are the changes in the illumination calculation in this case?

5.   Describe a method to calculate the illumination function in our local illumination model when the scene has spotlight sources.

6. The local illumination model does not calculate shadow regions. You can apply some tricks to place shadow in a scene where the illumination function is calculated using a local model.

   (a) How do you avoid the appearance of shadow by appropriately positioning the camera in the scene?

   (b) How do you use projective transformations to produce shadows of simple objects. What are the limitations of this method?

   (c) How do you use clipping operations to produce shadows? What are the difficulties of this method?

7. Show that the Gouraud interpolation method is dependent on the orientation of the polygon. What problems can this bring for the generated images? (Hint: consider rotational motions.)

8. Show that the Gouraud interpolation method on a triangle coincides with the interpolation method using barycentric coordinates.

9. Routines for intersection calculation are the essence of a ray tracing algorithm.

   (a) Determine the intersection point between a ray and a sphere of radius r and center at the origin.

   (b) Calculate the intersection point between a ray and a cylinder of equation $x^2 + y^2 = 1$.

   (c) Determine the intersection point between a ray and a triangle. (Hint: use barycentric coordinates.)

   (d) Determine the intersection point between a ray and an arbitrary polygon.

10. It is important to determine the intersection between a ray and a parallelepiped, as parallelepipeds are often used as bounding objects. Write a procedure to determine if a ray intersects a parallelepiped.

11. Based on the image in Figure 14.23, describe the pipeline of a 3D cartoon shading system.

Figure 14.23. Olaf rendered using cartoon shading. (Reprinted from *[Lake et al. 00]* *by permission of Intel Corporation. ©2000 Association for Computing Machinery, Inc. Reprinted by permission.* See Color Plate XXII.)

12. In the ray tracing acceleration algorithm using spatial subdivision by voxels (i.e., uniform subdivision), the same object can intersect several voxels. Write a procedure to prevent calculating the intersection between the ray and object more than once.

13. Let $\mathbf{n}$ be the normal vector to the surface at a point x, ω_i the incidence vector x, and $\mathbf{v}_t$ the transmission direction. From the Snell-Descartes law, we have $k_1(\omega_i \wedge \mathbf{n}) = k_2(\mathbf{v}_t \wedge \mathbf{n})$, where k_1 and k_2 are the indices of refraction. Show that

$$\mathbf{v}_t = \frac{k_1}{k_2}\left(\left(\sqrt{\langle \mathbf{n}, \omega_i\rangle^2 + \left(\frac{k_2}{k_1}\right)^2 - 1} - \langle \mathbf{n}, \omega_i\rangle\right)\mathbf{n} + \omega_i\right).$$

14. Describe an acceleration method by spatial subdivision using a *hierarchy of uniform grids*. Compare the advantages and disadvantages of this method with the grid method and with the subdivision method by octree.

15. From the *principle of Helmholtz*, the BRDF function is symmetrical; that is, $f_r(x, \omega_i, \omega_r) = f_r(x, \omega_r, \omega_i)$.

    (a) Explain the physical meaning of this symmetry.

    (b) Show that the Phong BRDF model does not satisfy the principle of Helmholtz.

    (c) What is the meaning of the previous item?

16. In some perfect diffuser, the decay of the reflected radiance with the angle of incidence θ_i is more accentuated than in the one from Lambert's law. How could you modify the Phong equation to take this fact into account?

17. Define the concept of shadows in a scene, including *umbra* and *penumbra*. Classify the types of shadow regions produced by each of the types of light sources studied in this chapter.

15 | Rasterization

The rasterization operation, which generates a virtual image of a scene, has two elements: a strategy for traversing the pixels on the virtual screen and a method of calculating the color value at each pixel (i.e., to perform the sampling of the shading function). Starting from the illumination calculation of the scene, we have the shading function that associates, to each point P on the virtual screen, the value of the luminous energy (color) arriving at that point along the direction of the center of projection (the observer's position).

15.1 Sampling

A pixel in the virtual screen covers a finite area determined by the screen dimensions and the image resolution. Usually several scene objects are projected in this area. These objects emit light (color) and different shades. The sampling process consists of choosing the most appropriate color to represent the scene at each pixel.

Reconstruction is closely related to sampling. The sampled image provides a discrete representation of the image. To visualize this image, we use a reconstruction method (to reconstruct the image either on a monitor or in a printer). Figure 15.1 illustrates the process. There are essentially two types of sampling: point sampling (shown in Figure 15.1) and area sampling.

15.2 Point Sampling

As we have mentioned previously, point sampling is a natural, simple method of representing functions: we choose a set of points x_i in the function domain, and each sample is the value $f(x_i)$ of the function at each point. When we use point sampling, the reconstruction is reduced to a method of sample interpolation.

Without taking proper care, the sampling and reconstruction process can result in rough errors. This is illustrated in Figure 15.2, where we perform point sampling of a sinusoid and use linear interpolation to reconstruct it. As the frequency of the sinusoid increases, the reconstructed signal becomes very different from the original one. Observe

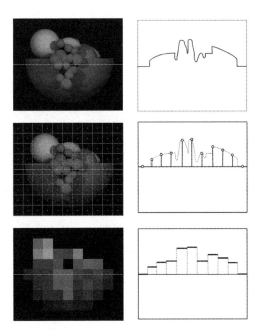

Figure 15.1. Sampling and reconstruction of a scene. Top left: the virtual screen with a projected scene, highlighting one scanline. Top right: the function graph of the associated shading on the scanline. Middle left: the pixels and the points at the center of the pixel where we calculate the shading function to obtain the color value of each pixel (point sampling). Middle right: the scanline samples. Bottom left: the reconstructed image. Bottom right: the scanline reconstruction. (*Left figures: ©Rosalee Wolfe. Used with permission.* See Color Plate XXIII.)

that the high frequencies disappear on the final reconstructed image. This is a general rule: very high frequencies disappear and are reconstructed as low frequencies, which do not exist on the original image (called spurious frequencies).

Note that in this example the reconstruction error comes from the fact that we used few samples in relation to the frequency of the signal. This leads us to two interesting questions: what is the ideal number of samples to be considered, once the signal is known?

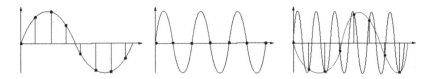

Figure 15.2. Point sampling and reconstruction.

Figure 15.3. The triangle disappears from the image, depending on its position at the pixel. Point sampling is at the center of the pixel.

Is there any interpolation method, associated to point sampling, that would allow an exact reconstruction of the original signal? We need Fourier theory to fully answer these queries, which is outside the scope of this book. For additional information, consult [Gomes and Velho 02].

Another point sampling problem is related to the presence of discontinuities in the signal: what is the correct sample value at a pixel where discontinuity exists? This is related to the previous problem: a discontinuity point introduces very high frequencies to the signal, making the use of point sampling prohibitive, under penalty of generating large errors in reconstruction.

To better illustrate this problem, consider a small triangle moving on the screen whose area is smaller than the area of the virtual pixel. As shown in Figure 15.3, this triangle appears and disappears during its motion, depending on its position at each pixel traversing its motion.

Sampling and reconstruction errors are generically known as *aliasing*. The various methods used to correct or minimize aliasing problems are called *anti-aliasing*. Separating sampling and reconstruction errors and thoroughly defining aliasing requires a detailed study of signal processing theory, which is beyond the scope of this book. For additional information, consult [Gomes and Velho 02] or [Gomes and Velho 97].

Given these problems we need more robust methods for calculating the value of the attribute function in a cell of a matrix representation. Area sampling is one of these methods.

15.3   Area Sampling

Area sampling consists of considering the virtual pixel as having a finite area and by taking, as a sample, a weighted average of the shading function at the pixel. The simplest way of formally understanding this method is to use concepts of linear algebra.

A unidimensional *sampling function* is a function $\phi\colon \mathbb{R} \to \mathbb{R}$, such that $\phi(0) = 1$, $\int_{\mathbb{R}} \phi(t)dt = 1$, and the translated functions $\phi_k(t) = \phi(t - k)$ form a linearly independent set.[1]

---

[1] Actually, linear independence is not necessary but will result in redundancy.

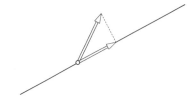

Figure 15.4. Projection and reconstruction.

This concept easily extends to multivariate sampling functions $\phi\colon \mathbb{R}^n \to \mathbb{R}$. In our case, we are particularly interested in the case of two variables, i.e., $n = 2$, which is the case of sampling the shading function to generate the virtual image.

Given a function f in a space of functions, the *representation* of f is given by the sequence $(f_k)_{k \in \mathbb{Z}}$, where

$$f_k = \int_{\mathbb{R}} f(t)\phi(t - k)dt.$$

A geometric interpretation of the representation is simple to obtain. Let us assume the set $\{\phi(t - k)\}$ forms an orthonormal basis. In this case we have

$$f_k = \int_{\mathbb{R}} f(t)\phi(t - k)dt = \langle f, \phi(t - k) \rangle, \tag{15.1}$$

and we can write

$$f(t) = \sum_{k} f_k \phi(t - k). \tag{15.2}$$

This equation provides the projection of the function f in the space generated by the functions $\phi(t - k)$ (see Figure 15.4).

The reconstruction of the function, from the representation sequence (f_k), is given by Equation (15.2). The reconstruction is exact if the function f belongs to the space generated by the functions $\phi(t - k)$.

The representation of f by the above sequence (f_k) is generically called *area sampling*. From the definition of f_k in Equation (15.1), each sample is a weighted average of f, using as weight the function ϕ, properly translated.

15.3.1   Haar Sampling

The simplest method of area sampling is *Haar sampling*. Consider a uniform partition of an interval $[a, b]$, where the length of each interval is 1. We define the kernel representation

$$\phi(x) = \begin{cases} 1 & \text{if } x \in (0, 1]; \\ 0 & \text{otherwise.} \end{cases}$$

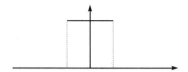

Figure 15.5. Haar basis.

The graph of ϕ is shown in Figure 15.5. The family $\phi(x-k)$ forms an orthonormal basis. The representation of a function f is given by the sequence (f_k), where

$$f_k = \langle f, \phi(x-k) \rangle = \int_k^{k+1} f(x)dx.$$

If the length of the interval is s instead of 1, we perform a change of scale of ϕ, obtaining

$$\phi^s(x) = \frac{1}{s}\phi(\frac{1}{s}).$$

In this case, the elements of the representation sequence of f are calculated by

$$f_k = \langle f, \phi^s(x-ks) \rangle = \frac{1}{s}\int_{ks}^{(k+1)s} f(x)dx.$$

The samples of f provide the average of function f in the interval of the partition. If the signal has large variations in the interval, it is more reasonable to take this average as a sampling instead of performing a point sampling. This is equivalent to using a low-pass filter in the function to be sampled, followed by point sampling the filtered function, which is, in this case, constant at the pixel. For this reason, several area sampling techniques are known as *pre-filtering techniques*.

This method is called *Haar representation*, which is a representation in an orthonormal basis. The function reconstruction is the one using the constant kernel discussed in Section 6.3.1 (see also Figure 15.6); therefore, the reconstruction is not exact.

As we saw in Chapter 6, the Haar representation extends to $\mathbb{R}^2$, thus providing a representation of an image. This representation is known as *unweighted area sampling*. The term *unweighted* comes from the fact that the weight function (i.e., the Haar function) is constant; therefore the sampling provides only the weighted average of the image function at the pixel.

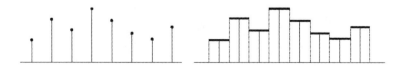

Figure 15.6. Haar representation and reconstruction.

Figure 15.7. Point and Haar representations.

Consider the rasterization of a subset in the plane. In this case, the color attribute is constant and the Haar representation calculates the percentage of the pixel area occupied by the object. Figure 15.7 shows the reconstruction of a gray color triangle, using point and Haar samplings.

In both cases, the reconstruction is performed with a constant kernel; however, in point sampling, the reconstruction coefficients are constant. Haar sampling results in a reconstruction that minimizes the staircase effect ("jaggies") at the boundaries of the high-frequency regions.

15.3.2 Weighted Area Sampling

Area sampling using Haar basis still presents problems. For example, consider a small triangle moving on the screen whose area is smaller than that of a pixel, as shown in Figure 15.8. The sample value at each pixel is the same, independent on the position of the triangle at the pixel. This problem occurs because the basis of the Haar representation is constant at the pixel. The problem can be avoided by using a nonconstant representation function. One could vary the intensity of the sample according to the distance between the barycenter of the triangle and the center of the pixel. A possibility is to use the triangular kernel (see Figure 15.9) defined by

$$h_1(x) = \begin{cases} 1 - |x| & \text{if } |x| \leq 1; \\ 0 & \text{if } |x| \geq 1. \end{cases}$$

Samples calculated with the triangular kernel constitute a weighted average, with more weight at the center of the pixel than at the boundaries. The representation basis formed

Figure 15.8. Sampling with a triangular kernel.

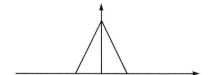

Figure 15.9. Kernel of a triangular representation.

by the kernel translations is not an orthonormal basis. What is more, the kernel support is formed by two representation cells. This fact, illustrated in Figure 15.8, is in itself important for sampling using this kernel because it provides a piecewise linear reconstruction of the underlying signal.

Area sampling using a nonconstant kernel in the representation cell is known as *weighted area sampling*. The extension of the triangular kernel to dimension 2 was described in Section 6.3.2.

15.3.3 Supersampling

A simple way of calculating area sampling using the Haar kernel consists of dividing each pixel into $n \times m$ subpixels, point sampling at each subpixel, and adding the area of the subpixels to obtain an approximation of the area of the objects at the pixel (see Figure 15.10(a)). The pixel color will be a percentage of the color of the given object by the value of the area in relation to the pixel area. (We will explore this fact later on.)

Another option is to perform supersampling: subdivide each pixel into $n \times m$ subpixels, perform a point sampling at each subpixel, and later calculate the average of the subsamples values to obtain the sampling value at the pixel (see Figure 15.10(b)). Whereas area sampling technique is a pre-filtering process using a low-pass filter, supersampling is a *post-filtering* technique because we initially perform a sampling at the subpixels, and only later compute the average. This average is a filtering of the samples at each subpixel (reconstruction filtering).

As the number of subpixels increases, the value obtained in the supersampling more closely approximates the value obtained by unweighted area sampling. A complete discussion of the relation between supersampling and area sampling can be found in [Fiume 89].

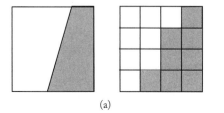 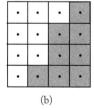

(a)   (b)

Figure 15.10. (a) Sampling by approximated area; (b) supersampling.

15.3.4   Stochastic Sampling

As we previously saw, performing point sampling on a signal with very high frequencies makes its high frequencies disappear and be reconstructed as low frequencies. The stochastic sampling algorithm seeks to avoid this problem by substituting high frequencies with noise so as to avoid the spurious low frequencies. Visual effects due to noise are perceptually more pleasant to the eye than effects generated by reconstructing high frequencies as low frequencies (see Figure 15.11).

The algorithm is quite simple. For each pixel on the image, we subdivide the pixel into subpixels; determine the geometric centroid of the pixel; perturb the centroid of each subpixel in a random way (uniform distribution or Gaussian are commonly used); perform point sampling at each perturbed centroid; and filter the sample values with some reconstruction filter (the most common strategy is to simply compute the average of the sampled pixels).

(a)   (b)

Figure 15.11. Sampling with random perturbation. (a) When the sampling frequency is adequate to perform a point sampling on the signal, stochastic sampling with a sufficiently small perturbation interval gives practically the same result as point sampling. (b) When the frequency of the signal is very high, the value obtained by stochastic sampling at a pixel is practically noise (it can assume any value of the signal in the illustration).

15.3.5   Analytical Sampling

Analytical sampling is a form of area sampling that directly explores the geometry of the objects at the pixel (see Figure 15.12). It was introduced by Ed Catmull in [Catmull 78], using the Weiler-Atherton algorithm for clipping the fragments.

In the case of a polygonal object, this algorithm contains the following steps: we clip each polygon, intersecting the pixel with relation to the pixel rectangle. Then we clip each polygon, contained in the pixel, with other polygons (here, we use the recursive clipping algorithm studied in Chapter 13). Next we exclude the nonvisible fragments. The final pixel color is given by the weighted average of the colors of each visible fragment, where the weight of each fragment is its area relative to the area of the pixel.

Note that the algorithm solves both the rasterization and visibility problems. Certainly, to use analytical sampling, we need a good 2D clipping algorithm for generic polygons (we can use one of the algorithms discussed in Chapter 12). We also need an efficient algorithm for calculating the area of polygons.

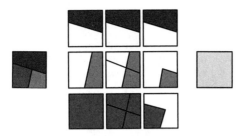

Figure 15.12. Analytical sampling of polygons. (See Color Plate XXIV.)

A more efficient option is to implement both the clipping and the area calculations for the particular case of triangles.

15.3.6 A-Buffer

In all of the above algorithms, with exception of analytical sampling, once we calculate the shading function we lose the information about pixel geometry. This information is important, as it can be used, for instance, to combine two or more images. In the case of analytical sampling, we can keep the exact information about the pixel geometry, but the storage cost is high.

The *A-buffer* sampling method, approximates the analytical sampling method and uses a data structure that allows us to store information about the pixel geometry without much storage cost. The method consists of subdividing each pixel into subpixels and then, for each polygon fragment at each subpixel, performing point sampling using one bit to indicate whether the geometry of that subpixel is empty (bit equals 0) or not (bit equals 1). Figure 15.13 shows one pixel of an image divided into 25 subpixels, with the corresponding matrix of bits. This matrix of bits is called a pixel *bitmask*. The bitmask represents the geometry of the fragment at the original pixel. In this case, the operations are reduced to bitmask operations.

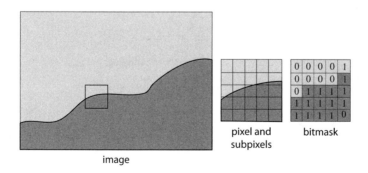

image   pixel and   bitmask
               subpixels

Figure 15.13. Bitmask of a pixel.

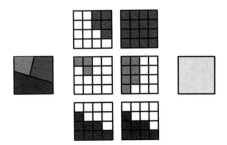

Figure 15.14. A-buffer of one pixel with three polygon fragments. (See Color Plate XXV.)

Figure 15.14 illustrates the use of the A-buffer for calculating the color of a pixel with the same geometry as the pixel in Figure 15.12.

15.4   Comments and References

A seminal work in rasterization and reconstruction with anti-aliasing was the doctoral dissertation of Frank Crow. In [Crow 81] we find a comparison of various anti-aliasing techniques. The A-buffer algorithm was introduced in [Carpenter 84]. Stochastic sampling techniques were introduced in computer graphics in [Cook 86], together with the distributed ray tracing method (see Chapter 14). This work implements a method for obtaining motion blur to avoid aliasing problems in temporal sampling.

15.4.1   Additional Topics

Certainly, in a more advanced course, the problem of aliasing and reconstruction of images should to be approached under the domain of Fourier transform.

Exercises

1.   Explain, in detail, why point sampling can cause aliasing.

2.   Discuss the relationships between filtering and area sampling.

3.   Explain the conditions under which area sampling eliminates aliasing.

4.   Explain the advantages of using A-buffer instead of applying supersampling only.

5.   Compare the techniques of stochastic and analytical sampling (i.e., based on a regular grid).

16 Mappings

Mapping techniques were introduced in the doctoral dissertation of Ed Catmull in 1974 [Catmull 74]. He developed a method called *texture mapping*, which applies an image onto a surface, like a decal (see Figure 16.1). When people hear about mapping, they usually immediately think of texture mapping.

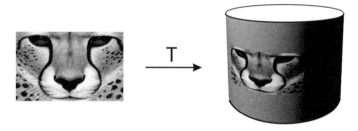

Figure 16.1. Texture mapping.

After Ed Catmull's work, several pioneering mapping applications appeared in both the literature and as special effects in movies. In general, work in mapping seeks to create the objects to be mapped (e.g., textures), to develop mapping techniques, and to perform the mapping calculation.

16.1 Mapping Graphics Objects

Given two graphics objects $\mathcal{O}_1 = (U, f)$ and $\mathcal{O}_2 = (V, g)$, a mapping of $\mathcal{O}_1$ in $\mathcal{O}_2 = (V, g)$ is a transformation $T \colon V \to U$. Notice, there is no error here: the definition really is inverted. Everything happens as in a change of coordinates: we want to map the coordinates of texture U in V, and therefore the transformation is from V to U. The object $\mathcal{O}_2 = (V, g)$ is called the *target object* and the object $\mathcal{O}_1 = (U, f)$ is called the *source object*. The mapping T defines one new attribute function $g^\star$ in the mapped object $\mathcal{O}_2$, given by $g^\star = f \circ T$, as indicated in Figure 16.2.

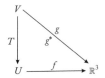

Figure 16.2. Mapping.

In this way, the new attribute of object $\mathcal{O}_2$ at a point $p \in V$ is the attribute of object $\mathcal{O}_1$ at a point $T(p)$; that is, $f(T(p))$. Generally we require the transformation T to be bijective, which guarantees that different points of V are not mapped in the same attribute. Besides, the bijectivity is useful in the mapping calculation, as we will see later on.

The new attribute function $g^\star$, obtained by the mapping T, can be combined with other attribute functions of the object $\mathcal{O}_2$ in several ways. This variety of combinations, together with the several choice possibilities of the mapping T, results in a great diversity of applications.

The dimension of the mapping is given by the dimension of the source graphics object $\mathcal{O}_1$; that is, the dimension of the geometric support U. The most common cases are in *2D mapping* where $\mathcal{O}_1$ is usually an image and in *3D mapping* where $\mathcal{O}_1$ is usually a 3D image (volumetric object). In animation and modeling, 1D mappings are used to change the attributes of curves. It is also common to use a nD mapping, where the dimension $n-1$ is used to define the texture to be mapped, and the extra dimension is used as the time parameter to change attributes of the animation. Of course, we do not have space or time here to treat all mapping types and applications; we therefore concentrate on 2D and 3D mappings.

16.1.1 2D Mapping

Consider the texture mapping shown in Figure 16.1. The target graphics object $\mathcal{O}_2$ is a cylinder and the source object ∞_1 is an image, as shown in Figure 16.3. This is a 2D mapping, specifically a *2D texture mapping*, which defines a new attribute function in the

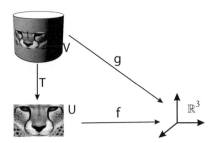

Figure 16.3. 2D mapping.

cylinder. The image in Figure 16.1 was obtained using this new attribute as the diffuse color of the surface; that is, the color component reflected in a diffuse way by the surface (with the Phong equation). This particular use of the new attribute has the visual effect of transferring the image to the surface.

16.1.2 3D Mapping

In the previous section we saw that the basic idea of mapping is to obtain a function defined in the geometric support of the target object, and to then use that function to obtain a new attribute function on this target object.

From a mathematical point of view, a very simple case happens when the geometric support V of the target object $\mathcal{O}_2 = (V, g)$ is a subset of $\mathbb{R}^3$; the source object $\mathcal{O}_1 = (U, f)$ is a 3D image $f: U \to \mathbb{R}^3$; and the condition $V \subset U$ is satisfied (see Figure 16.4(a)).

In this case we can define a mapping $T: V \to U$ for the inclusion transformation $T(p) = p$. Therefore the new attribute function $g^\star$ is simply the constraint $f|V$ (of the attribute function $f: V \to \mathbb{R}^n$ of the mapping volumetric object $\mathcal{O}_1$) to the set V. This is a typical case of a *3D mapping*.

In 3D mapping, the volumetric image being mapped on V can be interpreted as measuring a density of the ambient space. As in the case of the texture mapping of the cylinder, we can use the new attribute function to alter the diffuse color component of the mapped object. In this case, we have a *3D texture mapping*.

Figure 16.4(b) shows a piece of wood obtained by 3D texture mapping. Notice that by varying the parameters of the texture function, we obtain different types of pattern in the wood. As the texture is volumetric, if we "cut the wood," the texture will then appear in the interior parts as being an intrinsic part of the object.

In the example of 2D texture mapping previously seen, the image of the cheetah came from a real picture. On the other hand, the volumetric image in the above 3D texture mapping example was synthetically generated (procedural texture). This is a common characteristic of 3D mapping: the mapping is in itself very simple (usually, the inclusion

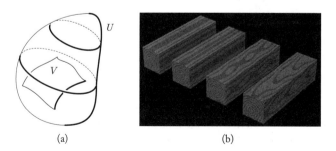

(a)   (b)

Figure 16.4. 3D texture mapping: (a) 3D mapping and (b) 3D wood texture (*[Wolfe 97]* ©*Rosalee Wolfe. Used with permission.* See Color Plate XXVII.)

function)—the difficulty lies in constructing the volumetric image to be mapped (i.e., constructing the 3D texture). We now discuss the problem of generating textures for several applications.

16.1.3   Creating Textures

To obtain good mappings, we need to create the source texture to be mapped. This process has a great artistic component associated with elaborated scientific methods. There are three basic methods of creating textures: scanning real images, synthesizing from real images, and defining textures algorithmically.

Scanning real images. A problem with using real scenes happens when we want to map a certain pattern of an existing texture. Consider, for instance, the problem of mapping a wood texture onto a surface. We can scan a texture of real wood, but the captured scale, can be very different from the scale of the surface to be mapped. We will have to perform significant scale changes, which can result in distortions that make the whole process unviable. A possible solution is to "glue," in the appropriate scale, small pieces of the scanned texture to obtain the surface texture. However this process of texture tiling creates a periodic texture and, in general, creates discontinuity problems at the boundaries of the collage.

The better solution is to obtain the desired texture by a process of texture synthesis. But this method is really related to image processing and therefore outside the scope of this book.

Defining algorithmically. The other method of obtaining textures, which works much better in any dimension, is the algorithmic, or procedural, representation method, which we mentioned in Chapter 10.

In the procedural method, a graphics object is represented by an algorithm in some virtual machine (for instance, the Turing machine):

$$\text{Object} = \text{algorithm}(\text{input}, \text{parameters}).$$

The algorithm input is generally a function or a set of points. The parameters allow us to control some characteristics of the object being represented. The semantics of the object are obtained in its reconstruction, which happens when running the algorithm.

Procedural methods are the most appropriate for representing objects with highly complex geometry and microgeometry (e.g., clouds, forests, fires, etc.) The methods have great flexibility, are easy to implement, and present good results. On the negative side, one would need training before one could develop the intuition to control the semantics of the objects.

Some might say that in addition to creating textures that come from scans or are algorithmically defined, textures can also be generated through *physical modeling*. But physical modeling methods are actually a type of procedural method that uses mathematical models from physics to construct the algorithm. Physical modeling is rarely used to generate textures because it is difficult to obtain and control the model, and the methods are computationally intensive.

The rest of this chapter is organized as follows: the next three sections focus on 2D mapping, covering a description of key methods, how to calculate them and examples of 2D mapping applications; the last three sections focus on covering methods to create procedural textures in both 2D and 3D.

16.2   2D Mapping Methods

As we previously saw, in 2D mapping the mapped object is generally a surface, and the mapping object is usually an image (2D texture). We will assume the mapped object is $\mathcal{O}_2 = (V, g)$, and the mapping object is $\mathcal{O}_1 = (U, f)$.

The most widely used methods for 2D mapping transformations are mapping by parameterization, mapping by projection, transformations of the plane, and mapping with an auxiliary surface.

16.2.1   Mapping by Parameterization

We take a parameterization $\varphi\colon U \to V$ of the surface defined in the support set U of the image. The mapping T of U in V is given by the inverse function $T = \varphi^{-1}\colon V \to U$. Therefore the new attribute function $g^\star$ is given by $g^\star = f \circ T = f \circ \varphi^{-1}\colon U \to \mathbb{R}^3$ (see Figure 16.5).

The texture mapping shown in Figure 16.1 was obtained using the parameterization of the cylinder by cylindrical coordinates: if the cylinder has height h and ray r, and the image has dimensions $u_0 \times v_0$. This parameterization is given by

$$\varphi(u,v) = \left(r \cos \frac{2\pi u}{u_0}, r \sin \frac{2\pi u}{u_0}, \frac{hv}{v_0} \right).$$

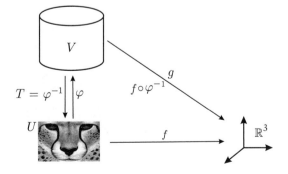

Figure 16.5. Mapping by parameterization.

The inverse of φ is given by

$$\varphi^{-1}(x, y, z) = \left(\frac{1}{2\pi} \arctan\left(\frac{y}{x}\right), \frac{v_0 z}{h} \right).$$

Note that mapping by parameterization requires that the surface of the object to be mapped has a "good" parameterization, which is not always easy to obtain.

16.2.2 Decal Mapping

Intuitively, decal mapping is equivalent to projecting an image onto surface V using a slide projector. For each point $p \in V$ in the support V of the object to be mapped, we take the ray r with origin at p whose direction is normal to V (see Figure 16.6(a)). If r intersects the image support U at a point q, we define $T(p) = q$. In other words, the mapping T is obtained by projecting each point $p \in V$ onto a point of the support U of the image, along the straight line passing through p, which is normal to surface V.

Note that in order to have a successful decal mapping, we need to obtain surfaces where the set of normal rays are "well behaved" so we can have a well-defined mapping (if, for instance, there is an intersection between two normals, the mapping is not well defined).

Another type of decal mapping consists of taking a family of parallel rays that are transversal to the surface U (the image support). For each ray r from this family, if the origin is a point $q \in U$ and the ray intersects the surface V at a point p, we define, as before, $T(p) = q$. In this case, the decal mapping can be seen as a particular case of 3D mapping. In fact, everything happens as if the texture of the image was extended into space in a constant way along the family of parallel rays (see Figure 16.6(b)).

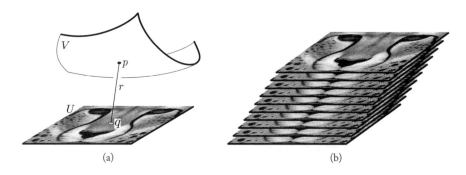

(a)   (b)

Figure 16.6. Decal mapping.

16.2.3   Mapping by Plane Transformations

Let U be a rectangular region on the plane, and consider an image $g\colon U \subset \mathbb{R}^2 \to \mathbb{R}^3$. If a transformation $T\colon U \to V$ of the rectangle U onto another region V of the plane exists, then we use T to map the image g onto the region V: for each point $p \in V$, the color attribute in p is given by the color attribute of the point $T^{-1}(p) \in U$ on image g. This process is called *mapping by deformation*, or warping, of image g (the rectangle of the image is transformed into the region V). Figure 16.7 shows the deformation of one image into a quadrilateral of the plane. The deformation of this figure was obtained by the (unique) projective transformation, taking the rectangle into the quadrilateral. This type of deformation is called *projective mapping*.

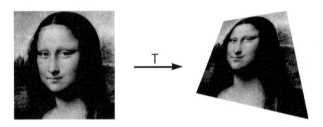

Figure 16.7. Projective mapping.

16.2.4   Mapping with Auxiliary Surfaces

The surfaces most used for decal mapping are planes, cylinders, and spheres, since these surfaces naturally have very good parameterizations. In the cylinder we have the cylindrical coordinates; in the sphere, we have the spherical coordinates

$$x = r \cos \varphi \cos \theta,$$
$$y = r \cos \varphi \sin \theta,$$
$$z = r \sin \varphi,$$

where θ and φ are the longitude and latitude angles, respectively. There are several other good parameterizations of the sphere; this is due to its importance in the construction of cartographic maps, such as the stereographic and Mercator projections, among others.

Mapping by parameterization requires a "good" parameterization on the surface of the object to be mapped, and decal mapping requires "well-behaved" lines normal to the surface of the object to be mapped. Surfaces we are mapping rarely meet these criteria, but we can instead find another surface that will approximate the actual surfaces we want to map but that will also allow us to use mapping by parameterization and decal mapping. We will need a surface M that has both a simple parameterization calculation and a well-behaved family of normal rays to allow the use of decal mapping. We therefore map the image in M by parameterization and then perform the definitive mapping of M onto the final graphics object using decal mapping. The auxiliary surfaces most used are planes, cubes, cylinders, and spheres. What type of auxiliary surface you use depends on the type of surface being mapped: use a plane when the surface to be mapped has an almost planar geometry, use a cylinder when the geometry has axial symmetry (e.g., a surface of revolution), etc. Later we will see several examples of this technique in action.

16.3   Calculating the 2D Mapping

Regardless of the method used, every 2D mapping requires one to map an image onto a region V on the plane. To study the challenges of calculating a 2D mapping, we will use the case of image deformation. We thus have a bijective transformation f of an image $f : U \subset \mathbb{R}^2 \to \mathbb{R}^3$ into a region V on the plane. U is the source and V the target object that should be mapped by U. The coordinates in U and V will be indicated by (x, y) and (u, v), respectively.

16.3.1   A Simple Example

Consider the case in which deformation f is a linear scaling of factor 2; that is, $f(x) = 2x$. By naively applying f to a digital image of resolution 1×4, we obtain the result shown in Figure 16.8.

As the application is expansive, there are some holes in the image, corresponding to pixels that were not painted. We need to reconstruct the target image at those pixels. If f is scaling by a factor of $1/2$, $f(x) = x/2$, we will have a contraction of the image and, in this case, several pixels will accumulate into the same pixel. Our problem will then be to decide what color this pixel should have. This example can be illustrated by taking the inverse transformation f^{-1} of Figure 16.8, which transforms 7 pixels in 4.

This problem represents a general one: when the transformation expands, there is a decrease of frequencies and holes appear; when the transformation contracts, there is a frequency increase and several pixels are mapped into a single pixel. In the second case, we need a good filtering technique to calculate the pixel value to prevent aliasing (i.e., high frequencies are wrongly constructed as low frequencies).

This example leads to two important conclusions: the expansion and contraction properties of the transformation play an important role in the mapping process, and problems result partially from the fact that we are working in the discrete domain. Notice that the application $x \mapsto 2x$, in the discrete domain, stops being bijective, and that its inverse is not even injective.

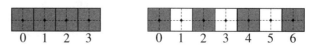

Figure 16.8. Scaling by a factor of 2.

16.3.2 Expansion and Contraction

Deformation of an application is measured by the distortion it causes along the distance between two points. More precisely, an application $f \colon U \to V$ is an *expansion*, if a constant $c > 1$ exists such that

$$|f(x) - f(y)| \geq c|x - y|, \quad \text{for all} \quad x, y \in U.$$

f is a *contraction*, if a constant $c < 1$ exists such that

$$|f(x) - f(y)| \leq c|x - y|, \quad \text{for all} \quad x, y \in U.$$

Of course, the condition $|f(x) - f(y)| = |x - y|$ means the application is an isometry and therefore preserves distances (it neither expands nor it contracts).

Example 16.1 (Linear transformations). Consider a linear transformation $T \colon \mathbb{R}^2 \to \mathbb{R}^2$. The image of the unit circle S^2 by T is an ellipse (see Figure 16.9). The deformation of S^2 in the ellipse provides us with information about the deformation in T: if the two rays of the ellipse OA and OB are greater than 1, the transformation is expansive; if they are less than 1, it is contractive; and if one of the rays is greater and one is less than 1, the transformation expands in one direction and contracts in the other. The largest value between the two rays of the ellipse is the *norm* $|T|$ of the linear transformation. Therefore, $|T| = \max\{|T(x)|; |x| = 1\}$. ❑

This example shows that the contraction and expansion phenomenon of an application can be anisotropic, even in the case of a linear transformation. In the general case, the contraction and expansion properties of a transformation can vary from one region to another in the application domain (observe this fact in the projective transformation of Figure 16.7).

A local estimate of the nature of the deformation can be obtained using the derivative f' of the transformation, which is given by its Jacobian matrix $J(f)$. If $f(x, y) = (u(x, y), v(x, y))$, then

$$J(f) = \begin{pmatrix} \frac{\partial u}{\partial x} & \frac{\partial v}{\partial x} \\ \frac{\partial u}{\partial y} & \frac{\partial v}{\partial y} \end{pmatrix}.$$

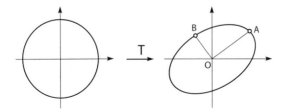

Figure 16.9. Image of a unit circle by a linear transformation.

If $J(f)$ is expansive at point p, then f is expansive in the neighborhood of p; if $J(f)$ is contractive, f is a contraction in the neighborhood. The same occurs if $J(f)$ contracts along one direction and expands along another.

The norm of the Jacobian matrix $J(f)$ can be calculated by

$$|J(f)| = \max\{|\operatorname{grad}(u)|, |\operatorname{grad}(v)|\}$$
$$= \max\left\{ \sqrt{\left(\frac{\partial u}{\partial x}\right)^2 + \left(\frac{\partial u}{\partial y}\right)^2}, \quad \sqrt{\left(\frac{\partial v}{\partial x}\right)^2 + \left(\frac{\partial v}{\partial y}\right)^2} \right\}. \qquad (16.1)$$

This spatial variation of the contraction and expansion properties of a transformation requires the use of filters whose kernels also vary spatially in both direction and position. These filtering methods are known as *space-invariant filters*. Given that locally a circle is approximately transformed into an ellipse, an interesting proposal consists of using a circular geometry for the pixels, together with anisotropic elliptic filters. There are several works on this subject.

16.3.3   Continuous Domain and Resampling

Besides expansion and contraction, we must also be aware of problems that can occur because we apply the transformation in the discrete domain; that is, we apply the transformation at each pixel considered as a point. In the continuous domain, the pixel is defined by the rectangle of the grid, exactly as in the image of the virtual screen. This way the image of the pixel, by the transformation f, is a curvilinear quadrilateral, as shown in Figure 16.10. We can approximate the curves on the edges by straight line segments, thus obtaining one linear quadrilateral (as indicated in dashed lines). With this approach, the image of the pixel is determined by the image of their four vertices.

To obtain the image in the continuous domain, we reconstruct it at the grid vertices of the matrix representation using an interpolation method. Here, the bilinear interpolation is a natural choice and works satisfactorily.

Once in the continuous domain, we apply the transformation in the image and soon afterward resample to obtain the final image in the discrete form. This method is illustrated in Figure 16.11. There are two possibilities for the processing order of the pixels: direct and inverse mappings.

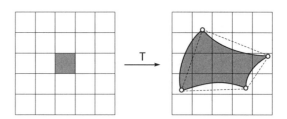

Figure 16.10. Image of a pixel in the continuous domain.

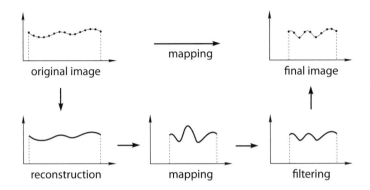

Figure 16.11. Transformation stages of an image: reconstruction, transformation, and resampling.

16.3.4 Direct Mapping

In the direct method, we apply the transformation to each pixel on the source image and paint the corresponding pixels on the target image. This process is illustrated in Figure 16.12(a). In the case of expansion, we rasterize the polygon and paint the pixels, taking into account the area occupied by the polygon at each pixel (analytical sampling); in the case of contraction, we join all the quadrilateral fragments at the pixel to perform the process of reconstructing the pixel color.

The problem of leaving holes in the image is attributed to direct mapping. However, as we saw above, this problem does not happen when working in the continuous domain. The main problem of direct mapping is that, in the case of a contraction, the image of the pixel is a polygon fragment at the pixel to be painted. Furthermore, the process of joining all the fragments is complex. Notice, this happens in the contraction, where the possibility of having aliasing problems is greater. The inverse mapping, covered below, solves this problem satisfactorily.

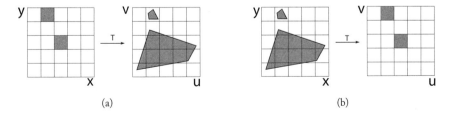

Figure 16.12. Processing order of the pixels: expansion and contraction in (a) direct and (b) inverse mapping.

16.3.5   Inverse Mapping

In the case of inverse mapping, processing starts at the target image. For each pixel $p \in V$, we calculate the inverse image $T^{-1}(p)$ and process this inverse image to calculate the color that will be attributed to pixel p. Notice in Figure 16.12(b), which illustrates inverse mapping, that when T is expansive the inverse mapping is a contraction and vice versa. Consequently, in the case of contractions, inverse mapping favors the filtering.

To process expansion, we take the barycenter of the quadrilateral $T^{-1}(p)$ and attribute to p the color of the closest pixel to the barycenter (notice, as we have an expansion, we can be less careful in the reconstruction).

To process contraction, we rasterize the quadrilateral $T^{-1}(p)$ to obtain the pixels that are mapped to pixel p, and we filter those pixels to obtain the color value at p.

With inverse mapping, even when working in the discrete domain, we will not have the problem of holes in the final image.

16.3.6   Prefiltering and Mipmap

As we previously described, performing the transformation calculation in the continuous domain is robust but computationally expensive. There are cheaper methods that are based on the following strategies: using inverse mapping, performing filtering in parallel with the transformation calculation, and precalculating and storing the filter.

One of those methods is the mipmap,[1] which was introduced by Lance Williams. The method aims at achieving a good balance between the quality of the final image and the processing time. It is easy to implement in hardware and is supported by several graphics devices (graphics workstations, video acceleration boards, etc.) and the OpenGL graphics system. Below are the important stages of this algorithm.

Type of mapping. Mipmapping uses inverse mapping and privileges filtering in the contraction regions, where the possibility of aliasing is more critical due to the increase of frequencies.

Simplification of the mapping properties. For filtering purposes, instead of considering the geometry of pixel $T^{-1}(p)$ (quadrilateral), the mipmap considers only the local expansion or contraction by a scaling factor of the ratio 2^n.

Prefiltering. To gain efficiency, the method performs prefiltering in several levels with a 2×2 filter (see Figure 16.13(a)). We can use the Gaussian, Bartlett, or even the Haar ("box") filters.

Prefiltering levels. If the image has a resolution of $2^n \times 2^n$, we have n filtering levels because, at each filtering, we obtain an image at a smaller scale by dividing its resolution by two. Each level tells us when the image should be smoothed out by the filter to reduce the high frequencies. This way, the filtered images can be structured using a pyramid, as

---

[1]"Mip" is the acronym of the Latin sentence "multum in parvo," which means "a lot in little space."

$\qquad$ (a) $\qquad\qquad\qquad\qquad\qquad\qquad$ (b)

Figure 16.13. Mipmap elements: (a) prefiltering in mipmap, (b) mipmap pyramid.

shown in Figure 16.13(b). This pyramidal structure is convenient because it saves space and facilitates access to the prefiltered image pixels. Each point q on the original image has a corresponding prefiltered point q_j at each level j of the pyramid.

Determination of the level. The filtering level we use is determined by the deformation (expansion or contraction) factor d of the transformation, which is given by the Jacobian norm in Equation (16.1). The level 0 (original image), corresponds to an expansion; the greater the contraction of the transformation, the higher the level.

The mipmap algorithm. Color attribute of each pixel $p \in V$ are obtained in the following way:

1. We calculate the expansion factor d or contraction of the transformation at the pixel $q = T^{-1}(p)$. The level is then given by $j = \text{floor}(d)$.

2. We calculate the pixels q_j and q_{j+1}, corresponding to the pixel q on the images of levels j and $j + 1$ of the pyramid.

3. We calculate the colors of $C(q_j)$ and $C(q_{j+1})$ at each level, performing a bilinear interpolation with the neighboring pixels.

4. The final value of color $C(q)$ of pixel $q = T^{-1}(p)$ is obtained by interpolating the color values $C(q_j)$ and $C(q_{j+1})$ of the previous item:

$$C(q) = \frac{d - \text{floor}(d)}{d}C(q_j) + \frac{\text{ceil}(d) - d}{d}C(q_{j+1}).$$

5. If the transformation expands into a neighborhood of q, we interpolate between the levels 0 (original image) and 1 of the table. Remember, there is a decrease of frequency in the expansion, and filtering is not critical.

16.4   Some 2D Mapping Applications

16.4.1   Texture Mapping

The colorization of an image depends on several factors, including the illumination model, the calculation of the illumination function, and the intrinsic surface characteristics of geometry, roughness, etc.

The geometry of an object has two different aspects. The *macrogeometry* defines the form of the object and is responsible for details of the object in a real scale, which make it identifiable in relation to our daily experience. The *microgeometry* takes care of the smaller details. The microgeometry of the object manifests itself perceptually through the pattern of reflected luminous energy, which characterizes the *texture* of its surface. If the same chair is modeled being made from two different kinds of wood, both models will have the same macrogeometry but different microgeometries.

Macrogeometry belongs to the area of geometric modeling. Our focus here is on the microgeometry. In principle, it would be possible to model the microgeometry of an object using geometric modeling methods together with elaborate models of light-object material interaction (see Chapter 14, [Blinn 77], and [Cook and Torrance 81]). However, this path leads to a significant increase in both the complexity of the scene and the illumination calculations. Instead, using textures to model the microgeometry allows us to increase the realism of images without increasing the geometric complexity of the scene. This fact is illustrated in Figure 16.14.

(a)

(b)

Figure 16.14. (a) Details created with texture mapping without geometry complexity. (*Image courtesy of Karin Eszterhas and 3DTotal.com, www.digitalgallery.dk, www.3dtotal.com.*) (b) Both images were obtained from the same scene containing 3,497 polygons; all the details of the image on the right were obtained using texture mapping. (*Image courtesy of Richard Tilbury and 3DTotal.com, www.richardtilburyart.com, www.3dtotal.com.* See Color Plate XXVI.)

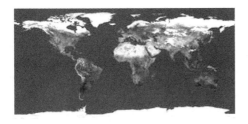

Figure 16.15. Texture mapping on a sphere. (See Color Plate XXVIII.)

Microgeometry alters the reflected luminous energy, thus creating a texture pattern. The texture governs everything we notice about the microgeometry of the object, making it possible to use a mapping function to alter the light pattern reflected from objects in order to simulate their microgeometry.

In texture mapping, we use a mapping function to map a texture function on the surface. Next, we use that function to alter the diffuse reflection of the object to obtain the simulation of the texture on the object surface. In Figure 16.15 we use sphere mapping by parameterization (Mercator projection), to obtain the image of a globe, starting from an image of the map of the earth.

16.4.2 Environment Mapping

We perceive the world by means of a projection onto our retina. Therefore, from a perceptual point of view, we do not need 3D models to visualize a scene. In fact, assume it is possible to photograph a scene by positioning the camera at any position and orientation. Also assume we are able to assemble all of the photos appropriately. Then the scene visualization can be done solely from the photographic montage.

The *plenoptic function* is essentially a photo montage. It is a function defined as $P = P(x, y, z, \theta, \varphi, \lambda, t)$ that associates, to each position (x, y, z) and direction (θ, φ), the value

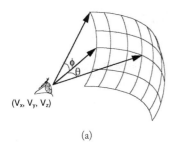
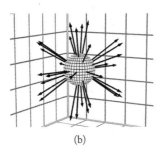

(a)   (b)

Figure 16.16. Elements of the plenoptic function: (a) plenoptic function sample, (b) beams of rays from p_0.

of the luminous energy of wavelength λ perceived by the observer. The plenoptic function also takes into account the temporary variations of those parameters. In other words, the plenoptic function captures, over time, all possible projections of the scene with the camera at different positions and orientations (see Figure 16.16(a)).

Two simplifying hypotheses will help us better understand the plenoptic function: the medium is nonparticipating, meaning the plenoptic function is constant along each ray, and time is fixed—in other words, we want one instantaneous exposure from the plenoptic function.

With these two hypotheses, for each position $p_0 = (x_0, y_0, z_0)$ in space, the plenoptic function is determined entirely by the beams of rays with origin at point p_0 (see Figure 16.16(b)). Therefore the plenoptic function is determined by the values in a sphere with radius $R > 0$ and center at p_0.

This sampling of the plenoptic function in a sphere is called *environment mapping*. Of course, instead of a sphere, we can use any other surface S whose solid angle is equal to 4π, with vertex at p_0, such that each ray in the beam, with center at p_0, intersects S at one point only. A cube meets these criteria, and a cylinder, despite not having a solid angle of 4π, is also used for environment mapping.[2] A surface satisfying these conditions is called a *plenoptic surface*.

You may be wondering what the relation is between this problem and 2D mapping. The answer is simple: to obtain an environment mapping we can take photographs of an environment and perform a texture mapping of the images obtained onto any of the surfaces mentioned in the previous paragraph.

Example 16.2 (Cylindrical environment mapping). There is an advantage to using a cylinder as a plenoptic surface because parameterization by cylindrical coordinates is an isometry between a region on the plane and the cylinder. This means that if the width of an image is

Figure 16.17. A stitched panoramic image and some of the photographs the image was stitched from. (*[Shenchang 95]* ©1995 *Association for Computing Machinery, Inc. Reprinted by permission.* See Color Plate XXIX.)

---

[2]Notice that the solid angle of the cylinder converges to 4π when its height increases.

 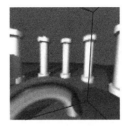

(a) Environment map   (b) Visualization

Figure 16.18. Virtual panorama with cubic mapping [Gomes et al. 98]. (a) Unfolded cubical environment map. (b) Cube reprojection in a given viewing direction. (*Reprinted from [Darsa et al. 98], courtesy of L. Darsa, B. Costa, and A. Varshney, with permission from Elsevier.*) In (b) we show parts of the cube edges for reference purposes only. (See Color Plate XXX.)

equal to the total length of the cylinder (i.e., $2\pi r$ in a cylinder with ray r), it can be mapped onto the cylinder without any distortions. An image satisfying these conditions is called a *cylindrical panorama*. Figure 16.17 displays such an image, constructed by "stitching" several images taken by rotating a camera fixed at a specific position.

We therefore obtain a cylindrical environment mapping by mapping a cylindrical panorama of width $2\pi r$ and height h onto a cylinder of ray r and height h. ❑

There are several methods for calculating environment mapping on a sphere, some of which we will discuss in the exercises. If the plenoptic surface used is the cube, we use six images of the scene, that have been obtained from the same point by rotating the camera by 90° so as to cover the whole environment. To compose the environment, we then map each image to one of the faces of the cube. Figure 16.18 shows the six faces of a cube and the image on each face of an environment mapping. Methods for calculating the transformations to change an environment mapping from one plenoptic surface to another can be found in [Greene 86].

Next we will discuss two important applications of environment mapping: reflection mapping and virtual panorama. These examples represent applications of texture mapping with auxiliary surfaces, and in this case the auxiliary surface is the plenoptic surface.

Reflection mapping. Consider a geometric object O in a certain scene. Let us take a plenoptic surface M and map the scene environment in M. The reflection mapping consists of using the environment mapping in a way that simulates reflection of the environment onto the object O. In other words, the reflection mapping provides a first-order approximation of the ray tracing method: to obtain a similar effect with ray tracing, we would need to trace secondary rays.

Once the environment mapping is created on the plenoptic surface, we only need to calculate the correct mapping to obtain the desired reflection effect. This mapping should be calculated having, as a basis, the reflected vector by the surface starting from the observer (position of the virtual camera), as shown in Figure 16.19.

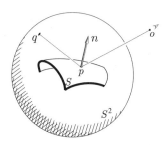

Figure 16.19. Calculating the reflection vector.

To calculate the value of the mapping at point $p \in O$, we consider the unit vector $\mathbf{n}$ normal to the surface O at point p, and the unit vector $\mathbf{v} = \overrightarrow{po}$ along the observer's direction. We take the specular reflection vector $\mathbf{r}$ associated to $\mathbf{n}$ and $\mathbf{v}$. The reflection vector is given by

$$\mathbf{r} = \mathbf{v} - 2\langle \mathbf{v}, \mathbf{n} \rangle \mathbf{n}.$$

Therefore, the parametric equation of the reflected ray is given by

$$\gamma(t) = p + \mathbf{r}t = p + (\mathbf{v} - 2\langle \mathbf{v}, \mathbf{n} \rangle \mathbf{n})t, \qquad t \geq 0.$$

The value of the mapping at point p is given by point q, where ray $\gamma(t)$ intersects plenoptic surface M.

Once the mapping is obtained, it should be used to alter the light specularly reflected by the object. Therefore, the attribute change should be made in the component of the specular reflection of the Phong equation.

The reflection mapping is simple to implement, computationally very efficient, and effective in several applications. There are two related applications of reflection mapping. One is using it to obtain a reflection of the environment for an approximation to the ray tracing method, as shown in Figure 16.20(a). The other application consists of modifying the specular reflection component with a texture of high and low frequencies to give a metallic look to the object. This effect is illustrated in Figure 16.20(b).

Virtual panoramas. Fundamentally, image-based rendering consists of reconstructing the plenoptic function starting from "scattered samples." This problem is very difficult in general, but solutions exist for several particular cases with important applications. One of those cases is the *virtual panorama*.

A virtual panorama has two basic ingredients: an environment map associated to a beam of rays with center at a point p_0, on a plenoptic surface S, and a camera model for visualizing the scene and reconstructing the plenoptic function, starting from the environment map. In other words, the virtual panorama performs the sampling and reconstruction of the plenoptic function for the case in which the position of the camera is fixed. The observer (virtual camera) is placed at the center of the plenoptic surface in which the

(a)   (b)

Figure 16.20. Examples of reflection mapping: (a) ray tracing approximation (*Courtesy of Castle Game Engine, http://castle-engine.sourceforge.net/*), (b) metal appearance (ⓒ*2011 Okino Computer Graphics, Inc.* See Color Plate XXXI).

environment is mapped. From this position, the observer can browse using the environment mapping. The constraint is that the observer cannot change positions, so navigation is restricted to changing the orientation of the camera and zooming (changing the focal distance of the camera). As there are no geometry details but only a mapping of the environment, the navigation can be performed in real time, even in computers with a simple configuration.

As we already know, the cube, sphere, and cylinder are the plenoptic surfaces most used for environment mapping. For each of these surfaces, we need an appropriate virtual camera model. In the sphere, we use a camera with spherical projection; in the cylinder, a camera with cylindrical projection and in the cube, we use the usual model described in Chapter 11. We leave the details of the spherical and cylindrical cameras for the exercises. Figure 16.18 shows environment mapping using a cube as an auxiliary surface.

16.4.3   Bump Mapping

In this technique, the mapped image is used to create a perturbation of the normal vector to the surface (see Figure 16.21). Assuming that (1) the image function of a grayscale image is given by $b(u,v) \in R$; (2) the surface is defined by a parameterization $p(u,v) \in R^3$; and (3) the normal vector at the point $p(u,v)$ is given by $N(u,v) \in R^3$, the calculation of the normal vector perturbation is simple:

$$q(u,v) = p(u,v);$$
$$N_1 = \frac{\partial q}{\partial u} \wedge \frac{\partial q}{\partial v};$$
$$N = \frac{N_1}{|N_1|}.$$

This technique was introduced in [Blinn 78].

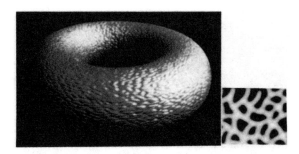

Figure 16.21. The result of bump mapping, obtained from the image at right, applied using mapping by parameterization. Hand drawn bump functions. (*[Blinn 78]* ©*1978 Association for Computing Machinery, Inc. Reprinted by permission.* See Color Plate XXXII.)

The displacement of the normals by the mapped image has the effect of transferring texture details of image b to the vector field that is normal to the surface; then, when we perform the illumination calculation, we have the illusion that those details are part of the geometry (remember, the normal vector is the only surface geometry component as part of the Phong equation). In Figure 16.22 we use bump mapping to obtain details of the geometry of a coin.

Figure 16.22. Face of a coin generated with bump mapping. The texture was mapped using decal mapping with orthogonal projection. (See Color Plate XXXIII.)

16.4.4 Displacement Mapping

One disadvantage of the bump mapping technique occurs when, by observing the silhouette of the surface, we loose details of the geometry being mapped.[3] This can create problems (e.g., in animation), but can be overcome using the intensities of the image to alter the geometry of the object. The change in the calculation in relation to bump mapping

---

[3]Silhouettes are view-dependent outlines of an object. A silhouette point p on the surface of an object is where the angle between the viewing direction and the normal vector at p is $90°$

            (a)   (b)

Figure 16.23. A deformed Utah teapot using the same texture for (a) bump and (b) displacement mappings. (*From [Wolfe 97], ©Rosalee Wolfe. Used with permission.* See Color Plate XXXIV.)

is simple:

$$p(u, v) = p(u, v);$$
$$Nb = \frac{\partial p}{\partial u} \wedge \frac{\partial p}{\partial v};$$
$$N = \frac{Nb}{|Nb|}.$$

In other words, instead of just changing the normal vector, we perform a displacement of the surface and then calculate the new normal vector. This technique is called *displacement mapping* and was introduced in [Cook 84].

Unlike bump mapping, displacement mapping actually deforms the geometry of a surface (see Figure 16.23). In some applications, simulation of the roughness of a surface can be obtained using bump mapping. When this technique fails (for instance, when we have to display the silhouette of the surface), the recommended technique is displacement mapping.

16.5   Noise Function

We will now move on from 2D mapping to an extremely powerful method for generating procedural textures in both 2D and 3D. The geometry of most natural objects is highly irregular, which makes it difficult to model with the simple use of deterministic procedures. Such irregularity is due to a certain degree of randomness in the geometry, as well as the existence of high and low frequencies at different scales—characteristics that are generally associated with *fractals*. Modeling objects with a high degree of irregularity involves three factors: frequency (determining the oscillation of the irregularity), amplitude (determining the magnitude of the irregularity), and scale (determining our perception of the irregularity).

16.5.1 Convolution Noise

Geofrey Gardner [Gardner 85] was a pioneer using these factors to generate textures related to natural objects. Gardner used discrete Fourier transform to obtain 2D textures of clouds by varying frequency and amplitude (spectral synthesis of textures). One of the limitations of the method is that Fourier transform does not allow scale variations. (We could obtain a variation of the three parameters by using Wavelet transform, but that is beyond the scope of this book.)

We can easily establish a naive procedure for constructing textures using the variation of the above three parameters. Let us take a random field random(i,j) that associates a pseudorandom scalar to each vertex (i, j) of an integer grid on a plane of order n:

for $i = 0, 1, \ldots, n$ **do**
  for $j = 0, 1, \ldots, n$ **do**
    Set_Pixel(random(x, y));
  end for
end for

This image is shown in Figure 16.24(a) for $n = 15$, where we used a pseudorandom variable with uniform distribution (white noise). To obtain this texture at different scales, we can perform successive filterings with a low-pass filter (i.e., to successively blur the image), as shown in Figure 16.24(b)–(d).

To vary both the frequency and amplitude, we generated another image with the same width and height, varying the grid resolution n, as well as the intensities of the pixels. Fig-

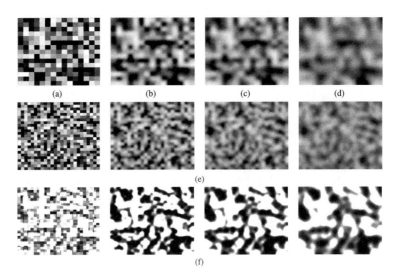

(a)   (b)   (c)   (d)

(e)

(f)

Figure 16.24. Texture with noise at different scales.

ure 16.24 repeats the above process for $n = 25$ (in this figure we maintain the same amplitude). To capture the information about resolution and frequency on the same image, we can add pairs of textures with resolution 15 and 25 (top and middle rows in Figure 16.24, respectively). We therefore obtain a sequence of textures at different scales as shown in Figure 16.24(e) and (f). This method of texture generation with noise is called *convolution noise*.

There are some problems in the convolution noise method. The process does not have a simple parameterization (which would make it possible for the user to control some characteristics of the texture sought). Additionally, the filtering process is computationally very costly.

16.5.2   Perlin Noise Function

While there are problems with the convolution noise method, the idea is a good one and is the basis for the construction of the *noise functions*. These functions were independently introduced by Ken Perlin [Perlin 85] and D. Peachey [Peachey 85], and are known as the *Perlin noise function*.[4]

The Perlin noise function allows us to control the frequency, amplitude, and scale parameters: it is related to ambient space, is present in several scales, is parameterizable, and has memory. Noise memory means the noise has the same value at each grid vertex. This property is important because if the noise is completely random, we will certainly obtain different textures at each new noise processing (in an animation, for instance, the texture would change in every frame).

There are three stages of the Perlin noise function: we create a grid in $\mathbb{R}^n$, make a pseudorandom field in the grid, and then reconstruct the pseudorandom field. Starting from a noise function, we construct another noise function with different frequencies, amplitudes, and scales. Perlin named this function *turbulence*.

Grid. To define the noise in dimension m, we need an integer grid in $\mathbb{R}^m$. For this, we define the partition J_n, in the interval $[0, n]$, by the integers $0, 1, 2, \ldots, n$, and we take the Cartesian product $J^m = [0, n]^m = J_n \times \cdots \times J_n$ of m copies of J_n. The parallelepiped $[0, n]^m$ is called the *grid domain*. We previously used this type of grid to obtain matrix representations of solid objects in $\mathbb{R}^m$. A generic grid vertex $(i_1, i_2, \ldots, i_m)$, $i_j = 0, 1, \ldots, m$ will be indicated by v_J, where J is a multi-index, $i_1 i_2 \ldots i_m$.

Random field. We have several options for defining a random field in the grid J^m, including using scalar, gradient, and scalar-gradient fields (see Figure 16.25).

Scalar noise. This method defines a random scalar field $N: J_m \to \mathbb{R}$, associating to each grid vertex $v_J \in J_m$ the random number $N(v_J)$ in the interval $[-1, 1]$. The noise generated by this method is called *scalar noise*.

---

[4]Ken Perlin received a Technical Achievement Oscar Award for the importance of his work in the film industry.

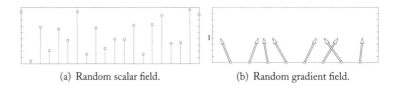

(a) Random scalar field.   (b) Random gradient field.

Figure 16.25. Two fundamental random fields.

Gradient noise. This method defines a random vector field $\mathbf{g} \colon J_m \to \mathbb{R}^{m+1}$ by placing $\mathbf{g}(v_J) = (\mathrm{N}(v_J), 1) \in \mathbb{R}^{m+1}$, where $\mathrm{N} \colon J_m \to S^{m-1}$ is a pseudorandom field, taking values in the unit sphere $S^{m-1} \subset \mathbb{R}^m$. (This field should have an approximately uniform distribution on the surface of the sphere.) The field $\mathbf{g}$ is called a *random gradient field*, and the associated noise is called *gradient noise.*

The gradient field $\mathbf{g}$ takes values in the set $S^{m-1} \times \{1\}$, which is contained in the "cone" $S^{m-1} \times \mathbb{R}$. For $m = 1$, a unidimensional noise, $S^{m-1} = \{-1, 1\}$. Thus we have only two possibilities for the gradient field $\mathbf{g}$: $(-1, 1)$ and $(-1, 1)$. In reality, we can replace the sphere S^{m-1} by the unit disk $D^m = \{x \in \mathbb{R}^m; |x| \le 1\}$. This was, in fact, our choice when implementing the 1D and 2D noises we used to produce the figures in this chapter.

Scalar-gradient noise. This method creates the hybrid scalar-gradient noise, defining a hybrid random field which associates, to each grid vertex v_J, a random scalar value $\mathrm{N}_1(v_J) \in [-1, 1]$ and a random vector $(\mathrm{N}_2(v_J), 1)$, where N_2 is a random gradient field on the sphere $S^{m-1} \subset \mathbb{R}^m$.

Reconstruction method. The reconstruction method defines the following function: Noise: $[0, n]^m \to \mathbb{R}$, of class $C^k, k \ge 0$, in the domain $[0, n]^m$ of the grid J^m, starting from the given random field.

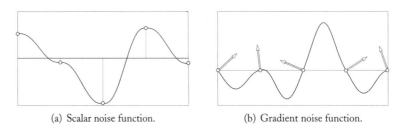

(a) Scalar noise function.   (b) Gradient noise function.

Figure 16.26. Noise function reconstruction. Note that in both instances, the gradient noise function has higher frequencies because the existence of zeros at the grid vertices forces the function to oscillate further more.

In the case of the random scalar field, the function $\text{Noise}(t)$ is obtained by interpolating the values $\text{N}(v_J)$ of the field, as shown in Figure 16.26(a). In the random gradient field, for each grid vertex v_J, we should have $\text{Noise}(v_J) = 0$, and the random vector $\mathbf{g}(v_J)$ should be normal to the graph $(p, \text{Noise}(p))$, $p \in \mathbb{R}^m$ in v_J. This fact is illustrated in Figure 16.26(b).

Having zeros at every grid vertex can result in problems with the appearance of periodic patterns in the noise function. This can be avoided by using a scalar-gradient noise function, which gathers properties of the two previous noises: the noise function interpolates the values of the scalar field, and in those points its graph is perpendicular to the gradient field. Of course, this hybrid noise function can be easily obtained by adding a scalar noise function to a gradient noise function.

16.5.3 Noise Function and Turbulence

A noise function, obtained by the process described in the previous section, will be indicated by $\text{Noise}(x)$. As we saw, this function is defined starting from a grid, a random field, and an interpolation method. By varying the scale—that is, simultaneously changing the frequency and the amplitude of the fundamental Noise—we obtain a family of fundamental noises. More precisely, by fixing a real number p, for each positive integer number i, we can define

$$\text{Noise}_i(x) = \frac{\text{Noise}(2^i x)}{p^i}. \tag{16.2}$$

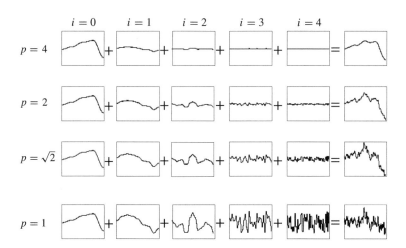

Figure 16.27. Perlin noise functions with different amplitudes. In each column we obtain five noise functions by keeping p constant and varying the frequency and the amplitude with Equation (16.2), taking $i = 0, 1, 2, 3, 4$. (*From http://freespace.virgin.net/ hugo.elias/ models/ m_perlin.htm.*)

As i increases, the factor 2^i reduces the scale of the Noise and, therefore, we have a resulting increase of the frequencies present in the function. On the other hand, the factor $1/p^i$ is the amplitude of the function. If $p > 1$, then as we increase the value of i, we reduce the scale and the amplitude; in other words, we increase the frequency and reduce the amplitude. This means the signal oscillates more with a smaller amplitude, and therefore there is an increase of the "fractal dimension." Geometrically, this corresponds to an increase of the irregularity. This fact is illustrated in Figure 16.27.

For each value of p, we define the *turbulence function* $T(x)$ by the sum

$$T(x) = \sum_{i=0}^{N-1} \text{Noise}_i(x) = \sum_{i=0}^{N-1} \frac{\text{Noise}(2^i x)}{p^i}.$$

The last column of Figure 16.27 shows the graph of four turbulence functions $T(x)$ for the values of $p = 4, 2\sqrt{2}$ and 1. In each case, we added five noise functions ($N = 5$).

16.6 Scalar Noise

We will now examine the reconstruction details of scalar noise. As we previously saw, in this case noise is constructed by interpolating the random scalar field $N(x)$. We will study the cases of 1D, 2D, and 3D noises, which can be generalized for n-dimensional noise.

16.6.1 Unidimensional Noise

In each interval $[j, j+1]$ of the grid, we have the values $N(j)$ and $N(j+1)$ of the random field N. We need an expression to calculate $\text{Noise}(x)$ for every $x \in [j, j+1]$. It will be enough to determine a function $h\colon [0,1] \to \mathbb{R}^n$, satisfying $h(0) = N(j)$ and $h(1) = N(j+1)$, because the noise function in the interval $[j, j+1]$ is defined by $\text{Noise}(x) = h(x - j)$. Depending on the differentiability class wanted for the noise function, the interpolating function h should satisfy additional boundary conditions involving its derivatives.

If $g\colon [0,1] \to \mathbb{R}$ is a function satisfying $g(0) = 1$ and $g(1) = 0$, we can define the interpolating function as

$$\begin{aligned} h(t) &= g(t)\,N(j) + (1 - g(t))\,N(j+1) \\ &= N(j+1) + g(t)(N(j) - N(j+1)). \end{aligned} \tag{16.3}$$

In this case, the obtained function h is of class C^k if and only if the kth derivatives of g in 0 and 1 coincide. There are many possibilities for the function g; we will study some of them here and leave others for the exercises at the end of the chapter.

Linear interpolation. In this case, we take $g(t) = 1 - t$ in Equation (16.3). The interpolating function $h\colon [0,1] \to \mathbb{R}^n$ is given by the classic formula

$$h(t) = (1 - t)\,N(j) + t\,N(j+1) = N(j) + t(N(j+1) - N(j)).$$

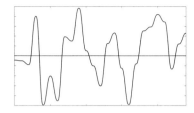

(a) Linear reconstruction.   (b) Polynomial reconstruction.

Figure 16.28. Different reconstructions of a scalar noise in the same random field.

We therefore have, for $x \in [j, j+1]$, $\mathrm{Noise}(x) = \mathrm{h}(x - j)$. The obtained noise function is of class C^0 only and is very computationally efficient, so, it can be used for interactive texture selection experiments where we have to generate textures in real time. Figure 16.28(a) displays the linear interpolation of a random scalar field in a unidimensional grid with 25 points.

Polynomial interpolation. A linear interpolation is actually a polynomial interpolation of degree 1. We can improve the differentiability class by using polynomials of higher degree. Ken Perlin's original work [Perlin 85] obtains cubic interpolation by using, for g in Equation (16.3), the polynomial $b_3(t) = 1 - (3t^2 - 2t^3) = 2t^3 - 3t^2 + 1$. We can easily verify that $g'(0) = g'(1) = 0$ and therefore the interpolating function is class C^1. (Since $g''(t) = 6 - 12t$, we can see the function is not class C^2.)

Perlin later published an article [Perlin 02] in which he recommended using, instead of b_3, the polynomial of degree 5, $b_5(t) = -6t^5 + 15t^4 - 10t^3 + 1$ to obtain an interpolating function of class C^2. The graph in Figure 16.28(b) shows the interpolation of the same random field used in the linear interpolation of Figure 16.28(a), taking $g(t) = b_5(t)$.

16.6.2 2D Noise

Let $J_n = \{0, 1, 2, \ldots, n\} \subset \mathbb{R}$ be an integer partition of the interval $[0, n] \subset \mathbb{R}$. The 2D grid, J^2 is obtained by the Cartesian product $J_n \times J_n \subset \mathbb{R}^2$. A generic vertex of this grid is indicated by $v_{ij} = (i, j)$, $i, j = 0, 1, \ldots, n$.

A cell $[i, i+1] \times [j, j+1]$ of the grid is denoted by C_{ij}. The vertices of this cell are given by $v_{ij,mn} = (i+m, j+n)$, $m, n = 0, 1$. That is: $v_{ij,00} = (i, j)$, $v_{ij,10} = (i+1, j)$, $v_{ij,01} = (i, j+1)$, and $v_{ij,11} = (i+1, j+1)$.

A random scalar field $\mathrm{N} \colon J^2 \to \mathbb{R}$ associates, to each grid vertex (i, j), a random real vector $\mathrm{N}(i, j) \in [0, 1]$. We need to determine a method to reconstruct the function in each grid cell C_{ij}. We will discuss three methods for reconstructing the noise function: bilinear interpolation, lofting, and Coons surface.

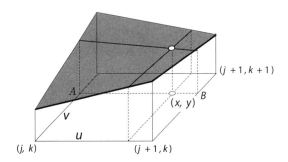

Figure 16.29. Interpolation in $\mathbb{R}^2$.

Reconstruction using bilinear interpolation. In this method, the value of the noise function $\text{Noise}(x, y)$ at a point (x, y) of the cell $C_{ij} = [i, i+1] \times [j, j+1]$ is calculated by making three successive linear interpolations (see Figure 16.29).

A linear interpolation, in the segment connecting vertex $v_{ij,00} = (i, j)$ to vertex $v_{ij,10} = (i+1, j)$, is defined by

$$\text{Noise}(x, j) = \text{Noise}(i, j)(1 - u) + \text{Noise}(i + 1, j)u,$$

where $u = x - i$. Next, another linear interpolation in the segment connecting the vertex $v_{ij,01} = (i, j+1)$ to vertex $v_{ij,11} = (i+1, j+1)$ is

$$\text{Noise}(x, j + 1) = \text{Noise}(i, j + 1)(1 - u) + \text{Noise}(i + 1, j + 1)u.$$

Finally, a linear interpolation along the segment connecting point (x, j) to point $(x, j+1)$, is

$$\text{Noise}(x, y) = \text{Noise}(x, j)(1 - v) + \text{Noise}(x, j + 1)v,$$

where $v = y - j$.

Reconstruction by lofting. As we saw in Chapter 8, this method reconstructs the surface starting from two boundary curves. Initially, we used one of the unidimensional interpolation methods (studied in the previous section) to reconstruct the noise on the side of the

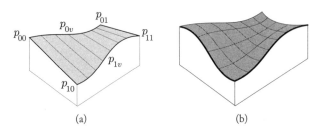

Figure 16.30. 2D scalar noise reconstruction: (a) lofting surface, (b) Coons surface.

cell connecting vertex $v_{ij,00} = (i, j)$ to vertex $v_{ij,10} = (i + 1, j)$ (see Figure 16.30(a)):

$$\text{Noise}(x, j) = h(u) = \text{N}(i + 1, j) + g(t)(\text{N}(i, j) - \text{N}(i + 1, j)).$$

Next we use the same interpolation method to reconstruct the unidimensional noise on the side of the cell connecting vertex $v_{ij,01} = (i, j + 1)$ to vertex $v_{ij,11} = (i + 1, j + 1)$:

$$\text{Noise}(x, j + 1) = h(u) = \text{N}(i + 1, j + 1) + g(t)(\text{N}(i, j + 1) - \text{N}(i + 1, j + 1)).$$

Finally, the noise function is reconstructed using linear interpolation along the y direction:

$$\text{Noise}(x, y) = \text{Noise}(x, j)(1 - v) + \text{Noise}(x, j + 1)v,$$

where $v = y - j$.

Interpolation with Coons patches. Bilinear interpolation and reconstruction by lofting can present problems. They accentuate directionality along the direction chosen for computing the linear interpolation in the interior of each cell. What is more, the differentiability class in the boundaries of the cell is, at the most, C^1.

We can avoid these problems using Coons surfaces (see Chapter 8) to perform the reconstruction. The Coons method constructs surfaces of class C^2 in the boundaries of the cell and does not have any directionality bias in the reconstruction inside each cell.

In this method, interpolation in cell $C_{ij} = [i, i + 1] \times [j, j + 1]$ is obtained as follows. We use four unidimensional linear interpolations to reconstruct the noise function in the cell edges:

$$(x, j) \mapsto f_{00,10}(x);$$
$$(x, j + 1) \mapsto f_{01,11}(x);$$
$$(i, y) \mapsto f_{00,01}(y);$$
$$(i + 1, y) \mapsto f_{10,11}(y).$$

The graphs of these functions are shown in Figure 16.30(a). The Coons method allows us to obtain a surface whose boundary is formed by these four graphs. (see Figure 16.30(b)).

16.6.3 3D Noise

In this case we have $J_n = \{0, 1, 2, \ldots, n\} \subset \mathbb{R}$. The 3D grid J^3 is obtained by the Cartesian product $J^3 = J_n \times J_n \times J_n$. A random scalar field $\text{N} \colon J^3 \to \mathbb{R}$ associates, to each grid vertex $v_{ijk} = (i, j, k)$, a random real vector $\text{N}(v_{ijk}) \in [-1, 1]$. A simple method of reconstructing the noise function consists of using trilinear interpolation. The procedure is analogous to the bilinear interpolation case but uses seven linear interpolations, as illustrated in Figure 16.31.

We can also use the lofting method by making one 2D reconstruction on two parallel faces of the cell, and a linear interpolation between them. We can also use the volumetric Coons method to perform the reconstruction of the 3D noise. Certainly the results of the Coons reconstruction are better than those of the previous two.

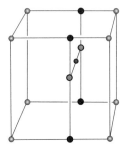

Figure 16.31. Reconstruction by trilinear interpolation.

16.7   Gradient Noise

We will now examine the reconstruction details of gradient noise in the cases of 1D, 2D, and 3D noise. The n-dimensional case is a natural extension and was left as an exercise.

16.7.1   Unidimensional Gradient Noise

Reconstruction of gradient noise uses the blending method. What is a blending method? A function $b_0 \colon [0, 1] \to \mathbb{R}$ of class C^∞, is said to be a *blending function* of order k, centered at 0, if the following conditions are satisfied:

1. $b_0(0) = 1$.

2. $b_0(1) = 0$.

3. $b_0^{(j)}(0) = b_0^{(j)}(1) = 0$, $j = 1, 2, \ldots, k$, where $b_0^{(j)}$ indicates its derivative of order j.

Likewise, we define a blending function centered at 1, b_1 by changing the conditions (1) and (2) by $b_1(0) = 0$ and $b_1(1) = 1$, respectively (the third condition is not affected). It is easy to see that if b_0 is a blending function centered at 0, then $b_1(t) = b_0(1 - t)$ is a blending function centered at 1, and vice versa.

Example 16.3. The polynomial $b_3(t) = 2t^3 - 3t^2 + 1$, used in the previous section for interpolation, is a blending function of order 1 with basis in 0. On the other hand, the polynomial of degree 5 $b_5(t) = -6t^5 + 15t^4 - 10t^3 + 1$ is a blending function of order 2 with basis in 0 (see Figure 16.32). ❑

As shown in Figure 16.32, if b_0 and b_1 are blending functions in $[0, 1]$ with basis 0 and 1, respectively, we can define the *blending* of two functions, if $h_0, h_1 \colon [0, 1] \to \mathbb{R}$, as being the function $h \colon [0, 1] \to \mathbb{R}$, given by

$$h(t) = b_0(t)h_0(t) + b_1(t)h_1(t). \tag{16.4}$$

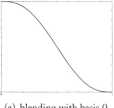 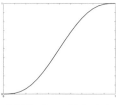 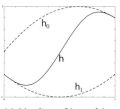

(a) blending with basis 0   (b) blending with basis 1   (c) blending of h_0 and h_1

Figure 16.32. Blending of two functions. (a) Graph of the function b_5. (b) Graph of the function $b_5(1-t)$, which is a blending function of order 2 with basis in 1. (c) The blending of function $h_0(t) = -9t^2 + 12t + 11$ with function $h_1(t) = 8t^2 - 8t + 1$. Here, we use $b_0(t) = b_5(t)$, and $b_1(t) = b_5(1-t)$, where b_5 is the polynomial defined in Example 16.3.

Theorem 16.4. The function h, resulting from the blending of functions h_0 and h_1, satisfies the following properties: $h(0) = h_0(0)$, $h(1) = h_1(1)$, $h'(0) = h_0'(0)$ and $h'(1) = h_1'(1)$.   ❑

The demonstration of the theorem has been left as an exercise. The theorem shows that blending produces a deformation of the functions h_0 and h_1 in the function h, preserving the value of h_0, its derivative at the initial point $t = 0$, and preserving the value of h_1 and its derivative at the final point $t = 1$ (this can be seen in Figure 16.32(c)). We also left as an exercise the demonstration that if the blending function is of order C^k, then $h^{(j)}(0) = h_0^{(j)}(0)$ and $h^{(j)}(1) = h_1^{(j)}(1)$, $j = 1, 2, \ldots, k$.

The extension of the concept of blending function, for an interval $[j, j+1]$ of the grid $J_n \subset \mathbb{R}$, is immediate. Besides, all the above results are valid, by replacing the interval $[0, 1]$ by an interval $[j, j+1]$; for this, it is enough to observe that if b_0 is a blending function in $[0, 1]$ with basis in 0, then $b_0^j(t) = b_0(t - j)$ is a blending function in the interval $[j, j+1]$ with basis in j. Similarly, if b_1 is a blending function in $[0, 1]$ with basis 1, then $b_1^j(t) = b_1^j(t - j)$ is a blending function in $[j, j+1]$ with basis in $j+1$. The *blending* of two functions, if $h_0, h_1 \colon [j, j+1] \to \mathbb{R}$, is the function $h \colon [j, j+1] \to \mathbb{R}$, given by

$$h(t) = b_0^j(t)h_0(t) + b_1^j(t)h_1(t). \tag{16.5}$$

As we previously saw, the reconstruction problem in the case of the gradient noise, is stated as the following. Given the random vector field $\mathbf{g}(j) = (N(j), 1) \in \mathbb{R}^2$, determine a noise function $\text{Noise}(x)$, that satisfies these two conditions: $\text{Noise}(j) = 0$ and the vector $\mathbf{g}(j) = (N(j), 1)$ is normal to the graph $(x, \text{Noise}(x))$ of the noise function for $x = j$, $j = 0, 1, \ldots, n$. (These conditions are illustrated in Figure 16.26(b).)

To solve this problem, consider the following procedure:

1. We obtain two blending functions b_j and b_{j+1} in $[j, j+1]$, with basis in j and $j+1$, respectively.

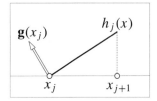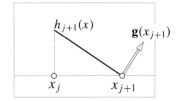

Figure 16.33. Functions h_j and h_{j+1}.

2. We obtain a function $h_j\colon [j, j+1] \to \mathbb{R}$, such that $h_j(j) = 0$ and the graph of h_j is perpendicular to vector $\mathbf{g}(j)$ at point $x = j$.

3. We obtain a function $h_{j+1}\colon [j, j+1] \to \mathbb{R}$, such that $h_{j+1}(j+1) = 0$ and the graph of h_{j+1} is perpendicular to the vector $\mathbf{g}(j+1)$ at point $x = j+1$.

4. We calculate the blending h of the functions h_j and h_{j+1}:

$$h(t) = h_j(t)b_j(t) + h_{j+1}(t)b_{j+1}(t).$$

We have left as an exercise to show, using Theorem 16.4, that function h, from this procedure, is a solution to the problem of reconstructing gradient noise.

As we already defined blending functions, to solve the problem of reconstructing the gradient noise function according to the above procedure, it is enough to determine the functions h_j and h_{j+1}. Ken Perlin uses, for this end, the linear functions: $h_j(x) = -\mathrm{N}(j)x$ and $h_{j+1}(x) = -\mathrm{N}(j+1)x$. Geometrically, the graph of the function h_j is the straight line passing through vertex j and is perpendicular to vector $\mathbf{g}(j)$ (see Figure 16.33). The same happens with function $h_{j+1}(x)$.

In Figure 16.34, we show the gradient noise function reconstructed by the above process using the random gradient field $\mathbf{g}(j) = (\mathrm{N}(j), 1)$, where the field $\mathrm{N}(j)$ is the same used to generate the scalar noise function shown in Figure 16.28. In this reconstruction, we

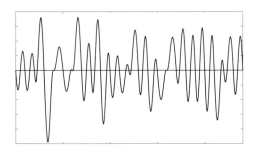

Figure 16.34. Gradient noise.

use the blending functions $b_0(t) = b_5(t)$ and $b_1(t) = b_5(1-t)$, where b_5 is the polynomial defined in Example 16.3. The noise function therefore has class C^2. Notice, once again, that by using the same grid and random field $N(j)$, the gradient noise presents higher frequencies.

16.7.2 2D Gradient Noise

Given the domain $I^2 = [0,1] \times [0,1]$ on the plane, we indicate the four vertices of the square by $v_{00} = (0,0)$, $v_{10} = (1,0)$, $v_{11} = (1,1)$ and $v_{01} = (0,1)$. A *2D blending function* of order k, with basis v_{00}, is a function $b_{00}\colon I^2 \to \mathbb{R}$, such that $b_{00}(0,0) = 1$; b_{00} is canceled out in the edges $x = 1$ and $y = 1$, that is, $b_{00}(1,y) = b_{00}(x,1) = 0$; and the derivatives up to order k of b_{00} are canceled out in the edges $x = 1$ and $y = 1$. Figure 16.35(a) shows the graph of a 2D blending function centered at $(0,0)$.

Note that edges $x = 1$ and $y = 1$, where b_{00} and their derivatives are canceled out, are the edges that do not contain the vertex $(0,0)$. This makes it easy to generalize the concept of the blending function in order to define functions with basis at vertices v_{10}, v_{11}, and v_{01} of the unit square. The graph of those functions is shown in Figure 16.35(b)–(d).

Starting from a blending function b_{00}, centered at v_{00}, we can obtain blending functions centered at the other three vertices of I^2: $b_{10}(x,y) = b_{00}(1-x,y)$, $b_{11}(x,y) = b_{00}(1-x,1-y)$, and $b_{01}(x,y) = b_{00}(x,1-y)$.

Example 16.5. This is an example of how to construct 2D blending functions starting from 1D blending functions.

If $b_0\colon [0,1] \to \mathbb{R}$ is a 1D blending function in the interval $[0,1]$, centered at 0, then $b_{00}(x,y) = b_0(x)b_0(y)$ is a 2D blending function centered at $v_{00} = (0,0)$. In this way, we obtain the four blending functions of the grid $[0,1]^2$:

1. $b_{00}(x,y) = b_0(x)b_0(y)$, centered at $v_{00} = (0,0)$.

2. $b_{10}(x,y) = b_1(x)b_0(y) = b_0(1-x)b_0(y)$, centered at $v_{10} = (1,0)$.

3. $b_{11}(x,y) = b_1(x)b_1(y) = b_0(1-x)b_0(1-y)$, centered at $v_{11} = (1,1)$.

4. $b_{01}(x,y) = b_0(x)b_1(y) = b_0(x)b_0(1-y)$, centered at $v_{01} = (0,1)$.

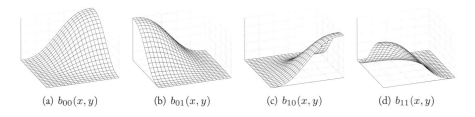

(a) $b_{00}(x,y)$ (b) $b_{01}(x,y)$ (c) $b_{10}(x,y)$ (d) $b_{11}(x,y)$

Figure 16.35. Blending functions in $[0,1] \times [0,1]$.

Starting from the 1D blending functions previously studied, we can obtain several 2D blending functions. The blending functions shown in Figure 16.35, were obtained through this process by taking $b_0(x)$ as being the 1D blending function of degree 5 from Example 16.3: $b_0(t) = b_5(t) = -6t^5 + 15t^4 - 10t^3 + 1$. (Other examples of blending functions are in the exercises.) ❏

Given four functions $h_{00}, h_{10}, h_{11}, h_{01} \colon I^2 \to \mathbb{R}$, we define the blending as being the function $h \colon I^2 \to \mathbb{R}$ given by

$$h(x,y) = \sum_{i,j=0}^{1} h_{ij}(x,y)b_{ij}(x,y) \tag{16.6}$$
$$= (h_{00}b_{00} + h_{01}b_{01} + h_{10}b_{10} + h_{11}b_{11})(x,y),$$

where each b_{ij} is a blending function centered at the vertex $v_{ij} = (i,j)$. We left as an exercise the statement and demonstration of the 2D analog of Theorem 16.4 with the properties of the function h in the above equation.

This concept can be extended to obtain blending functions in a cell $C_{ij} = [i, i+1] \times [j, j+1]$ of the 2D grid J^2. The four vertices of that cell will be indicated by $v_{ij,mn} = (i+n, j+n)$, $m, n = 0, 1$. As in the unidimensional case, if b_{mn}, $m, n = 0, 1$, is a blending function in $[0,1] \times [0,1]$, centered at the vertex $v_{(mn)}$, then $b_{ij,mn}(x,y) = b_{mn}(x-i, y-j)$ is a blending function in the cell C_{ij}, centered at the vertex $(i+m, j+n)$.

As in the unidimensional case, the problem of reconstructing the 2D gradient noise function is stated in the following way. Given the random vector field $\mathbf{g}(i,j) = (\mathrm{N}(i,j), 1) \in \mathbb{R}^3$, $\mathrm{N}(i,j) \in S^2$, the noise function $\mathrm{Noise}(x,y)$ should satisfy the following two conditions: $\mathrm{Noise}(i,j) = 0$, and the vector $\mathbf{g}(i,j) = (\mathrm{N}(i,j), 1)$ is normal to the graph $(x, y, \mathrm{Noise}(x,y))$ of the noise function at the grid vertices: $(x,y) = (i,j)$.

Figure 16.36 illustrates the problem, showing, in a cell, the graph of the noise function together with the gradient field.

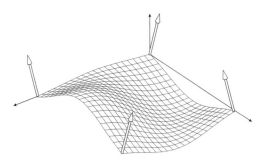

Figure 16.36. Noise function and gradient field.

Using the concept of 2D blending, the solution to the reconstruction problem is similar to the unidimensional case. In other words, in each grid cell $C_{ij} = [i, i+1] \times [j, j+1]$, we use the following procedure:

1. We obtain four blending functions $b_{ij,00}$, $b_{ij,01}$, $b_{ij,10}$, and $b_{ij,11}$ centered, respectively, at vertices $v_{ij,mn}$, $m, n = 0, 1$; that is, (i, j), $(i, j+1)$, $(i+1, j)$, and $(i+1, j+1)$.

2. For each cell vertex $v_{ij,mn}$, $m, n = 0, 1$, we obtain a function $h_{ij,mn} \colon C_{ij} \to \mathbb{R}$, such that $h_{ij,mn}(v_{ij,mn}) = 0$ and the graph of $h_{ij,mn}$ is perpendicular to the vector $\mathbf{g}(v_{ij,mn})$ at the vertex $v_{ij,mn}$.

3. The noise function $\text{Noise}(x, y)$ in the cell is obtained by performing blending of the four functions $h_{ij,mn}$ of the previous item, according to Equation (16.6).

We left, as an exercise, the demonstration of the statement in the last item above.

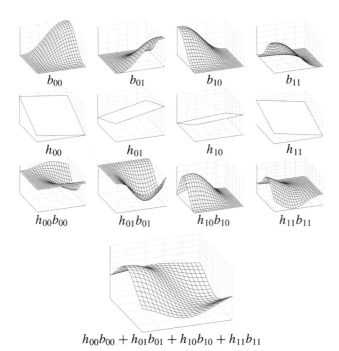

$$h_{00}b_{00} + h_{01}b_{01} + h_{10}b_{10} + h_{11}b_{11}$$

Figure 16.37. Calculating the 2D noise function in a cell. Top row: the four blending functions, b_{ij}, at each of the vertices. Second row: the linear functions h_{ij} associated to the gradient field at each vertex of the cell. Third row: the product $b_{ij}h_{ij}$. Bottom: the final blending given by the sum $\sum_{i,j} h_{ij}b_{ij}$.

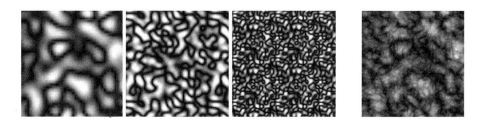

Figure 16.38. Noise obtained after adding noises up.

To calculate $\text{Noise}(x, y)$, we can use the blending function from Example 16.5. The problem is reduced to obtain the functions $h_{ij,mn}$. As in the unidimensional case, Ken Perlin uses linear functions:

$$h_{ij,mn}(x, y) = -\langle \text{N}(v_{ij,mn}), (x, y) \rangle.$$

Geometrically, the graph of $h_{ij,mn}$ is a plane passing through vertex $v_{ij,mn}$, which is perpendicular to the vector $\mathbf{g}(v_{ij,mn})$ at this vertex. Figure 16.37 illustrates the calculation of the noise function in a cell.

Example 16.6. The last image of Figure 16.38 displays a 2D texture created using a turbulence function, obtained by adding three gradient noises, as shown in the three images on the left. ❏

16.7.3   3D Gradient Noise

Given a 3D cell $I^3 = [0, 1] \times [0, 1] \times [0, 1]$ in $\mathbb{R}^3$, we indicate the eight vertices of the cube by $v_{mno} = (m, n, o)$, $m, n, o = 0, 1$. Explicitly, we have $v_{000} = (0, 0, 0)$, $v_{100} = (1, 0, 0)$, $v_{010} = (0, 1, 0)$, $v_{001} = (0, 0, 1)$, $v_{110} = (1, 1, 0)$, $v_{011} = (0, 1, 1)$, $v_{101} = (1, 0, 1)$, and $v_{111} = (1, 1, 1)$. A *2D blending function* of order k, centered at the vertex v_{mno}, is a function $b_{mno} \colon I^3 \to \mathbb{R}$, such that

1. $b_{mno}(m, n, o) = 1$;

2. b_{mno} is canceled out in the faces $x = (m + 1)\%2$, $y = (n + 1)\%2$, and $z = (o + 1)\%2$. (The operation % indicates the remainder of the division by 2.) That is, $b_{mno}((m + 1)\%2, y, z) = b_{mno}(x, (n + 1)\%2, z) = b_{mno}(x, y, (o + 1)\%2) = 0$;

3. The derivatives up to order k of b_{mno} are canceled out in the faces $x = (m + 1)\%2$, $y = (n + 1)\%2$, and $z = (o + 1)\%2$ of the cell.

Starting from a blending function b_{000}, centered at (000), we can obtain blending functions centered at the other vertices of the cube I^3:

$$b_{100}(x, y, z) = b_{000}(1 - x, 0, 0),$$
$$b_{010}(x, y, z) = b_{000}(0, 1 - y, 0),$$
$$b_{001}(x, y, z) = b_{000}(0, 0, 1 - z),$$
$$b_{110}(x, y, z) = b_{000}(1 - x, 1 - y, 0),$$
$$b_{011}(x, y, z) = b_{000}(0, 1 - y, 1 - z),$$
$$b_{101}(x, y, z) = b_{000}(1 - x, 0, 1 - z),$$
$$b_{111}(x, y, z) = b_{000}(1 - x, 1 - y, 1 - z).$$

The example below shows how to construct 3D blending functions, starting from 1D blending functions.

Example 16.7. If $b_0 \colon [0, 1] \to \mathbb{R}$ is a 1D blending function in the interval $[0, 1]$, centered at 0, then $b_{000}(x, y, z) = b_0(x)b_0(y)b_0(z)$ is a 3D blending function centered at $v_{000} = (0, 0, 0)$. In this way, we obtain the eight blending functions of the cell $[0, 1]^3$:

$$b_{000}(x, y, z) = b_0(x)b_0(y)b_0(z), \qquad \text{centered at } v_{000} = (0, 0, 0),$$
$$b_{010}(x, y, z) = b_0(x)b_0(1 - y)b_0(z), \qquad \text{centered at } v_{010} = (0, 1, 0),$$
$$b_{001}(x, y, z) = b_0(x)b_0(y)b_0(1 - z), \qquad \text{centered at } v_{001} = (0, 0, 1),$$
$$b_{110}(x, y, z) = b_0(1 - x)b_0(1 - y)b_0(z), \qquad \text{centered at } v_{110} = (1, 1, 0),$$
$$b_{011}(x, y, z) = b_0(x)b_0(1 - y)b_0(1 - z), \qquad \text{centered at } v_{011} = (0, 1, 1),$$
$$b_{101}(x, y, z) = b_0(1 - x)b_0(y)b_0(1 - z), \qquad \text{centered at } v_{101} = (1, 0, 1),$$
$$b_{100}(x, y, z) = b_0(1 - x)b_0(y)b_0(z), \qquad \text{centered at } v_{100} = (1, 0, 0),$$
$$b_{111}(x, y, z) = b_0(1 - x)b_0(1 - y)b_0(1 - z), \quad \text{centered at } v_{111} = (1, 1, 1),$$

Starting from the 1D blending functions previously studied, we can obtain several 3D blending functions. ❏

Given eight real functions h_{000}, h_{100}, h_{010}, h_{001}, h_{110}, h_{011}, h_{101}, and h_{111}, defined in the interval I^3, we define the blending as the function $h \colon I^3 \to \mathbb{R}$, given by

$$h(x, y, z) = \sum_{m,n,o=0}^{1} h_{mno}(x, y, z)b_{mno}(x, y, z), \tag{16.7}$$

where each b_{mno} is a blending function centered at the vertex $v_{mno} = (m, n, o)$. We left as an exercise the statement and demonstration of the 3D application of Theorem 16.4.

This concept can be extended to obtain blending functions in a cell $C_{ijk} = [i, i + 1] \times [j, j + 1] \times [k, k + 1]$ of the 3D grid J^3. The eight vertices of this cell will be indicated by $v_{ijk,mno} = (i + n, j + n, k + o)$, $m, n, o = 0, 1$. As in the unidimensional and 2D

446

cases, if b_{000} is a blending function in unit cell $[0,1]^3$ centered at vertex $(0,0,0)$, then $b_{ijk,000}(x,y,z) = b_{ijk}(x-i, y-j, z-k)$ is a blending function in the cell C_{ijk}, centered at vertex (i,j,k).

As in the 2D case, the problem of reconstructing the 3D gradient noise function is stated as the following. Given a random vector field $\mathbf{g}(i,j,k) = (\mathrm{N}(i,j,k),1) \in \mathbb{R}^4$, the noise function $\mathrm{Noise}(x,y,z)$ should satisfy these two conditions: $\mathrm{Noise}(i,j,k) = 0$, and the vector $\mathbf{g}(i,j,k)$ is normal to the graph $(x,y,z,\mathrm{Noise}(x,y,z))$ of the noise function at the grid vertices; that is, for $(x,y,z) = (i,j,k)$.

Solving this 3D case is similar to our solution for the 2D case. In each grid cell $C_{ijk} = [i,i+1] \times [j,j+1] \times [k,k+1]$, we use the following procedure:

1. We obtain eight blending functions: $b_{ijk,000}(x,y,z)$, $b_{ijk,100}(x,y,z)$, $b_{ijk,010}(x,y,z)$, $b_{ijk,001}(x,y,z)$, $b_{ijk,110}(x,y,z)$, $b_{ijk,011}(x,y,z)$, $b_{ijk,101}(x,y,z)$, and $b_{ijk,111}(x,y,z)$, centered at the vertices $v_{ijk,mno}$, $m,n,o = 0,1$, of the cube defined by the cell.

2. For each cell vertex $v_{ijk,mno}$, $m,n = 0,1$, we obtain a function $h_{ijk,mno}: C_{ijk} \to \mathbb{R}$, such that $h_{ijk,mno}(v_{ijk,mno}) = 0$, and the graph of $h_{ijk,mno}$ is perpendicular to the vector $\mathbf{g}(v_{ijk,mno})$ at the vertex $v_{ijk,mno}$.

3. The noise function $\mathrm{Noise}(x,y,z)$ in the cell is obtained by blending the eight functions $h_{ijk,mno}$ of the previous item, according to the Equation (16.7).

We left as an exercise the demonstration of the statement in the last item above.

To calculate $\mathrm{Noise}(x,y,z)$, we can use the blending function of Example 16.7. The problem is therefore reduced to obtaining the functions $h_{ijk,mno}$. As in the 2D case, we can use the linear functions

$$h_{ijk,mno}(x,y,z) = -\langle \mathrm{N}(v_{ijk,mno}), (x,y,z)\rangle.$$

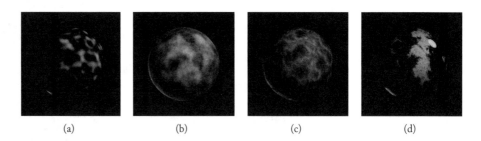

(a)　　　　(b)　　　　(c)　　　　(d)

Figure 16.39. Spheres with 3D textures defined with the Perlin noise function. (©*2001 Ken Perlin.*) (a) Applying noise itself to modulate surface color. (b) Using a texture that consists of a fractal sum of noise calls: $\sum 1/f(\mathrm{noise})$. (c) Using a fractal sum of the absolute value of noise: $\sum 1/f(|\mathrm{noise}|)$. (d) Using the turbulence texture from (c) to do a phase shift in a stripe pattern, created with a sine function of the x coordinate of the surface location: $\sin(x + \sum 1/f(|\mathrm{noise}|)$. (See Color Plate XXXV.)

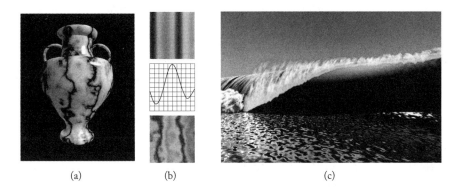

(a)   (b)   (c)

Figure 16.40. Procedural textures. (a) 3D marble vase (*[Perlin 85] ©1985 Association for Computing Machinery, Inc. Reprinted by permission.*) (b) Marble texture obtained by using a sinusoid. (c) Water textures applied to a breaking wave model. (*Figure courtesy of Manuel Gamito and Ken Musgrave.* See Color Plate XXXVI.)

Geometrically, the graph of $h_{ijk,mno}$ is a hyperplane of $\mathbb{R}^4$ passing through the vertex $v_{ijk,mno}$, which is perpendicular to the vector $\mathbf{g}(v_{ijk,mno})$ in that vertex.

Figure 16.39 shows examples of the variety of texture patterns we can obtain with the Perlin noise function. In Figure 16.40(a), we show a marble vase where the texture of the marble was obtained with a 3D texture mapping. The 3D texture was constructed using the Perlin noise function in a quite ingenious way: a turbulence function T was used to modify the phase of a sinusoid,

$$f_{\text{marble}}(x, y, z) = \text{marble\_color}(\sin(x + \text{T}(x, y, z)),$$

as illustrated in the 3D marble vase. Figure 16.40(b) illustrates the texture that would be obtained by using the sinusoid without the phase change:

$$f_{\text{marble}}(x, y, z) = \text{marble\_color}(\sin(x)).$$

In the breaking wave of Figure 16.40(c), the spray and the foam are modeled with procedural density functions. A procedural texture is also applied on the top region of the wave. Notice this texture follows naturally the downward movement of the wave. The water surface is modeled using bump mapping (Section 16.4.3) where the texture mapped simulating the small-scale ripples is obtained using noise.

16.8   Comments and References

Mapping techniques were introduced by Ed Catmull's doctoral dissertation [Catmull 74]. In this chapter, we attempted to give a unified view of the area, but development of mapping techniques since Catmull's pioneering work happened quickly and in a nonintegrated way.

Bump mapping was introduced by Jim Blinn in [Blinn 78]. Reflection mapping was introduced by Jim Blinn and Martin Newell in [Blinn and Newell 76]. In this work, auxiliary cylindrical surfaces were used. The Perlin noise function was originally published in [Perlin 85]. The technique was simultaneously developed by Edwin Peachey [Peachey 85].

16.8.1   Additional Topics

Sampling and reconstruction is crucial to mapping. Here, we studied the classic method of mipmap. In a more advanced course, this problem could be covered much more broadly.

The synthesis of textures, from a real image, is another important and interesting topic. Along the line of procedural textures, it would be interesting to study other procedural methods of texture creation and their use in modeling natural phenomena (fire, water, clouds, etc.) A broad reference on the subject is [Ebert et al. 02]. As we saw in Chapter 10, the 3D noise function can be used in modeling 3D objects.

Texture mapping techniques are just the tip of the iceberg of image-based rendering, and area that models environments starting from images and using very simple geometries. For details about this area, consult [Gomes et al. 98] or [McMillan 97]. One could devote an entire seminar to many of the techniques in this area.

Exercises

1.   Show that the interpolation operation is a particular case of blending in the case of two functions being constant.

2.   When we map an image onto a sphere, we have a deformation of the image. Various mappings (stereographic, Mercator projections, etc.) try to minimize such distortion. Do you think it is possible to obtain a mapping of an image onto a sphere without distortion? Why? What about in the case of a cylinder?

3.   Explain, in detail, how to use the Perlin noise function to generate the image in Figure 16.41.

4.   Besides the noise functions covered in this chapter, there are other interesting variants.

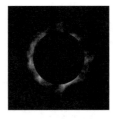

Figure 16.41. A procedural texture simulating flames. (©*2001 Ken Perlin.* See Color Plate XXXVII.)

    (a) Describe, in detail, how to define a *simplicial noise function*, by replacing the grid by a volumetric triangulation in space.

    (b) What are the advantages of a simplicial noise function in relation to a noise function using a matrix grid?

    (c) Describe, in detail, how to introduce a noise function associated to a scattered set of points in space (*scattered noise function*).

5. Determine the equations for projective mapping of an image on the plane.

6. Describe a method to guarantee that the pseudorandom vector field $N \colon J_m \to S^{m-1}$ used to obtain the gradient noise will be approximately evenly distributed on the unit disk.

7. Consider, two functions h_j and h_{j+1}, both of class C^∞, defined in the interval $[x_j, x_{j+1}]$, and let b_0 be a blending function.

    (a) If the derivatives of the function b_0 satisfy $b_0^{(k)}(x_j) = b_0^{(k)}(x_{j+1}) = 0$, $k = 0, 1, \ldots, n$, then the differentiability class of the function h, obtained by the blending, is at least equal to n.

    (b) Determine the differentiability class of the Perlin blending gradient, where the blending function is the cubic polynomial $b_0(t) = 3t^2 - 2t^3$.

    (c) Repeat the previous item for the polynomial of degree 5, $b_0(t) = 6t^5 - 15t^4 + 10t^3$.

8. Define an interpolation of the noise function using the function

$$g(t) = \frac{1 - \cos(\pi t)}{2}.$$

(Trigonometrical interpolation).

    (a) What is the differentiability class of this interpolation?

    (b) Show that g is a blending function.

9. A cubic interpolation of the noise function can be obtained using four consecutive grid vertices, j_0, j_1, j_2, and j_3, to perform the interpolation in the interval $[j_1, j_2]$. Take

$$
\begin{aligned}
A &= (N(j_3) - N(j_2)) - (N(j_0) - N(j_1)); \\
B &= N(j_0) - N(j_1) - A; \\
C &= N(j_2) - N(j_0); \\
D &= N(j_1).
\end{aligned}
$$

and define the interpolating polynomial by

$$h(t) = At^3 + Bt^2 + Ct + D.$$

    (a) Prove that $h(0) = N(j_1)$ and $h(1) = N(j_2)$.

    (b) Show the interpolation is of class C^2.

10. Show that the rotation of an image can be performed by consecutively using a horizontal shear (i.e., along the x-axis) followed by a vertical shear (i.e., along the y-axis).

11. Study the spherical reflection mapping of OpenGL and give a detailed description using the concepts of this chapter. What are the disadvantages of this mapping?

12. Consider an image U on the plane and a quadrilateral V.

    (a) Determine a deformation mapping of U in V by performing a bilinear interpolation.

    (b) Calculate the inverse mapping.

    (c) Compare this mapping with the projective mapping.

13. Define Mercator projection in the sphere and discuss its advantages and disadvantages in relation to the parameterization with spherical coordinates. Do the same for the stereographic projection.

14. Describe how mapping can be used to simulate the effect of refraction (*refraction mapping*).

15. Describe how mapping can be used to simulate the effect of transparency of an object (*transparency mapping*).

16. Using the z-buffer algorithm, describe how mapping can be used to generate shadows of the objects in a scene (*shadow mapping*).

17. Discuss the problem of using the cylinder as an auxiliary surface in reflection mapping. Discuss methods for overcoming this problem.

18. Using texture mapping, describe a method to simulate the highlight from the Phong illumination method.

(a) 360° sphere mapping   (b) Reflection mapping

Figure 16.42. Sphere mapping in the teapot. (*Image appears in online Panda3D Manual, Panda3D open source 3D game engine, http://panda3d.org.* See Color Plate XXXVIII.)

19. *Sphere mapping* is an environment mapping widely used in hardware implementations and in the OpenGL graphics system. This mapping uses an image of the environment, taken with a 360° fisheye lens, as shown in Figure 16.42(a). An environment mapping is created by mapping that image onto a hemisphere of the sphere using orthogonal decal mapping. In Figure 16.42(b), we show the effect of the reflection mapping associated to this environment map.

   (a) Fill out the details of reflection mapping.

   (b) Does the reflection mapping depend on the observer's position?

20. One of the problems with cubic reflection mapping (when the auxiliary surface is a cube) happens when the object is polygonal and the reflection vector, calculated at each vertex of one of the polygons, associates points on different faces of the cube. How can this problem be solved?

21. This exercise defines some nonprojective camera models.

   (a) Define a virtual camera model using cylindrical projection.

   (b) Describe the mapping of the virtual screen of the camera in (a) to the graphics device screen.

   (c) Define a virtual camera model using spherical projection.

   (d) Describe the mapping of the virtual screen of the camera in (c) to the graphics device screen. (Hint: you should take into account the geometry of the virtual screen.)

   (e) Discuss the problem of uniform sampling with cylindrical and spherical cameras.

22. Outline at least five distinct ways of defining a bump mapping operation using an RGB image instead of a grayscale one.

17 | Composition

This chapter deals with post-processing operations. These are operations between images after the rasterization process, such as operations that combine several images to form a single one (this is an important operation used, for instance, when we have very complex scenes or when we want to combine real images with virtual ones).

17.1 The Alpha Channel

In Chapter 15, the considerations established about sampling a pixel on the image led us immediately to the concept of a pixel geometry. Usually, if a pixel does not have geometry—that is, if no scene object is projected onto the pixel—a *background color* (usually black) is attributed to it.

Another option is to make empty pixels transparent: if the pixel does not have geometry, it is transparent; if the entire pixel region is occupied by points from one or more graphics objects, it is opaque; if only parts of the pixel area are occupied by the geometry of graphics objects, it is partially transparent.

In this way, transparency becomes one more attribute of an image—an attribute that can be stored as an additional channel to the existing three R, G, B color channels. This channel is called the *alpha channel*, and is represented by the Greek letter α. Figure 17.1 displays an image and its α channel. Notice we have $\alpha = 0$ and $\alpha = 1$ in the transparent and opaque pixels, respectively. In other pixels (in the boundary between regions of opacity and transparency), α assumes intermediate values. Therefore, the value of the α channel in a pixel measures the percentage of the fragment area (the area filled with geometric objects at the pixel), in relation to the pixel area. In other words, it provides the probability that a point in the pixel will belong to a fragment of an object.

One application of the alpha channel is in the combination of two images by a process generically called *composition*. There a number of instances in which an image g can be divided into two parts: foreground f and background b. In this case, the final image is obtained by linear interpolation using the α channel:

$$g = \alpha f + (1 - \alpha)b.$$

Figure 17.1. Image and alpha channel. (See Color Plate XXXIX.)

The resulting image g is equal to the back image b at the transparent pixels ($\alpha = 0$) and to the front image f at the opaque pixels ($\alpha = 1$). In pixels with partial transparency ($0 < \alpha < 1$), g is a combination of the front and background images. Figure 17.2 displays a foreground-background composition operation using the α channel.

The alpha channel is much more than a way to combine two images. If we store an image with its values pre-multiplied by alpha, we eliminate three multiplication operations for each pixel and therefore have a significant computational gain. What is more, this image will occupy less space because it now has only three color channels. Pixels with $\alpha = 0$ have a null color value and therefore do not even need to be stored. In this case, the image becomes a 2D planar graphics object $\mathcal{O} = (U, f)$, with color attribute f, whose geometric support U is not necessarily a rectangle on the plane. The transparency of pixels produces an attenuation of the intensities at the boundary of the image, allowing for a smoother reconstruction of the high frequencies, thus avoiding the staircase (jaggie) visual effect. Images with pre-multiplied alpha are called *sprites*. To facilitate operations with sprites, we associate a rectangle (bounding box) to each sprite.

Figure 17.2. Composition with the alpha channel. (See Color Plate XL.)

17.2   Composition and Pixel Geometry

The image composition method we described in the previous section uses the concept of pixel transparency without taking into account its geometry. In this section, we will discuss an even better solution to the composition problem that takes into account the pixel geometry.

17.2.1   Composition with the Alpha Channel

As we saw in the previous section, the α channel is obtained by discretizing the transparency function. This function is spatially quantized using the same discretization grid as the image, and the transparency information at each pixel is quantized using the same number of bits from the quantization of the color channels of the image. The resulting image is appended to the color components of the image, constituting the α channel of the image. This new channel becomes an integral part of the image for which it was calculated. In this way, we obtain a quite homogeneous representation of the image, together with its transparency function.

We previously studied an image composition operator that uses the transparency function of the pixel to overlap two images. By taking into account the pixel geometry, we can use the transparency function of the pixel to define other image composition operators.

When we perform the composition of two images f and g to obtain an image h, we should combine, for each pixel, not just the color information of images f and g, but also the information of the transparency channels α_f and α_g of those images. The combination of these channels results in the α channel of the final image, which is of great importance for future combinations of this image with other ones.

While the α channel provides the percentage of each pixel's area occupied by objects, it does not store any information about the geometry of these objects (i.e., the pixel geometry). Consequently, when we combine the α channel of two images, we have to establish certain assumptions.

Consider the simplest case: f and g each have only a polygon fragment at the pixel. Figure 17.3(a), (b), and (c) show the three possible configurations of the combination geometry at a pixel from two images: the color regions of f and g overlap either completely, partially, or not at all.

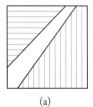 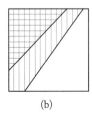 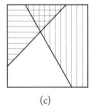 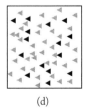

(a)	(b)	(c)	(d)

Figure 17.3. Configurations of the pixel geometry: (a) no overlap, (b) total overlap, (c) partial overlap, (d) statistical distribution of the microgeometry of the pixels.

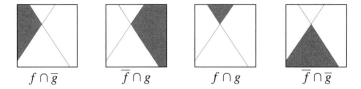

$$f \cap \overline{g} \qquad \overline{f} \cap g \qquad f \cap g \qquad \overline{f} \cap \overline{g}$$

Figure 17.4. Partition defined by the polygon fragments.

The configuration of partial overlap, which is the most probable, is known as *generic position*. In the operations, we will assume this position always happens. This is equivalent to admitting the microgeometry of the pixels has a statistical distribution (Figure 17.3(d)). In this case, the polygon fragments on each image f and g determine a partition of the pixel into four regions: $f \cap \overline{g}$, $\overline{f} \cap g$, $f \cap g$, and $\overline{f} \cap \overline{g}$, where the bar indicates the complement set in relation to the pixel region. In Figure 17.4, we show the pixel together with the four sets of that partition.

Different configurations of the pixel geometry can be obtained by making Boolean operations between sets of those regions. For instance, the pixel geometry of image f is given by $[f \cap \overline{g}] \cup [f \cap g]$.

If α_f and α_g represent the values of the transparency functions of f and g, respectively, at a pixel, then we will assume the percentage of intersection area between the two regions is given by the product $\alpha_f \alpha_g$. This assumption seems plausible, given that the product $\alpha_f \alpha_g$ indicates the probability of a point belonging to the intersection between the two regions. We can have crude errors when performing successive image compositions, but usually such errors are irrelevant.

As the percentage of covered area by the region $f \cap g$ is $\alpha_f \alpha_g$, we can easily calculate the percentage of the pixel area corresponding to other partition regions in Figure 17.4. These values are given in Table 17.1. From this point, we can calculate the transparency value associated to several Boolean operations with the pixel fragments. The pixel area only covered by the polygon fragment on the image f is given by $f - (f \cap g)$; therefore, the corresponding transparency of this polygon fragment on image f is given by $\alpha_f - \alpha_f \alpha_g$. Similarly, the percentage covered only by the geometric fragment of image g is $\alpha_g - \alpha_f \alpha_g$. The percentage of pixel area that is covered neither by the image fragments f nor by the image g, is given by $(1 - \alpha_f)(1 - \alpha_g)$.

Region	Transparency values
$\overline{f} \cap \overline{g}$	$(1 - \alpha_f)(1 - \alpha_g)$
$f \cap \overline{g}$	$\alpha_f(1 - \alpha_g)$
$\overline{f} \cap g$	$(1 - \alpha_f)\alpha_g$
$f \cap g$	$\alpha_f \alpha_g$

Table 17.1. Percentage of pixel area corresponding to partition regions in Figure 17.4.

Of course, we do not have unicity. For instance, the transparency given by the pixel fragment on image f is obviously given by α_f. However, we have $f = (f \cap \bar{g}) \cup (f \cap g)$ and therefore the transparency given by the fragment on image f can also be calculated by $\alpha_f(1 - \alpha_g) + \alpha_f\alpha_g$.

17.2.2 Composition with Bitmask

We have described two extreme methods of combining two images: one was using analytical sampling to store the fragments of objects in each pixel and to perform the operations in an exact way; the other was replacing the geometry of each pixel with the alpha channel and performing the operations in a roughly approximate way.

There is also a more moderate option that is more efficient than using the analytical method and far more precise than using the transparency function. This method consists of representing the pixel geometry with a bitmask, as described in Section 15.3.6.

We divide the pixel into subpixels, attributing one transparency bit to each subpixel. If the bit is 1, the pixel is opaque in that region (i.e., it does not have any degree of transparency); if the bit is 0, the subpixel is transparent (see Figure 17.5).

When we use the bitmask process, the composition operation happens in four stages. We calculate first the bitmasks, then the percentage of each image, then the color, and finally the alpha channel.

Calculating the bitmask. In the first stage, the bitmasks of the front and back images are combined, one bit at time, by means of logic operations in a way that divides the pixel decomposition into subregions. The logic operations used are defined by the Boolean operators *and* and *not*. More specifically, by making M_f and M_g the bitmask of a generic pixel on the images f and g, respectively, we can write:

$$\begin{aligned}
\bar{f} \cap \bar{g} &= (\text{not } M_f) \text{ and } (\text{not } M_g); \\
f \cap \bar{g} &= M_f \text{ and } (\text{not } M_g); \\
\bar{f} \cap g &= (\text{not } M_f) \text{ and } M_g; \\
f \cap g &= M_f \text{ and } M_g.
\end{aligned} \qquad (17.1)$$

Calculating the percentage of each image. The percentage of color A_f on image f at the final pixel is given by calculating the quotient between the area of the pixel fragment on

Figure 17.5. A bitmask.

the final image and the area of the total pixel fragment of image f. The same result is valid to perform the percentage calculations A_g of image g. In terms of bitmask, the values are obtained by

$$A_f = \frac{\text{number of bits of } f \text{ 'on'}}{\text{total number of subpixels in the fragments of } f},$$

and

$$A_g = \frac{\text{number of bits of } g \text{ 'on'}}{\text{total number of subpixels in the fragments of } g}, \tag{17.2}$$

where the numbers of bits "on" in f (or g) means the number of bits equal to 1 in the fragments of image f (or g) that belong to the final image.

Calculating the color. In this stage we use the percentages of color contribution calculated in the previous stage for each image, A_f and A_g, to obtain the color of each pixel in the final image r. This is done by taking the weighted average

$$r = A_f f + A_g g. \tag{17.3}$$

Calculating the alpha channel. Similarly to the previous stage, we use the percentages A_f and A_g to calculate the alpha channel of the final image r, by the equation

$$\alpha_r = A_f \alpha_f + A_g \alpha_g. \tag{17.4}$$

Example 17.1 (Overlap operation). Consider the composition operation we just saw, in which we placed image f (front) over image g. For this, consider the polygon fragments at the pixel shown in Figure 17.6. We want to perform the overlap operation of image f on image g.

Every fragment of image f at the pixel exists in the final pixel: therefore, $A_f = 1$. In the case of image g, the fragment present on the final image is given by the Boolean operation $\overline{f} \cap g$; therefore,

$$A_g = \frac{\text{Area}(\overline{f} \cap g)}{\text{Area}(g)}. \tag{17.5}$$

With a bitmask, the calculation of A_g is performed using the discrete operations in (17.1) together with Equation (17.2). More precisely, the discrete calculation of the area

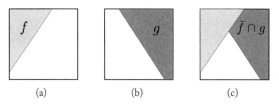

Figure 17.6. Overlap operation. (a) Front image. (b) Back image. (c) Overlap of f on g.

percentage A_g of Equation (17.5) is given by

$$A_g = \frac{\text{number of bits } [(\text{not } M_f) \text{ and } M_g]}{\text{number of bits } M_g}.$$

Starting with A_f and A_g, which measure, respectively, the percentage of fragment area coming from images f and g, the value of the pixel color is given by Equation (17.3). Similarly, the transparency value (α channel) is calculated with Equation (17.4).   ❏

17.3   Composition Algebra

As we saw above, composition using a bitmask is analogous to composition using a transparency channel. We split the geometry of the pixel decomposition into four subsets: $\bar{f} \cap \bar{g}$, $f \cap \bar{g}$, $\bar{f} \cap g$, and $f \cap g$ so we can proceed to calculating the transparency and color information of the composed image. The difference with the α channel method is that in this case the pixel geometry is known and therefore no assumptions need to be made. In this way, we have the result with the discretization precision given by the bitmask.

The final color attribute of the pixel at the composed image comes from the weighted average of the colors in the fragments of images f and g given by Equation (17.3). In addition to the colors of images f and g, we can also attribute, to a given fragment, the color 0, in the case where we want to eliminate any color information in this geometric fragment of the pixel. Table 17.2 gives the possible color choices in each pixel fragment.

Region	Possible Colors
$\bar{f} \cap \bar{g}$	0
$f \cap \bar{g}$	0 or f
$\bar{f} \cap g$	0 or g
$f \cap g$	0, f, or g

Table 17.2. Possible color choices in each pixel fragment.

17.3.1   Composition Operators

It is possible to define 12 operators involving the fragments ($2 \times 2 \times 3$). To illustrate the various composition operations, we will use the images f and g shown on the left in Figure 17.7. Image f is a torus, and image g is a checkered floor. Both images are synthetically generated. We also show in the figure the α channel of each of the images.

On the right of Figure 17.7 we show, in different tones of gray, the partitions $\bar{f} \cap \bar{g}$, $f \cap \bar{g}$, $\bar{f} \cap g$, and $f \cap g$, determined by the elements of images f and g in the supporting set. The black color indicates the image background, where we do not have elements f or g. Note that region $g \cap \bar{f}$ is not connected.

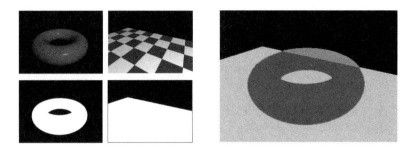

Figure 17.7. Images used to illustrate the composition operations (left; see Color Plate XLI.) Partition of the supporting set of images f and g (right).

The over operator. This is the overlap operator we saw in Section 17.1 and in Section 17.3. On the resulting image, $r = f \text{over} g$, the color of the elements of image f at the pixel always has predominance over the color in g. In terms of polygon fragments, the pixel geometry of image r is given by $f \cup (\overline{f} \cap g)$. The table in Figure 17.8 gives the color values of each fragment in the pixel geometry. Note that the polygon fragment of image f contributes 100% to the color of the pixel; that is, $A_f = 1$. The percentage of color contribution on image g is given by the relative area from the intersection $\overline{f} \cap g$ between the fragments $\overline{f}$ and g; that is,

$$A_g = \frac{\text{Area}(\overline{f} \cap g)}{\text{Area}(g)}.$$

Of course the operator is not commutative.

Using the bitmask method, we obtain

$$M_r = M_f \text{or} [(\text{not } M_f) \text{and } M_g];$$
$$A_f = 1;$$
$$A_g = \frac{\text{number of bits } [(\text{not } M_f) \text{ and } M_g]}{\text{number of bits } M_g}.$$

Region	Color
$f \cap \bar{g}$	f
$\bar{f} \cap g$	g
$f \cap g$	f
$\bar{f} \cap \bar{g}$	0

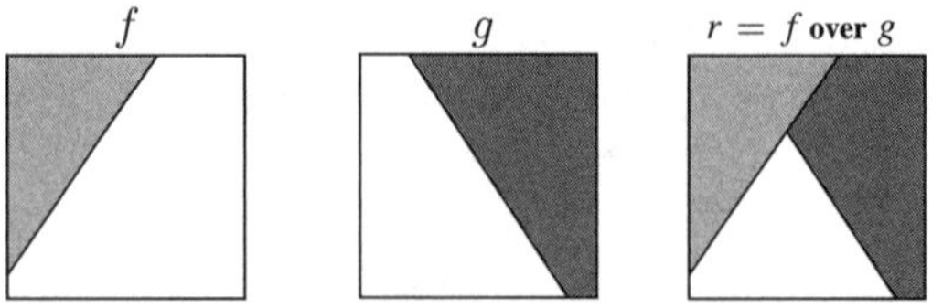

Figure 17.8. Pixel geometry of the **over** operator.

Figure 17.9. The **over** operator.

Figure 17.9 shows the effect of the **over** operator applied to images f and g; we also show the α channel of the resulting image, obtained by combining the α channel of each image, according to Equation (17.4).

The in operator. In the operation $r = f\mathbf{in}g$, we consider only the information on image f that is contained in image g. In terms of fragments, the pixel geometry on image r is given by $f \cap g$, as illustrated in Figure 17.10. The table in this figure shows the color values of each polygon fragment on the final image r. The percentage of pixel area with color information from image f is given by the relative area of the fragment of $f \cap g$ in relation to the fragment g; that is,

$$A_f = \frac{\text{Area}(f \cap g)}{\text{Area}(g)}.$$

The percentage of area occupied by the color of image g is $A_g = 0$. Note that we are interested only in the part of image f contained in image g; therefore, the color of image g does not have any influence on the pixel color of the final image. Here, the operator is also not commutative.

In terms of bitmask, we have the following result:

$$
\begin{aligned}
M_r &= M_f \text{ and } M_g; \\
A_f &= \frac{\text{number of bits } (M_f \text{and } M_g)}{\text{number of bits } M_g}; \\
A_g &= 0.
\end{aligned}
$$

Region	Color
$f \cap \bar{g}$	0
$\bar{f} \cap g$	0
$f \cap g$	f
$\bar{f} \cap \bar{g}$	0

f g $r = f\,\mathbf{in}\,g$

Figure 17.10. Pixel geometry of the **in** operator.

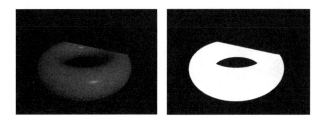

Figure 17.11. The **in** operator.

Figure 17.11 shows the result of operator **in** on images f and g, as well as the α channel of the final image, obtained by the combination of the α channel of each image.

The out operator. The result of the operation $r = f\,\mathbf{out}\,g$ consists of considering the region of image f outside the region delimited by the elements of image g. In terms of fragments, the pixel geometry of the final image r is given by the intersection $f \cap \bar{g}$ of fragments f and $\bar{g}$, as illustrated in Figure 17.12. The table in this figure shows the colors in each of the fragments. The percentage of pixel area with color attributes from image f is given by

$$A_f = \frac{\text{Area}(f \cap \bar{g})}{\text{Area}(f)}.$$

As image g does not contribute to the final pixel color, we have $A_g = 0$. Of course, the operator is not commutative.

From the bitmask method, we obtain:

$$
\begin{aligned}
M_r &= M_f \text{ and (not } M_g); \\
A_f &= \frac{\text{number of bits } (M_f \text{and (not } M_g))}{\text{number of bits } M_f}; \\
A_g &= 0.
\end{aligned}
$$

The composition operation **out** is illustrated in Figure 17.13, where we also show the α channel of the resulting image.

Region	Color
$f \cap \bar{g}$	f
$\bar{f} \cap g$	0
$f \cap g$	0
$\bar{f} \cap \bar{g}$	0

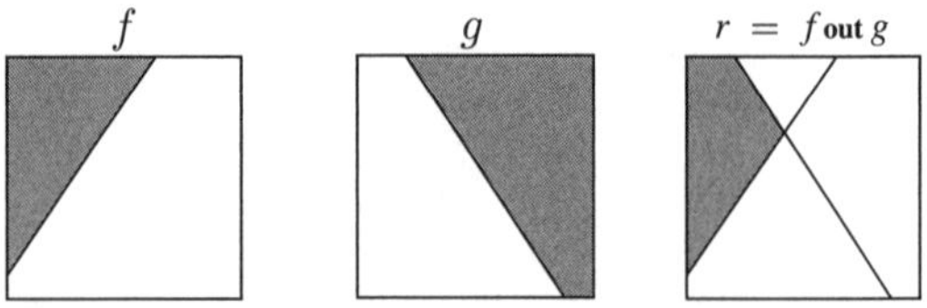

Figure 17.12. Pixel geometry of the **out** operator.

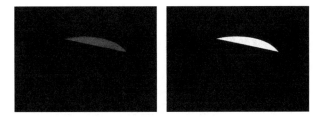

Figure 17.13. The **out** operator.

The atop operator. The image r resulting from the operation $r = f\mathbf{atop}g$ consists of overlapping the color of image f on the regions where elements from image g have color. In terms of fragments, the pixel geometry of image r is given by $(f \cap g) \cup (\overline{f} \cap g)$, as illustrated in Figure 17.14. The figure also shows the color table in the various geometric fragments of the pixel of image r. The percentage of pixel area occupied with color information from image f is given by

$$A_f = \frac{\text{Area}(f \cap g)}{\text{Area}(f)}.$$

The percentage for image g is given by

$$A_g = \frac{\text{Area}(\overline{f} \cap g)}{\text{Area}(g)}.$$

Using the discretization of the transparency channel with the bitmask, we obtain:

$$
\begin{aligned}
M_r &= M_g; \\
A_f &= \frac{\text{number of bits } (M_f \text{ and } M_g)}{\text{number of bits } M_f}; \\
A_g &= \frac{\text{number of bits } [(\text{not } M_f) \text{ and } M_g]}{\text{number of bits } M_g}.
\end{aligned}
$$

The result of applying the **atop** operator on f and g is shown in Figure 17.15. In this figure, we also show the α channel of the resulting image, which of course, coincides with the α channel of image g.

Region	Color
$f \cap \bar{g}$	0
$\bar{f} \cap g$	g
$f \cap g$	f
$\bar{f} \cap \bar{g}$	0

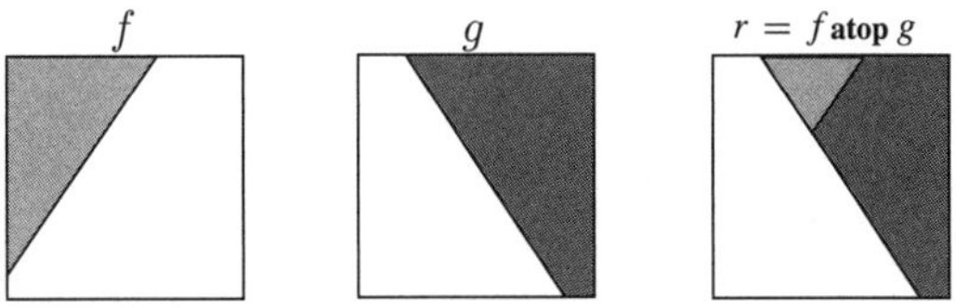

Figure 17.14. Pixel geometry of the **atop** operator.

Figure 17.15. The **atop** operator.

The xor operator. The image r resulting from the application of the operator $r = f\mathbf{xor}g$, contains all the elements belonging to images f or g, with exception to the elements that simultaneously belong to the two images. In terms of operations with geometric fragments at each pixel, the pixel geometry on r is given by the Boolean operation called *symmetric difference*, between the polygon fragments of images f and g. That is, $f\mathbf{xor}g = (f - g) \cup (g - f)$, as illustrated in Figure 17.16.

This figure also provides a table with the color value of each fragment. The percentage of pixel area with the color of image f is given by

$$A_f = \frac{\text{Area}(f \cap \bar{g})}{\text{Area}(f)},$$

and the percentage of the area of the pixel with the color of image g is

$$A_g = \frac{\text{Area}(g \cap \bar{f})}{\text{Area}(g)}.$$

Observe the operator **xor** is commutative.

In the bitmask process, we obtain the following parameter values:

$$M_r = [M_f \text{and} (\text{not } M_g)] \text{ or } [(\text{not } M_f) \text{and } M_g];$$
$$A_f = \frac{\text{number of bits } [M_f \text{ and } (\text{not } M_g)]}{\text{number of bits } M_f};$$
$$A_g = \frac{\text{number of bits } [(\text{not } M_f) \text{ and } M_g]}{\text{number of bits } M_g}.$$

Region	Color
$f \cap \bar{g}$	f
$\bar{f} \cap g$	g
$f \cap g$	0
$\bar{f} \cap \bar{g}$	0

Figure 17.16. Pixel geometry of the **xor** operator.

Figure 17.17. The **xor** operator.

In Figure 17.17, we show the result of the operation $f\,\mathbf{xor}\,g$. In this figure, we also show the α channel of the resulting image.

The set operator. The operator $f\,\mathbf{set}\,g$ considers only the color information originating from f, therefore ignoring the pixel color of image g. This way, we have $A_f = 1$ and $A_g = 0$. In Figure 17.18, we show the color values on each region of the pixel partition, and we illustrate the operation at the pixel geometry level. Of course, the operator is not commutative.

Using bitmask, we have $M_r = M_f$, $A_f = 1$, and $A_g = 0$.

Region	Color
$f \cap \bar{g}$	f
$\bar{f} \cap g$	0
$f \cap g$	f
$\bar{f} \cap \bar{g}$	0

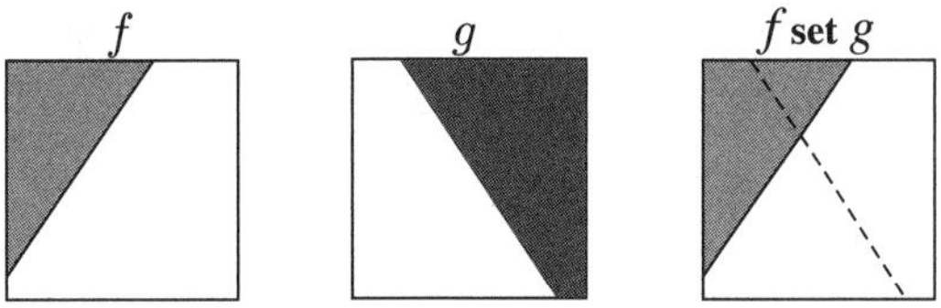

Figure 17.18. Pixel geometry of the **set** operator.

17.3.2 Unary Operator

The attentive reader will notice that while we said we would describe twelve operators, we have defined only six: **over, in, out, atop, xor,** and **set.** We can add to this list the unary operator **clear**, which is defined below. It happens that, with exception of the operators **clear** and **xor**, the other operators should be double-counted, as they are noncommutative and can therefore be performed in the reverse—giving us twelve operators total.

The clear operator. The $\mathbf{clear}(f)$ operator sets to zero (i.e., background color) the alpha value of each pixel, independent of geometry information. As a consequence, it makes the pixel completely transparent (see Figure 17.19). In terms of fragments, it excludes all the polygon fragments of the pixel. We therefore have $A_f = 0$. The operator **clear** transforms the pixel into a pixel of the type $(0,0,0,0)$. Consider the difference between a pixel $(0,0,0,0)$ and a pixel $(0,0,0,\alpha)$, with $\alpha > 0$. In the first case, we have a pixel of null, transparent color, and in the second case we have a pixel of null, nontransparent color. Using a bitmask, we have $A_f = 0$, $A_g = 0$, and $M_r = 0$.

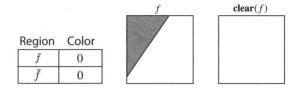

Region	Color
f	0
$\bar{f}$	0

Figure 17.19. Pixel geometry of the **clear** operator.

17.4   Composition of Images and Visibility

We usually associate the image composition operations with the post-processing stage, but they are also useful in solving the visibility problem in 3D scene visualization.

17.4.1   Revisiting the Painter's Algorithm

The **over** operator, in particular, allows us to perform image composition of individual scene objects along the visibility order, from the most distant to the closest to the camera. This process is the basis of the painter's algorithm (see Section 13.4.1).

When there is no interference in the visibility relations among scene objects, we can split them in relation to the distance to the observer and sort them in depth (along the z direction in the camera coordinate system). Therefore the scene is decomposed by objects in layers, $I_1, I_2, \ldots, I_n$, and split by planes, which are then combined in the correct order with the operator **over**:

$$I = I_1 \text{ over } I_2 \text{ over } \cdots \text{ over } I_n.$$

This mechanism is proposed as a general architecture for visualization systems in [Potmesil and Hoffert 87]. Its advantage is the independence among layers, which can be processed in parallel using different methods since composition is performed in the last stage of the process.

17.4.2   Solving Cycles in the Painter's Algorithm

This visualization method does not work if there is interference in the visibility relation in the scene objects. This interference is caused by the existence of cycles in the visibility graph, which indicates the priority order of objects in the scene.

As we saw in Section 13.4.1, in the original painter's algorithm the problem of cycles in the visibility graph is solved by subdividing the objects so as to eliminate interferences.

Another approach to solving this problem was proposed in [Snyder and Lengyel 98]. The idea is to solve the cycles by gathering, using composition operators. In other words, each cycle in the visibility graph is transformed into a correct image from the point of view of the camera in relation to the visibility.

To show how the algorithm works, consider the visibility graphs in Figure 17.20. The graph of Figure 17.20(c) contains a cycle, indicating we cannot define a priority order

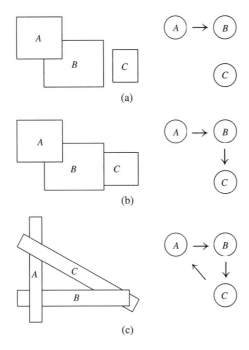

Figure 17.20. Example visibility graphs with and without cycles.

for the scene objects (different parts of an object have dependencies in two points of the graph).

In this method, the cycles are resolved using composition operators that place a part of the object in front and other parts behind in the visibility relation. For example, consider the cycle in Figure 17.20(c). Clearly, the operation A **over** B **over** C would produce a wrong result because C interferes with A (indicated by $C \rightarrow A$ in the graph).

One simple modification using composition operators would produce the correct result (see Figure 17.21): $(A$ **out** $C) + (B$ **out** $A) + (C$ **out** $B)$. Another method of obtaining the same result is to use the sequence $(C$ **atop** $A)$ **over** B **over** C, as also illustrated in Figure 17.21.

The composition operators are used to extract the parts of an object interfering with other objects, according to the visibility graph. Generically speaking, we have the following composition expression:

$$\sum_{i=1}^{n} = I_i \mathbf{out}_{j \mid O_j \rightarrow O_i} I_j.$$

To implement the painter's algorithm using image composition, it is enough to calculate the visibility graph and construct the corresponding composition expression.

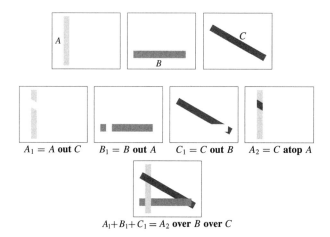

$$A_1 = A \text{ out } C \qquad B_1 = B \text{ out } A \qquad C_1 = C \text{ out } B \qquad A_2 = C \text{ atop } A$$

$$A_1 + B_1 + C_1 = A_2 \text{ over } B \text{ over } C$$

Figure 17.21. Resolving cycles with composition operations. (See Color Plate XLII.)

17.5   Comments and References

Image composition methods with an alpha channel were initially proposed in [Porter and Duff 84]. The image composition operations can be used effectively to create special effects in 3D computer graphics. An example is the combination between real and synthetic images, which is extensively used in virtual worlds. Composition operators are also used in digital painting systems.

17.5.1   Additional Topics

Composition methods find vast applications in image-based rendering and modeling. Another important topic to be studied consists of the use of composition in virtual reality environments.

Exercises

1.   Define a composition operation between two images when each of the images has, besides color, the depth information (depth buffer).

2.   Outline and discuss the relationships between composition and anti-aliasing techniques.

3.   Give examples of how to use composition to implement an algorithm for rendering visible opaque surfaces.

4.   Show how we could modify the solution from the previous exercise to consider translucent surfaces.

5.   Discuss the relationship between A-buffer and composition techniques.

This chapter complements Chapters 5 and 14 and provides background material for Chapter 19. We discuss radiometry and photometry in more depth to provide a more focused view of these topics at an introductory level.

18.1 Radiometry and Illumination

Radiometry treats the measurement of several quantities related to the transfer of radiant energy. The main radiometric quantities are *radiant potency* (or *flux*), *radiance*, *irradiance*, *radiosity*, and *intensity*.

While radiometry deals with physical quantities, photometry deals with perceptual quantities. Photometry treats the measurement of the quantities of luminous perception. These main quantities are *luminous intensity*, *luminous potency*, *brightness illuminance*, and *luminance*.

18.1.1 Solid Angle

The concept of a *solid angle*, which is widely used in defining radiometric quantities, is extremely important for describing flows in a certain direction through a particular region of a surface. To understand a solid angle, we will start with the concept of a planar angle. But in order to understand planar angles, we need to introduce *spherical projection*.

Given a sphere $S^{n-1}(r)$ of radius r, in $\mathbb{R}^n$ with center at the origin, the spherical projection is the transformation $p \colon \mathbb{R}^n - \{0\} \to S^{n-1}(r)$ defined by $p(x) = rx/|x|$ (see Figure 18.1(a)). Geometrically, this transformation projects the point x onto the surface of the sphere along the ray exiting at the origin 0 of $\mathbb{R}^n$ and passing through x.

Now, consider the angle formed by object A when observed from viewpoint O (Figure 18.1(b)). To measure this angle (the planar angle), we take a unit circle S^1 with center at O and perform the spherical projection of the object on the circle obtaining the arch ℓ. The angle is measured by the length of the arch ℓ. Notice the angle is a measure of the apparent length of the object seen by an observer on the plane ("flat land"), located at the

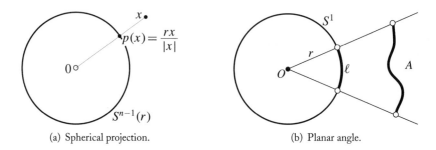

(a) Spherical projection.   (b) Planar angle.

Figure 18.1. The concept of planar angle.

point O. We could use a circle with radius r, but in this case we divide the length of the projected arch by the radius of the circle.

The notion of angle can be extended to measure the apparent area of objects in space (Figure 18.2(a)). This extension results in the concept of a solid angle. The definition is similar to a planar angle, replacing the plane by the space $\mathbb{R}^3$, and the circle by the unit sphere. More precisely, consider a subset A in space, and a point O, which is the observation origin. The measure of the solid angle ω, determined by A, is obtained by taking a unit sphere S^2 with center at O and performing the spherical projection $p(A)$ of the set A on the sphere S^2. The area of $p(A)$ is the measure of ω.

We can take a sphere of radius r with center at O, but in this case we divide the projected area by the square of the radius. In other words, the measure of the solid angle is given by

$$\omega = \frac{\text{Area}(p(A))}{r^2}. \tag{18.1}$$

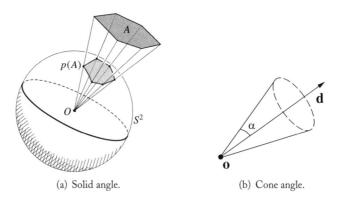

(a) Solid angle.   (b) Cone angle.

Figure 18.2. The concept of solid angle.

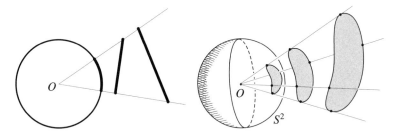

Figure 18.3. Perceptually congruent segments and surfaces.

The spherical projection of the set A on the unit sphere $S^2 \subset \mathbb{R}^3$ determines a *cone* of vertex O and basis A. This cone is the solid formed by the set of all the rays in space with origin at O, and passing through the points of set A, as illustrated in Figure 18.2(b). The cone generalizes, for 3D space, the geometric notion of the angular region in planar geometry. Equation (18.1) is a measure of this 3D angular region.

You can avoid confusion if you remember that both planar and the solid angles are adimensional quantities. The reason is that in the first case, we divide the arch length by the radius of the circle, and in the second case we divide an area by the square of the radius length. However, we use the *radian* (rd) to indicate the measure of a planar angle, and the *stereoradian* (sr) as a unit of measure of a solid angle. These units of measure are just symbolic—they avoid notational confusion and explicitly show that a certain quantity depends on an angle, thus providing a comfort zone when working with angular measures.

Perceptual congruence and projected area. Sets that determine the same angle, when observed from a certain point of view, are called *perceptually congruent* (Figure 18.3). The solid angle is the extension of this concept for 3D. Figure 18.3 shows perceptually congruent surfaces observed from point O.

We say two perceptually congruent surfaces have the same *apparent area* or *projected area*—the projection of the real area on the plane perpendicular to the viewing ray. Luminous phenomena depend on the observer's position, and we are interested in the projected

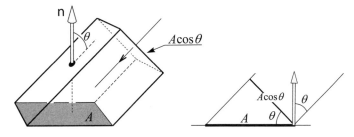

Figure 18.4. Projected area.

area. The viewing projection is conical, but if the area is very small in relation at the observer's distance, we can assume the projection is parallel (see Figure 18.4):

$$\text{Projected area} = A \cos \theta. \tag{18.2}$$

In other words, the projected area is equal to the product of the real area A and the cosine of the angle that the viewing ray makes with the vector that is normal to surface A.

Element of the solid angle. The infinitesimal elements of area and solid angle, $dA, d\omega$, etc. are important because, as we will see, radiometric quantities are defined by the rates of variation (derivatives) of the radiant energy entering or leaving a surface. In the discrete domain (implementation), those infinitesimal values are approximated by small intervals or volumes ($\Delta A, \Delta \omega$, etc.).

Consider an area element dA of a surface A in space, and let $x \in A$ be a point of that area element at a distance r from the origin O (Figure 18.5). This area element determines an element of solid angle $d\omega$ with origin at the point O toward the direction $\mathbf{u}$ defined by the origin O and by point x. From the definition of solid angle, the measure of $d\omega$ is given by

$$d\omega = \frac{\text{Area}(p(dA))}{r^2}, \tag{18.3}$$

where p is the spherical projection of $d(A)$ on the sphere of radius r with center at point O. As we are treating infinitesimal quantities, we can assume dA is a planar surface, and replace the spherical projection by a parallel projection along the direction $\mathbf{u}$. From Equation (18.2), if the normal $n(x)$ to the surface at point x makes an angle θ with the direction $\mathbf{u}$, the value of the projected area is given by $dA \cos \theta$. Therefore, from Equation (18.3) we obtain

$$d\omega = \frac{dA \cos \theta}{r^2}. \tag{18.4}$$

Note that a solid angle element simultaneously specifies a direction (ray) and an area (measure of the solid angle). The direction $\mathbf{u}$ is entirely specified by a unit vector, that is, a point in the unit sphere S^2, and this is defined by the spherical coordinates (θ, ϕ). Any quantity g, depending on both a position x in space and a direction $\mathbf{u}$, is defined in the Cartesian product $\mathbb{R}^3 \times S^2$. This set is called *phase space*.

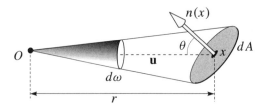

Figure 18.5. Element of the solid angle.

The geometry of the area dA defining a solid angle element can be arbitrary. In the case of a sphere, it is natural to represent the solid angle element by a pyramid instead of a cone. We illustrate this fact in the example below.

Example 18.1 (Element of the solid angle in spherical coordinates). An important particular case is the expression of an infinitesimal element of the solid angle of a sphere. Consider the sphere of radius r with spherical coordinates θ (azimuth) and ϕ (longitude), as shown in Figure 18.6(a). We will determine the solid angle element at a point with coordinates (ϕ, θ). The planar angle element of a meridian passing through the point is $d\theta$ and therefore the corresponding arch element is $rd\theta$; the planar angle element of the equator is $d\phi$. By projecting the point (ϕ, θ) on the plane of the equator, we obtain the projected ray $r \sin \theta$, so the corresponding arch element is $r \sin \theta d\phi$. The area element dA of the sphere is the product of those two arch elements:

$$dA = (rd\theta)(r \sin \theta d\phi) = r^2 \sin \theta d\theta d\phi.$$

Using Equation (18.4), the value of the measure of the infinitesimal solid angle element $d\omega$ in spherical coordinates is

$$d\omega = \frac{dA}{r^2} = \sin \theta d\theta d\phi. \tag{18.5}$$

This example is useful in calculating radiant energy and illumination. By integrating the above volume element over the entire sphere, we obtain the solid angle of the sphere:

$$\Omega = \int_0^\pi \in_0^{2\pi} \sin \theta d\theta d\phi = 4\pi.$$

Therefore, the solid angle of a hemisphere is 2π.   ❏

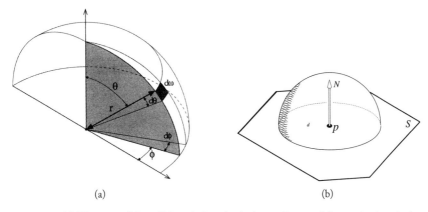

Figure 18.6. (a) Element of the solid angle in spherical coordinates; (b) superior hemisphere.

Illumination hemispheres. Radiometric quantities are based on the radiant energy either intercepting or emanating from a surface. To measure this energy, it is useful to introduce the concept of superior and inferior hemispheres.

Given a surface S and a point $p \in S$, we consider a unit sphere in space with center at p (Figure 18.6(b)). The sphere is divided by the surface into two hemispheres. The hemisphere on the side pointing toward the normal to the surface is called the *superior hemisphere*; the other is the *inferior hemisphere*. The superior hemisphere is indicated by Ω^+; the inferior by Ω^-. (When there is no doubt about the hemisphere being referenced, we just use Ω.)

The superior hemisphere is used to measure the radiant energy entering or leaving the surface (incident and reflected energy). The inferior hemisphere is useful if surface is translucent, because it captures information about the energy leaving the surface (transmitted energy).

18.1.2 Radiometric Quantities

We use the letter Q to represent radiant energy. As with every form of energy, *radiant energy* is measured in joules (J), an SI unit.

Flux or radiant potency. Light travels so quickly that rather than think of it in terms of speed, it is more appropriate to think of it in terms of *flux*, which is, by definition, the potency of the radiant energy: radiant energy per second. The flux is indicated by Φ. The flux unit is therefore *joule/second*, called *watts*, w ($1\text{w} = 1\text{J/s}$). It is important to measure the flux per area unit. This quantity is called *flux density* and is measured in w/m^2.

Radiance. It is common to measure the radiant energy based on the flow of energy passing through a real or imaginary surface. Consider a beam of photons flowing toward a direction $\mathbf{u}$ through the surface element dA confined in the element of the solid angle $d\omega$. We assume all the photons are contained in the element of the spatial volume $dAds$, where $ds = cdt$, with c being the speed of the photons (i.e., the speed of light in the vacuum; see Figure 18.7).

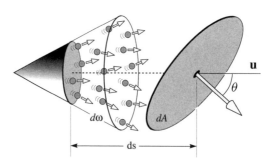

Figure 18.7. Beam of photons in the element of the volume $dAds$.

If x is a point in the volume element $dAds$, the content of particles in the volume element $dAdsd\omega$ in the phase space will be indicated by

$$n(x, \mathbf{u})dA \cos \theta dsd\omega.$$

As $ds = cdt$, we can still write

$$cn(x, \mathbf{u})dA \cos \theta d\omega dt.$$

Each photon has an energy given by $\hbar\nu$, where $\hbar$ is the *Planck constant* and ν is the frequency of the photon. Therefore, the radiant energy transported by the photons in the volume element $dA \cos \theta d\omega ds$ is given by

$$\hbar\nu cn(x, \mathbf{u})dA \cos \theta d\omega dt.$$

The element of the radiant energy dQ, transported within the period d in the volume element $dA \cos \theta d\omega$, is obtained by integrating the above expression in the phase space; that is,

$$dQ = \left(\int_\omega \int_A \hbar\nu cn(x, \mathbf{u})dA \cos \theta d\omega \right) dt.$$

We then conclude the radiant potency is given by

$$\phi = \frac{dQ}{dt} = \int_\omega \int_A \hbar\nu cn(x, \mathbf{u})dA \cos \theta d\omega,$$

and therefore

$$\frac{d^2\phi}{dA \cos \theta d\omega} = \hbar\nu cn(x, \mathbf{u}).$$

The quantity of the above equation is called *radiance* and is indicated by the symbol L. We therefore have

$$L = \frac{d^2\phi}{dA \cos \theta d\omega} \qquad \Leftrightarrow \qquad d^2\phi = LdAd\omega \cos \theta. \qquad (18.6)$$

The unit of measure of radiance is $w/m^2 sr$. From the definition of radiance, we have

$$L = \frac{d}{d\omega} \left(\frac{d\phi}{dA \cos \theta} \right).$$

The expression between parentheses measures the density of the radiant flux ϕ on the projected surface toward the normal direction; therefore, radiance measures the variation of the radiant flux density in relation to the solid angle toward a certain direction.

More precisely, if $x \in \mathbb{R}^3$ is a point in space on a surface S and $\mathbf{u}$ is a direction, the *radiance* $L(x, \mathbf{u})$ is the flux at the point x per area unit perpendicular to $\mathbf{u}$ (i.e., projected

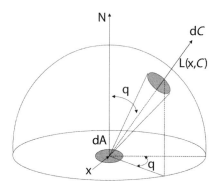

Figure 18.8. Radiance emitted by a surface.

area toward the direction of propagation), and per unit of solid angle. Figure 18.8 illustrates the concept (the sphere was drawn only as reference to the normal direction of ray propagation.)

By integrating the expression $d^2\phi = LdAd\omega\cos\theta$, given by Equation (18.6), in a certain region R of a surface (i.e., integral in dA), and in the region of corresponding solid angle (i.e., integral in $d\omega$), we obtain the total flow of radiant energy emitted by region R within the specified solid angle.

Still, by integrating the same expression $d^2\phi = LdAd\omega\cos\theta$ in relation to the solid angle element in the entire superior hemisphere Ω, we obtain

$$d\Phi = \left(\int_\Omega L(x,\theta,\phi)\cos\theta d\omega\right)dA,$$

or

$$\frac{d\Phi}{dA} = \int_\Omega L(x,\theta,\phi)\cos\theta d\omega, \qquad (18.7)$$

which is an expression for the density of radiant flux in the entire superior hemisphere of the surface.

Irradiance and radiosity. It is common to distinguish between flow arriving and flow leaving a surface in a certain direction. Arriving flow is called *irradiance*, and departing flow is called *radiant exitance* or *radiosity*.[1] We will use the subindex i ("inside") to indicate radiometric quantities associated to the radiant energy entering the superior hemisphere of illumination (i.e., the energy arrives at the surface), and o ("outside") to indicate quantities associated to the radiant energy leaving the superior hemisphere (i.e., radiant energy which leaves the surface).

---

[1] *Radiosity* is a term used in thermal engineering but not in illumination. The reason is that the first computer graphics works on global illumination were based on the heat transfer studies.

Given a point x on a surface A and a direction $\mathbf{u} = (\theta_i, \phi_i)$, the *irradiance* E at x toward the direction $\mathbf{u}$ is the density of total incident flux on the surface in the specified direction. In other words, $E = d\Phi/dA$. Using Equation (18.7) we can write

$$E = \frac{d\Phi}{dA} = \int_\Omega L_i(x, \theta, \phi) \cos\theta d\omega. \qquad (18.8)$$

This equation states that the density of total incident flux in the surface is the integral of the incident radiance, in the direction $\mathbf{u}$, across the entire superior hemisphere.

In particular, by using spherical coordinates in the superior hemisphere, we obtain, with Equation (18.5),

$$E = \int_\Omega L_i(x, \theta, \phi) \cos\theta d\omega = \int_0^\pi \int_0^{2\pi} L_i(x, \theta, \phi) \cos\theta \sin\theta d\theta d\phi.$$

Deriving the Equation (18.8) in relation to the solid angle ω, we obtain

$$\frac{dE}{d\omega} = L(x, \theta, \phi) \cos\theta \quad \Rightarrow \quad dE = L(x, \theta, \phi) \cos\theta d\omega. \qquad (18.9)$$

The *radiosity* B at a certain point x of a surface in a direction (θ_r, ϕ_r) is the flux density emanating from the surface at point x toward the specified direction.[2] To determine radiosity, we use the irradiance equations previously given and replace the incident radiance L_i with the radiance exiting the surface L_r.

$$B = \frac{d\Phi}{dA} = \int_\Omega L_r(x, \theta, \phi) \cos\theta d\omega.$$

By using the expression of the solid angle element in the sphere (Equation (18.5)), we obtain

$$B = \int_\Omega L_r(x, \theta, \phi) \cos\theta d\omega = \int_0^\pi \int_0^{2\pi} L_r(x, \theta, \phi) \cos\theta \sin\theta d\theta d\phi.$$

Radiant intensity. Consider a point light (i.e., a source whose geometric dimensions are negligible) emitting energy in all directions. The radiant flux emitted toward a certain direction can be measured by the infinitesimal radiant flux in a certain solid angle element. This gives us the definition of *radiant intensity* (or simply *intensity*):

$$I = \frac{d\phi}{d\omega}.$$

The unit of measure of intensity is w/sr.

---

[2]Radiometers are instruments that measure the intensity of radiant energy. The most common radiometers measure either radiance or irradiance.

Let us now consider a surface located at a distance d from the light source, so that its normal vector makes an angle θ with the illumination direction. The solid angle element of this surface is given by $d\omega = dA \cos\theta / r^2$. If E is the irradiance in the area element dA of the surface, we have

$$E = \frac{d\phi}{dA} = \frac{d\phi \cos\theta}{r^2 d\omega} = \frac{1}{r^2} \frac{d\phi \cos\theta}{d\omega} = \frac{I \cos\theta}{d^2}. \qquad (18.10)$$

This is the attenuation law of the radiant intensity with the square of the distance.

18.1.3 Spectral Radiometric Quantities

Light sources, in general, emit radiant energy in many regions of the light spectrum. The sun, for instance, emits both the visible and invisible radiation (gamma, ultraviolet rays, etc.). A light source can be characterized by the amount of emitted radiant energy for each wavelength of the energy spectrum. Thus every color can be defined by its spectral distribution, i.e., a function that associates to each wavelength the corresponding radiant energy.

In computer graphics we do not calculate all the spectral illumination due to computational costs. Instead we usually sample the spectrum in three different wavelengths corresponding to low, average, and high frequencies (which correspond to the red, green, and blue colors). We then work with the calculated quantities in each of those wavelengths (for more details, see Chapter 5).

The *spectral radiant energy* Q_λ measures the total radiant energy per unit of wavelength λ:

$$Q_\lambda = \frac{dQ}{d\lambda}.$$

The unit of measure is joules/nanometers (J/nm). This is always the standard to define a spectral radiometric quantity: we take the derivative in relation to the wavelength. (The unit of measure is always divided by nm.)

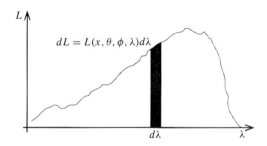

Figure 18.9. Spectral radiance.

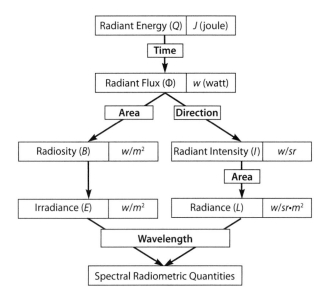

Figure 18.10. Radiometric quantities.

Spectral radiance at a point x of a surface in a direction (θ, ϕ), $L_\lambda = L(x, \theta, \phi, \lambda)$, is given by

$$L_\lambda = \frac{d^3\phi}{dA d\omega \cos\theta d\lambda}, \qquad (18.11)$$

and its unit of measure is $w/m^2 srnm$. The total radiance would be expressed as an integral over the visible spectrum (Figure 18.9).

$$L(x, \theta, \phi) = \int_{\lambda_{\min}}^{\lambda_{\max}} L(x, \theta, \phi, \lambda) d\lambda. \qquad (18.12)$$

Figure 18.10 summarizes the radiometric quantities covered in this section.

18.2 BRDF

In the physical world, the appearance of objects depends on our viewing position, on the angle of the light source, or even the color and intensity of the light illuminating the objects. To simulate this in computer graphics we need mathematical models. A key piece of this modeling is the *bidirectional reflectance distribution function* (BRDF).

The bidirectional reflectance distribution function relates incident energy at a point x with the reflected energy. The main elements influencing the radiance emitted by a surface S include illumination, reflectance properties of the surface, and the observer's position.

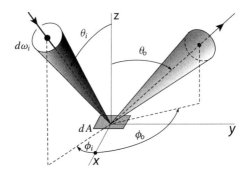

Figure 18.11. Incident and reflected fluxes in an area element.

We could also include the medium of propagation in this list, but we will assume the medium is nonparticipative.

Consider an incident radiant flux at a point x on a surface S in a direction $\mathbf{u}_i = (\theta_i, \phi_i)$, as an element of the solid angle $d\omega_i$ (see Figure 18.11). We indicate by dA the surface area element and by $\mathbf{n}$ the unit normal vector at point x. The irradiance at point x, in the direction $\mathbf{u}_i$, and in the cone of the solid angle element $d\omega_i$, is indicated by dE_i. The radiance reflected in the direction $\mathbf{u}_r = (\theta_r, \phi_r)$, originated at dE_i, will be indicated by dL_r. It is common to use the Greek letter ω to indicate a direction (θ, ϕ). In this case, dw would be the element of the solid angle in that direction (this notation is good but should be used carefully because ω often represents a solid angle). Using this notation, the incidence and reflection directions are indicated, by ω_i and ω_r, respectively.

We define the BRDF, by the quotient

$$f_r(x, \omega_i, \omega_r) = \frac{dL_r(x, \omega_r)}{dE_i(x, \omega_i)}, \tag{18.13}$$

between the radiance of the reflected flux and the irradiance of the incident flux. The unit of measure of S is $1/sr$.

Remember that the choice of notation is arbitrary. Mathematical notation is widely consistent, but notation in radiometry is not so uniform, especially when it comes to BRDF. It is common to find k or ρ_{bd}, instead of f_r; also, it is common to use the notation $f_r(x, \omega \to \omega')$, where the subindices in the variables are eliminated and the arrow is used to indicate ω is the direction of incidence, and ω' the direction of reflection. Another notation places the indices i and r in the hemisphere Ω; in this way, if $\omega \in \Omega_i$ (Ω_r), we know we are dealing with an incidence (reflection) direction. You may also find as subindex the letter o ("outgoing") instead of r ("reflected"); in fact, we feel the letter o is more correct, because one can have radiant energy leaving a surface without being reflected (i.e., the object's own energy).

The BRDF quantitatively expresses the connection between the irradiance (incident flux per area unit) and the radiance (incident flux per area unit per unit of solid angle) reflected by the surface. To simplify, we assume the reflection point is the same as the incidence point x (true except when there is dispersal), and the irradiance depends only on the direction of incidence (θ_i, ϕ_i) (in reality it can also depend, for instance, on wavelength). Therefore, the BRDF only depends on the directions of incidence and reflection (which is why it's called "bidirectional"). We therefore have

$$f_r = f_r(x, \theta_i, \phi_i, \theta_r, \phi_r) = f_r(x, \omega_i, \omega_r).$$

The expression $dE_i = L_i(\omega_i) \cos \theta d\omega_i = L_i(\omega_i)\langle \mathbf{n}, \omega_i \rangle d\omega_i$ from Equation (18.9) allows us to write f_r in terms of both the radiance and reflected radiance:

$$f_r(\omega_i, \omega_r) = \frac{dL_r}{dE_i} = \frac{dL_r}{dL_i \cos \theta d\omega_i} = \frac{dL_r}{dL_i \langle \mathbf{n}, \omega_i \rangle d\omega_i}. \tag{18.14}$$

This equation shows a curious fact: unlike the common sense, the BRDF can assume arbitrarily large values. (Of course L_r is smaller than L_i, unless the surface is a light source, however the cosine factor in the denominator forces the BRDF to not be limited within the interval $[0, 1]$.) Notice however, from Equation (18.14) that we have

$$\frac{dL_r(x, \omega_r)}{dL_i(x, \omega_i)} = f_r(x, \omega_i, \omega_r) \cos \theta d\omega_i = f_r(x, \omega_i, \omega_r)\langle \mathbf{n}, \omega_i \rangle d\omega_i. \tag{18.15}$$

As the total radiance leaving the superior hemisphere is less than or equal to the radiance entering that hemisphere, by integrating both members of this equation in the entire superior hemisphere Ω, we can conclude that

$$\int_\Omega f_r(x, \omega_i, \omega_r)\langle \mathbf{n}, \omega_i \rangle d\omega_i \leq 1.$$

This inequality is as BRDF's *law of energy conservation*: the total energy reflected in the superior hemisphere cannot exceed the total incident energy.

18.2.1   The Radiance Equation

We can now write the radiance equation of a surface. In physical terms, the total radiance exiting a surface has two different sources: the object's own energy (i.e., emitted) and the reflected energy.

$$\text{Emanated radiance} = \text{Emitted radiance} + \text{Reflected radiance}. \tag{18.16}$$

From Equation (18.15), the total radiance reflected L_r is given by the integral

$$L_r(x, \omega_0) = \int_\Omega f_r(x, \omega_i, \omega_r) L_i(x, \omega_i)\langle \mathbf{n}, \omega_i \rangle d\omega_i \tag{18.17}$$

in the superior hemisphere Ω. This equation is called *radiance equation*. We can use operators to write the radiance equation. We define the linear operator $= K$,

$$Kg(x,\omega_r) = K(g)(x,\omega_r) = \int_\Omega f_r(x,\omega_i,\omega_r)g(x,\omega_i)\langle \mathbf{n},\omega_i\rangle d\omega_i. \qquad (18.18)$$

Equation (18.17) can be written as

$$L_r = KL_i.$$

(Equation (18.18) shows that the BRDF f_r is in fact a distribution in the sense of functional analysis, therefore it can be called "function distribution.")

Indicating emitted radiance by L_e, and using the expression in Equation (18.16) together with the radiance equation, we can write

$$L_r(x,\omega_0) = L_e(x,\omega_0) + \int_\Omega f_r(x,\omega_i,\omega_r)L_i(x,\omega_i)\cos\theta_i d\omega_i, \qquad (18.19)$$

or,

$$L_r = L_e + KL_i.$$

This equation represents the total radiance of the energy leaving a surface from the superior hemisphere.

18.2.2 BRDF Calculation

There are three basic methods for calculating the BRDF: analytical, geometric, and experimental methods. The analytical method uses a theoretical illumination model, given by an expression of a function which is defined analytically (for example, the Phong model, see Chapter 14). The geometric method uses a geometric modeling of the surface's microgeometry. An example of this method is the BRDF of Cook-Torrance, which uses the geometry model with microfacets. The experimental method uses equipment to measure values of the BRDF of the surface in a laboratory and then reconstructs the function from those samples. The BRDF can be extended to define transmittance as well as reflectance, through the bidirectional reflectance and transmittance distribution function (BRTDF).

18.3 Photometry

Photometry measures the luminous information produced when the sensors of our viewing system[3] are excited by radiant energy. The luminous information measured by photometry can be either monochromatic or spectral radiation.

Photometry is thus the perceptual manifestation of radiometry. In this way, each radiometric quantity corresponds to one photometric quantity. Historically, however, this

---

[3]Photometry treats a more varied range of sensors beyond the human eye.

relation was not so directly established. Photometry is much older than radiometry (ancient astronomers classified the stars according to their luminous information, which they called *quantities*). The association of light with electromagnetic radiation and radiance is a more recent accomplishment.

Photometers are instruments that measure the value of photometric quantities. (The most well-known example is the photometer in cameras that evaluates the illumination of the environment in order to get the right exposure.)

18.3.1 It All Began with a Candle

The basic unit of measure in photometry is the *candela* (*cd*), which is associated with the photometric quantity called *luminous intensity*. Originally, the candela definition was established in relation to a "standard candle": one candela is the luminous intensity of a standard candela (the Latin word for "candle"). Over time, the definition of a candela went through several changes to bring it up to date technologically. The current definition is the luminous intensity, in a given direction, of one light source which emits monochromatic radiation in the frequency 540×10^{12} hertz, and whose radiant intensity is $1/683 \text{w sr}^{-1}$. This frequency, which corresponds to a wavelength of 555nm, was chosen because our perception of the illumination from a light source depends on the wavelength and not just on the radiant potency. Quantifying this phenomenon is the essence of photometry.

By observing the candela definition, we can see it measures the luminous information associated with radiant intensity (radiant flux per solid angle). The candela is one of the seven fundamental units of measure in the international system of units (SI).

As the candela is a unit per solid angle, it does not vary with distance from the light source. It is a convenient unit for measuring the luminous energy of a light source, but not for measuring the luminous energy arriving at objects in a scene. For this, we need other photometric quantities associated with radiometric ones.

18.3.2 Function of Luminous Efficiency

In this section, we will show the quantitative correspondence between radiometric and photometric quantities. The human eye is sensitive to radiation within 380nm and 770nm of the electromagnetic spectrum (for more details, see Chapter 5). The response of our visual system to luminous stimuli is not the same for every wavelength; that is, two radiant fluxes with same radiant potency can generate different luminous information (as when the radiance emitted by them is concentrated within different degrees of the spectrum). To give an extreme example, recall that a luminous source with a radiant potency of 200w, in the infrared region of the spectrum, does not have any luminous information. (Why does a fluorescent lamp of 20w illuminate more than an incandescent one of 60w?).

In 1924, the Comission Internacionale of L'Eclairage (CIE), the international commission that sets illumination standards, performed an experiment to measure the sensibility of the human eye in different degrees of the spectrum. We will briefly describe this experiment, which will lead us to the definition of the function of luminous efficiency.

We begin with a sampling sequence $(\lambda_1, \lambda_2, \ldots, \lambda_n)$ of wavelengths in the visible spectrum. We then take a monochromatic light source of reference ℓ_r, in one of the wavelengths of the sequence, with a radiant potency of 1w. We consider monochromatic light sources ℓ_i in each wavelength λ_i of the sequence, called test sources. For each λ_i, we make an exposition for a reference light observer in parallel with the test light source ℓ_i. The observer will notice different luminous information coming from the two sources. We then vary the radiant flux of the test source until the observer does not perceive any difference in the luminous information of both sources.

After performing the experiment in every point of the sequence, we take the quotient:

$$K_i = \frac{\text{Radiance of } \ell_r}{\text{Radiance of } \ell_i}.$$

Intuitively, K is a measure of the attenuation value of the luminous perception, when we vary the wavelength of the luminous source. The *function of luminous efficiency* $V(\lambda)$ is obtained by normalizing this function. That is, if $K_{\max}$ is the maximum value of K, we set

$$V_i = V(\lambda_i) = \frac{K_i}{K_{\max}}.$$

By interpolating the obtained values of V_i, we obtain the function $V(\lambda)$ defined in the entire visible spectrum.

To conclude the definition, we need to determine the value of $K_{\max}$. Note that to convert the radiant intensity I to the luminous intensity I_v, we use the expression

$$I_v(\lambda) = K_{\max} V(\lambda) I(\lambda). \tag{18.20}$$

The wavelength λ, where K assumes the maximum value (i.e., in which $V(\lambda) = 1$) is $\lambda = 555$nm. From the candela definition given in Section 18.3.1, we know that, for $I(555) = 1/683$ lm/sr, we have $I_v(555) = 1$cd. Placing these values in the Equation (18.20), we obtain $K_{\max} = 683$lm/w. (Of course, this value of $K_{\max}$ was determined in a way that makes the radiometric units of measure compatible with the classic units of photometry.) With the calculation of $K_{\max}$, we conclude the definition of the function of luminous efficiency. Its graph is shown in Figure 18.12.

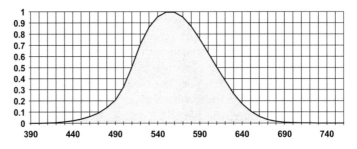

Figure 18.12. Function of luminous efficiency.

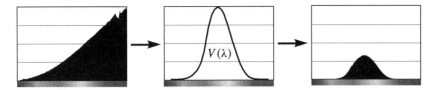

Figure 18.13. Filtering by the function of luminous efficiency. (See Color Plate XLIII.)

The graph clearly shows that the eye is more sensitive to radiant energy with wavelength in the middle band of the spectrum (as we saw, the maximum value is reached in 555 nm).

Finally, if X is a radiometric quantity and X_v its corresponding photometric quantity,[4] we have the conversion equation we were looking for:

$$X_v(\lambda) = 683V(\lambda)X(\lambda). \tag{18.21}$$

Example 18.2. Consider two laser pointers, both with potency of 0.005w, one of them emitting a laser of wavelength $\lambda = 635$nm and the other with wavelength $\lambda = 670$nm. Which one has greater luminous intensity? We have

$$I_{635} = 0.005\text{w} \cdot \text{V}(635) \cdot 683\frac{\text{lm}}{\text{w}} = 0.005 \cdot 0.032 \cdot 683\text{lm} = 0.11\text{lm}.$$

$$I_{670} = 0.005\text{w} \cdot \text{V}(670) \cdot 683\frac{\text{lm}}{\text{w}} = 0.005 \cdot 0.265 \cdot 683\text{lm} = 0.74\text{lm}.$$

Therefore, the laser of greater wavelength is seven times more luminous than the other. (The values of $V(635)$ and $V(670)$ are obtained by consulting a table of the function of luminous efficiency $V(\lambda)$.) ❑

If we have a nonmonochromatic, spectral radiometric quantity $X(\lambda)$, the corresponding photometric quantity X_v is obtained by the integral

$$X_v = 680 \int_{\lambda_{\min}}^{\lambda_{\max}} X(\lambda)V(\lambda)d\lambda. \tag{18.22}$$

Notice this integral performs a weighted average of the radiometric quantities with the function of luminous efficiency.

The function of luminous efficiency works as a filter to convert radiometric quantities into photometric ones:

Radiometric Quantity $\rightarrow V(\lambda) \rightarrow$ Photometric Quantity

This fact is illustrated in Figure 18.13.

---

[4]If X is a radiometric quantity, the corresponding photometric quantity will always be indicated by the notation X_v (v for visual).

18.3.3 Photometric Quantities

In this section we will study other photometric quantities and their relationship to the corresponding radiometric quantities. We will leave the reader with the task of developing analogous results for the spectral photometric ones.

In the SI standard unit, the emitted radiation energy is measured in joules. However, if the wavelength of the radiation is within the visible spectrum, then the energy is called luminous energy and its unit is the *talbot* (however, this unit is very little used).

Luminous intensity. We initially defined luminous intensity so we could introduce the candela and describe luminous efficiency. Let us make a revision.

The *luminous intensity* I_v of a light source measures the amount of luminous flux emitted in a given direction within a certain region determined by a solid angle. Therefore, it is an appropriate measure for directional lights. In infinitesimal terms, luminous intensity measures the derivative of the flux in relation to a solid angle:

$$I = \frac{d\phi}{d\omega}.$$

The corresponding radiometric quantity is the radiant intensity I, whose unit of measure is w/sr. The unit of measure of the luminous intensity is the candela (cd), (1cd = 1lm/sr). The conversion from radiant intensity for luminous intensity comes from Equation (18.21):

$$I_v(\lambda) = 683V(\lambda)I(\lambda).$$

The luminous intensity does not decrease when we distance ourselves from the light source. Therefore, it is an appropriate measure for specifying luminous information about the light source. But it does not provide useful information about the light arriving at the objects.

Luminous flux. *Luminous flux* ϕ_v is the photometric quantity associated with radiant flux (or radiant potency). Its unit of measure is the *lumen* (lm), defined as the luminous flux of a light source with radiant flux of $1/683$ w to a frequency of 540×10^{12} hertz (wavelength of 555nm).

From Section 18.3.1, we know the frequency of 540×10^{12} hertz corresponds to the wavelength of 555nm, where the eye reaches the maximum efficiency. In fact, by reading again the candela definition given in that section, and by combining with the lumen definition given above, we conclude that an isotropic light source has 1 lm, if its luminous intensity per stereoradian is equal to one candela. (In particular we have cd = lm/sr.)

The conversion from radiant flux ϕ to luminous flux ϕ_v comes from Equation (18.21):

$$\phi_v(\lambda) = 683V(\lambda)\phi(\lambda).$$

Notice the luminous flux does not take into account the propagation direction of the radiation or the area of the emitting source. For this reason, it is very appropriate for point light sources. What is more, unlike the luminous intensity, the luminous flux decays with distance from the light source. This quantity is therefore more appropriate when we want to effectively measure the luminous energy arriving at the objects of the scene. (This is the quantity used by light designers.)

Luminance. Luminance is the best known the photometric quantity; in fact, its name is used generically to refer to the luminous information of a color (which is why we have avoided using the term "luminance" in this chapter.)

Most light sources are not points; instead they have area. For these light sources, called sources with extension, we define *luminance* L_v by the second derivative of the luminous flux regarding both the area and the solid angle in a certain direction. That is,

$$L_v = \frac{d^2\phi}{dA\cos\theta d\omega}.$$

The luminance therefore corresponds to radiance L in radiometry. Conversion from radiant flux ϕ to luminous flux ϕ_v comes from the Equation (18.21):

$$L_v(\lambda) = 683V(\lambda)L(\lambda).$$

The unit of measure is $\text{lm/m}^2\text{sr}$.

Luminance is widely used in television engineering, where color systems are based on the chrominance-luminance decomposition (see Chapter 5). In fact, we have the familiar equation to calculate the "NTSC luminance" associated to a trichromatic color of (R, G, B) coordinates:

$$Y = 0.299R + 0.587G + 0.114B.$$

This expression is obtained using the conversion Equation (18.21) for each of the colors R, G, and B.

Illuminance. Illuminance E_v is the photometric quantity corresponding to irradiance E. It is a measure of the concentration of the luminous flux incident on a surface; that is, it measures the density of luminous flux arriving at the surface. In infinitesimal terms, the illuminance is defined as the derivative of the luminous flux in relation to the area

$$E = \frac{d\phi_v}{dA}.$$

The unit of measure is lm/m^2, called lux (lx).

The vast majority of the commercial photometers measure luminance or illuminance of light sources.

18.4   Summary

For the photometric quantities, we can devise a diagram similar to the one in Figure 18.10. The table below provides a summary of the radiometric and photometric quantities studied in this chapter, with their units of measure and symbols used. Recall that, in the international system of measures (SI), the fundamental measure of the illumination area is the candela.

Radiometry	Units	Photometry	Units
Radiant Energy (Q)	J (joule)	Luminous Energy (Q_v)	talbot
Radiant Flux (Φ)	w (watt)	Luminous Flux (Φ_v)	lm (lumen)
Radiosity (B)	w/m^2	Luminous Excitement (M_v)	lm/m^2 (lux)
Irradiance (E)	w/m^2	Illuminance (E_v)	lm/m^2 (lux)
Radiant Intensity (I)	w/sr	Luminous Intensity (I_v)	lm/sr (candela)
Radiance (L)	w/sr·m^2	Luminance (L_v)	$lm/m^2 \cdot sr$

There are several other units of measure for photometric quantities besides the ones introduced in this chapter (e.g., foot-candle, phot, lambert, etc.), but we chose to not include them because many are in disuse and they do not contribute to a better understanding of photometry.

18.5   Comments and References

In this chapter, all topics were discussed in the classic spirit of engineering methods using "infinitesimal elements." A rigorous mathematical handling (which would follow the line of study of the quantities in the phase space, as we will do in the introduction of radiance in Chapter 19), would demand the use of differential forms, tensorial calculus, and measurement theory—topics too extensive to be considered here.

A very important text in colorimetry, radiometry, and photometry is [Wyszecki and Stiles 00]. There are specific books on colorimetry and photometry; a good reference is [Walsh 58]. You can also find plenty of bibliographical material on photometry and radiometry in the extensive literature of books on optics. A good reference is [Klein and Furtak 86].

19 The Illumination Equation

The illumination of a 3D scene is the result of the interaction between the light sources and surfaces of the environment. In Chapter 14 we examined an illumination equation that expressed the local interaction of light with a surface. In that case, the value of the resulting energy (color) is given explicitly by the equation. However, that equation is limited and cannot describe all the phenomena of luminous energy propagation in a scene. In this chapter, we will study an illumination equation that describes such phenomena and the computational methods to numerically solve that equation. Fundamental prerequisites for this chapter include Chapters 14 and 18.

19.1 Illumination Model

The illumination equation describes the propagation of radiant energy in ambient space. As we saw in Chapter 14, according to the particle model of light, each photon carries a certain amount of radiant energy. At any point in time, a photon can be characterized by its position and by the direction of its motion. In this way, the state of a photon is given by a point $s \in \mathbb{R}^3$ and a vector $\omega \in S^2$. The space $\mathbb{R}^3 \times S^2$ of the pairs (s, ω) is called *phase space*.

Illumination is the result of photon motion, so it is convenient to measure the flux of the radiant energy. Radiant energy per unit of time is also called *radiant potency*. We will use the notation Φ to indicate the radiant flux. We can consider the radiant flux as being defined in the phase space. In this case, if $dAd\omega$ is the volume element in the phase space, the expression $\Phi(s, \omega, t)dAd\omega$ is the number of photons passing through an area dA in the neighborhood of point s under an infinitesimal element of solid angle $d\omega$, along the direction ω, and at the instant of time t (see Figure 19.1).

Light transports radiant energy; therefore, the illumination equation can be considered a problem in the area of *transport theory*, which studies the distribution of abstract particles in space and time. It is in this context that we obtain an expression of the illumination equation.

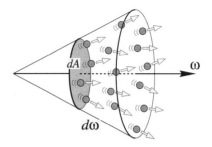

Figure 19.1. Radiant flux in phase space.

To develop a global illumination equation, we will start with two hypotheses that will simplify the problem:

1. The system is in equilibrium, meaning that conditions of light propagation in the environment do not change during the simulated time interval. (This condition implies the flux is constant at every point in the scene; that is, $\partial \Phi / \partial t = 0$.)

2. The medium is nonparticipative, meaning that the only important phenomena happen on the surface of the scene objects. (This hypothesis assumes the simulation is done in a vacuum.)

19.1.1 Transport of Radiant Energy

The transport of radiant energy between two points in a vacuum is given by the equation

$$\Phi(r, \omega) = \Phi(s, \omega), \tag{19.1}$$

where r, s are mutually visible points along the direction ω (see Figure 19.2(a)). While the equation is valid for any pair of points from the ambient space satisfying the visibility condition, we are interested on points $r, s \in M = \cup M_i$ of the surfaces in the scene.

Given a point r, we find a point $s \in M$ using the *visibility function for surfaces*, $\nu : \mathbb{R}^3 \times S^2 \to \mathbb{R}$, which returns the distance of the closest visible point on a surface in the scene:

$$\nu(r, \omega) = \inf \{\alpha > 0 : (r - \alpha\omega) \in M\}.$$

The point s is then given by $s = r - \nu(r, \omega) \, \omega$.

At point p on a surface, the illumination function potentially depends on the radiant energy flux arriving from all directions. For this reason we use the concept of an *illumination hemisphere* Ω at the point p, which is defined by an imaginary sphere of radius one with center at p.

The luminous energy irradiating at point p in direction ω defines a solid angle in the hemisphere Ω_o; similarly, the luminous energy arriving at point p defines a solid angle in the illumination hemisphere Ω_i. In this way, the illumination hemisphere contains every exchange of energy between surface point p and the environment (see Figure 19.2(b)).

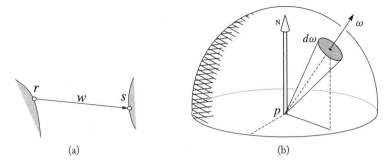

Figure 19.2. The illumination hemisphere. (a) Transport of radiant energy between two points. (b) Exchange of energy between point p and the environment.

19.1.2 Boundary Conditions

Equation (19.1) describes the *transport conditions* of radiant energy in the vacuum. To formulate the illumination problem fully, we need to specify the *boundary conditions*. When luminous rays reach surfaces, two boundary conditions exist: explicit and implicit.

Explicit condition. In the explicit condition, flux leaving surface point s in direction ω is independent of the incident flux. This illumination function is given by

$$\Phi(s,\omega) = \mathcal{E}(s,\omega), \tag{19.2}$$

where $\mathcal{E}$ specifies the *emissivity function* of the surface.

Implicit condition. In the implicit condition, flux leaving surface point s in direction ω depends on the incident flux in the illumination hemisphere:

$$\Phi(s,\omega) = \int_{\Omega_i} k(s,\omega' \to \omega)\Phi(s,\omega')d\omega', \tag{19.3}$$

where $\omega \in \Omega_o$, $\omega' \in \Omega_i$, and k is the bidirectional reflectance function of the surface. The law of conservation of energy states that the radiant energy emitted in Ω_o has to be smaller than the incident radiant energy in Ω_i.

19.1.3 Radiance Equation

We now introduce the boundary conditions into the transport equation. At each point $r \in M$, we want to obtain the contribution of radiant energy coming from all points $s \in M$ that are visible inside the illumination hemisphere Ω_i at r. We therefore split the transport equation between two points r and s:

$$\Phi(r,\omega) = \Phi(s \to r, \omega), \tag{19.4}$$

where the notation $a \to b$ means the transport from a to b.

We now place on the right side of the equality the two boundary conditions (19.2) and (19.3):

$$\Phi(r,\omega) = \mathcal{E}(s,\omega) + \int_{\Omega_i} k(s,\omega' \to \omega)\Phi(s,\omega')d\omega', \tag{19.5}$$

where $r, s \in M$, $\omega \in \Omega_o$, $\omega' \in \Omega_i$, and s is determined by the visibility function ν.

Equation (19.5), called the *transport equation*, is a *Fredholm integral equation of the second kind*, which has been the object of much study.

The transport equation describes radiant energy flux in terms of the number of photons (irradiance). What is now needed is to obtain the total amount of radiant energy, or *radiance*, which is the flux Φ multiplied by the energy E of the transported photons $L = E\Phi$, where $E = \hbar\, c/\lambda$, with c being the speed of the electromagnetic radiation in the vacuum (speed of light), λ the wavelength, and $\hbar$ the *Plank constant*. As is explained in Chapter 18, radiance L is the radiant flux on a surface along a certain direction ω,

$$L(r,\omega) = \frac{d^2\Phi(r,\omega)}{d\omega dS \cos\theta_s}, \tag{19.6}$$

where $d\omega dS \cos\theta_s$ is the projected differential area.

The *radiance equation*, or *illumination equation*, is given by

$$L(r,\omega) = L_E(r,\omega) + \int_{\Omega_i} k(s,\omega' \to \omega)L(s,\omega')\cos\theta_s\, d\omega' \tag{19.7}$$

This equation is also known as the *rendering equation*, or *temporal invariant equation of monochrome radiance in a vacuum*. Its geometry is illustrated in Figure 19.3(a).

A modified version of Equation (19.7) describes the irradiated energy starting from point r in direction ω^o in terms of the incident energy in the illumination hemisphere at r:

$$L(r,\omega^o) = L_E(r,\omega^o) + \int_{\Omega_i} k(r,\omega \to \omega^o)L(s,\omega)\cos\theta_r\, d\omega, \tag{19.8}$$

where $\omega^o \in \Omega_o$ and θ_r is the angle between the normal and the surface at r and $\omega \in \Omega_i$. The geometry of the equation is illustrated in Figure 19.3(b). This form of the radiance equation will be used in the elaboration of several methods for calculating the illumination.

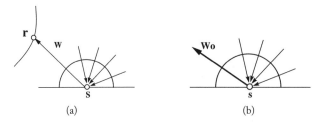

Figure 19.3. Geometry of the rendering equation.

19.1.4   Numerical Approximation

The illumination equation should be solved in a numerical and approximated way. To study the solution of this integral equation, we will use the notation of the operators. As a space of functions, we define the integral operator $\mathcal{K}\colon E \to E$, placing

$$(\mathcal{K}f)(x) = \int k(x,y)f(y)dy.$$

The function k is called a *kernel* of the operator. We indicate that the operator $\mathcal{K}$ is applied to a function $f(x)$ by $(\mathcal{K}f)(x)$. Using the notation of operators, the illumination equation is written in the form

$$L(r,\omega) = L_E(s,\omega) + (\mathcal{K}L)(s,\omega), \qquad (19.9)$$

or $L = L_E + KL$.

Part of the difficulty of solving Equation (19.9) comes from the fact that the unknown function, L, appears on both sides of the equality, inside and outside the integral. A strategy for an approximated solution to the problem is to use the method of successive substitutions. Notice that function L is defined in a recursive way; consequently, Equation (19.9) provides an expression for L. The basic idea consists of substituting L by its expression in the right side of the equality, obtaining

$$L = L_E + K(L_E + KL)$$
$$= L_E + KL_E + K^2L,$$

where the exponent indicates the successive application of the operator K to a function f, i.e., $(\mathcal{K}^2 f)(x) = (\mathcal{K}(\mathcal{K}f))(x)$.

Repeating the substitution process $n + 1$ times gives us

$$L = L_E + KL_E + \cdots + K^n L_E + K^{n+1}L$$
$$= \sum_{i=0}^{n} K^i L_E + K^{n+1}L.$$

This recurrence relation provides a way to approximately calculate L. Ignoring the residual term of order $n + 1$, $K^{n+1}L$, we have

$$L \approx L_n = \sum_{i=0}^{n} K^i L_E. \qquad (19.10)$$

The substitution method, applied to the illumination function calculation, has an intuitive, physical interpretation. Notice that the term L_E corresponds to the radiant energy emitted by the light sources. The integral operator K models the propagation of the reflected light on the surfaces. Therefore, KL_E corresponds to the illumination of the light sources, which are reflected directly by the surfaces. Its successive application models the propagation of the reflected light n times in the scene.

19.1.5   Methods for Calculating the Illumination

We have shown a developed methodology that performs calculations of the illumination function in an approximate way. In practice, this strategy translates itself into two methods used for calculating the illumination: local and global methods.

Local methods. The approximation given by $L_1 = L_E + K L_E$, corresponds to the direct contribution of the light sources. These methods use the local model of illumination, which we studied in Chapter 14.

Global methods. In a global method the approximation is given by

$$L_n = \sum_{i=0}^{n} K^i L_E.$$

This corresponds to the direct contribution of the light sources as well as the indirect contribution from the reflection on the surfaces. This class of methods uses the global illumination model studied in this chapter. The two most important forms of implementation are the methods of ray tracing and radiosity. In the next sections, they will be presented in detail.

19.2   Ray Tracing Method

The ray tracing method provides a solution for calculating the global illumination, by sampling the path of light rays in the scene. The basic idea consists of following the rays coming from the scene and arriving on the virtual screen. The most appropriate name for this method would be *reverse ray tracing*. The ray tracing method, due to its characteristics, is very appropriate to model specular reflection (and transmission) phenomena, which are dependent on the virtual camera.

In reverse ray tracing, the illumination integral is calculated by probabilistic sampling, using Monte Carlo integration. In this section, we will see that integration is not necessary in the case of perfectly specular surfaces.

19.2.1   Photon Transport

To solve the illumination equation using the ray tracing method, we will calculate the transport of photons in the scene using the geometric optics model. The goal is to follow the path of those particles carrying radiant energy.

We will focus on particle p. The path of this particle in the scene corresponds to a series of states $\{s_0, s_1, \ldots s_n\}$, where each state is associated with the attributes of p in stage t of the simulation, such as its position, direction, and energy.

A particle has an existence interval, or life span. This interval is determined by events associated to p: its creation, in the initial state s_0; and its extinction, in the final state s_n.

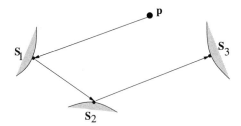

Figure 19.4. Trajectory of a particle.

The illumination equation describes the transport of radiant energy, which is equivalent to the propagation of photons (light particles) in the environment. Therefore, in the context of ray tracing, it is convenient to formulate the problem through a stochastic transport equation describing the state change of the particles,

$$\ell(t) = g(t) + \int k(s \rightarrow t)\ell(s)ds. \tag{19.11}$$

In this equation, ℓ, g and k are interpreted as probability distributions. More precisely: $\ell(s_i)$ is the probable number of particles *existing* in state s_i; $g(s_i)$ is the probable number of particles *created* in state s_i; and $k(s_i \rightarrow s_j)$ is the probability that a particle will move from state s_i to state s_j.

We want to calculate the illumination function on the surfaces of the scene. We associate the states of the particles with events related to the decomposition of the surfaces $M = \cup M_i$. A particle is in state s_i, when its path arrives at the surface M_i (see Figure 19.4).

We want to estimate ℓ, given g and k in the Equation (19.11). For this we will use Monte Carlo methods for calculating the integral value.

It is possible to follow the history of the particles moving forward or retreating along its path. In the context of visualization, we begin with the particles on the image plane and then register the path followed since their creation (when they are emitted by the light sources). By beginning at the state t_n and tracing back the history of the particle, we have

$$\ell(t_n) \approx g(t_n) + \int k(s \rightarrow t_n)\ell(s)ds.$$

We know $g(t_n)$ and we need to estimate the integral value

$$\int k_n(s)\ell(s)ds,$$

where $k_n(s)$ indicates the probability of a particle to arrive at the state t_n, coming from a previous state s.

Using Monte Carlo methods, we can estimate $\ell(t_{n-1})$ by performing a random sampling

$$
\begin{aligned}
\ell(t_n) &\approx g(t_n) + \int k_n(s)\ell(s)ds \\
&= g(t_n) + \ell(t_{n-1}) \\
&= g(t_n) + g(t_{n-1}) + \int k(t_{n-1} \to r)\ell(r)dr.
\end{aligned}
$$

Continuing with the process, we obtain an approximate estimate of the probability distribution $\ell(t_n)$

$$
\ell(t_n) \approx g(t_n) + g(t_{n-1}) + \dots.
$$

The result is in accordance with the methodology for calculating the illumination equation, as developed in the previous section.

The stochastic transport equation has the same structure of the radiance equation

$$
L(r, \omega^o) = L_E(r, \omega^o) + \int_\Omega k(r, \omega \to \omega^o) L(s, \omega) \cos \theta_r \, d\omega.
$$

To solve this equation through a probabilistic approach, we use two techniques that efficiently estimate the integral value by the Monte Carlo method: stratification and importance sampling.

Stratification. This technique consists of performing a partition of the illumination hemisphere $\Omega = \cup \, \Pi_m$ into *strata*

$$
D_m = \{\omega; \omega \in \Pi_m\}
$$

so the function $L(s, \omega)$, for $\omega \in \Pi_m$, presents a small variation in each element of the partition.

Importance sampling. This technique consists of performing the sampling of $L(s, \omega)$ in each stratum in a way to select more representative samples. This is done through the *importance function*:

$$
g_m : \mathbb{R}^3 \times S^2 \to [0, 1].
$$

Incorporating these two techniques in the radiance equation, we obtain

$$
L(r, \omega^o) = L_E(r, \omega^o) + \sum_{m=1}^{M} \int_{D_m} k(r, \omega \to \omega^o) \frac{L(s, \omega) \cos \theta_r}{g_m(r, \omega)} g_m(r, \omega) \, d\omega,
$$

where each stratum D_i is given by a set of directions $\omega \in \Pi_i$. We divide L by g_m and later we multiply the result by g_m, to avoid introducing distortions in the equation.

The points on the visible surfaces in each stratum can be obtained by using the visibility function $N(r, \omega) = r - \nu(r, \omega) \ \omega$. With this, contribution of each stratum D_i can be determined:

$$\int_{D_i} = \int_{\omega \in \Pi_i} k(r, \omega \to \omega^o) L(N(r, \omega), \omega) \cos \theta d\omega.$$

In short, the general schema of the ray tracing method for calculating the illumination consists of the following steps: we choose the stratification $\{D_m\}$, with $\Omega = \cup \Pi_m$; then we determine the visibility $N(s, \omega)$ of the strata D_m; and finally we estimate the illumination integral in each stratum D_m.

The stratification and the importance function should be based on both local information on the surface and global information in the scene. A good choice is to divide it into two strata: direct and indirect illuminations. With direct illumination, the stratum is calculated based on knowledge about the light sources. With indirect illumination, the stratum is calculated based on the bidirectional reflectance function of the surface.

19.2.2   Revisiting Classic Ray Tracing

The classic ray tracing algorithm uses two hypotheses that simplify the problem: point light sources and perfect specular surfaces. With these hypotheses, the reflectance function corresponds to a Dirac delta distribution and the stratification is reduced to a discrete set

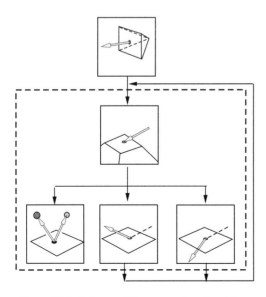

Figure 19.5. Schema of the ray tracing algorithm.

of directions. The illumination equation is then reduced to

$$L(r, \omega) = \sum_{\omega^l} \left[k_l(r, \omega \to \omega^l) L_E(s_l, \omega^l) \right] + \left[k_r L(s_r, \omega^r) + k_t L(s_t, \omega^t) \right],$$

where $s_i = N(r, \omega^i) = r - \nu(r, \omega^i)\, \omega^i$, for $i = l, s, t$, given by the rays ω_l along the direction of the light sources, and by the reflected and refracted rays, ω^r and ω^t, respectively.

Notice the first part of the equation corresponds to the direct illumination of the light sources, while the second part corresponds to the indirect specular illumination. The second part is calculated in a recursive way. Figure 19.5 illustrates the schema used in the ray tracing algorithm.

19.3   Radiosity Method

The radiosity method provides a solution for the calculation of global illumination based on a discretization of the surfaces in the scene. The basic idea consists of decomposing the surfaces into polygonal elements and then calculate the exchange of energy among all those elements. The radiosity method is particularly suitable to model interactions of diffuse reflection, which are independent of the virtual camera. In radiosity, the illumination integral is calculated using finite elements with the Galerkin method.

We start from Equation (19.8), describing the radiance $L(r, \omega^o)$ propagated from point r along the direction ω^o, as a function of the emitted and incident radiant energies at the point:

$$L(r, \omega^o) = L_E(r, \omega^o) + \int_{\Omega_i} k(r, \omega \to \omega^o) L(s, \omega) \cos \theta_r \, d\omega.$$

We determine the transport of energy arriving at r along the direction ω, using the visibility function ν such that $s = r + \nu(r, s)\omega$. Besides, when $s \in dS$ is in a distant surface, the solid angle $d\omega$ can be written in the following way:

$$d\omega = \frac{dS \cos \theta_s}{\|r - s\|^2},$$

where θ_s is the angle between the normal at dS and the vector $(r - s)$.

In order to place this expression in the illumination equation, we have to guarantee the integration will only be evaluated at the points on the visible surfaces. In this way, we define the *visibility test function*

$$V(r, s) = \begin{cases} 1 & \text{if } s = r - \nu(r, s - r)(s - r), \\ 0 & \text{otherwise.} \end{cases}$$

We can change the integration domain from solid angles in the illumination hemisphere, to areas in the visible surfaces. This is achieved by replacing the expression of the

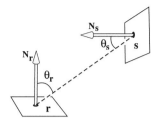

Figure 19.6. Integration domain for areas in the visible surfaces.

solid angle in the equation and introducing the visibility function. With this, we have

$$L(r, \omega^o) = L_E(r, \omega^o) + \int_M k(r, \omega \to \omega^o) L(s, \omega) G(r, s) \, d\omega$$

where the function

$$G(r, s) = \frac{\cos \theta_s \cos \theta_r}{\|r - s\|^2} V(r, s)$$

only depends on the geometry (see Figure 19.6).

The discretization method consists of (1) dividing the surfaces by polygonal patches $M_i = \cup m_k$ (finite elements) and (2) defining a basis of functions $\{b_j\}_{j \in J}$ that generates an approximation space on the surfaces in the scene. The projection of the solution $L(r, \omega)$ in that space can be written as a linear combination of the functions of the basis

$$\widehat{L}(r, \omega) = \sum_j L_j b_j(r, \omega).$$

By calculating the projection of the equation in this space of functions, we have

$$\langle \widehat{L}, \, b_i \rangle = \langle L_E, \, b_i \rangle + \left\langle \int_M k(r, \omega) G(r, s) \widehat{L}, \, b_i \right\rangle.$$

By replacing the expression of $\widehat{L}$ in the equation, we have

$$\left\langle \sum_j L_j b_j, \, b_i \right\rangle = \langle L_E, \, b_i \rangle + \left\langle \int_M k(r, \omega) G(r, s) \sum_j L_j b_j, \, b_i \right\rangle.$$

Rearranging the terms in L_j and removing the sum of the internal product results in

$$\langle L_E, \, b_i \rangle = \left\langle \sum_j L_j b_j, \, b_i \right\rangle - \left\langle \int_M k(r, \omega) G(r, s) \sum_j L_j b_j, \, b_i \right\rangle,$$

$$\langle L_E, \, b_i \rangle = \sum_j L_j \left[\langle b_j, \, b_i \rangle - \left\langle \int_M k(r, \omega) G(r, s) b_j, \, b_i \right\rangle \right].$$

Notice we can indicate the above expression in the matrix form $L_E = KL$.

The classic radiosity method makes the following assumptions to simplify the problem:

1. surfaces are opaque: there is no propagation by transmission.

2. there is Lambertian reflectance, with perfectly diffuse surfaces.

3. radiance and irradiance are constant in each element.

The diffuse reflection implies that the bidirectional reflectance function $k(s, \omega \to \omega')$ is constant in all directions and, therefore, does not depend on ω. Thus we can replace it by a function $\rho(s)$ that is outside the integral:

$$\int_M k(s, \omega \to \omega') L(s, \omega) G(r, s) \, d\omega = \rho(s) \int_M L(s, \omega) G(r, s) \, d\omega.$$

With this, we can also perform the following substitution to transform radiance into radiosity $L\pi = B$.

By considering a piecewise constant illumination function we know we can adopt the Haar basis $\{b_i\}$ for the approximation space of the finite elements.

$$b_i(r) = \left\{ \begin{array}{ll} 1 & r \in M_i, \\ 0 & \text{otherwise.} \end{array} \right.$$

Besides, as the functions of the Haar basis are disjunct, we have

$$\langle b_i, \, b_j \rangle = \delta_{ij} A_i.$$

By combining the above data, the illumination integral expressed in the Haar basis now is

$$\left\langle \int_M k(r, \omega) G(r, s) b_j, \, b_i \right\rangle = \frac{\rho_i}{\pi} \int_{M_i} \int_{M_k} G(i, k) \, dk di = \rho_i A_i F_{i,k},$$

where

$$F_{i,k} = \frac{1}{A_i} \int_{M_i} \int_{M_k} \frac{\cos \theta_i \cos \theta_k}{\pi \|i - k\|^2} V(i, k) dk di$$

is the *form factor*, which represents the percentage of radiant energy leaving element i and arriving at element j.

Using the fact that

$$\langle L_E, b_i \rangle = \int_{M_i} L_E(s) ds = E_i A_i,$$

and substituting in the equation $L \mapsto \frac{B}{\pi}$, we have

$$E_i A_i = \sum_k B_k \left(\delta_{ik} A_i - \rho_i A_i F_{i,k} \right),$$

$$E_i A_i = B_i A_i - \rho_i \sum_k B_i A_i F_{i,k},$$

or dividing both members by A_i and rearranging the terms

$$B_i = E_i + \rho_i \sum_k B_i F_{i,k}.$$

This equation is called the *classic radiosity equation*. In reality, we have a system of n equations for the radiosities B of n elements of the discretization. In matrix form

$$
\begin{aligned}
B &= E + FB, \\
(I - F)B &= E;
\end{aligned}
$$

that is

$$
\begin{pmatrix}
1 - \rho_1 F_{11} & \cdots & -\rho_1 F_{1n} \\
-\rho_2 F_{21} & \cdots & -\rho_2 F_{2n} \\
\vdots & \ddots & \vdots \\
-\rho_n F_{n1} & \cdots & 1 - \rho_n F_{nn}
\end{pmatrix}
\begin{pmatrix}
B_1 \\ B_2 \\ \vdots \\ B_n
\end{pmatrix}
=
\begin{pmatrix}
E_1 \\ E_2 \\ \vdots \\ E_n
\end{pmatrix}.
$$

Given that the discretization of the surfaces into finite elements is established by a polygonal mesh, we have $F_{ii} = 0$ and the diagonal of the matrix is equal to 1.

We want to find the numerical solution of the system given by

$$B = (I - F)^{-1} E.$$

When the linear system is very large, inverting the matrix becomes impracticable. Thus, we need to look for alternative methods that efficiently obtain the solution. The three methods discussed are *classic matrix*, *progressive*, and *hierarchical methods*.

19.3.1 Classic Matrix Methods

This method is based on the iterative solution of linear systems, a classic topic in computational linear algebra. In this class of methods, given the linear system $Mx = y$, we want to generate a series of approximated solutions x^k, converging toward the solution x when $k \to \infty$.

The approximation error at stage k is given by

$$e^k = x - x^k,$$

and the residue r^k, caused by the approximation $Mx^k = y + r^k$, is

$$r^k = y - Mx^k.$$

We want to minimize the residue r^k at each stage k. To express the residue in terms of the error, we subtract the equality $y - Mx = 0$ of r^k

$$r^k = (y - Mx^k) - (y - Mx) = M(x - x^k) = Me^k.$$

The basic idea of the iterative methods is to refine the approximation x^k, producing a better approximation x^{k+1}. The Southwell method, which consists of seeking a transformation which makes, in the next stage, the residue r_i^{k+1} of one of the elements x_i^{k+1}, equal to zero. Therefore, we select the element x^i with a residue of larger magnitude, and we calculate x_i^{k+1}, satisfying

$$
\begin{aligned}
r_i^{k+1} &= 0, \\
y_i + \sum_j M_{ij} x_j^{k+1} &= 0,
\end{aligned}
$$

given that only the component i of vector x is altered, we have $x_j^{k+1} = x_j^k$, for $j \neq i$. The new value of x_i^{k+1} is

$$
x_i^{k+1} = \frac{1}{M_{ii}} \left(y_i - \sum_{i \neq j} M_{ij} x_j^k \right) = x_i^k + \frac{r_i^k}{M_{ii}} = x_i^k + \Delta x_i^k.
$$

The new residue can then be calculated

$$
\begin{aligned}
r^{k+1} &= y - M x^{k+1} \\
&= y - M(x^k + \Delta x^k) \\
&= y - M x^k - M \Delta x^k \\
&= r^k - M \Delta x^k.
\end{aligned}
$$

However, the vector $\Delta x^k = x^{k+1} - x^k$ has all the components equal to zero, except for Δx_i^k. Then

$$
r_j^{k+1} = r_j^k - \frac{K_{ji}}{K_{ii}} r_i^k.
$$

Notice, we only use a column of the matrix to update the vector of residues. This is indicated below

$$
\begin{pmatrix} x \\ x \\ x \\ x \end{pmatrix} = \begin{pmatrix} x \\ x \\ x \\ x \end{pmatrix} + \begin{pmatrix} \vdots \\ x \\ \vdots \end{pmatrix} \begin{pmatrix} \cdots & x & \cdots \\ \cdots & x & \cdots \\ \cdots & x & \cdots \\ \cdots & x & \cdots \end{pmatrix}.
$$

19.3.2   Progressive Radiosity

The algorithm of progressive radiosity uses a variant of the Southwell method, which results in good approximations to the solution with few iterations. In the context of the illumination problem, we can interpret the residue R_i^k as being the radiant energy of the element M_i, in the stage k, not yet propagated into the scene.

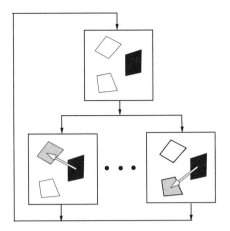

Figure 19.7. Progressive radiosity.

With this physical interpretation, we can see that the Southwell method consists of transferring the nondistributed radiant energy of an element M_i, for all the other elements M_j, with $j \neq i$. Those elements will then spread the received energy in a subsequent stage.

In this case, $M = (I - F)$. We know that $M_{ii} = 1$ and $M_{ij} = -\rho_j F_{ij}$. We then have the following rule for updating the vector of residues:

$$R_j^{k+1} = R_j^k + (\rho_j F_{ij}) R_i^k,$$

and $R_i^{k+1} = 0$. Besides, in the progressive radiosity, we also update the vector and the radiosities of the elements B_j, for $j \neq i$.

$$B_j^{k+1} = B_j^k + (\rho_j F_{ij}) R_i^k.$$

The method of progressive radiosity achieves, in a few iterations, a good approximation of the illumination function. This is because, at each stage, it chooses the element with the largest accumulated energy to be transferred to other elements in the scene. Figure 19.7 shows the schema of processing the progressive radiosity algorithm.

19.3.3 Hierarchical Radiosity

The method of hierarchical radiosity explores the sparse structure of the system of luminous energy exchange among the surfaces in the scene. This method applies techniques inspired by the solution to the problems of gravitational simulation among heavenly bodies (n-body). The idea is to treat interactions between distant and close objects, respectively, at low and high resolutions. Such a strategy allows one to reduce the computational cost of the method, from quadratic to linear, in the number of elements used in the discretization of the scene.

The hierarchical radiosity applies multiresolution meshes to represent the radiosity function. The interactions among elements can happen at arbitrary levels in the hierarchy. In this method, the hierarchical representation commonly adopted to represent the multiresolution mesh is the quadtree (see Chapter 9) and each element of the scene geometry corresponds to a quaternary tree.

Each instance of the transport of luminous energy between the elements of the quadtree is represented in a data structure called the transport link. It stores the amount of radiance transported between the elements, as well as an error estimate of the approximation. The transport link is equivalent to the form factor in traditional radiosity algorithms.

The hierarchical radiosity algorithm is separated in two stages: refinement of the transport links; and solution of the radiosity system.

The first stage determines the optimal level of interaction between the elements of the hierarchy, based on the approximation error. The second stage solves the problem of energy transfer in the system.

Refinement of the transport link. The refinement process begins with a geometric representation of the scene in its lower resolution. The refinement of the structure is determined by the error in the energy transport between each pair of elements in the scene. When this error is larger than a certain tolerance, the link should be refined. Three options exist to refine the link between two elements A e B.

- ❑ Subdivide A into subelements k_A and replace the link $A \leftrightarrow B$ by new links between B and k_A.

- ❑ Subdivide B into subelements k_B and replace the link $A \leftrightarrow B$ by new links between A and k_B.

- ❑ Subdivide A and B into subelements k_A and k_B, respectively, and replace the link $A \leftrightarrow B$ by new links between k_A and k_B.

The decision of which option to pick is based on the relative error of each element.

Solution of the radiosity system. The solution of the hierarchical radiosity equation is achieved in two steps: gather and push-pull (hierarchical-consistency).

- ❑ *Gather.* The radiance of each element of the hierarchy is gathered for all the links of this element with the rest of the scene, resulting in the irradiance transported to the container element. The irradiance of all the links of an element is accumulated, thus obtaining the total irradiance of the element.

- ❑ *Hierarchical Consistency.* To obtain a solid representation of the radiosity at the leaves of the quadtree, the irradiance of each element are propagated from the parents to their children, producing the total irradiance accumulated at each leaf. This irradiance is then converted into radiosity at the leaf elements, by using the reflectance function of the surface. The resulting radiosities are then propagated from the children to their parents, filling out the entire hierarchical structure with the correct values of the illumination function.

19.4   Comments and References

Global illumination models were initially proposed in the context of ray tracing. The pioneering work in this direction was by Turner Whitted [Whitted 80] and Roy Hall [Hall 89]. These models aim at performing a global sampling by using local illumination models (see Chapter 14).

Later, other global illumination models were developed using the radiosity method. The pioneering work in this area is [Goral et al. 84]. Among the works in this area, we can cite [Cohen and Greenberg 85, Hanrahan et al. 91, Chen 89, Chen 90]. Methods combining ray tracing and radiosity were also developed [Wallace et al. 87].

Another global illumination method is photon mapping, which distributes the luminous energy from the light sources [Ward 94, Jensen 01]. Photon mapping is a two-pass global illumination algorithm; it is an efficient alternative to pure Monte Carlo ray tracing techniques. Rays from both the light source and from the camera are traced independently until some termination criterion is met; those rays are then coupled in a second step to produce a radiance value.

The general formulation of global illumination is given by the *rendering equation*, proposed by Jim Kajiya [Kajiya 86]. Another relevant study was done by Jim Arvo [Arvo and Kirk 90].

19.4.1   Additional Topics

A more detailed study of the functional analysis methods and their applications in the modeling and solution of illumination problems should be explored. The methods of tonal application—transformation of radiant quantities to the color space of the monitor—are important in this area, and several studies exist on the theme. Another interesting topic is the illumination calculation on surfaces whose material admits subsurface scattering. To approach this topic, it is necessary to extend the BRDF function, given that the light is reflected at a different point from where the incidence of the luminous ray took place.

Bibliography

[Albuquerque 99] A. L. P. Albuquerque. "Cenários Virtuais com um Estudo de Sincronismo de Câmera (Virtual Sets with a Camera Synchronization Study)." Master's thesis, (in Portuguese) Department of Informatics, PUC-Rio, Brazil, 1999.

[Alexa 02] M. Alexa. "Linear Combination of Transformations." *ACM Transactions on Graphics* 21:3 (2002), 380–387.

[Allgower and Schmidt 85] E. Allgower and P. H. Schmidt. "An Algorithm for Piecewise Linear Approximation of an Implicitly Defined Manifold." *SIAM Journal of Numerical Analysis* 22:2 (1985), 322–346.

[Amanatides 84] J. Amanatides. "Ray Tracing with Cones." In *Proceedings of the 11th Annual Conference on Computer Graphics and Interactive Techniques, SIGGRAPH '84*, pp. 129–135, 1984.

[Anastacio et al. 09] F. Anastacio, P. Prusinkiewicz, and M. C. Sousa. "Sketch-Based Parameterization of L-systems using Illustration-Inspired Construction Lines and Depth Modulation." *Computers & Graphics* 33:4 (2009), 440–451.

[Andersson and Stewart 10] L.-E. Andersson and N. F. Stewart. *Introduction to the Mathematics of Subdivision Surfaces*. Philadelphia, PA: Society for Industrial and Applied Mathematics, 2010.

[Arvo and Kirk 90] J. Arvo and D. Kirk. "Particle Transport and Image Synthesis." In *Proceedings of the 17th Annual Conference on Computer Graphics and Interactive Techniques, SIGGRAPH '90*, pp. 63–66, 1990.

[Assarsson and Möller 00] U. Assarsson and T. Möller. "Optimized View Frustum Culling Algorithms for Bounding Boxes." *Journal of Graphics Tools* 5:1 (2000), 9–22.

[Atherton et al. 78] P. Atherton, K. Weiler, and D. Greenberg. "Polygon Shadow Generation." In *Proceedings of the 5th Annual Conference on Computer Graphics and Interactive Techniques, SIGGRAPH '78*, pp. 275–281, 1978.

[Badler et al. 91] Norman I. Badler, Brian A. Barsky, and David Zeltzer, editors. *Making Them Move: Mechanics, Control, and Animation of Articulated Figures*. San Francisco, CA: Morgan Kaufmann Publishers, 1991.

[Badler et al. 93] Norman I. Badler, Cary B. Phillips, and Bonnie Lynn Webber. *Simulating Humans: Computer Graphics Animation and Control*. New York: Oxford University Press, 1993.

[Baumgart 75] B. Baumgart. "Winged-Edge Polyhedron Representation for Computer Vision." http://www.baumgart.org/winged-edge/winged-edge.html, 1975. National Computer Conference.

[Blinn and Newell 76] J. F. Blinn and M. E. Newell. "Texture and Reflection in Computer Generated Images." *Communications of the ACM* 19:10 (1976), 542–546.

[Blinn 77] J. F. Blinn. "Models of Light Reflection For Computer Synthesized Pictures." In *Proceedings of the 4th Annual Conference on Computer Graphics and Interactive Techniques, SIGGRAPH '77*, pp. 192–198, 1977.

[Blinn 78] J. F. Blinn. "Simulation of Wrinkled Surfaces." In *Proceedings of the 5th Annual Conference on Computer Graphics and Interactive Techniques, SIGGRAPH '78*, pp. 286–292, 1978.

[Bloomenthal and Wyvill 97] J. Bloomenthal and B. Wyvill, editors. *Introduction to Implicit Surfaces*. San Francisco, CA: Morgan Kaufmann Publishers, 1997.

[Bouknight and Kelly 70] W. J. Bouknight and K. C. Kelly. "An Algorithm for Producing Half-Tone Computer Graphics Presentations with Shadows and Movable Light Sources." In *Proceedings of the May 5–7, 1970, Spring Joint Computer Conference, AFIPS '70 (Spring)*, pp. 1–10, 1970.

[Bouknight 70] W. J. Bouknight. "A Procedure for Generation of Three-Dimensional Half-Toned Computer Graphics Presentations." *Communications of the ACM* 13:9 (1970), 527–536.

[Bouthors et al. 08] A. Bouthors, F. Neyret, N. Max, E. Bruneton, and C. Crassin. "Interactive Multiple Anisotropic Scattering in Clouds." In *Proceedings of the 2008 Symposium on Interactive 3D Graphics and Games, I3D '08*, pp. 173–182, 2008.

[Bresenham 65] J. E. Bresenham. "Algorithm for Computer Control of a Digital Plotter." *IBM Systems Journal* 4 (1965), 25–30.

[Carpenter 84] L. Carpenter. "The A-buffer, an Antialiased Hidden Surface Method." In *Proceedings of the 11th Annual Conference on Computer Graphics and Interactive Techniques, SIGGRAPH '84*, pp. 103–108, 1984.

[Catmull 74] E. E. Catmull. "A Subdivision Algorithm for Computer Display of Curved Surfaces." Ph.D. thesis, Department of Computer Science, University of Utah, 1974.

[Catmull 78] E. E. Catmull. "A Hidden-Surface Algorithm with Anti-Aliasing." In *Proceedings of the 5th Annual Conference on Computer Graphics and Interactive Techniques, SIGGRAPH '78*, pp. 6–11, 1978.

[Chen 89] S. E. Chen. "A Progressive Radiosity Method and its Implementation in a Distributed Processing Environment." Master's thesis, Program of Computer Graphics, Cornell University, 1989.

[Chen 90] S. E. Chen. "Incremental Radiosity: An Extension of Progressive Radiosity to an Interactive Image Synthesis System." In *Proceedings of the 17th Annual Conference on Computer Graphics and Interactive Techniques, SIGGRAPH '90*, pp. 135–144, 1990.

[Coelho 98] L. C. G. Coelho. "Modelagem de Cascas com Interseções Paramétricas (Shell Modeling with Parametric Intersections)." Ph.D. thesis, (in Portuguese) Department of Informatics, PUC-Rio, Brazil, 1998.

[Cohen and Greenberg 85] M. F. Cohen and D. P. Greenberg. "The Hemi-Cube: A Radiosity For Complex Environments." In *Proceedings of the 12th Annual Conference on Computer Graphics and Interactive Techniques, SIGGRAPH '85*, pp. 31–40, 1985.

[Comba and Naylor 96] J. Comba and B. Naylor. "Conversion of Binary Space Partitioning Trees to Boundary Representation." In *Proceedings of Theory and Practice of Geometric Modeling 96, (Blaubeuren II)*. Springer Verlag, 1996.

[Cook and Torrance 81] R. L. Cook and K. E. Torrance. "A Reflectance Model for Computer Graphics." In *Proceedings of the 8th Annual Conference on Computer Graphics and Interactive Techniques, SIGGRAPH '81*, pp. 307–316, 1981.

[Cook et al. 84] R. L. Cook, T. Porter, and L. Carpenter. "Distributed Ray Tracing." In *Proceedings of the 11th Annual Conference on Computer Graphics and Interactive Techniques, SIGGRAPH '84*, pp. 137–145, 1984.

[Cook et al. 87] R. L. Cook, L. Carpenter, and E. Catmull. "The Reyes Image Rendering Architecture." In *Proceedings of the 14th Annual Conference on Computer Graphics and Interactive Techniques, SIGGRAPH '87*, pp. 95–102, 1987.

[Cook 84] R. L. Cook. "Shade Trees." In *Proceedings of the 11th Annual Conference on Computer Graphics and Interactive Techniques, SIGGRAPH '84*, pp. 223–231, 1984.

[Cook 86] R. L. Cook. "Stochastic Sampling in Computer Graphics." *ACM Transactions on Graphics* 5:1 (1986), 51–72.

[Coons 74] S. Coons. "Surface Patches and B-Spline Curves." In *Computer Aided Geometric Design*, edited by R. Barnhill and R. Riesenfeld. Academic Press, 1974.

[Craig 89] J. J. Craig. *Introduction to Robotics: Mechanics and Control*, Second edition. Boston, MA: Addison-Wesley Longman Publishing, 1989.

[Crow 81] F. C. Crow. "A Comparison of Antialiasing Techniques." *IEEE Computer Graphics and Applications* 1:1 (1981), 40–48.

[Crow 82] F. C. Crow. "A More Flexible Image Generation Environment." In *Proceedings of the 9th Annual Conference on Computer Graphics and Interactive Techniques, SIGGRAPH '82*, pp. 9–18, 1982.

[Cyrus and Beck 78] M. Cyrus and J. Beck. "Generalized Two- and Three-Dimensional Clipping." *Computers & Graphics* 3:1 (1978), 23–28.

[da Silva 98] F. W. da Silva. "Um Sistema de Animação Baseado em Movimento Capturado (A Motion Capture Based Animation System)." Master's thesis, (in Portuguese) COPPE/Sistemas, Federal University of Rio de Janeiro (UFRJ), Brazil, 1998.

[Darsa et al. 98] L. Darsa, B. Costab, and A. Varshneyc. "Walkthroughs of Complex Environments using Image-based Simplification." *Computers & Graphics* 22:1 (1998), 55–69.

[de Berg et al. 97] M. de Berg, M. van Kreveld, M. Overmars, and O. Schwarzkopf. *Computational Geometry: Algorithms and Applications*. Secaucus, NJ: Springer-Verlag New York, 1997.

[de Figueiredo and Stolfi 96] L. H. de Figueiredo and J. Stolfi. "Adaptive Enumeration of Implicit Surfaces with Affine Arithmetic." *Computer Graphics Forum* 15:5 (1996), 287–296.

[do Carmo 75] M. P. do Carmo. *Differential Geometry of Curves and Surfaces*. Upper Saddle River, NJ: Prentice-Hall, 1975.

[Dorst et al. 07] L. Dorst, D. Fontijine, and S. Mann. *Geometric Algebra for Computer Science: An Object-Oriented Approach to Geometry (The Morgan Kaufmann Series in Computer Graphics)*. San Francisco, CA: Morgan Kaufmann Publishers, 2007.

[Eastman and Weiss 82] C. M. Eastman and S. F. Weiss. "Tree Structures for High Dimensionality Nearest Neighbor Searching." *Information Systems* 7:2 (1982), 115–122.

[Ebert et al. 02] D. S. Ebert, F. K. Musgrave, D. Peachey, K. Perlin, and S. Worley. *Texturing and Modeling: A Procedural Approach*, Third edition. San Francisco, CA: Morgan Kaufmann Publishers, 2002.

[Farin 93] G. Farin. *Curves and Surfaces for Computer Aided Geometric Design (3rd ed.): A Practical Guide.* San Diego, CA: Academic Press Professional, 1993.

[Fiume 89] E. L. Fiume. *The Mathematical Structure of Raster Graphics.* San Diego, CA: Academic Press Professional, 1989.

[Floyd and Steinberg 75] R. W. Floyd and L. Steinberg. "An Adaptive Algorithm for Spatial Gray Scale." In *Proceedings of ID (Society of Information Displays): Digest of Technical Papers*, pp. 36–37, 1975.

[Foley et al. 96] J. D. Foley, A. van Dam, S. K. Feiner, and J. F. Hughes. *Computer Graphics (2nd ed. in C): Principles and Practice.* Boston, MA: Addison-Wesley Longman Publishing, 1996.

[Fowler et al. 92] D. R. Fowler, H. Meinhardt, and P. Prusinkiewicz. "Modeling Seashells." In *Proceedings of the 19th Annual Conference on Computer Graphics and Interactive Techniques, SIGGRAPH '92*, pp. 379–387, 1992.

[Fuchs et al. 80] H. Fuchs, Z. M. Kedem, and B. F. Naylor. "On Visible Surface Generation by A Priori Tree Structures." In *Proceedings of the 7th Annual Conference on Computer Graphics and Interactive Techniques, SIGGRAPH '80*, pp. 124–133, 1980.

[Fuchs et al. 83] H. Fuchs, G. D. Abram, and E. D. Grant. "Near Real-Time Shaded Display of Rigid Objects." In *Proceedings of the 10th Annual Conference on Computer Graphics and Interactive Techniques, SIGGRAPH '83*, pp. 65–72, 1983.

[Galbraith et al. 02] C. Galbraith, P. Prusinkiewicz, and B. Wyvill. "Modeling a Murex Cabritii Sea Shell with a Structured Implicit Surface Modeler." *The Visual Computer* 18:2 (2002), 70–80.

[Gardner 85] G. Y. Gardner. "Visual Simulation of Clouds." In *Proceedings of the 12th Annual Conference on Computer Graphics and Interactive Techniques, SIGGRAPH '85*, pp. 297–304, 1985.

[Ghazanfarpour and Hasenfratz 98] D. Ghazanfarpour and J.-M. Hasenfratz. "A Beam Tracing with Precise Antialiasing for Polyhedral Scenes." *Computers & Graphics* 22:1 (1998), 103–115.

[Glassner 89] A. S. Glassner, editor. *An Introduction to Ray Tracing.* London, UK: Academic Press, 1989.

[Gomes and Velho 95] J. Gomes and L. Velho. "Abstraction Paradigms for Computer Graphics." *The Visual Computer* 11:5 (1995), 227–239.

[Gomes and Velho 97] J. Gomes and L. Velho. *Image Processing for Computer Graphics*, First edition. Secaucus, NJ: Springer-Verlag New York, 1997.

[Gomes and Velho 02] J. Gomes and L. Velho. *Computer Graphics: Image (in Portuguese).* (Segunda edição). Série de Computação e Matemática, Instituto de Matemática Pura e Aplicada (IMPA), Rio de Janeiro, Brazil, 2002.

[Gomes et al. 93] J. Gomes, C. Hoffman, V. Shapiro, and L. Velho. "Modeling in Computer Graphics." In *ACM SIGGRAPH 1993 Courses, SIGGRAPH '93*. New York: ACM, 1993.

[Gomes et al. 96] J. Gomes, L. Darsa, B. Costa, and L. Velho. "Graphical Objects." *The Visual Computer* 12:6 (1996), 269–282.

[Gomes et al. 98] J. Gomes, L. Darsa, B. Costa, and L. Velho. *Warping and Morphing of Graphical Objects*. San Francisco, CA: Morgan Kaufmann Publishers, 1998.

[Gooch and Gooch 01] B. Gooch and A. Gooch. *Non-Photorealistic Rendering*. Natick, MA: AK Peters Ltd, 2001.

[Goral et al. 84] C. M. Goral, K. K. Torrance, D. P. Greenberg, and B. Battaile. "Modelling the Interaction of Light Between Diffuse Surfaces." In *Proceedings of the 11th Annual Conference on Computer Graphics and Interactive Techniques, SIGGRAPH '84*, pp. 213–222, 1984.

[Gouraud 71] H. Gouraud. "Continuous Shading of Curved Surfaces." *IEEE Transactions on Computers* 20:6 (1971), 623–629.

[Green and Hatch 95] D. Green and D. Hatch. "Fast Polygon-Cube Intersection Testing." In *Graphics Gems V*, edited by Paul Heckbert, pp. 375–379, 1995.

[Greene 86] N. Greene. "Applications of World Projections." In *Proceedings of Graphics Interface '86*, pp. 108–114, 1986.

[Greene 94] N. Greene. "Detecting Intersection of a Rectangular Solid and a Convex Polyhedron." In *Graphics Gems IV*, edited by Paul Heckbert, pp. 74–82, 1994.

[Greiner and Hormann 98] G. Greiner and K. Hormann. "Efficient Clipping of Arbitrary Polygons." *ACM Transactions on Graphics* 17:2 (1998), 71–83.

[Group 68] Mathematical Applications Group. "3-D Simulated Graphics Offered by Service Bureau." *Datamation* 13.

[Haines 03] E. Haines. "Ray Tracing News Guide." http://tog.acm.org/resources/RTNews/html/, 2003.

[Hall 89] R. A. Hall. *Illumination and Color in Computer Generated Imagery*. New York: Springer-Verlag New York, 1989.

[Hanrahan et al. 91] P. Hanrahan, D. Salzman, and L. Aupperle. "A Rapid Hierarchical Radiosity Algorithm." In *Proceedings of the 18th Annual Conference on Computer Graphics and Interactive Techniques, SIGGRAPH '91*, pp. 197–206, 1991.

[Harary and Tal 11] G. Harary and A. Tal. "The Natural 3D Spiral." *Computer Graphics Forum (Proc. Eurographics '11)* 30:2 (2011), 237–246.

[Heckbert and Hanrahan 84] P. S. Heckbert and P. Hanrahan. "Beam Tracing Polygonal Objects." In *Proceedings of the 11th Annual Conference on Computer Graphics and Interactive Techniques, SIGGRAPH '84*, pp. 119–127, 1984.

[Heckbert 82] P. S. Heckbert. "Color Quantization for Frame Buffer Display." In *Proceedings of the 9th Annual Conference on Computer Graphics and Interactive Techniques, SIGGRAPH '82*, pp. 297–307, 1982.

[Higham 96] N. J. Higham. *Accuracy and Stability of Numerical Algorithms*. Philadelphia, PA: Society for Industrial and Applied Mathematics (SIAM), 1996.

[Hoffmann 89] C. M. Hoffmann. *Geometric and Solid Modeling: An Introduction*. San Francisco, CA: Morgan Kaufmann Publishers, 1989.

[Hoffmann 03] G. Hoffmann. "Basic Principles for Computer Graphics and Mechanics.", 2003. Course Documents, Department of Mechanical Engineering University of Applied Sciences Emden/Leer, Germany. Available online (http://www.fho-emden.de/~hoffmann/howww41a.html).

[Jensen 01] H. W. Jensen. *Realistic Image Synthesis Using Photon Mapping*. Natick, MA: A K Peters, 2001.

[Kajiya 86] J. T. Kajiya. "The Rendering Equation." In *Proceedings of the 13th Annual Conference on Computer Graphics and Interactive Techniques*, SIGGRAPH '86, pp. 143–150, 1986.

[Kaufman 94] A. Kaufman. "Voxels as a Computational Representation of Geometry." In *ACM SIGGRAPH 1994 Courses*, SIGGRAPH '94. ACM, 1994.

[Kay and Kajiya 86] T. L. Kay and J. T. Kajiya. "Ray Tracing Complex Scenes." In *Proceedings of the 13th Annual Conference on Computer Graphics and Interactive Techniques*, SIGGRAPH '86, pp. 269–278, 1986.

[Klein and Furtak 86] M. Klein and T. Furtak. *Optics*. Second edition, New York: John Wiley and Sons, 1986.

[Kolb et al. 95] C. Kolb, D. Mitchell, and P. Hanraran. "A Realistic Camera Model for Computer Graphics." In *Proceedings of the 22nd Annual Conference on Computer Graphics and Interactive Techniques*, SIGGRAPH '95, pp. 317–324, 1995.

[Lake et al. 00] A. Lake, C. Marshall, M. Harris, and M. Blackstein. "Stylized Rendering Techniques for Scalable Real-Time 3D Animation." In *Proceedings of the 1st International Symposium on Non-photorealistic Animation and Rendering*, NPAR '00, pp. 13–20. New York: ACM, 2000.

[Liang and Barsky 84] Y.-D. Liang and B. Barsky. "A New Concept and Method for Line Clipping." *ACM Transactions on Graphics* 3.

[Lima 83] E. L. Lima. *Analysis Course, Volume 2 (in Portuguese)*. Projeto Euclides, Instituto de Matemática Pura e Aplicada (IMPA), Rio de Janeiro, Brazil, 1983.

[Lima 99] E. L. Lima. *Linear Algebra (in Portuguese)*. (Terceira edição). Instituto de Matemática Pura e Aplicada (IMPA), Rio de Janeiro, Brazil: Coleção Matemática Universitária, 1999.

[Lorensen and Cline 87] W. E. Lorensen and H. E. Cline. "Marching Cubes: A High Resolution 3D Surface Construction Algorithm." In *Proceedings of the 14th Annual Conference on Computer Graphics and Interactive Techniques*, SIGGRAPH '87, pp. 163–169, 1987.

[McMillan 97] L. McMillan. "An Image-Based Approach to Three-Dimensional Computer Graphics." Ph.D. thesis, University of North Carolina, 1997.

[Murray et al. 94] R. Murray, Z. Li, and S. S. Sastry. *A Mathematical Introduction to Robotic Manipulation*, First edition. Boca Raton, FL: CRC Press, 1994.

[Newell et al. 72] M. E. Newell, R. G. Newell, and T. L. Sancha. "A Solution to the Hidden Surface Problem." In *Proceedings of the ACM annual conference - Volume 1*, ACM '72, pp. 443–450, 1972.

[Nicholl et al. 87] T. M. Nicholl, D. T. Lee, and R. A. Nicholl. "An Efficient New algorithm for 2-D Line clipping: Its Development and Analysis." In *Proceedings of the 14th Annual Conference on Computer Graphics and Interactive Techniques*, SIGGRAPH '87, pp. 253–262, 1987.

[Olsen et al. 09] L. Olsen, F. F. Samavati, M. C. Sousa, and J. Jorge. "Sketch-Based Modeling: A Survey." *Computer & Graphics* 33:1 (2009), 85–103.

[Omohundro 89] S. M. Omohundro. "Five Balltree Construction Algorithms." Technical Report TR-89-063, International Computer Science Institute, Berkeley, CA, 1989.

[Ostromoukhov and Hersch 95] V. Ostromoukhov and R. D. Hersch. "Artistic Screening." In *Proceedings of the 22nd Annual Conference on Computer Graphics and Interactive Techniques, SIGGRAPH '95*, pp. 219–228, 1995.

[Peachey 85] D. R. Peachey. "Solid Texturing of Complex Surfaces." In *Proceedings of the 12th Annual Conference on Computer Graphics and Interactive Techniques, SIGGRAPH '85*, pp. 279–286, 1985.

[Penna and Patterson 86] A. Penna and R. Patterson. *Projective Geometry and Its Applications to Computer Graphics.* Upper Saddle River, NJ: Prentice-Hall, 1986.

[Perlin and Hoffert 89] K. Perlin and E. M. Hoffert. "Hypertexture." In *Proceedings of the 16th Annual Conference on Computer Graphics and Interactive Techniques, SIGGRAPH '89*, pp. 253–262, 1989.

[Perlin 85] K. Perlin. "An Image Synthesizer." In *Proceedings of the 12th Annual Conference on Computer Graphics and Interactive Techniques, SIGGRAPH '85*, pp. 287–296, 1985.

[Perlin 02] K. Perlin. "Improving Noise." *ACM Transactions on Graphics (Proc. SIGGRAPH '02)* 21:3 (2002), 681–682.

[Perwass 09] C. Perwass. *Geometric Algebra with Applications in Engineering*, First edition. Springer Publishing Company, Incorporated, 2009.

[Phong 75] B.-T. Phong. "Illumination for Computer Generated Pictures." *Communications of the ACM* 18:6 (1975), 311–317.

[Porter and Duff 84] T. Porter and T. Duff. "Compositing Digital Images." In *Proceedings of the 11th Annual Conference on Computer Graphics and Interactive Techniques, SIGGRAPH '84*, pp. 253–259, 1984.

[Potmesil and Hoffert 87] M. Potmesil and E. M. Hoffert. "FRAMES: Software Tools for Modeling, Rendering and Animation of 3D Scenes." In *Proceedings of the 14th Annual Conference on Computer Graphics and Interactive Techniques, SIGGRAPH '87*, pp. 85–93, 1987.

[Prusinkiewicz and Lindenmayer 96] P. Prusinkiewicz and A. Lindenmayer. *The Algorithmic Beauty of Plants.* New York: Springer-Verlag New York, 1990 (second printing 1996).

[Raup 62] D. M. Raup. "Computer as Aid in Describing Form in Gastropod Shells." *Science* 138:3537 (1962), 150–152.

[Reeves 83] W. T. Reeves. "Particle Systems: A Technique for Modeling a Class of Fuzzy Objects." *ACM Transactions on Graphics* 2:2 (1983), 91–108.

[Requicha 80] A. A. G. Requicha. "Representations for Rigid Solids: Theory, Methods, and Systems." *ACM Computing Surveys* 12:4 (1980), 437–464.

[Roberts 66] L. G. Roberts. "Homogeneous Matrix Representation and Manipulation of N-Dimensional Constructs." Technical Report MS-1045, Lincoln Lab, MIT, Lexington, MA, 1966.

[Roth 82] S. D. Roth. "Ray Casting for Modeling Solids." *Computer Graphics and Image Processing* 18:2 (1982), 109–144.

[Samet 90] H. Samet. *The Design and Analysis of Spatial Data Structures*. Boston, MA: Addison-Wesley Longman Publishing, 1990.

[Samet 05] H. Samet. *Foundations of Multidimensional and Metric Data Structures (The Morgan Kaufmann Series in Computer Graphics and Geometric Modeling)*. San Francisco, CA: Morgan Kaufmann Publishers, 2005.

[Shenchang 95] E. C. Shenchang. "QuickTime VR: An Image-Based Approach to Virtual Environment Navigation." In *Proceedings of the 22nd Annual Conference on Computer Graphics and Interactive Techniques, SIGGRAPH '95*, pp. 29–38, 1995.

[Shoemake 85] K. Shoemake. "Animating Rotation with Quaternion Curve." In *Proceedings of the 12th Annual Conference on Computer Graphics and Interactive Techniques, SIGGRAPH '85*, pp. 245–254, 1985.

[Shumacker et al. 69] R. A. Shumacker, R. Brand, M. Gilliland, and W. Sharp. "Study for Applying Computer-Generated Images to Visual Simulation." Technical Report AFHRL-TR-69-14, U.S. Air Force Human Resources Lab., 1969.

[Sims 90] K. Sims. "Particle Animation and Rendering using Data Parallel Computation." In *Proceedings of the 17th Annual Conference on Computer Graphics and Interactive Techniques, SIGGRAPH '90*, pp. 405–413, 1990.

[Smith 78] A. R. Smith. "Color Gamut Transform Pairs." In *Proceedings of the 5th Annual Conference on Computer Graphics and Interactive Techniques, SIGGRAPH '78*, pp. 12–19, 1978.

[Smith 84] A. R. Smith. "The Viewing Transformation." Technical Memo No. 84, Pixar, 1984.

[Snyder and Lengyel 98] J. Snyder and J. Lengyel. "Visibility Sorting and Compositing without Splitting for Image Layer Decomposition." In *Proceedings of the 25th Annual Conference on Computer Graphics and Interactive Techniques, SIGGRAPH '98*, pp. 219–230, 1998.

[Sutherland and Hodgman 74] I. E. Sutherland and G. W. Hodgman. "Reentrant Polygon Clipping." *Communications of the ACM* 17:1 (1974), 32–42.

[Sutherland et al. 74] I. E. Sutherland, R. F. Sproull, and R. A. Shumacker. "A Characterization of Ten Hidden-Surface Algorithms." *ACM Computing Surveys* 6:1 (1974), 1–55.

[Thibault and Naylor 87] W. Thibault and B. Naylor. "Set Operations on Polyhedra using Binary Space Partitioning Trees." In *Proceedings of the 14th Annual Conference on Computer Graphics and Interactive Techniques, SIGGRAPH '87*, pp. 153–162, 1987.

[Ulichney 87] R. Ulichney. *Digital Halftoning*. Cambridge, MA: MIT Press, 1987.

[Vatti 92] B. R. Vatti. "A Generic Solution to Polygon Clipping." *Communications of the ACM* 35:7 (1992), 56–63.

[Velho and Gomes 91] L. Velho and J. Gomes. "Digital Halftoning with Space Filling Curves." In *Proceedings of the 18th Annual Conference on Computer Graphics and Interactive Techniques, SIGGRAPH '91*, pp. 81–90, 1991.

[Velho et al. 97] L. Velho, J. Gomes, and M. R. Sobreiro. "Color Image Quantization by Pairwise Clustering." In *Proceedings of the X Brazilian Symposium of Computer Graphics and Image Processing (SIBGRAPI '97)*, pp. 203–210. IEEE Computer Society, 1997.

[Velho et al. 02] L. Velho, J. Gomes, and L. H. Figueiredo. *Implicit Objects in Computer Graphics*. New York: Springer-Verlag, 2002.

[Wallace et al. 87] J. R. Wallace, M. F. Cohen, and D. P. Greenberg. "A Two-Pass Solution to the Rendering Equation: A Synthesis of Ray Tracing and Radiosity Methods." In *Proceedings of the 14th Annual Conference on Computer Graphics and Interactive Techniques, SIGGRAPH '87*, pp. 311–320, 1987.

[Walsh 58] J. T. Walsh. *Photometry.* New York: Dover Publications, 1958.

[Ward 94] G. J. Ward. "The RADIANCE Lighting Simulation and Rendering System." In *Proceedings of the 21st Annual Conference on Computer Graphics and Interactive Techniques, SIG-GRAPH '94*, pp. 459–472, 1994.

[Warnock 69] J. Warnock. "A Hidden-Surface Algorithm for Computer Generated Half-Tone Pictures." Technical Report TR 4–15, NTIS AD-733 671, Department of Computer Science, University of Utah, 1969.

[Warren and Weimer 01] J. Warren and H. Weimer. *Subdivision Methods for Geometric Design: A Constructive Approach*, First edition. San Francisco, CA: Morgan Kaufmann Publishers, 2001.

[Watkins 70] G. S. Watkins. "A Real-Time Visible Surface Algorithm." Technical Report UTEC-CS-70-101, Department of Computer Science, University of Utah, 1970.

[Weiler and Atherton 77] K. Weiler and K. Atherton. "Hidden Surface Removal using Polygon Area Sorting." In *Proceedings of the 4th Annual Conference on Computer Graphics and Interactive Techniques, SIGGRAPH '77*, pp. 214–222, 1977.

[Weiler 82] K. Weiler. "The Radial Edge Structure: A Topological Representation for Non-Manifold Geometric Boundary Modeling." In *Geometric Modeling for CAD Applications*, edited by M. J. Wozny, H. W. McLauhlin, and J. L. Encamcao, pp. 3–36. Amsterdam: North-Holland, 1982.

[Whitted 80] T. Whitted. "An Improved Illumination Model for Shaded Display." *Communications of the ACM* 23:6 (1980), 343–349.

[Wolfe 97] R. Wolfe. "Teaching Texture Mapping Visually." *Computer Graphics* 31:4 (1997), 66–70.

[Wyszecki and Stiles 00] G. Wyszecki and W. S. Stiles. *Color Science: Concepts and Methods, Quantitative Data and Formulae, 2nd Edition*, Second edition. New York: John Wiley & Sons, 2000.

[Yamamoto 98] J. K. Yamamoto. "A Review of Numerical Methods for the Interpolation of Geological Data." *Anais da Academia Brasileira de Ciências* 70:1 (1998), 91–116.

[Zaretskiy et al. 10] Y. Zaretskiy, S. Geiger, K. S. Sorbie, and M. Foerster. "Efficient Flow and Transport Simulations in Reconstructed 3D Pore Geometries." *Advances in Water Resources* 33:12 (2010), 1508–1516.

Index